MATHEMATICAL PROGRAMMING *for* ECONOMICS AND BUSINESS

MATHEMATICAL ECONOMICS

Roger C. Pfaffenberger

THE IOWA STATE

PROGRAMMING for AND BUSINESS

David A. Walker

UNIVERSITY PRESS *Ames* IOWA

Roger C. Pfaffenberger, associate professor of Management Science and Statistics, University of Maryland, holds the M.S. and Ph.D. degrees in statistics from Texas A & M University.

David A. Walker, financial economist, Federal Deposit Insurance Corporation, and professorial lecturer, Georgetown University, holds the M.S. and Ph.D. degrees in economics from Iowa State University.

Composed by Science Press
Printed by The Iowa State University Press

First edition, 1976

Library of Congress Cataloging in Publication Data

Pfaffenberger, Roger C 1943–
Mathematical programming for economics and business.

Bibliography: p.
Includes index.
1. Programming (Mathematics) I. Walker, David A., 1941– joint author. II. Title.
QA402.5.P46 1976 519.7'02'433 75-23045
ISBN 0-8138-1055-8

To

OUR PARENTS; ANN and JANELLE; and RON

CONTENTS

PREFACE

This book is intended for use in an introductory graduate course in mathematical programming for M.S. and Ph.D. students in such applied programs as business, economics, industrial engineering, and related areas. Numerous applications, most of which deal with economic and business problems, are used to illustrate the ideas developed in the text.

Use of operations research methods as an aid to managerial decision making has increased dramatically. Many management and business decisions involve constrained optimization processes; consequently, interest in mathematical programming (the optimization of constrained functions) has grown accordingly. This growth has been reinforced by the availability and efficiency of high-speed digital computers. Most of the early applications of mathematical programming to the resolution of business problems involved linear programming, i.e., optimizing a linear function of a set of decision variables subject to constraints (usually depicting resource restrictions) that assume linear functional forms.

For a number of reasons, linear programming continues to be the most widely used mathematical programming technique. First, the theory of linear programming is relatively easy to comprehend, and many excellent and highly readable books are available on the subject. Second, the method used to solve linear programming problems guarantees the optimal solution and is easily programmed. Third, even though many applied optimization problems are extremely complex and inherently nonlinear, the linear programming model has proved to be useful as an approximating problem in numerous instances.

Nonlinear mathematical programming models and techniques are being employed more frequently as the users of this analytic tool become more sophisticated and availability of nonlinear computer codes increases. Unlike linear programming, no single algorithm or method is available to solve a nonlinear problem. The solution technique chosen often depends on the characteristics of the particular problem. Due to this and the difficulties of dealing with nonlinear functions mathematically, understanding and use of nonlinear techniques requires more rigorous mathematical training than that necessary for linear programming and; hence the nonlinear approach is being adopted more slowly than linear programming.

This book develops the theory of classical mathematical programming in a somewhat rigorous fashion. Since linear programming continues to be widely used, an extended treatment is given. Examples are used to illustrate and explain the theoretical ideas developed. The important mathematical programming methods are presented, but coverages of techniques is by no means exhaustive. Many new approaches to solving mathematical programming problems are constantly being proposed in the literature, particularly in the areas of nonlinear programming and discrete programming where the optimal values of the decision variables are restricted to integer values, which is a common occurrence in many business and economic applications.

The methods covered have been chosen because of their wide applicability and acceptance among users of mathematical programming. Sufficient theory is developed to provide a basic understanding of the techniques so that they may be properly used and perhaps extended or amended by those with particular interests and requisite mathematical skills. Differential and integral calculus and linear algebra are required to assimilate this material. For full appreciation of Chapter 10, an introductory course in statistics would be helpful.

The authors wish to express their appreciation to associates, reviewers, and students who have helped to modify and improve this treatment of mathematical programming. Several faculty members of the Management Science Department at The Pennsylvania State University have provided numerous suggestions. Professor G. K. Nelson was an enthusiastic supporter of the project from its inception.

The authors would like to express their appreciation to Ann Pfaffenberger, who patiently did much of the typing, and to Elaine Gabriletto and Brenda Tolliver, who participated in the final manuscript preparation. They are grateful to the editorial staff of the Iowa State University Press for assistance and cooperation in preparation of this book. Special thanks are due to Nancy Bohlen, associate editor, for her outstanding editorial contribution to the refinement and improvement of the text material. The authors are also grateful to the Economics Research Unit of the Federal Deposit Insurance Corporation for support.

MATHEMATICAL PROGRAMMING *for* ECONOMICS AND BUSINESS

1 CHARACTERISTICS AND TYPES OF MODELS

1.1 MATHEMATICAL PROGRAMMING PROBLEM

The problem of optimizing a numerical function of one or more variables when they are constrained in some manner is called a mathematical programming problem. Specifically, the purpose of such a problem is to determine the values of n variables $x_1, x_2, \ldots, x_n$ that optimize the function

$$z = f(x_1, x_2, \ldots, x_n) \tag{1.1}$$

subject to the constraints

$$g_i(x_1, x_2, \ldots, x_n)\{\leq \; = \; \geq\} b_i \qquad i = 1, 2, \ldots, m \tag{1.2}$$

It is usually assumed that the values of the n variables cannot be negative numerically. The nonnegativity restrictions on the variables may be stated as

$$x_j \geq 0 \qquad j = 1, 2, \ldots, n \tag{1.3}$$

Also, it is usually desired to determine the optimal value (minimum or maximum) of the function z in (1.1), which is called the objective function.

The formulation of business and economic questions as mathematical programming problems has resulted in the successful resolution of many complex real-world optimization situations. Most of the applications of mathematical programming to business and economics involve the maximization of revenues or profits and the minimization of costs. As an illustration of a typical business problem that can be formulated as a mathematical programming model, consider the following example.

Example 1.1

Suppose a firm has L dollars available for capital investment in its next planning period. To simplify matters for illustrative purposes, suppose that n different projects are competing for these funds. Assume that the jth project requires an investment of c_j dollars and m_j man-hours and yields a profit of p_j dollars. In the coming period M man-hours are available to work on the selected projects.

The firm desires to determine which projects should be undertaken to maximize profits in the coming period, while not exceeding capital allocations of the allotted L dollars. Assume that a project must either be undertaken or not; it is not possible to change the amount of investment in the jth project once it has been selected. Additionally, suppose that only one unit of each project may be undertaken.

The situation may be formulated as a mathematical programming problem by introducing integer-valued variables d_j that can take on only the values 0 or 1. By maximizing the function $z = \sum_{j=1}^{n} p_j d_j$ subject to the constraints

$$\sum_{j=1}^{n} c_j d_j \leq L \qquad \sum_{j=1}^{n} m_j d_j \leq M$$

$$0 \leq d_j \leq 1 \qquad j = 1, 2, \ldots, n$$

where d_j must be integer valued, the projects to select for maximizing profits will be determined.

Numerous generalizations of the model in Example 1.1 exist, some of which will be discussed in conjunction with the subject matter presented in the following section.

1.2 MODEL CLASSIFICATION

A real-world optimization problem may be classified in five ways.

1. The functional relationships in the problem may be known (deterministic) or uncertain (probabilistic).
2. The functions $f(x_1, x_2, \ldots, x_n)$ and $g_i(x_1, x_2, \ldots, x_n)$, $i = 1, 2, \ldots, m$, in (1.1) and (1.2) may be linear in $x_1, x_2, \ldots, x_n$; or at least one function in the set may be nonlinear.
3. The functions may be continuously differentiable (smooth) or nondifferentiable (nonsmooth).
4. The variables $x_1, x_2, \ldots, x_n$ in the mathematical programming problem may be continuous or may be restricted to integer values.
5. The optimization may take place at a fixed point in time (static) or during an interval of time (dynamic).

The model specified by (1.1)–(1.3) and most mathematical programming models are deterministic; given $x_1, x_2, \ldots, x_n$, the values of $f, g_1, g_2, \ldots, g_m$ are uniquely determined. Most of the current applications of mathematical programming to business and economic problems assume that all model functions are linear. There is a very simple reason for this. The simplex method, devised by Dantzig in 1947 [1963], is an extremely efficient procedure for solving linear programming problems. When this method is programmed on a computer, it is possible to solve linear problems involving hundreds of variables and thousands of constraints. If one or more of the functions is nonlinear, the problem is almost always more difficult to solve than linear ones. Thus, even though the real-world problem may be complex and inherently highly nonlinear, successful modeling of it may be possible by using many variables and constraints in a linear formulation.

Most algorithms devised to solve mathematical programming problems require that the functions in the model be continuously differentiable; thus all functions typically must be smooth. Some algorithms are designed to produce integer solutions for the variables in a linear problem. However, if one or more of the functions in the model is nonlinear, no efficient algorithm has been developed to solve the programming problem and to guarantee that the values of $x_1, x_2, \ldots, x_n$ will all be integer valued. Finally, the optimization of the objective function $f(x_1, x_2, \ldots, x_n)$ in (1.1) is almost always considered at one point in time, i.e., the model is static.

In most instances when mathematical programming methods are applied to a real-world problem, the functions are inherently nonlinear, nonsmooth, and probabilistic; the variables of the problem must be integer valued; and the optimization of the objective function associated with the problem should be conducted dynamically. However, the model chosen to represent the problem is invariably a linear-smooth-deterministic-continuous-static model. It is the hope of the model builder in these instances that the solution to the simplified mathematical programming model will furnish usable information in learning more about the real-world problem.

The problem in Example 1.1 is linear in the variables $d_1, d_2, \ldots, d_n$; the functions are smooth and are determined; the variables must be integer valued; and the optimization process is static; i.e., it is a linear-smooth-deterministic-integer-static model.

Since the optimization problem in Example 1.1 concerns functions specified in a future time period, a generalization to a more appropriate probabilistic model is easily envisaged. The profit for the jth project p_j is being forecast from information at the present time. It would seem more realistic to consider the p_j coefficients as random variables. The objective function $z = \sum_{j=1}^{n} p_j d_j$ is now specified probabilistically. That is, z is associated with some probability distribution determined by the joint distribution of $p_1, p_2, \ldots, p_n$. This introduces the difficulty of defining the maximization of z since it is a variable. A common solution to this dilemma is to replace each p_j with its expected value. Since expected values of random variables are constants, the objective function is converted from a probabilistic quantity to a deterministic one.

Probabilistic elements may be introduced into the model in other ways. For example, the number of man-hours M available in the next period may be viewed as a random variable. Alternatively, it may well be that the man-hours constraint $\sum_{j=1}^{n} m_j d_j \leq M$ needs only to be satisfied probabilistically as follows:

$$\Pr\left(\sum_{j=1}^{n} m_j d_j \leq M\right) \geq 0.95 \tag{1.4}$$

Problems with the constraints of (1.4) have been solved, but only in a limited number of cases. When the m_j are assumed to be normally distributed random variables, Charnes and Cooper [1959] give a very elegant solution to the mathematical programming problem if all functions are linear and all variables are continuous, for instance.

Nonlinearity may occur in the model if the projects are related in some way. For example, it may be that if projects 1 and 2 are both undertaken, the profit enjoyed by participating in these projects may exceed $p_1 + p_2$ due to use of some of the same materials on both. The objective function would have to be adjusted to reflect nonadditive behavior of the first two variables.

The optimization may be required over time if, for example, amounts of the investments may be altered in later periods. A dynamic optimization model would be required in this case. Thus an important element in employing mathematical programming methods to solve real-world problems is the specification of the model. A mathematical programmer learns quickly to use the simplest manageable model that adequately represents the real-world problem. Generalizations to more complex models generally follow with greater ease if approximate solutions to the problem determined from simple models are available.

1.3 SPECIFIC MODELS OF INTEREST

The best known mathematical programming model is the *linear programming* model. All functions in (1.1) and (1.2) are linear in the n variables $x_1, x_2, \ldots, x_n$. The model may be written as

$$\text{optimize} \quad z = f(x_1, x_2, \ldots, x_n) = \textstyle\sum_{j=1}^{n} c_j x_j \tag{1.5}$$

subject to

$$g_i(x_1, x_2, \ldots, x_n) = \textstyle\sum_{j=1}^{n} a_{ij} x_j \{\le \ = \ \ge\} b_i \qquad i = 1, 2, \ldots, m \tag{1.6}$$

$$x_j \ge 0 \qquad j = 1, 2, \ldots, n \tag{1.7}$$

where the c_j's, b_i's, and a_{ij}'s are known constants.

The linear programming model has been successfully used to solve a variety of business and economic problems. The reader is directed to the Bibliography for references to specific examples.

The model specified by (1.5)–(1.7) is called an *integer linear programming* model if, additionally, the variables $x_1, x_2, \ldots, x_n$ are required to be integer valued. Many integer linear programming problems have been proposed, and algorithms for solving this class of mathematical programming models have been developed.

If in the model,

$$\begin{aligned} &\text{optimize} \quad z = f(x_1, x_2, \ldots, x_n) \\ &\text{subject to} \quad g_i(x_1, x_2, \ldots, x_n)\{\le \ = \ \ge\} b_i \qquad i = 1, 2, \ldots, m \\ &\qquad\qquad\qquad\qquad\qquad x_j \ge 0 \qquad j = 1, 2, \ldots, n \end{aligned}$$

at least one function in the set $f, g_1, g_2, \ldots, g_m$ is nonlinear, it is called a *nonlinear programming* model. As noted earlier, a nonlinear problem is generally much more difficult to solve than a linear one. Many algorithms have been developed to alleviate this, among which are those by Rosen [1960] and Fiacco and McCormick [1964], but often the best solution to a nonlinear problem cannot be guaranteed by the algorithm.

If the nonlinear programming model contains the additional restriction that the variables $x_1, x_2, \ldots, x_n$ must be integer valued, then it is called an *integer nonlinear programming* model. No efficient algorithm has been developed to solve problems of this type, although they occur quite frequently in business and economics optimization modeling.

A special case of the general nonlinear programming model, which has received a great deal of attention, is the quadratic programming model. In this model the objective function is quadratic in $x_1, x_2, \ldots, x_n$ and the constraints are linear. Specifically, the model is

$$\begin{aligned} &\text{optimize} \quad z = \textstyle\sum_{j=1}^{n} c_j x_j + \sum_{i=1}^{n} \sum_{j=1}^{n} d_{ij} x_i x_j \\ &\text{subject to} \quad \textstyle\sum_{j=1}^{n} a_{ij} x_j \leq b_i \qquad i = 1, 2, \ldots, m \\ &\qquad\qquad\qquad x_j \geq 0 \qquad j = 1, 2, \ldots, n \end{aligned}$$

where the c_j's, d_{ij}'s, and a_{ij}'s are known constants.

The quadratic programming model has been successfully employed for studying models of the firm operating in a market characterized by imperfect competition. If the firm sells its outputs in an imperfectly competitive market, its total revenue is assumed to be a quadratic function. If the firm buys inputs in that same market, its cost of production is a quadratic function.

2 LINEAR PROGRAMMING

2.1 INTRODUCTION

Linear optimization problems are the simplest and most thoroughly explored mathematical programming problems. The objective is to optimize a linear function of one or more variables when the solution must satisfy one or more linear constraints placed on the variables. Applications to economic and business problems and algorithms for linear programming have been studied for years. Dantzig is well known for his participation in these developments. In one of the most thorough presentations of linear programming, he [1963, Ch. 2] summarizes the origins and achievements in linear optimization problems. Some of these are mentioned in this chapter.

The interdependence of economic activities probably was first stressed in 1758 by François Quesnay in the *Tableau èconomique.* Phillips [1955] has shown the conversion of the *Tableau èconomique* to a simple Leontief input-output model. One of the applications of linear programming to economic analysis is presented in Section 2.8 and utilizes a Leontief model as the set of constraints.

Linear programming may be considered as a way to delineate a problem where limited resources are to be allocated among competing activities in an optimal manner. The functions specifying how the resources are limited and how their allocation is to be completed must all be linear. As an illustration, suppose that the Acme Machine Shop is considering the production of two new products. The first will yield a profit of \$10 per unit and the second will yield \$20 per unit produced. Each product requires production time on two machines, the pressing machine and the stamping machine. The first requires 2 hours per unit on each machine. The second requires 3 hours per unit on the pressing machine and 1 hour per unit on the stamping machine. The pressing machine is run 18 hours per day and the stamping machine 10 hours. If other restrictions are neglected (such as coordinating production on the two machines), how many units of each product should be produced each day so that profit is maximized?

Before responding to this question, consider the following mathematical statement of it, where x_1 and x_2 are the number of units produced of the first and second product respectively,

$$\begin{aligned} \text{maximize} \quad & z = f(\mathbf{x}) = 10x_1 + 20x_2 && (2.1)\\ \text{subject to} \quad & 2x_1 + 3x_2 \leq 18 && (2.2)\\ & 2x_1 + x_2 \leq 10 && (2.3)\\ & x_1, x_2 \geq 0 && (2.4) \end{aligned}$$

The function in (2.1) is an expression of profit in terms of x_1 and x_2. The constraints (2.2) and (2.3) reflect the time restrictions in hours placed on the production of the two products on the pressing and stamping machines respectively. The constraints in (2.4) express the fact that a negative number of units of either product cannot be produced.

The problem as stated in (2.1)–(2.4) is a linear programming problem. The word "linear" in linear programming is certainly appropriate; i.e., a linear function is optimized subject to restrictions expressed as linear inequalities (restrictions may occasionally be expressed as equalities) in the decision variables. The word "program" refers to the way these problems are usually solved, i.e., by a program of steps that will lead to an optimal solution of the problem.

The most important and widely used program to solve a linear programming problem is called the simplex algorithm, which is presented in Section 2.5. Other methods used to solve this type of optimization problem are presented in this chapter. The graphic method is used in Section 2.3 to solve (2.1)–(2.4).

2.2 LINEAR PROGRAMMING MODEL

The linear programming problem may be stated as

$$\text{optimize} \quad z = f(x_1, x_2, \ldots, x_n) = c_1x_1 + c_2x_2 + \cdots + c_nx_n$$

subject to

$$g_1(x_1, x_2, \ldots, x_n) = a_{11}x_1 + a_{12}x_2 + \cdots + a_{1n}x_n \; \{\leq \; = \; \geq\} b_1$$

$$g_2(x_1, x_2, \ldots, x_n) = a_{21}x_1 + a_{22}x_2 + \cdots + a_{2n}x_n \ \{\leq \ = \ \geq\} b_2$$
$$\vdots \qquad\qquad\qquad\qquad \vdots$$
$$g_m(x_1, x_2, \ldots, x_n) = a_{m1}x_1 + a_{m2}x_2 + \cdots + a_{mn}x_n \{\leq \ = \ \geq\} b_m$$
$$x_1, x_2, \ldots, x_n \geq 0 \qquad (2.5)$$

The a_{ij}'s, b_i's, and c_j's are assumed to be known constants. The variables $x_1, x_2, \ldots, x_n$ are called decision variables. The function f to be optimized is called the objective function; it may be maximized with respect to the decision variables $x_1, x_2, \ldots, x_n$ (such as in maximizing profit) or minimized (such as in minimizing loss). The optimization of f is carried out so that the m constraints $g_1, g_2, \ldots, g_m$ are satisfied. The constraints may be cast either as inequalities in either direction ($\leq$ or $\geq$) or as equalities ($=$). The nonnegative restrictions on the decision variables in (2.5) are consistent with most applications of the linear programming model to business and economic problems. For example, if the decision variables represent the number of units of products to produce with limited resources, they cannot assume negative values.

Several examples follow that illustrate formulation of a linear programming model and its diverse applicability.

Example 2.1. Diet Problem

The diet problem was the first economic problem to be solved using linear programming. A particular form of this classic will now be described.

A team of scientists in the year 2000 are concerned with the production of a cheap but nutritious food product for the mass population. It is decided that the product will be composed of three elements—soylent green, red, and blue. Each pound of the final product must contain at least 2000 calories of energy and 1000 units of vitamins. The vitamin and calorie composition of each element is given in Table 2.1. The cost per pound of each element is \$3 for green, \$1 for red, and \$2 for blue. The objective is to minimize the total cost subject to the calorie and vitamin restrictions.

Table 2.1. Composition of Soylent Green, Red, and Blue

Element	Calories/lb	Vitamins/lb
Green	3200	800
Red	1600	1000
Blue	2400	1200

Let x_1, x_2, and x_3 be the proportion of soylent green, red, and blue respectively to be used in making 1 pound of the food product. Formulation as a linear programming problem is

$$\begin{aligned} \text{minimize} \quad & z = f(x_1, x_2, x_3) = 3x_1 + 1x_2 + 2x_3 \\ \text{subject to} \quad & 3200x_1 + 1600x_2 + 2400x_3 \geq 2000 \\ & 800x_1 + 1000x_2 + 1200x_3 \geq 1000 \\ & x_1 + x_2 + x_3 = 1.00 \\ & x_1, x_2, x_3 \geq 0 \end{aligned}$$

Example 2.2. Blending Problem

A producer of 1-pound cans of mixed nuts must determine how much of two mixtures *A* and *B* to blend so that his profit is maximized. The composition of each mixture and supply of each type of nut is given in Table 2.2. The costs of cashews, peanuts, and almonds are \$1.60, \$0.80, and \$1.00 per pound respectively. Mixture *A* sells for \$2 per pound, while mixture *B* sells for \$1.80 per pound. Any nuts left over after the mixtures have been blended may be sold at cost.

Table 2.2. Composition (in %) of Mixtures *A* and *B*

Mixture	Cashew	Peanut	Almond
A	0.25	0.50	0.25
B	0.25	0.75	0
Supply (lb)	375	1000	300

Let x_1 and x_2 be the number of cans of mixtures *A* and *B* to be prepared respectively. The linear programming problem is

$$\begin{aligned} \text{maximize} \quad & z = f(x_1, x_2) = p_1 x_1 + p_2 x_2 \\ \text{subject to} \quad & 0.25x_1 + 0.25x_2 \leq 375 \\ & 0.50x_1 + 0.75x_2 \leq 1000 \\ & 0.25x_1 \leq 300 \\ & x_1, x_2 \geq 0 \end{aligned}$$

where

$$p_1 = \$2.00 - [0.25(\$1.60) + 0.50(\$0.80) + 0.25(\$1.00)] = \$0.95$$
$$p_2 = \$1.80 - [0.25(\$1.60) + 0.75(\$0.80)] = \$0.80$$

Example 2.3. Transportation Problem

The Block Company has three factories that supply four of its warehouses with units of one of its products. The number of units produced by each factory and demanded by each warehouse and the cost of shipping one unit from each factory to each warehouse are shown in Table 2.3.

Table 2.3. Shipping Costs per Unit and Marginal Units Produced and Demanded

	Warehouse				
Factory	1	2	3	4	Production
1	\$5	\$7	\$9	\$6	45
2	7	5	6	10	30
3	6	8	12	7	25
Demand	30	10	40	20	100

The Block Company is interested in determining the least expensive means of shipping units from the three factories to the four warehouses, so that each factory ships all units it produces and the demand at each warehouse is satisfied.

Let x_{ij} = number of units shipped from the ith factory to the jth warehouse

c_{ij} = cost of shipping one unit from the ith factory to the jth warehouse

a_i = number of units produced at the ith factory

b_j = number of units demanded by the jth warehouse

The problem may now be stated as a linear programming problem:

$$
\begin{aligned}
\text{minimize} \quad & z = f(x_{11}, x_{12}, \ldots, x_{34}) = \textstyle\sum_{i=1}^{3} \sum_{j=1}^{4} c_{ij} x_{ij} \\
\text{subject to} \quad & \textstyle\sum_{j=1}^{4} x_{ij} = a_i \qquad i = 1, 2, 3 \\
& \textstyle\sum_{i=1}^{3} x_{ij} = b_j \qquad j = 1, 2, 3, 4 \\
& x_{ij} \geq 0 \qquad i = 1, 2, 3; j = 1, 2, 3, 4
\end{aligned}
$$

The first set of three equations insures that the number of units produced by each factory is distributed to the warehouses. The second set of four equations insures that the warehouse demands are satisfied.

Example 2.4. Advertising Allocation

The Exo-Lent advertising firm has under contract a client who wishes to expend \$100,000 for advertising one of its products in the New York metropolitan area. The firm can purchase radio time for the client at \$200 per spot, TV time at \$1000 per spot, and newspaper advertisements at \$400 per insertion. The client asks that at least \$15,000 be spent on TV advertising. Based upon the experience of the firm, the client agrees that no more than 25% of the budget should be used for newspaper advertisements and that newspaper allocation should not exceed 50% of the TV expenditure.

The payoff from the advertising is a measure of the size of the audience reached. Based on past experience of the firm with a similar product, a radio spot is given 40 audience points, a TV spot 160 points, and a newspaper insertion 300 points. The firm wishes to determine the number of dollars to allocate to each advertising medium so that the number of total audience points is maximized.

Let x_1, x_2, and x_3 denote the dollars (in thousands) spent on the radio spots, TV spots, and newspaper insertions respectively. The linear programming problem becomes

$$\begin{aligned}
\text{maximize} \quad z &= f(x_1, x_2, x_3) \\
&= (40/200)x_1 + (160/1000)x_2 + (300/400)x_3 \\
\text{subject to} \quad & x_1 + x_2 + x_3 = 100 \\
& x_2 \geq 15 \\
& x_1 \leq 25 \\
& x_1 - 0.5x_2 \leq 0 \\
& x_1, x_2, x_3 \geq 0
\end{aligned}$$

Since the unit of measure of the decision variables is \$1000 dollars, the costs c_1, c_2, and c_3 in the objective function must be expressed in terms of points per dollar.

The above examples suggest the versatility of linear programming in representing constrained optimization problems. Some limitations restrict its use, however.

1. All functions in the model must be linear in the decision variables. Numerous optimization models in business and economics require a nonlinear objective function or require one or more of the constraints to be nonlinear. The delineation and methods of solving

nonlinear constrained optimization problems are presented in Chapters 3 and 4.

2. Decision variables may enter the model only in an additive way. If two variables interact, nonlinear functions would normally be required to model the interaction effect.
3. Linear programming will produce real-numbered solutions. If integer solutions are required, such as in Examples 2.2 and 2.3, the integer linear programming procedures described in Chapter 6 should ordinarily be used. One might think that linear programming methods could be used, if integer solutions are required, by rounding the solution values to whole numbers. Unfortunately, the integer solution obtained by rounding the values of the variables produced by linear programming may not be the optimal solution to the corresponding integer linear programming problem, as will be seen in Chapter 6.
4. The linear programming model is deterministic. Coefficients in the objective function and in the constraints are considered to be constants. If the coefficients are allowed to be random variables or if the constraint inequalities need only to be satisfied probabilistically, the linear programming model does not apply.

Linear programming has been successfully applied to literally thousands of real-world problems where, in many, one or more of the assumptions listed above did not completely hold. In these cases a more sophisticated model might have produced better results; yet linear programming is easily applied, and much theoretical and computational work on resolving problems by this method has been performed and published. With the availability of high-speed digital computers and computer program codes for solving linear programming problems, it is natural to first attempt to model a constrained optimization problem by linear programming. If the linear model does not prove to be satisfactory, then others, such as those discussed in Chapters 3–7, may be tried.

The general linear programming problem may be stated in matrix notation as

$$\text{optimize } f(\mathbf{x}) = \mathbf{c}'\mathbf{x}$$
$$\text{subject to } \mathbf{A}\mathbf{x}\,\{\leq\ =\ \geq\}\,\mathbf{b} \qquad \mathbf{x} \geq \mathbf{0}$$

where

$$\mathbf{c}' = [c_1, c_2, \ldots, c_n]$$
$$\mathbf{b}' = [b_1, b_2, \ldots, b_m]$$
$$\mathbf{x}' = [x_1, x_2, \ldots, x_n] \qquad (2.6)$$

and

$$\mathbf{A} = \begin{bmatrix} a_{11} & a_{12} & \cdots & a_{1m} \\ a_{21} & a_{22} & \cdots & a_{2m} \\ \vdots & & & \vdots \\ a_{n1} & a_{n2} & \cdots & a_{nm} \end{bmatrix} \tag{2.7}$$

The matrix form is a concise way of stating the model and will be used later in the chapter to present theoretical aspects of linear programming and solution methods.

2.3 GRAPHIC SOLUTION OF LINEAR PROGRAMMING PROBLEMS

If the linear programming problem involves only two variables, a simple graph is often the most efficient solution method. Consider the problem introduced in (2.1)–(2.4) and represented graphically in Figure 2.1. The shaded area represents the constraint region—the set of points (x_1, x_2) that simultaneously satisfy the linear inequalities (2.2)–(2.4). Three contours (lines) of the objective function ($z = 80$,

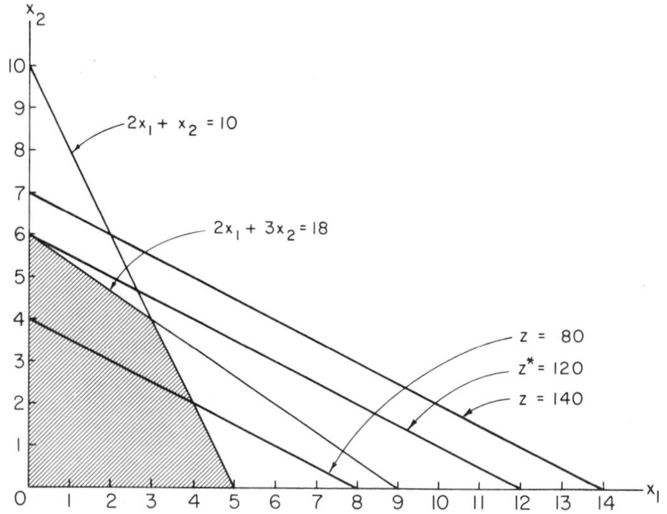

Fig. 2.1. Graphic solution of the linear programming problem in (2.1)–(2.4).

120, and 140) are shown in Figure 2.1. The maximum value of the objective function in the linear programming problem is found by moving the line $10x_1 + 20x_2 = z$ up by increasing values of z until the line contacts the perimeter of the constraint region. The largest value of z that produces a line still in contact with the constraint region is $z = 120$. The point at which the contact is made by the line $10x_1 + 20x_2 = 120$ is $(x_1 = 0, x_2 = 6)$. The solution to this linear programming problem is therefore $x_1^* = 0$, $x_2^* = 6$ (produce no units of the first product and 6 units of the second) and $z^* = 120$ (the maximum profit is \$120). The asterisks will be used to denote optimal solutions for the decision variables and the objective function.

From Figure 2.1 it should be clear that the solution to the linear programming problem must occur at a corner point or along an edge of the constraint region. The point at which the objective function contour will last contact the constraint region as z is increased in Figure 2.1 occurs at the corner point $(x_1 = 0, x_2 = 6)$. If the slope of the objective function had been equivalent to the slope of the line $2x_1 + 3x_2 = 18$ associated with the first constraint (2.2), the solution to the problem would be any pair of x_1 and x_2 values that occurs on the edge of the constraint region, as illustrated in Figure 2.2. In this case, there is not a unique maximizing point for the linear programming problem.

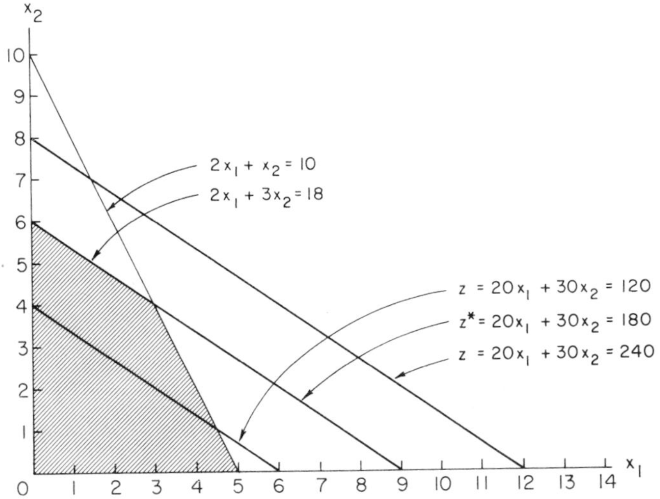

Fig. 2.2. Graphic solution for maximizing $z = 20x_1 + 30x_2$ subject to constraints (2.2)–(2.4).

Example 2.5

By graphic methods, solve the problem

$$\begin{aligned} \text{minimize} \quad & z = 10x_1 + 15x_2 \\ \text{subject to} \quad & x_1 + 4x_2 \geq 8 \\ & 4x_1 + x_2 \geq 12 \\ & x_1, x_2 \geq 0 \end{aligned}$$

The solution to the problem is illustrated graphically in Figure 2.3. The constraint region encompasses the entire first quadrant except for the area not shaded in Figure 2.3. By decreasing the value of z, it is seen that the minimum value is $z^* = 46.62$, which occurs at the point ($x_1^* = 2.66$, $x_2^* = 1.33$). Notice that the maximum of $z = 10x_1 + 15x_2$ is unbounded, since larger and larger values of x_1 and x_2 would continue to increase the magnitude of z.

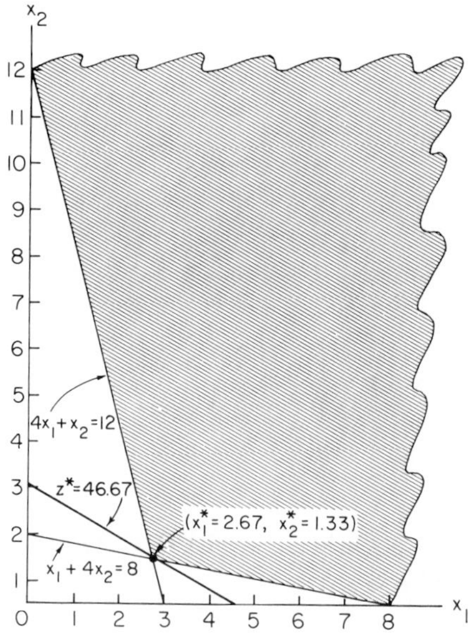

Fig. 2.3. Graphic solution to the linear programming problem in Example 2.5.

The graphic solution method is ill-suited for solving complex linear programming problems involving three or more variables, since graphing functions with three variables (3-space) is tedious and with four or more variables it becomes impossible. The graphic method does illustrate two very important features of linear programming problems, however. First, the solution will occur at a corner point (or along an edge if the solution is not unique) of the constraint region. Second, the constraint region forms a convex set of points. A set of points S is convex if for any two points in the set (say a and b) all points on the line connecting them are also in S. Figure 2.4 illustrates three sets that are convex and two that are not convex (nonconvex).

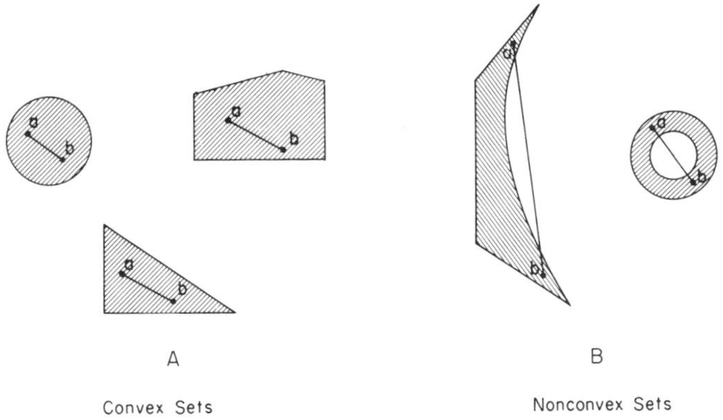

Fig. 2.4. Illustration of convex (A) and nonconvex (B) sets.

These two features (corner point solutions and convex set constraint regions) can be used to construct highly efficient solution methods to linear programming problems, since only corner points are contenders for the optimal solution and they can be identified by moving along the linear edges of the convex constraint region. Before these methods can be fully explained, it will be necessary to first present and discuss several linear programming definitions and theorems.

2.4 LINEAR PROGRAMMING DEFINITIONS AND THEOREMS

Suppose that the linear programming model contains all three types of constraints as shown below:

$$\begin{aligned}
\text{optimize} \quad & z = c_1x_1 + c_2x_2 + \cdots + c_nx_n \\
\text{subject to} \quad & \textstyle\sum_{j=1}^{n} a_{ij}x_j \le b_i \qquad i = 1, 2, \ldots, k & (2.8)\\
& \textstyle\sum_{j=1}^{n} a_{ij}x_j \ge b_i \qquad i = k + 1, \ldots, s & (2.9)\\
& \textstyle\sum_{j=1}^{n} a_{ij}x_j = b_i \qquad i = s + 1, \ldots, m \\
& x_j \ge 0 \qquad j = 1, 2, \ldots, n
\end{aligned}$$

Many of the definitions and theorems relating to linear programming depend on the model being in such a form that all constraints are represented as equalities. The first set of inequalities (2.8) can be transformed to equalities by writing the ith inequality as

$$\sum_{j=1}^{n} a_{ij}x_j + x_{n+i} = b_i \qquad 1 \le i \le k$$

The new variable x_{n+i} has taken up the slack in the "less than or equal to" inequality and is therefore called a slack variable. The second set of inequalities (2.9) can be transformed in a similar manner by writing the ith inequality as

$$\sum_{j=1}^{n} a_{ij}x_j - x_{n+i} = b_i \qquad k + 1 \le i \le s$$

The new variable x_{n+i} is called a surplus variable, since it represents the amount by which the ith "greater than or equal to" inequality exceeds b_i.

In matrix form the linear programming problem may now be written as

$$\text{optimize} \quad z = \bar{\mathbf{c}}'\bar{\mathbf{x}} \tag{2.10}$$
$$\text{subject to} \quad \bar{\mathbf{A}}\bar{\mathbf{x}} = \mathbf{b} \tag{2.11}$$
$$\bar{\mathbf{x}} \ge \mathbf{0} \tag{2.12}$$

where

$$\bar{\mathbf{c}}' = [c_1, c_2, \ldots, c_n; c_{n+1}, \ldots, c_{n+k}; c_{n+k+1}, \ldots, c_{n+s}]$$
$$= [\mathbf{c}', \mathbf{c}'_{sl}, \mathbf{c}'_{su}] \tag{2.13}$$
$$\bar{\mathbf{x}}' = [x_1, x_2, \ldots, x_n; x_{n+1}, \ldots, x_{n+k}; x_{n+k+1}, \ldots, x_{n+s}]$$
$$= [\mathbf{x}', \mathbf{x}'_{sl}, \mathbf{x}'_{su}] \tag{2.14}$$

$$\bar{\mathbf{A}} = \begin{bmatrix} a_{11} & a_{12} & \cdots & a_{1n}; & 1 & 0 & \cdots & 0; & 0 & 0 & \cdots & 0 \\ a_{21} & a_{22} & \cdots & a_{2n}; & 0 & 1 & \cdots & 0; & 0 & 0 & \cdots & 0 \\ \vdots & & & & & & & & & & & \\ a_{k1} & a_{k2} & \cdots & a_{kn}; & 0 & 0 & \cdots & 1; & 0 & 0 & \cdots & 0 \\ a_{k+1,1} & a_{k+1,2} & \cdots & a_{k+1,n}; & 0 & 0 & \cdots & 0; & 1 & 0 & \cdots & 0 \\ \vdots & & & & & & & & & & & \\ a_{s1} & a_{s2} & \cdots & a_{sn}; & 0 & 0 & \cdots & 0; & 0 & 0 & \cdots & 1 \\ a_{s+1,1} & a_{s+1,2} & \cdots & a_{s+1,n}; & 0 & 0 & \cdots & 0; & 0 & 0 & \cdots & 0 \\ \vdots & & & & & & & & & & & \\ a_{m1} & a_{m2} & \cdots & a_{mn}; & 0 & 0 & \cdots & 0; & 0 & 0 & \cdots & 0 \end{bmatrix}_{m \times (n+s)} = [\mathbf{A}, \mathbf{A}_s] \tag{2.15}$$

and where the $(1 \times n)$ vectors $\mathbf{c}'$ and $\mathbf{x}'$, the $(1 \times m)$ vector $\mathbf{b}'$, and the $(m \times n)$ matrix $\mathbf{A}$ are given by (2.6) and (2.7).

It is natural to write the linear programming problem in this form since it is noted in Section 2.3 that only corner points need to be considered in solving the problem. Corner point solutions, if they exist, can be determined by solving the system of linear equations in (2.11) for $\bar{\mathbf{x}}$. To formalize what represents a solution to the linear programming problem in relation to the set of m linear equations in the $n + s$ variables, a series of definitions and theorems will now be presented.

Definition 2.1 A *feasible solution* to a linear programming problem is a vector $\bar{\mathbf{x}}' = [x_1, x_2, \ldots, x_{n+s}]$, which satisfies the m equality constraints in (2.11) and the nonnegativity restriction (2.12).

Definition 2.2 An *optimal solution* to a linear programming problem is a feasible solution that makes the problem's objective function an optimum (either a minimum or a maximum, whichever is requested in the problem).

From among the feasible solutions, the requirement is to select one that provides the optimal value of the objective function. It may be that more than one feasible solution gives the same optimal value for the objective function, as mentioned in Section 2.3. To determine an optimum, it would be sufficient to check each and every possible feasible solution, but fortunately this is not necessary.

To see how the number of contenders for an optimal solution can be decreased to a manageable set, consider the system of equations in (2.11):

$$\bar{\mathbf{A}}_{m \times (n+s)} \bar{\mathbf{x}}_{(n+s) \times 1} = \mathbf{b}_{m \times 1}$$

First, under what circumstances will a solution for $\bar{\mathbf{x}}$ exist? From results in linear algebra, it is well known that at least one solution exists if and only if the rank[1] of $\bar{\mathbf{A}}$ equals the rank of $[\bar{\mathbf{A}}, \mathbf{b}]$, the $m \times (n + s + 1)$ matrix formed by augmenting the column vector $\mathbf{b}$ to $\bar{\mathbf{A}}$. If the ranks are equal, either of two cases must have occurred. The first case occurs when the rank of $\bar{\mathbf{A}}$ is $n + s$, in which case the system of equations will possess a unique solution. If the rank of $\bar{\mathbf{A}}$ is less than $n + s$, the second case occurs in which there will be an infinite number of solutions.

---

1. The rank of a matrix is the greatest number of rows or columns (the number is equal) that can be used to form a nonzero determinant. See Hadley [1961] and Hohn [1958].

If the first case occurs, $m \geq n + s$. If $m > n + s$, there must be $m - (n + s)$ redundant equations; these may be eliminated from the system without altering the unique solution. As will be seen, this case can present computational difficulties in solving linear programming problems unless it is recognized in the model construction phase of the solution process.

If the ranks of $\bar{\mathbf{A}}$ and $[\bar{\mathbf{A}}, \mathbf{b}]$ are both $r < n + s$, any set of r variables may be solved in terms of the remaining $n + s - r$ variables. Equivalently, the $n + s - r$ variables can be assigned arbitrary values, and the solution for the remaining r variables can be found. To illustrate these results, consider the following examples where the number of variables $n + s$ has been set at two.

Example 2.6. $(m = 1)$

$$2x_1 + 4x_2 = 8$$

In this example, the ranks of $\bar{\mathbf{A}} = [2, 4]$ and $[\bar{\mathbf{A}}, \mathbf{b}] = [2, 4, 8]$ are both one; x_1 may be solved for in terms of x_2 $(x_1 = 4 - 2x_2)$, or x_2 may be solved for in terms of x_1, $(x_2 = 2 - 0.5x_1)$.

Example 2.7. $(m = 2)$

Illustration 2.7.1:
$$\begin{aligned} 2x_1 + 4x_2 &= 8 \\ 2x_1 + 2x_2 &= 6 \end{aligned}$$

The ranks of $\bar{\mathbf{A}}$ and $[\bar{\mathbf{A}}, \mathbf{b}]$ are both two; the unique solution is $(x_1 = 2, x_2 = 1)$.

Illustration 2.7.2:
$$\begin{aligned} 2x_1 + 4x_2 &= 8 \\ 4x_1 + 8x_2 &= 16 \end{aligned}$$

The ranks of $\bar{\mathbf{A}}$ and $[\bar{\mathbf{A}}, \mathbf{b}]$ are both one; one variable (say x_1) may be solved for in terms of the other $(x_1 = 4 - 2x_2)$.

Illustration 2.7.3:
$$\begin{aligned} 2x_1 + 4x_2 &= 8 \\ x_1 + 2x_2 &= 2 \end{aligned}$$

The rank of $\bar{\mathbf{A}}$ is one while the rank of $[\bar{\mathbf{A}}, \mathbf{b}]$ is two; no solution exists.

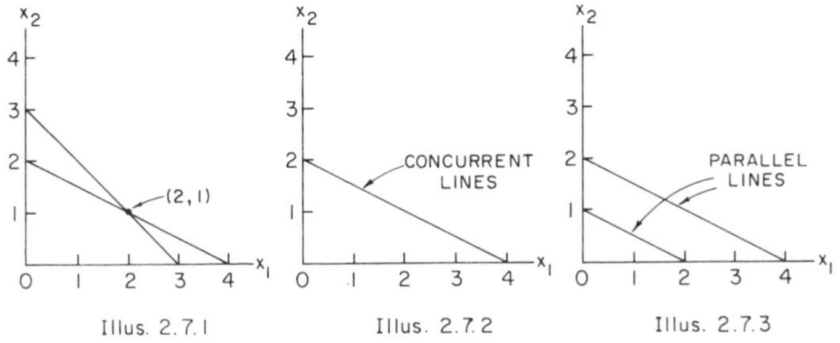

Fig. 2.5. Graphic presentation of the illustrations in Example 2.7.

The three illustrations in Example 2.7 are shown graphically in Figure 2.5.

Example 2.8. $(m = 3)$

Illustration 2.8.1:
$$\begin{aligned} x_1 + x_2 &= 2 \\ 2x_1 + x_2 &= 3 \\ 3x_1 + x_2 &= 4 \end{aligned}$$

The ranks of $\bar{\mathbf{A}}$ and $[\bar{\mathbf{A}}, \mathbf{b}]$ are both two; a unique solution exists $(x_1 = 1, x_2 = 1)$.

Illustration 2.8.2:
$$\begin{aligned} x_1 + x_2 &= 2 \\ 2x_1 + 2x_2 &= 4 \\ 4x_1 + 4x_2 &= 8 \end{aligned}$$

The ranks of $\bar{\mathbf{A}}$ and $[\bar{\mathbf{A}}, \mathbf{b}]$ are both one; one variable (say x_1) may be expressed in terms of the other $(x_1 = 2 - x_2)$.

Illustration 2.8.3:
$$\begin{aligned} x_1 + x_2 &= 2 \\ 2x_1 + x_2 &= 3 \\ 3x_1 + 2x_2 &= 6 \end{aligned}$$

The rank of $\bar{\mathbf{A}}$ is two while the rank of $[\bar{\mathbf{A}}, \mathbf{b}]$ is three; no solution exists.

The three illustrations in Example 2.8 are shown graphically in Figure 2.6. The problems in Examples 2.6–2.8 do not illustrate all the graphic possibilities for the cases presented (e.g., three parallel lines in

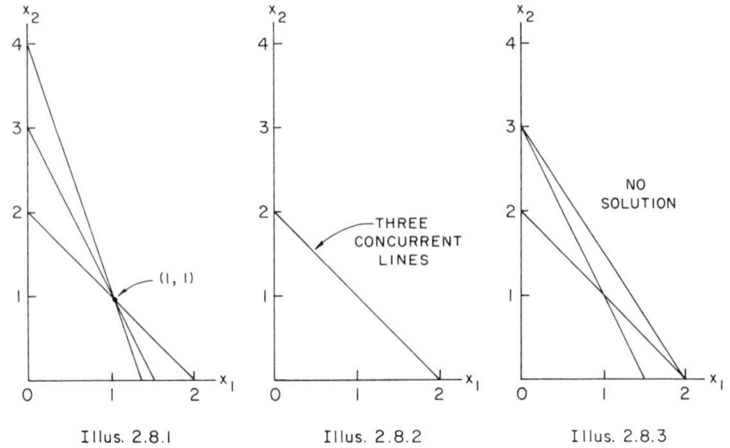

Fig. 2.6. Graphic presentation of the illustrations in Example 2.8.

Example 2.8—no solution), but they do indicate the variety of possibilities. Normally, the case in Example 2.6 ($m < n + s$) will occur in linear programming.

Definition 2.3 A *basic feasible solution* to a linear programming problem is a feasible solution with no more than m positive variables [the remaining variables, if any, must be zero to satisfy the nonnegativity restriction (2.12)]. The variables that are positive in the vector $\bar{\mathbf{x}}$ correspond to the linearly independent columns of the matrix $\bar{\mathbf{A}}$.

Definition 2.4 A *basic feasible optimal solution* to a linear programming problem is a basic feasible solution that optimizes the objective function z.

Definition 2.5 A *degenerate basic feasible solution* is a basic feasible solution with fewer than m positive variables. If exactly m variables are positive, the basic feasible solution is nondegenerate.

For a nondegenerate linear programming problem, none of the basic feasible solutions will have a row of coefficients in $\bar{\mathbf{A}}$ that is a linear combination of the other rows. A basic feasible solution to a nondegenerate linear programming problem will have exactly m positive variables and $n + s - m$ variables set equal to zero.

Example 2.9

Find the set of basic feasible solutions to the following linear programming problem:

$$\begin{aligned} \text{maximize} \quad & z = 10x_1 + 15x_2 \\ \text{subject to} \quad & x_1 + 4x_2 \leq 20 \\ & 4x_1 + x_2 \leq 35 \\ & x_1, x_2 \geq 0 \end{aligned}$$

The constraint set is first converted to the equality form as in (2.11) by using two slack variables x_3 and x_4:

$$\begin{aligned} x_1 + 4x_2 + x_3 \quad\quad &= 20 \\ 4x_1 + x_2 \quad\quad + x_4 &= 35 \end{aligned}$$

In this form, $m = 2, n + s = 2 + 2 = 4$, and

$$\bar{\mathbf{A}} = \begin{bmatrix} 1 & 4 & 1 & 0 \\ 4 & 1 & 0 & 1 \end{bmatrix} \qquad [\bar{\mathbf{A}}, \mathbf{b}] = \begin{bmatrix} 1 & 4 & 1 & 0 & 20 \\ 4 & 1 & 0 & 1 & 35 \end{bmatrix}$$

Since the ranks of $\bar{\mathbf{A}}$ and $[\bar{\mathbf{A}}, \mathbf{b}]$ are both two, two of the variables may be solved for in terms of the other two. There are six possible solutions to the system of equations with $4 - 2 = 2$ variables set equal to zero; these are shown in Table 2.4. The first four solutions are basic feasible since they satisfy the requirement that $m = 2$ variables are positive and $n + s - m = 4 - 2 = 2$ are zero. The last two solutions are not basic feasible since they contain negative values of variables.

Table 2.4. Set of Solutions to Example 2.9 Equations with Two Variables Set Equal to Zero

Solution	x_1	x_2	x_3	x_4
1	8	3	0	0
2	0	0	20	35
3	8.75	0	11.25	0
4	0	5	0	30
5	20	0	0	−45
6	0	35	−120	0

The optimal basic feasible solution to the problem in Ex-

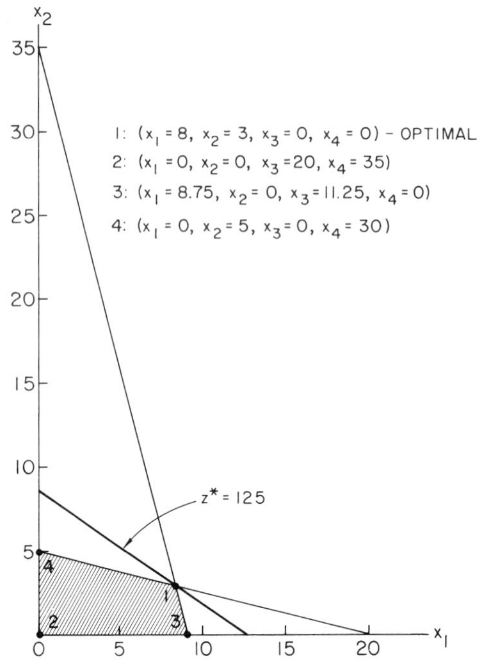

Fig. 2.7. Four basic feasible solutions to the problem in Example 2.9.

ample 2.9 is ($x_1^* = 8, x_2^* = 3, x_3^* = 0, x_4^* = 0$), since among the basic feasible solutions it maximizes the objective function $f^*(8, 3) = 125$.

The four basic feasible solutions and the optimal basic feasible solution are shown in Figure 2.7. Notice that the four basic feasible solutions correspond to the corner points of the constraint region. Thus the role of basic feasible solutions to a linear programming problem is of paramount importance, for these solutions correspond to corner points of the constraint region and the optimal solution will occur at a corner point. To insure that only basic feasible solutions need to be considered when solving a linear programming problem, certain conditions must be satisfied. These conditions and guarantees of optimality follow from a set of linear programming theorems. Unfortunately, some of these theorems and many computational methods will break down if degenerate basic feasible solutions exist, for example. Before turning to the theorems, consider the following problem in which degeneracy exists.

Example 2.10

$$\begin{aligned} \text{maximize} \quad z &= 10x_1 + 15x_2 \\ \text{subject to} \quad x_1 + x_2 &= 2 \\ 3x_1 + 2x_2 &\le 6 \\ x_1, x_2 &= 0 \end{aligned}$$

The equality constraint set is

$$\begin{aligned} x_1 + x_2 \qquad &= 2 \\ 3x_1 + 2x_2 + x_3 &= 6 \end{aligned}$$

It is clear that there exists a degenerate basic feasible solution ($x_1 = 2, x_2 = x_3 = 0$). This solution contains less than $m = 2$ positive variables. The reason for this may be seen in Figure 2.8; the inequality constraint $3x_1 + 2x_2 \le 6$ is redundant. The shaded region represents the set of points satisfying this constraint in the first quadrant.

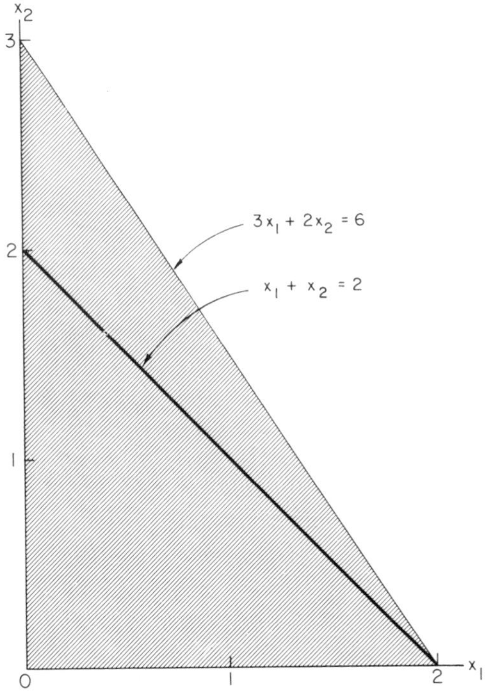

Fig. 2.8. Graph of constraints in the Example 2.10 problem.

The line $x_1 + x_2 = 2$ is contained entirely in this region in the first quadrant. Therefore, the constraint $3x_1 + 2x_2 \leq 6$ could be dropped. The basic feasible solutions to the remaining constraint are $(x_1 = 2, x_2 = 0)$ and $(x_1 = 0, x_2 = 2)$ since $m = 1$ and $n + s = 2$.

As suggested by Example 2.10, degenerate linear programming problems are rather pathological; they rarely occur in practice. In Section 2.5, when the simplex solution algorithm is presented, more will be said about the potential difficulties caused by degeneracy and how the problem can be identified in the simplex calculations.

Theorems 2.1–2.4 are stated and discussed, but not formally proved. For the interested reader, references are given where detailed proofs of the theorems may be found.

Theorem 2.1 The set of feasible solutions to a linear programming problem is convex.

This theorem is based upon the following facts:

1. A linear function is both convex and concave (see Appendix B for definitions of convex and concave functions and convex sets).
2. The set of points X such that $g(\mathbf{x}) \leq \mathbf{b}$, $\mathbf{x} \geq \mathbf{0}$ forms a convex set of points if $g(\mathbf{x})$ is a convex function for all positive values of $\mathbf{x}$. Additionally, the set of points X such that $h(\mathbf{x}) \geq \mathbf{b}$, $\mathbf{x} \geq \mathbf{0}$ is convex if $h(\mathbf{x})$ is a concave function for all positive values of $\mathbf{x}$.
3. The intersection of convex sets is a convex set.

These facts may be applied to a linear programming problem as follows. The function $g_i(\mathbf{x})$, representing the left-hand side of the ith constraint, where

$$g_i(\mathbf{x}) = \sum_{j=1}^{n} a_{ij}x_j\{\leq \; = \; \geq\}b_i \qquad 1 \leq i \leq m$$

is a linear function and therefore is both convex and concave in $\mathbf{x}$. The set of points X such that $g_i(\mathbf{x})\{\leq \; = \; \geq\}b_i$ forms a convex set for the ith constraint by fact (2) (if equality holds, $g_i(\mathbf{x}) = b_i$, then the set of convex points is composed of all points on this line). The intersection of all convex sets formed from each of the m constraints is a convex set by fact (3). This convex set is the set of all feasible solutions to the linear programming problem.

Definition 2.6 A vector point $\mathbf{x}$ is called a *corner point* of a convex set if and only if no other points $\mathbf{x}_1$ and $\mathbf{x}_2$ exist in the set such that

$$\mathbf{x} = \lambda \mathbf{x}_1 + (1 - \lambda)\mathbf{x}_2 \qquad 0 < \lambda < 1$$

If the vector $\mathbf{x}$ has two components, the condition for $\mathbf{x}$ to be a corner point is that two points $\mathbf{x}_1$ and $\mathbf{x}_2$ in the set do not exist such that

$$\mathbf{x} = \begin{bmatrix} \mathbf{x}_1 \\ \mathbf{x}_2 \end{bmatrix} = \lambda \begin{bmatrix} x_{11} \\ x_{12} \end{bmatrix} + (1 - \lambda) \begin{bmatrix} x_{21} \\ x_{22} \end{bmatrix} = \lambda \mathbf{x}_1 + (1 - \lambda)\mathbf{x}_2$$

for $0 < \lambda < 1$. For this two-variable case, the geometric interpretation is that no two points in the set exist such that the corner point $\mathbf{x}$ lies on a line connecting them. Examples of noncorner and corner points for the two-dimensional case are illustrated in Figure 2.9.

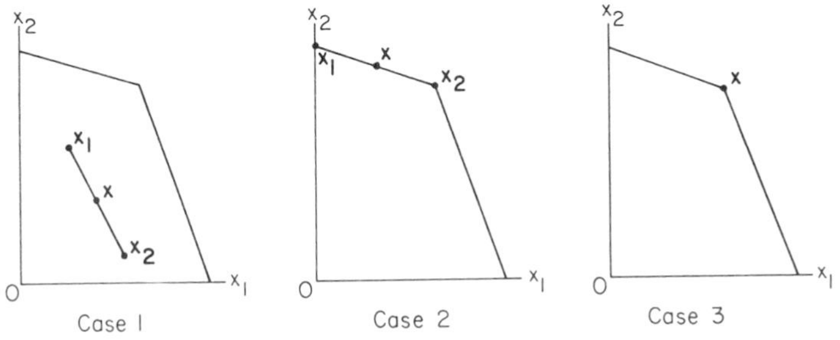

Fig. 2.9. Illustration of two noncorner points (Cases 1 and 2) and a corner point (Case 3).

Theorem 2.2 The optimal solution to a linear programming problem occurs at a corner point of the convex set of feasible solutions.

This may be seen to be true for the two-variable case by observing Figure 2.10. Various convex sets formed from linear inequalities are illustrated. The asterisk in each set indicates the point at which the maximum occurs for the objective function z, where the contours of z correspond to increasing values as they move away from the origin toward larger values of x_1 and x_2. Proofs of Theorems 2.1 and 2.2 may be found by consulting Hadley [1962, pp. 83–93].

The first two theorems simply reiterate what has been said; namely, the optimum of a linear programming problem will occur at a

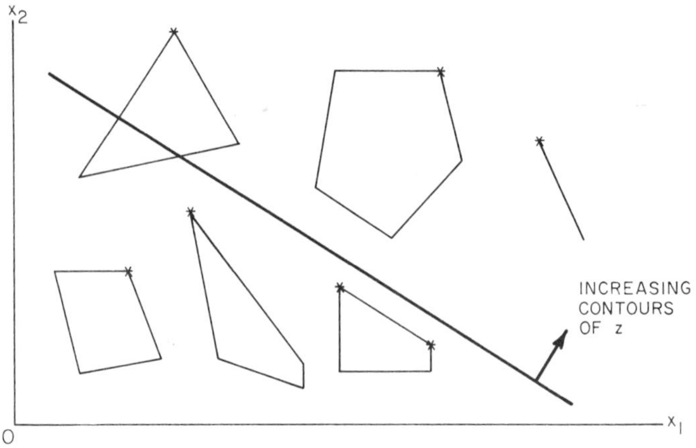

Fig. 2.10. Corner point solutions to linear programming problems.

corner point of the constraint region. The basic feasible solutions must now be tied in with these corner points.

Theorem 2.3 Suppose that $\bar{\mathbf{x}}_0$ is a basic feasible solution. Then $\bar{\mathbf{x}}_0$ is a corner point in the convex set of feasible solutions.

By definition, if $\bar{\mathbf{x}}_0$ is a basic feasible solution, no more than m components of $\bar{\mathbf{x}}_0$ are positive and these components correspond to linearly independent columns of the matrix $\bar{\mathbf{A}}$. The rows or columns of a square matrix are linearly independent if the determinant they can form has a value different from zero. The rank of an $n \times m$ matrix is the largest possible number of linearly independent rows or columns that occur. The row rank (number of independent rows) and column rank (number of independent columns) of a matrix are equal. Assume that $r = m$ components of $\bar{\mathbf{x}}_0$ are positive and these are associated with the first r columns of $\bar{\mathbf{A}}$, denoted by $\bar{\mathbf{a}}_1, \bar{\mathbf{a}}_2, \ldots, \bar{\mathbf{a}}_r$. If the r independent column vectors are not the first r in the set of $n + s$ column vectors in $\bar{\mathbf{A}}$, the vectors can be appropriately renumbered so that they become the first r column vectors.

Theorem 2.3 may be verified by contradiction. Assume that $\bar{\mathbf{x}}_0$ is not a corner point in the set. This implies that there are two points $\bar{\mathbf{x}}_1$ and $\bar{\mathbf{x}}_2$ in the set, such that for $0 < \lambda < 1$, $\bar{\mathbf{x}}_0 = \lambda\bar{\mathbf{x}}_1 + (1 - \lambda)\bar{\mathbf{x}}_2$. However, the last $n + s - r$ components of $\bar{\mathbf{x}}_0$ are zero; and since λ cannot be zero or one, it follows that the last $n + s - r$ components of $\bar{\mathbf{x}}_1$ and $\bar{\mathbf{x}}_2$ are zero. Thus the $1 \times (n + s)$ row vectors $\bar{\mathbf{x}}_1'$ and $\bar{\mathbf{x}}_2'$ must appear as

$$\bar{\mathbf{x}}_1' = [x_{11}, x_{12}, \ldots, x_{1r}; 0 \cdots 0]$$
$$\bar{\mathbf{x}}_2' = [x_{21}, x_{22}, \ldots, x_{2r}; 0 \cdots 0]$$

Since $\bar{\mathbf{x}}_0, \bar{\mathbf{x}}_1$, and $\bar{\mathbf{x}}_2$ are feasible points, it follows that

$$x_{01}\bar{\mathbf{a}}_1 + x_{02}\bar{\mathbf{a}}_2 + \cdots + x_{0r}\bar{\mathbf{a}}_r = \mathbf{b}$$
$$x_{11}\bar{\mathbf{a}}_1 + x_{12}\bar{\mathbf{a}}_2 + \cdots + x_{1r}\bar{\mathbf{a}}_r = \mathbf{b}$$
$$x_{21}\bar{\mathbf{a}}_1 + x_{22}\bar{\mathbf{a}}_2 + \cdots + x_{2r}\bar{\mathbf{a}}_r = \mathbf{b}$$

By subtracting the second equation from the first and the third from the first, it follows that

$$(x_{01} - x_{11})\bar{\mathbf{a}}_1 + (x_{02} - x_{12})\bar{\mathbf{a}}_2 + \cdots + (x_{0r} - x_{2r})\bar{\mathbf{a}}_r = \mathbf{0}$$
$$(x_{01} - x_{21})\bar{\mathbf{a}}_1 + (x_{02} - x_{22})\bar{\mathbf{a}}_2 + \cdots + (x_{0r} - x_{2r})\bar{\mathbf{a}}_r = \mathbf{0}$$

But since $\bar{\mathbf{a}}_1, \bar{\mathbf{a}}_2, \ldots, \bar{\mathbf{a}}_r$ are linearly independent vectors,

$$x_{0j} - x_{1j} = 0 \qquad x_{0j} - x_{2j} = 0 \qquad j = 1, 2, \ldots, r$$

This implies that $\bar{\mathbf{x}}_0 = \bar{\mathbf{x}}_1 = \bar{\mathbf{x}}_2$; i.e., $\bar{\mathbf{x}}_0$ cannot be written as a linear combination of $\bar{\mathbf{x}}_1$ and $\bar{\mathbf{x}}_2$. Therefore, $\bar{\mathbf{x}}_0$ is a corner point in the convex set of feasible solution

Theorem 2.4 Suppose that $\bar{\mathbf{x}}_0$ is a corner point in the convex set of feasible solutions. Then $\bar{\mathbf{x}}_0$ is a basic feasible solution.

Assume that the first r components of $\bar{\mathbf{x}}_0$ are nonzero so that $\bar{\mathbf{x}}_0$ may be written as a $1 \times (n + s)$ row vector in the following form: $\bar{\mathbf{x}}_0' = [x_{01}, x_{02}, \ldots, x_{0r}; 0 \cdots 0]$. To show that $\bar{\mathbf{x}}_0$ is a basic feasible solution, it is necessary to verify that the r nonzero components in $\bar{\mathbf{x}}_0$ correspond to linearly independent columns of $\bar{\mathbf{A}}$ and that $r = m$.

First, it will be shown that the r columns in $\bar{\mathbf{A}}$ corresponding to the nonzero components in $\bar{\mathbf{x}}_0$ are independent. This result may be verified by contradiction; assume that the first r columns of $\bar{\mathbf{A}}$, denoted by $\bar{\mathbf{a}}_1, \bar{\mathbf{a}}_2, \ldots, \bar{\mathbf{a}}_r$, are linearly dependent. It must then be true that there exists a set of constants $d_1, d_2, \ldots, d_r$, not all zero, such that

$$d_1\bar{\mathbf{a}}_1 + d_2\bar{\mathbf{a}}_2 + \cdots + d_r\bar{\mathbf{a}}_r = \mathbf{0} \tag{2.16}$$

Now select a constant e sufficiently small so that

$$x_{0i} + ed_i > 0 \qquad x_{0i} - ed_i > 0 \qquad i = 1, 2, \ldots, r$$

and write

$$\bar{\mathbf{x}}_1 = \bar{\mathbf{x}}_0 + e\mathbf{d} = \begin{bmatrix} x_{01} + ed_1 \\ x_{02} + ed_2 \\ \vdots \\ x_{0r} + ed_r \\ 0 \\ \vdots \\ 0 \end{bmatrix} \qquad \bar{\mathbf{x}}_2 = \bar{\mathbf{x}}_0 - e\mathbf{d} = \begin{bmatrix} x_{01} - ed_1 \\ x_{02} - ed_2 \\ \vdots \\ x_{0r} - ed_r \\ 0 \\ \vdots \\ 0 \end{bmatrix} \tag{2.17}$$

where the $1 \times (n + s)$ row vector $\mathbf{d}'$ is given by

$$\mathbf{d}' = [d_1, d_2, \ldots, d_r; 0 \cdots 0] \tag{2.18}$$

If e is chosen small enough, $\bar{\mathbf{x}}_1 \geq 0$ and $\bar{\mathbf{x}}_2 \geq 0$. Furthermore, the $\bar{\mathbf{x}}_1$ and $\bar{\mathbf{x}}_2$ points will belong to the convex set of feasible solutions, since

$$\bar{\mathbf{A}}\bar{\mathbf{x}}_1 = \bar{\mathbf{A}}(\bar{\mathbf{x}}_0 + e\mathbf{d}) = \bar{\mathbf{A}}\bar{\mathbf{x}}_0 + e\bar{\mathbf{A}}\mathbf{d} = \mathbf{b} + e(\mathbf{0}) = \mathbf{b}$$
$$\bar{\mathbf{A}}\bar{\mathbf{x}}_2 = \bar{\mathbf{A}}(\bar{\mathbf{x}}_0 - e\mathbf{d}) = \bar{\mathbf{A}}\bar{\mathbf{x}}_0 - e\bar{\mathbf{A}}\mathbf{d} = \mathbf{b} - e(\mathbf{0}) = \mathbf{b}$$

where $\bar{\mathbf{A}}\mathbf{d} = \mathbf{0}$ by (2.16) and (2.18). From (2.17) it is clear that $\bar{\mathbf{x}}_0 = 0.50\bar{\mathbf{x}}_1 + 0.50\bar{\mathbf{x}}_2$. Since $\bar{\mathbf{x}}_0$ can be written as a linear combination of two different points in the convex set of feasible solutions, it cannot be a corner point. This is a contradiction to the assumption that $\bar{\mathbf{x}}_0$ is a corner point. Therefore, the r columns of $\bar{\mathbf{A}}$ associated with the nonzero components of $\bar{\mathbf{x}}_0$ must be linearly independent.

Since the r column vectors $\bar{\mathbf{a}}_1, \bar{\mathbf{a}}_2, \ldots, \bar{\mathbf{a}}_r$ must be linearly independent, r cannot be greater than m, for each vector contains exactly m components. Thus no more than m components of the corner point $\bar{\mathbf{x}}_0$ may be positive. It follows that $\bar{\mathbf{x}}_0$ is a basic feasible solution. If the corner point $\bar{\mathbf{x}}_0$ has less than m positive components, it can be associated with a degenerate basic feasible solution.

Theorems 2.3 and 2.4 insure that every basic feasible solution to a linear programming problem is a corner point of the convex set of feasible solutions and that every corner point is a basic feasible solution. If the problem is nondegenerate, each basic feasible solution will correspond to one and only one corner point and vice versa. This will not be true, however, if degenerate basic feasible solutions exist. In this case a number of different degenerate basic feasible solutions may correspond to the same corner point.

The reader is directed to Hadley [1962, pp. 95–104] for further discussions related to Theorems 2.3 and 2.4.

In summary, the following results have been developed regarding the solution of a linear programming problem:

1. The optimal solution occurs at a corner point of the convex set of feasible solutions.
2. In the absence of degeneracy, there is a one-to-one correspondence between members of the corner point set and the members of the set of basic feasible solutions.
3. The optimal solution never will have more than m variables different from zero.
4. In the absence of degeneracy, every extreme point will have m linearly independent column vectors in $\bar{\mathbf{A}}$ associated with it.

These results imply for nondegenerate problems that only a finite number of points, the corner points, need to be considered in determining the optimal solution to a linear programming problem. The fact that corner points correspond to basic feasible solutions means that the contending corner points for the solution can be generated by the sets of m linearly independent vectors from the $n + s$ vectors $\bar{\mathbf{a}}_1, \bar{\mathbf{a}}_2, \ldots, \bar{\mathbf{a}}_{n+s}$.

Since the set of corner points is equivalent to the set of basic feasible solutions and these can be determined by solving systems of linear equations where $n + s - m$ of the variables are set to zero, there is a temptation to consider using the method of solving systems of linear equations (the complete description method) to find the optimal solution to a linear programming problem. Recall that this method was used to optimize the problem in Example 2.9; the amount of calculations required to find the optimum was minimal. However, consider applying this method in general. Assuming that $n + s > m$, there will be at most $(n + s)!/m!(n + s - m)!$ basic feasible solutions. In Example 2.9, $n + s = 4$, $m = 2$, and there will be at most $4!/(2!)(2!) = 6$ basic feasible solutions. In fact there were four, since two of the six possible solutions formed by setting two variables to zero were not feasible. It is not uncommon to encounter linear programming problems that involve several hundred variables and literally thousands of constraints. Even with a problem involving 50 variables and 200 constraints, it becomes out of the question to solve all $(50 + s)!/(200!)(50 + s - 200)!$ systems of equations.

By introducing the concept of basic feasible solutions and relating them to corner points in the constraint region, the number of points that needs to be considered in solving a linear programming problem has been reduced from an infinite number (all feasible solutions) to a finite number (the basic feasible solutions). Unfortunately, there may be too many basic feasible solutions to consider when the complete description method is used as a solution method. A more efficient solution method must be devised.

The readers familiar with optimization methods used in economics will probably recall that the method of Lagrangian multipliers can be used to solve constrained optimization problems. Consider the linear programming problem

$$\text{maximize} \quad z = \textstyle\sum_{j=1}^{n+s} c_j x_j \tag{2.19}$$
$$\text{subject to} \quad \textstyle\sum_{j=1}^{n+s} a_{ij} x_j = b_i \qquad i = 1, 2, \ldots, m \tag{2.20}$$
$$x_j \geq 0 \qquad j = 1, 2, \ldots, n + s \tag{2.21}$$

where any necessary slack or surplus variables (s in number) have been introduced so that the constraints may be written in equality form (2.20). The Lagrangian function is formed by augmenting to the objective function z the constraints together with Lagrangian multipliers. The nonnegativity conditions in (2.21) may be converted to equality conditions by adding $n + s$ new variables $(y_1, y_2, \ldots, y_{n+s})$ and writing $x_j = y_j^2, j = 1, 2, \ldots, n + s$.

By using $m + n + s$ Lagrangian multiplier variables $\lambda_1, \lambda_2, \ldots, \lambda_m; \theta_1, \theta_2, \ldots, \theta_{n+s}$, the Lagrangian function $g(\mathbf{x}, \mathbf{y}, \boldsymbol{\lambda}, \boldsymbol{\theta})$ may now be formed:

$$g(\mathbf{x}, \mathbf{y}, \boldsymbol{\lambda}, \boldsymbol{\theta}) = \sum_{j=1}^{n+s} c_j x_j + \sum_{i=1}^{m} \lambda_i \left(b_i - \sum_{j=1}^{n+s} a_{ij} x_j \right) + \sum_{j=1}^{n+s} \theta_j (x_j - y_j^2) \tag{2.22}$$

This function may be maximized with respect to the $3n + 3s + 3m$ variables by methods of differential calculus. The partial derivatives of $g(\mathbf{x}, \mathbf{y}, \boldsymbol{\lambda}, \boldsymbol{\theta})$ are taken with respect to each variable and set to zero; the following system of equations results:

$$\frac{\partial g}{\partial x_j} = c_j - \sum_{i=1}^{m} \lambda_i a_{ij} - \theta_j = 0 \qquad j = 1, 2, \ldots, n + s \tag{2.23}$$

$$\frac{\partial g}{\partial \lambda_i} = b_i - \sum_{j=1}^{n+s} a_{ij} x_j = 0 \qquad i = 1, 2, \ldots, m \tag{2.24}$$

$$\frac{\partial g}{\partial \theta_j} = x_j - y_j^2 = 0 \qquad j = 1, 2, \ldots, n + s \tag{2.25}$$

$$\frac{\partial g}{\partial y_j} = -2y_j\theta_j = 0 \qquad j = 1, 2, \ldots, n + s \tag{2.26}$$

These results can be rewritten as

$$c_j - \sum_{i=1}^{m} \lambda_i a_{ij} - \theta_j = 0 \qquad j = 1, \ldots, n + s$$

$$\sum_{j=1}^{n+s} a_{ij} x_j = b_i \qquad i = 1, \ldots, m$$

$$x_j \theta_j = 0 \qquad j = 1, \ldots, n + s$$

or in matrix form

$$\bar{\mathbf{A}}'\boldsymbol{\lambda} + \boldsymbol{\theta} = \bar{\mathbf{c}}$$

$$\bar{\mathbf{A}}\bar{\mathbf{x}} = \mathbf{b}$$

$$x_j \theta_j = 0 \qquad j = 1, \ldots, n + s$$

From (2.26) if $y_j = 0$, then $x_j = 0$ by observing the condition in (2.25); and if $\theta_j = 0$, then $x_j > 0$. The solution to the linear programming problem can be determined, therefore, by solving the system of equations in (2.23)–(2.26) for each of the 2^{n+s} cases as reflected in (2.26) ($x_j = 0$ or $x_j > 0$). The computations can get out of hand very quickly, even for problems with a modest number of variables.

The Lagrangian method will be fully presented in Chapter 3, where a computational scheme using the method for solving nonlinear problems will be discussed. Additionally, the equivalence between the constrained maximization problem (2.19)–(2.21) and the unconstrained maximization problem (2.22) will be demonstrated.

Since neither the Lagrangian method nor the complete description method (setting $n + s - m$ variables equal to zero and solving the resulting systems of equations as in Example 2.9) are satisfactory computational tools for solving linear programming problems, we might ask why such methods are studied at all in this context. The complete description method will be used extensively in Chapter 10 where stochastic programming problems are studied. The Lagrangian method can be effectively used to demonstrate additional features of linear programming. In Section 2.6 the method will be used, for example, to develop important results in the duality theory of linear programming.

The computational method used most often in solving linear programming problems (for good reason as will be seen) is the simplex algorithm presented in the following section.

2.5 SIMPLEX ALGORITHM

The simplex algorithm is an efficient iterative procedure that will identify the basic feasible solutions in such a way that each new solution is at least as good as the previous one in terms of optimizing the objective function.

In this section the computational aspects are presented and problems are solved by the simplex algorithm to illustrate its application. In Section 2.6 the theory of the simplex algorithm is developed to demonstrate why a linear programming problem can be solved by this method so efficiently.

Consider the problem introduced in the previous section where two slack variables x_3 and x_4 have been added:

$$\begin{aligned} \text{maximize} \quad & z = 10x_1 + 20x_2 + 0x_3 + 0x_4 && (2.27)\\ \text{subject to} \quad & 2x_1 + 3x_2 + x_3 = 18 && (2.28)\\ & 2x_1 + x_2 + x_4 = 10 && (2.29)\\ & x_i \geq 0 \quad i = 1, 2, 3, 4 \end{aligned}$$

Notice that the slack variables in the objective function (2.27) are assigned zero costs ($c_3 = 0$ and $c_4 = 0$) since no profit will be earned if they are nonzero.

Equations (2.28) and (2.29) can be written in the following form:

$$\begin{bmatrix} 2 \\ 2 \end{bmatrix} x_1 + \begin{bmatrix} 3 \\ 1 \end{bmatrix} x_2 + \begin{bmatrix} 1 \\ 0 \end{bmatrix} x_3 + \begin{bmatrix} 0 \\ 1 \end{bmatrix} x_4 = \begin{bmatrix} 18 \\ 10 \end{bmatrix}$$

From observation of these equations a very convenient basic feasible solution is evident: $x_1 = x_2 = 0$ and $x_3 = 18$, $x_4 = 10$. This is a basic feasible solution since exactly $m = 2$ of the $n + s = 2 + 2 = 4$ variables are positive. This solution is illustrated in Table 2.5. The first four columns in the table contain each variable and its coefficients in the two equations. The fifth column contains the variables that are positive in the solution; these will be referred to as *basic variables*. The last column gives the right-hand sides (18, 10) of the two equations. The basic variable and index columns give the basic feasible solution ($x_3 = 18$, $x_4 = 10$). All other variables not in this column are equal to zero and are called *nonbasic variables* ($x_1 = 0$ and $x_2 = 0$). Notice that the basic

Table 2.5. Equations in (2.28) and (2.29) and a Basic Feasible Solution

x_1	x_2	x_3	x_4	Basic variable	Index
2	3	1	0	x_3	18
2	1	0	1	x_4	10

variables x_3 and x_4 correspond to the columns in Table 2.5 that contain a one and a zero.

This basic feasible solution is not very attractive, since it provides a zero value for the objective function; $z = 10(0) + 20(0) + 0(18) + 0(20) = 0$. Thus another basic feasible solution must be found. In the simplex algorithm this is accomplished from the information contained in Table 2.5 in the following manner.

To find another basic feasible solution, a different set of basic variables must be chosen. That is, one of the nonbasic variables x_1 or x_2 must become basic, while one of the basic variables x_3 or x_4 becomes nonbasic; this process will generate a new basic feasible solution. The problem is to select a nonbasic variable to become basic so that the new solution provides a value of the objective function that is at least as good as the one corresponding to the present solution.

In deciding which nonbasic variable to select for addition to the basic set, it seems reasonable to choose the one that will give the greatest per unit increase in the objective function z when all other nonbasic variables are held at zero. First consider x_1. If $x_1 = 1$,

$$2x_1 + 3x_2 + x_3 + 0x_4 = 18 \tag{2.30}$$

and it is clear that $x_3 = 18 - 2(1) = 16$ for this equation to be satisfied. That is, x_3 must be reduced by 2 units. From the second equation,

$$2x_1 + x_2 + 0x_3 + x_4 = 10 \tag{2.31}$$

it follows that x_4 would have to be reduced from 10 to $x_4 = 10 - 2(1) = 8$, a reduction of 2 units. Each unit increase in x_1 will increase the objective function by $10(1) + 20(0) + 0(-2) + 0(-2) = 10$ units.

Now consider adding x_2 to the set of basic variables. From (2.30), it is apparent that a one-unit increase in x_1 results in a three-unit decrease in x_3; i.e., $x_3 = 18 - 3(1) = 15$. From (2.31) x_4 must be decreased by one unit; i.e., $x_4 = 10 - 1(1) = 9$. Thus each unit increase in x_2 will result in an increase in z of $10(0) + 20(1) + 0(-3) + 0(-1) = 20$ units.

Since x_2 gives the greatest per unit increase in z, x_2 is selected to enter the set of basic variables. It now becomes necessary to determine how many units of x_2 can be employed and which basic variable will become nonbasic. From (2.30) it is apparent that x_2 can be at most 6 if this equation is satisfied, since no variables can assume negative values. From (2.31) x_2 can be at most 10 for this equation to be satisfied. Since both equations must be satisfied by the new basic feasible solution, x_2 can be at most 6; (2.30) is more restrictive. By making $x_2 = 6$, the slack variable x_3 in (2.30) must become zero. Thus the basic variable in the present solution, which becomes nonbasic in the new solution x_3, has also been identified in the process of determining which constraint equation is more restrictive as one attempts to increase x_2.

At this point it is known that $x_2 = 6$ and $x_3 = 0$ will be part of the new solution. The complete new basic feasible solution can be determined by applying the Gauss-Jordon elimination process to the equations in Table 2.5. The new set of equations generated by this process should result in a column containing a one and a zero for x_2 since it has been chosen to become a basic variable. The column corresponding to x_2 is called the *pivot column.* Since the first constraint equation is more restrictive, the one in the pivot column should appear in the first row; this is called the *pivot row.* The pivot row and column for introducing x_2 as the new basic variable and dropping x_3 from the basic set in Table 2.5 are shown in Table 2.6. The coefficient $a_{12} = 3$ at the intersection of the pivot row and column is called the *pivot element.* In the new set of equations generated by applying the Gauss-Jordon process, this element must be a one while the other coefficients in the pivot column become zeroes. To make the pivot element a one, the first equation in Table 2.6 (2.30) is divided by 3 resulting in

$$(2/3)x_1 + x_2 + (1/3)x_3 + 0x_4 = 6 \tag{2.32}$$

(The pivot row and column in each simplex tableau are designated by boldface type.) Notice that these equations are equivalent; i.e., any solution satisfying (2.30) must satisfy (2.32) and vice-versa.

Table 2.6. Pivot Row and Column for Applying the Gauss-Jordan Elimination Process

x_1	**Pivot column** $\mathbf{x_2}$	x_3	x_4	Basic variable	Index	
2	**[$\mathbf{a_{12} = 3}$]**	**1**	**0**	$\mathbf{x_3}$	**18**	**Pivot row**
2	**1**	0	1	x_4	10	

The second equation in Table 2.6 (2.31) must be changed so that the coefficient of x_2 becomes zero. The Gauss-Jordon process accomplishes this by subtracting 1/3 times (2.30) from (2.31):

$$\begin{array}{r} 2x_1 + x_2 + 0x_3 + x_4 = 10 \\ -[(2/3)x_1 + x_2 + (1/3)x_3 + 0x_4 = 6] \\ \hline (4/3)x_1 + 0x_2 - (1/3)x_3 + x_4 = 4 \end{array} \tag{2.33}$$

The new second equation (2.33) is not equivalent to the present equation (2.30). For example, the solution ($x_1 = 0, x_2 = 5, x_3 = 0, x_4 = 5$) satisfies (2.31), but not (2.33). However, the Gauss-Jordon procedure guarantees that the set of solutions satisfying the present equations,

$$\begin{aligned} 2x_1 + 3x_2 + x_3 + 0x_4 &= 18 \\ 2x_1 + x_2 + 0x_3 + x_4 &= 10 \end{aligned} \tag{2.34}$$

is equivalent to the set of solutions satisfying the new pair of equations generated by the process,

$$\begin{aligned} (2/3)x_1 + x_2 + (1/3)x_3 + 0x_4 &= 6 \\ (4/3)x_1 + 0x_2 + (-1/3)x_3 + x_4 &= 4 \end{aligned} \tag{2.35}$$

In tabular form the new equations (2.35) contain an obvious new basic feasible solution ($x_1 = 0, x_2 = 6, x_3 = 0, x_4 = 4$) as illustrated in Table 2.7. The reader should check that this new solution satisfies both pairs of equations in (2.34) and (2.35). An objective function value of $z = 10(0) + 20(6) + 0(0) + 0(4) = 120$ corresponds to this new basic feasible solution.

Table 2.7. New Basic Feasible Solution

x_1	x_2	x_3	x_4	Basic variable	Index
2/3	1	1/3	0	x_2	6
4/3	0	−1/3	1	x_4	4

The process of moving from the basic feasible solution portrayed in Table 2.5 to the new basic feasible solution in Table 2.7 by employing the Gauss-Jordan elimination process is the essence of the simplex algorithm. A summary of the steps employed in moving from Table 2.5 to Table 2.7 follows.

1. To identify a new basic feasible solution, add one presently nonbasic variable to the basic set and move a present basic variable to the nonbasic set.
2. Select the nonbasic variable to add to the basic set by identifying the one that contributes the greatest per unit increase in the objective function when the remaining nonbasic variables are held at the zero level.
3. Select the basic variable to drop from the basic set by identifying which equation is most restrictive as units of the newly selected basic variable are added. The basic variable corresponding to the most restrictive equation is the one dropped from the basic set.
4. Apply the Gauss-Jordan method to the present equations in such a way that the column of coefficients associated with the new basic variable contains a one in the pivot row and zeroes in the remaining positions in the pivot column.

Tables 2.5 and 2.7 are called tableaus in the simplex process; they contain the necessary information taken together with the objective function to move from one basic feasible solution to the next.

Since it has not been determined that the best basic feasible solution has been encountered in the tableau in Table 2.7, the search for a better basic feasible solution is undertaken. The two nonbasic variables are x_1 and x_3. If one unit of x_1 is added to the solution, from the two equations (2.35) in Table 2.7 it is apparent that x_2 would have to become (16/3), $[x_2 = 6 - (2/3)(1)]$, a reduction of (2/3)'s of a unit from its present value and x_4 would become (8/3), $[x_4 = 4 - (4/3)(1)]$, a reduction of (4/3)'s of a unit from its present value. The net increase in the objective function by adding one unit of x_1 is $10(1) + 20(-2/3) + 0(0) + 0(-4/3) = -3\ 1/3$. Thus a unit increase in x_1 results in a 3 1/3 unit decrease in the objective function; therefore it is not desirable to make x_1 a basic variable.

If one unit of x_3 is added to the solution, from (2.35) it is apparent that x_2 would be decreased by 1/3 of a unit $[x_2 = 6 - (1/3)(1) = 5\ 2/3]$ from its present value of 6, and x_4 would be increased by 1/3 of a unit $[x_4 = 4 - (-1/3)(1) = 4\ 1/3]$ from its present value of 4. The net increase in the objective function by adding one unit of x_3 is therefore $10(0) + 20(-1/3) + 0(1) + 0(1/3) = -6\ 2/3$. A unit increase in x_3 is not advisable since it will decrease the objective function by 6 2/3 units.

If either of the two nonbasic variables x_1 or x_3 is made basic, the objective function will decrease in value. Therefore, the present solution $(x_1 = 0, x_2 = 6, x_3 = 0, x_4 = 4)$ cannot be improved upon; it is the best feasible solution. This optimal solution is the same one identified by the graphic solution method illustrated in Figure 2.1.

The simplex algorithm solution to this problem is illustrated in Figure 2.11. The algorithm began with the basic feasible solution ($x_1 = 0, x_2 = 0, x_3 = 18, x_4 = 10$) and moved in one step to the optimal basic feasible solution ($x_1^* = 0, x_2^* = 6, x_3^* = 4, x_4^* = 0$).

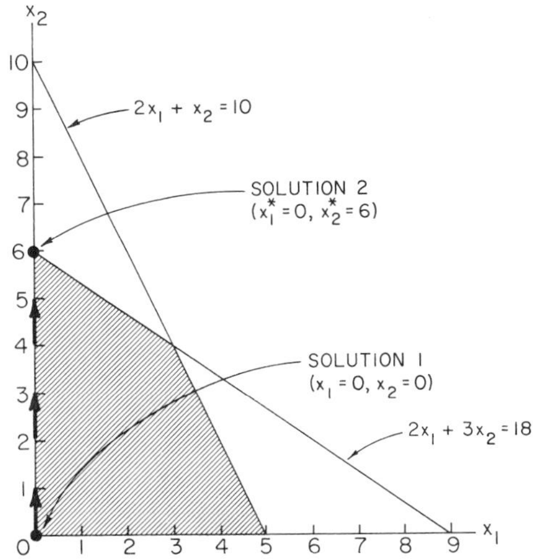

Fig. 2.11. Simplex algorithm solution pattern.

The simplex algorithm will move from corner point to corner point in its generation of new basic feasible solutions in such a way that each newly encountered solution is at least as good as the previous one; this property will be proved in Section 2.6. In this way not all basic feasible solutions need to be enumerated. In the above problem it was not necessary to identify and use the basic feasible solution ($x_1 = 5, x_2 = 0, x_3 = 8, x_4 = 0$), for example. Only one step or iteration of the simplex algorithm is required to identify the optimal basic feasible solution in this problem. Ordinarily, the simplex algorithm will require more than one iteration, but will only need to generate a small subset of the set of all basic feasible solutions.

Pivot Row and Column Rules

Suppose there are two variables $n = 2$ and three "less than or equal to" constraints $m = 3$ in a linear programming problem, and at the tth iteration in the algorithm the corresponding tableau is the one illustrated in Table 2.8.

Table 2.8. Typical Tableau at the *t*th Iteration

Cost coefficient:	c_1	c_2	c_3	c_4	c_5	Basic	
Variable:	x_1	x_2	x_3	x_4	x_5	variable	Index
	a_{11}^t	a_{12}^t	1	0	0	x_3	b_1^t
	a_{21}^t	a_{22}^t	0	1	0	x_4	b_2^t
	a_{31}^t	a_{32}^t	0	0	1	x_5	b_3^t

Note: a_{ij}^t, b_i^t represent the coefficients and index values at the *t*th simplex iteration.

Assume that the present solution $(x_1 = 0, x_2 = 0, x_3 = b_1^t, x_4 = b_2^t, x_5 = b_3^t)$ is not optimal and either x_1 or x_2 will become a basic variable in the $(t + 1)$st tableau. The per unit increase in z resulting from a one-unit increase in x_1 can be determined as follows. From the three equations in the tableau,

$$\begin{aligned} a_{11}^t x_1 + a_{12}^t x_2 + x_3 \qquad\qquad &= b_1^t \\ a_{21}^t x_1 + a_{22}^t x_2 \qquad + x_4 \qquad &= b_2^t \\ a_{31}^t x_1 + a_{32}^t x_2 \qquad\qquad + x_5 &= b_3^t \end{aligned}$$

if x_1 increases by one unit, x_3, x_4, and x_5 must decrease by a_{11}^t, a_{21}^t, and a_{31}^t units when x_2 is held at zero. Thus the per unit increase in z is $c_1(1) + c_2(0) + c_3(-a_{11}^t) + c_4(-a_{21}^t) + c_5(-a_{31}^t)$, or in vector notation,

$$c_1 - [c_3, c_4, c_5]\begin{bmatrix} a_{11}^t \\ a_{21}^t \\ a_{31}^t \end{bmatrix} = c_1 - z_1 \tag{2.36}$$

where

$$z_1 = [c_3, c_4, c_5]\begin{bmatrix} a_{11}^t \\ a_{21}^t \\ a_{31}^t \end{bmatrix}$$

Notice that the scalar z_1 is the inner product of the vector of cost coefficients associated with the *basic variables* x_3, x_4, x_5 and the *column coefficients* corresponding to x_1 in the tableau. The quantity $c_1 - z_1$ is called the *simplex criterion.* It represents the increase in the objective function experienced by adding one unit of the nonbasic variable x_1 to the solution when the other nonbasic variables remain at zero.

The simplex criterion for the other nonbasic variable in the tableau is

$$c_2 - z_2 = c_2 - [c_3, c_4, c_5]\begin{bmatrix} a^t_{12} \\ a^t_{22} \\ a^t_{32} \end{bmatrix}$$

Consider the partial tableau illustrated in Table 2.9. The simplex criterion for the nonbasic variable x_2 is

$$c_2 - [c_1, c_5, c_3]\begin{bmatrix} a^t_{12} \\ a^t_{22} \\ a^t_{32} \end{bmatrix} = c_2 - z_2$$

Notice that the cost coefficients of the basic variables must correspond to the appropriate elements of the column vector associated with x_2. The reader should verify this by the procedure used to write (2.36).

Table 2.9. Computing the Simplex Criterion

Cost coefficient:	$[c_1]$	c_2	$[c_3]$	c_4	$[c_5]$	Basic
Variable:	x_1	x_2	x_3	x_4	x_5	variable
	(1)	$[a^t_{12}]$	0	a^t_{14}	0	x_1
	0	$[a^t_{22}]$	0	a^t_{24}	(1)	x_5
	0	$[a^t_{32}]$	(1)	a^t_{34}	0	x_3

Once the pivot column has been determined, the pivot row is selected in the following way. Suppose that x_1 in Table 2.8 corresponds to the pivot column. The fraction formed by dividing the index by the element in the same row in the pivot column determines how many units of the entering variable x_1 may be added to the solution without having the constraint associated with the row violated. The pivot row is selected by determining which fraction r_i is the smallest *nonnegative* number, where r_i is the index divided by the element in the pivot column corresponding to the ith row. The minimum r_i determines the maximum amount of x_1 that may be brought into the basis such that *all* constraints are satisfied. If the fraction is infinite or negative, it need not be considered, since the associated constraint will not restrict the nonbasic variable chosen to join the set of basic variables. This will be shown in Section 2.6.

In Table 2.10, the computation of the fractions r_1, r_2, and r_3 is shown, given that x_1 is the entering variable. Thus the criterion for selecting the pivot row is to determine

$$\min_{r_i \geq 0} (r_i = b_i^t / a_{ij}^t) \qquad i = 1, 2, \cdots m \tag{2.37}$$

where j is the index of the pivot column. The row in which the minimum r_i occurs is the pivot row.

Table 2.10. Determination of the Pivot Row

Pivot column $\mathbf{x}_1$	x_2	x_3	x_4	x_5	Index	Fraction
$\mathbf{a}_{11}^t$	a_{12}^t	1	0	0	b_1^t	$r_1 = b_1^t / a_{11}^t$
$\mathbf{a}_{21}^t$	a_{22}^t	0	1	0	b_2^t	$r_2 = b_2^t / a_{21}^t$
$\mathbf{a}_{31}^t$	a_{32}^t	0	0	1	b_3^t	$r_3 = b_3^t / a_{31}^t$

Note: a_{ij}^t, b_i^t represent the coefficients and index values at the tth simplex iteration.

The simplex algorithm for maximizing a linear objective function subject to linear constraints and nonnegativity restrictions on the variables may now be delineated in the following steps:

1. Convert all inequality constraints to equality constraints by adding slack or surplus variables.
2. Structure the problem so that a basic feasible solution is evident and write the problem in tableau form.
3. Determine the simplex criterion for each nonbasic variable.
4. Select as the pivot column the one associated with the variable that has the greatest positive simplex criterion. If none of the simplex criteria are greater than zero, the optimal basic feasible solution has been encountered.
5. Select as the pivot row the one having the smallest fraction r_i in (2.37) that is greater than or equal to zero.
6. Using the Gauss-Jordan elimination process, convert the pivot column to such a form that a one appears in the pivot row and zeroes in other positions of this column.
7. Go to step 3 and continue the process until an optimal basic feasible solution is encountered in step 4.

Example 2.11

Solve by using the simplex algorithm:

$$\begin{aligned}
\text{maximize} \quad & z = 2x_1 + 4x_2 + 2.5x_3 \\
\text{subject to} \quad & 3x_1 + 4x_2 + 2x_3 \leq 60 \\
& 2x_1 + x_2 + 2x_3 \leq 40 \\
& x_1 + 3x_2 + 3x_3 \leq 30 \\
& x_1, x_2, x_3 \geq 0
\end{aligned}$$

The problem is restructured by adding three slack variables x_4, x_5, and x_6:

$$\begin{array}{ll} \text{maximize} & z = 2x_1 + 4x_2 + 2.5x_3 + 0x_4 + 0x_5 + 0x_6 \\ \text{subject to} & 3x_1 + 4x_2 + 2x_3 + x_4 = 60 \\ & 2x_1 + x_2 + 2x_3 + x_5 = 40 \\ & x_1 + 3x_2 + 3x_3 + x_6 = 30 \\ & x_1, x_2, x_3 \geq 0 \end{array}$$

Table 2.11. Initial Tableau for Example 2.11

Cost coefficient:	2	**4**	2.5	0	0	0	Basic	
Variable:	x_1	$\mathbf{x_2}$	x_3	x_4	x_5	x_6	variable	Index
	3	**4**	2	1	0	0	x_4	60
	2	**1**	2	0	1	0	x_5	40
	1	**[3]**	**3**	**0**	**0**	**1**	$\mathbf{x_6}$	**30**
$c_j - z_j$:	2	4	2.5	0	0	0		

The initial tableau appears in Table 2.11. The addition of the three slack variables provides a convenient set of basic variables. The initial basic feasible solution is ($x_1 = x_2 = x_3 = 0$, $x_4 = 60$, $x_5 = 40$, and $x_6 = 30$). The simplex criterion is now computed for the three nonbasic variables x_1, x_2, and x_3:

$$x_1\colon c_1 - z_1 = 2 - [0,0,0]\begin{bmatrix} 3 \\ 2 \\ 1 \end{bmatrix} = 2 - 0 = 2$$

$$x_2\colon c_2 - z_2 = 4 - [0,0,0]\begin{bmatrix} 4 \\ 1 \\ 3 \end{bmatrix} = 4 - 0 = 4$$

$$x_3\colon c_3 - z_3 = 2.5 - [0,0,0]\begin{bmatrix} 2 \\ 2 \\ 3 \end{bmatrix} = 2.5 - 0 = 2.5$$

The maximum $c_j - z_j$ occurs when $j = 2$; thus x_2 is chosen to become the new basic variable. The pivot column is shown in Table 2.11. To determine the pivot row, the fractions formed by dividing the index elements by the corresponding row element in the pivot column are calculated: $r_1 = 60/4 = 15$, $r_2 = 40/1 = 40$, $r_3 = 30/3 = 10$. Since r_3 is the minimum, the third row in the tableau in Table 2.11 is the pivot row. The Gauss-Jordan elimination method is employed to convert the pivot column to the form where a one appears in the pivot row and zeroes elsewhere. The new tableau is illustrated in Table 2.12.

Table 2.12. First Tableau for Example 2.11

Cost coefficient:	2	4	2.5	0	0	0	Basic	
Variable:	$\mathbf{x}_1$	x_2	x_3	x_4	x_5	x_6	variable	Index
	[5/3]	**0**	**−2**	**1**	**0**	**−4/3**	$\mathbf{x}_4$	**20**
	5/3	0	1	0	1	−1/3	x_5	30
	1/3	1	1	0	0	1/3	x_2	10
$c_j - z_j$:	2/3	0	−3/2	0	0	−4/3		

The third row of the new tableau is generated by dividing the third row of the initial tableau in Table 2.11 by 3. In the basic variable column, the new basic variable x_2 replaces the old one x_6. The second row of the new tableau is generated by subtracting 1/3 times the third row from the second in the initial tableau. The first row of the new tableau is generated by subtracting 4/3 times the third row from the first in the initial tableau. The new basic feasible solution is ($x_1 = x_3 = x_6 = 0$, $x_4 = 20$, $x_5 = 30$, $x_2 = 10$), which corresponds to an objective function value of $z = 2(0) + 4(10) + 2.5(0) + 0(20) + 0(30) + 0(0) = 40$.

In the new tableau, step 3 of the simplex algorithm is applied to see whether or not the optimal basic feasible solution has been encountered:

$$x_1\colon c_1 - z_1 = 2 - [0,0,4]\begin{bmatrix}5/3\\5/3\\1/3\end{bmatrix} = 2/3$$

$$x_3\colon c_3 - z_3 = 2.5 - [0,0,4]\begin{bmatrix}-2\\1\\1\end{bmatrix} = -3/2$$

$$x_6\colon c_6 - z_6 = 0 - [0,0,4]\begin{bmatrix}-4/3\\-1/3\\1/3\end{bmatrix} = -4/3$$

Since there is a positive simplex criterion value, the process must continue. The new basic variable will be x_1, so that column 1 becomes the pivot column.

The pivot row fractions are: $r_1 = 20/(5/3) = 12$, $r_2 = 30/(5/3) = 18$, $r_3 = 10/(1/3) = 30$. The minimum nonnegative fraction is 12; thus the first row in the tableau in Table 2.12 becomes the pivot row. The new (second) tableau generated by applying the Gauss-Jordan method to the first tableau in Table 2.12 is shown in Table 2.13. The new basic feasible solution is ($x_3 = x_4 = x_6 = 0, x_1 = 12, x_5 = 10, x_2 = 6$), which corresponds to an objective function value of $z = 2(12) + 4(6) + 2.5(0) + 0(0) + 0(10) + 0(0) = 48$.

Table 2.13. Second Tableau for Example 2.11

Cost coefficient:	2	4	2.5	0	0	0	Basic	
Variable:	x_1	x_2	x_3	x_4	x_5	x_6	variable	Index
	1	0	−6/5	3/5	0	−4/5	x_1	12
	0	0	3	−1	1	1	x_5	10
	0	1	7/5	−1/5	0	9/15	x_2	6
$c_j - z_j$:	0	0	−7/10	−2/5	0	−4/5		

The simplex criterion is again computed for the nonbasic variables in the second tableau in Table 2.13:

$$x_3\colon c_3 - z_3 = 2.5 - [2, 0, 4]\begin{bmatrix} -6/5 \\ 3 \\ 7/5 \end{bmatrix} = -7/10 < 0$$

$$x_4\colon c_4 - z_4 = 0 - [2, 0, 4]\begin{bmatrix} 3/5 \\ -1 \\ -1/5 \end{bmatrix} = -2/5 < 0$$

$$x_6\colon c_6 - z_6 = 0 - [2, 0, 4]\begin{bmatrix} -4/5 \\ 1 \\ 9/15 \end{bmatrix} = -4/5 < 0$$

The process terminates, since the simplex criterion is negative for each nonbasic variable. The optimal basic feasible solution is ($x_1^* = 12, x_2^* = 6, x_3^* = 0, x_4^* = 0, x_5^* = 10, x_6^* = 0$). The maximum value of the objective function is $z^* = 48$.

Notice that in the optimal solution the slack variable x_5 is not at the zero level. This implies that the second constraint of the problem ($2x_1 + x_2 + 2x_3 \leq 40$) is not tight at the optimal solution. This is easily verified by substituting for x_1, x_2, and x_3 the optimal basic feasible solution values of 12, 6, and 0 respectively: $2(12) + 1(6) + 2(0) = 30 < 40$. The slack in this constraint is equal to ten units.

The Minimizing Problem

The simplex algorithm has been described for solving a linear programming problem in which the objective function is to be maximized. If a linear programming problem must be solved for the minimum of a constrained linear objective function, the above algorithm procedure may be used by making use of the fact that the following two linear programming problems are equivalent:

$$\begin{aligned} \text{minimize} \quad & w = \bar{\mathbf{c}}'\bar{\mathbf{x}} \\ \text{subject to} \quad & \bar{\mathbf{A}}\bar{\mathbf{x}} = \mathbf{b} \qquad \bar{\mathbf{x}} \geq \mathbf{0} \end{aligned}$$

and

$$\begin{aligned} \text{maximize} \quad & z = -w = -\bar{\mathbf{c}}'\bar{\mathbf{x}} \\ \text{subject to} \quad & \bar{\mathbf{A}}\bar{\mathbf{x}} = \mathbf{b} \qquad \bar{\mathbf{x}} \geq \mathbf{0} \end{aligned}$$

That is, a function may be minimized by maximizing its negative.

ARTIFICIAL VARIABLES

Consider the following problem:

$$\begin{aligned} \text{minimize} \quad & z = 3x_1 + x_2 + 2x_3 \\ \text{subject to} \quad & 8x_1 + 2x_2 + 3x_3 \geq 5 \\ & 4x_1 + 5x_2 + 6x_3 \geq 5 \\ & x_1, x_2, x_3 \geq 0 \end{aligned} \tag{2.38}$$

The problem can be converted to a maximization problem with equality constraints by multiplying the objective function in (2.38) by a negative one and by adding two surplus variables x_4 and x_5 respectively. The maximization problem is

$$\text{maximize} \quad z = -3x_1 - x_2 - 2x_3 + 0x_4 + 0x_5 \tag{2.39}$$

$$\text{subject to} \quad 8x_1 + 2x_2 + 3x_3 - x_4 = 5 \tag{2.40}$$

$$4x_1 + 5x_2 + 6x_3 - x_5 = 5 \tag{2.41}$$

$$x_i \geq 0 \qquad i = 1, 2, \cdots, 5 \tag{2.42}$$

The "greater than or equal to" constraints introduce a new problem. When these are converted to equality constraints, no convenient basic feasible solution is evident. Notice that the solution ($x_1 = x_2 = x_3 = 0$,

$x_4 = x_5 = -5$) is infeasible since the latter two variables violate the nonnegativity restriction in (2.42).

One way to obviate this problem is to use artificial variables. One new variable is added to each equation in (2.40) and (2.41). Their addition to the equality constraint set will generate a convenient basic feasible solution. Since the two variables are being artificially employed to gain a basic feasible solution initially, they should not occur in the optimal basic feasible solution; i.e., each artificial variable should drop out at some simplex iteration and never become a basic variable again, since the simplex method iterates to an optimal solution. One method that will force the artificial variables to become nonbasic is called the big *M* method, in which large negative cost coefficients are attached to the artificial variables so that they become unattractive as basic variables, since the desire is to maximize the objective function. To see how this method is employed, the problem in (2.39)–(2.42) is solved in Example 2.12, where x_6 and x_7 are the artificial variables with cost coefficients of -10^5.

Example 2.12

Solve by using the simplex algorithm

$$\begin{aligned}
\text{maximize} \quad & w = -z = -3x_1 - x_2 - 2x_3 + 0x_4 \\
& \qquad + 0x_5 - 10^5x_6 - 10^5x_7 \\
\text{subject to} \quad & 8x_1 + 2x_2 + 3x_3 - x_4 \qquad + x_6 \qquad = 5 \\
& 4x_1 + 5x_2 + 6x_3 \qquad - x_5 \qquad + x_7 = 5 \\
& x_i \geq 0 \qquad i = 1, 2, \cdots, 7
\end{aligned}$$

The initial tableau is shown in Table 2.14. The simplex criterion is now calculated for each nonbasic variables:

$$x_1\text{: } c_1 - z_1 = -3 - [-10^5, -10^5]\begin{bmatrix} 8 \\ 4 \end{bmatrix} = 12(10^5) - 3$$

$$x_2\text{: } c_2 - z_2 = -1 - [-10^5, -10^5]\begin{bmatrix} 2 \\ 5 \end{bmatrix} = 7(10^5) - 1$$

$$x_3\text{: } c_3 - z_3 = -2 - [-10^5, -10^5]\begin{bmatrix} 3 \\ 6 \end{bmatrix} = 9(10^5) - 2$$

$$x_4\text{: } c_4 - z_4 = 0 - [-10^5, -10^5]\begin{bmatrix} -1 \\ 0 \end{bmatrix} = -10^5$$

$$x_5\text{: } c_5 - z_5 = 0 - [-10^5, -10^5]\begin{bmatrix} 0 \\ -1 \end{bmatrix} = -10^5$$

Table 2.14. Initial Tableau for Example 2.12

Cost coefficient:	−3	−1	−2	0	0	-10^5	-10^5	Basic	
Variable:	x_1	x_2	x_3	x_4	x_5	x_6	x_7	variable	Index
	[8]	2	3	−1	0	1	0	x_6	5
	4	5	6	0	−1	0	1	x_7	5
$c_j - z_j$:	$12(10^5) - 3$	$7(10^5) - 1$	$9(10^5) - 2$	-10^5	-10^5	0	0		

Since x_1 has the largest simplex criterion, column 1 becomes the pivot column. The pivot row fractions are $r_1 = 5/8$, $r_2 = 5/4$. Since r_1 is the smallest nonnegative fraction, row 1 becomes the pivot row. The first tableau is shown in Table 2.15.

Table 2.15. First Tableau for Example 2.12

Cost coefficient:	−3	−1	−2	0	0	-10^5	-10^5	Basic	
Variable:	x_1	x_2	x_3	x_4	x_5	x_6	x_7	variable	Index
	1	1/4	3/8	−1/8	0	1/8	0	x_1	5/8
	0	4	[9/2]	1/2	−1	−1/2	1	x_7	5/2
$c_j - z_j$:	0	$4(10^5) - 1/4$	$4.5(10^5) - 7/8$	$0.5(10^5) - 3/8$	-10^5	$-1.5(10^5) + 3/8$	0		

The simplex criteria for the nonbasic variables in the first tableau are

$$x_2 \colon c_2 - z_2 = -1 - [-3, -10^5] \begin{bmatrix} 1/4 \\ 4 \end{bmatrix} = 4(10^5) + (3/4) - 1$$

$$x_3 \colon c_3 - z_3 = -2 - [-3, -10^5] \begin{bmatrix} 3/8 \\ 9/2 \end{bmatrix} = 4.5(10^5) + (9/8) - 2$$

$$x_4 \colon c_4 - z_4 = 0 - [-3, -10^5] \begin{bmatrix} -1/8 \\ 1/2 \end{bmatrix} = 0.5(10^5) - 3/8$$

$$x_5 \colon c_5 - z_5 = 0 - [-3, -10^5] \begin{bmatrix} 0 \\ -1 \end{bmatrix} = -10^5$$

$$x_6 \colon c_6 - z_6 = -10^5 - [-3, -10^5] \begin{bmatrix} 1/8 \\ -1/2 \end{bmatrix} = -1.5(10^5) + 3/8$$

Since x_3 is associated with the largest positive simplex criterion, the third column in the first tableau becomes the pivot column. The pivot row fractions are $r_1 = (5/8)/(3/8) = 5/3$, $r_2 = (5/2)/(9/2) = 5/9$. Since r_2 is the smaller nonnegative fraction, the second row becomes the pivot row. The second tableau is shown in Table 2.16.

Table 2.16. Second Simplex Tableau for Example 2.12

Cost coefficient:	−3	**−1**	−2	0	0	-10^5	-10^5	Basic	
Variable:	x_1	$\mathbf{x_2}$	x_3	x_4	x_5	x_6	x_7	variable	Index
	1	**−1/12**	0	−1/6	1/12	1/6	−1/12	x_1	5/12
	0	**8/9**	**1**	**1/9**	**−2/9**	**−1/9**	**2/9**	$\mathbf{x_3}$	**5/9**
$c_j - z_j$:	0	19/36	0	−5/18	−7/36	$-10^5 + 5/18$	$-10^5 + 7/36$		

The simplex criteria for the nonbasic variables in the second tableau are

$$x_2: c_2 - z_2 = -1 - [-3, -2]\begin{bmatrix} -1/12 \\ 8/9 \end{bmatrix} = 19/36$$

$$x_4: c_4 - z_4 = 0 - [-3, -2]\begin{bmatrix} -1/6 \\ 1/9 \end{bmatrix} = -5/18$$

$$x_5: c_5 - z_5 = 0 - [-3, -2]\begin{bmatrix} 1/12 \\ -2/9 \end{bmatrix} = -7/36$$

$$x_6: c_6 - z_6 = -10^5 - [-3, -2]\begin{bmatrix} 1/6 \\ -1/9 \end{bmatrix} = -10^5 + 5/18$$

$$x_7: c_7 - z_7 = -10^5 - [-3, -2]\begin{bmatrix} -1/12 \\ 2/9 \end{bmatrix} = -10^5 + 7/36$$

Since x_2 is associated with the largest positive simplex criterion, the second column in the second tableau becomes the pivot column. The pivot row fractions are $r_1 = (5/12)/(-1/12) = -5$, $r_2 = (4/9)/(8/9) = 1/2$. Since the divisor in r_1 is negative, it is dropped from consideration. Thus the second row is the pivot row. The third tableau is shown in Table 2.17.

Table 2.17. Second Simplex Tableau for Example 2.12

Cost coefficient:	−3	−1	−2	0	0	-10^5	-10^5	Basic	
Variable:	x_1	x_2	x_3	x_4	x_5	x_6	x_7	variable	Index
	1	0	3/32	−5/32	1/16	+5/32	−1/16	x_1	15/32
	0	1	9/8	1/8	−1/4	−1/8	1/4	x_2	5/8
$c_j - z_j$:	0	0	−19/32	−11/32	−1/16	$-10^5 + 11/32$	$-10^5 + 1/16$		

The simplex criteria for the nonbasic variables in the second tableau are

$$x_3\colon c_3 - z_3 = -2 - [-3, -1]\begin{bmatrix}3/32\\9/8\end{bmatrix} = -19/32$$

$$x_4\colon c_4 - z_4 = 0 - [-3, -1]\begin{bmatrix}-5/32\\1/8\end{bmatrix} = -11/32$$

$$x_5\colon c_5 - z_5 = 0 - [-3, -1]\begin{bmatrix}1/16\\-1/4\end{bmatrix} = -1/16$$

$$x_6\colon c_6 - z_6 = -10^5 - [-3, -1]\begin{bmatrix}5/32\\-1/8\end{bmatrix} = -10^5 + 11/32$$

$$x_7\colon c_7 - z_7 = -10^5 - [-3, -1]\begin{bmatrix}-1/16\\1/4\end{bmatrix} = -10^5 + 1/16$$

Since all the simplex criteria are negative, the optimal basic feasible solution has been determined. The optimal solution is ($x_1^* = 15/32, x_2^* = 5/8, x_3^* = 0, x_4^* = x_5^* = x_6^* = x_7^* = 0$).

Notice that the artificial variables are driven out of the solution by the second tableau. The cost coefficients corresponding to these variables do not have to be of such great magnitude (10^5). The idea in the big *M* method is to make these coefficients sufficiently large negative numbers so that the artificial variables readily drop out of the set of basic variables in the first few simplex iterations.

Linear programming problems involving equality constraints can also be structured using artificial variables so that an initial basic feasible solution is readily apparent, as Example 2.13 illustrates.

Example 2.13

Find an initial basic feasible solution for the problem

$$\begin{aligned}\text{maximize} \quad & z = 4x_1 + 10x_2\\ \text{subject to} \quad & 5x_1 + 6x_2 \le 30\\ & 8x_1 + 12x_2 \ge 24\\ & x_1 - x_2 = 0\\ & x_1, x_2 \ge 0\end{aligned}$$

The problem can be restructured as

maximize

$$z = 4x_1 + 10x_2 + 0x_3 + 0x_4 - 10^5x_5 - 10^5x_6$$

$$\text{subject to}\quad \begin{aligned} 5x_1 + 6x_2 + x_3 &= 30 \\ 8x_1 + 12x_2 - x_4 + x_5 &= 24 \\ x_1 - x_2 + x_6 &= 0 \\ x_i \geq 0 \quad i = 1, 2, \cdots, 6 \end{aligned}$$

where x_3 is a slack variable, x_4 is a surplus variable, and x_5 and x_6 are artificial variables. The initial tableau is shown in Table 2.18.

Table 2.18. Initial Tableau for Example 2.13

Cost coefficient:	4	10	0	0	-10^5	-10^5	Basic	
Variable:	x_1	x_2	x_3	x_4	x_5	x_6	variable	Index
	5	6	1	0	0	0	x_3	30
	8	12	0	-1	1	0	x_5	24
	1	-1	0	0	0	1	x_6	0
$c_j - z_j$:	$4 + 9(10^5)$	$10 + 11(10^5)$	0	10^5	0	0		

Degeneracy and Cycling

The definition of a degenerate basic feasible solution has been given (see Definition 2.5). If such solutions are encountered in solving a linear programming problem by using the simplex method, computational difficulties may arise that might result in the basic feasible optimal solution not being identified by the simplex algorithm. Such difficulties are known as *cycling*; the simplex algorithm may cycle through the same chain of basic feasible solutions so that the tableau computations may not terminate, although a finite number of basic feasible solutions exist.

To illustrate the difficulty that degeneracy may present in conjunction with solving a problem by the simplex method, two example problems are presented. In the first, no cycling occurs and the basic feasible optimal solution is identified, although a degenerate solution is encountered. In the second, cycling occurs and the simplex method fails to identify the basic feasible optimal solution.

Example 2.14

$$\begin{aligned} \text{maximize} \quad & z = x_1 + x_2 \\ \text{subject to} \quad & -x_1 + x_2 \le 0 \\ & 2x_1 - x_2 \le 2 \\ & x_2 \le 2 \\ & x_1, x_2 \ge 0 \end{aligned}$$

Adding three slack variables x_3, x_4, and x_5, the problem becomes

$$\begin{aligned} \text{maximize} \quad & z = x_1 + x_2 + (0)x_3 + (0)x_4 + (0)x_5 \\ \text{subject to} \quad & -x_1 + x_2 + x_3 = 0 \\ & 2x_1 - x_2 + x_4 = 2 \\ & x_2 + x_5 = 2 \\ & x_i \ge 0 \quad i = 1, 2, \ldots, 5 \end{aligned}$$

The initial tableau is illustrated in Table 2.19.

Table 2.19. Initial Tableau for Example 2.14

Cost coefficient:	**1**	1	0	0	0	Basic	
Variable:	$\mathbf{x_1}$	x_2	x_3	x_4	x_5	variable	Index
	−1	1	1	0	0	x_3	0
	[2]	**−1**	**0**	**1**	**0**	$\mathbf{x_4}$	**2**
	0	1	0	0	1	x_5	2
$c_j - z_j$:	1	1	0	0	0		

Notice that the initial solution ($x_1 = 0$, $x_2 = 0$, $x_3 = 0$, $x_4 = 2$, $x_5 = 2$) is degenerate, since fewer than $m = 3$ variables are positive. The simplex criteria for the nonbasic variables are

$$x_1\colon c_1 - z_1 = 1 - [0, 0, 0] \begin{bmatrix} -1 \\ 2 \\ 0 \end{bmatrix} = 1$$

$$x_2\colon c_2 - z_2 = 1 - [0, 0, 0] \begin{bmatrix} 1 \\ -1 \\ 1 \end{bmatrix} = 1$$

There is a tie for the entering variable. A commonly used tie-breaker rule is to select the variable with the smallest subscript.

Using this rule causes x_1 to enter the basis. The departing variable fraction is $r_2 = 2/2 = 1$; r_1 and r_3 are not computed since $a_{11} = -1 \leq 0$ and $a_{31} = 0 \leq 0$. Thus the departing variable is x_5. Applying the Gauss-Jordan elimination process to the initial tableau results in the first tableau, shown in Table 2.20. The new basic feasible solution is ($x_2 = x_4 = 0, x_1 = x_3 = 1, x_5 = 2$) and is nondegenerate.

Table 2.20. First Tableau for Example 2.14

Cost coefficient:	1	1	0	0	0	Basic	
Valuable:	x_1	$\mathbf{x_2}$	x_3	x_4	x_5	variable	Index
	0	**[1/2]**	**1**	**1/2**	**0**	$\mathbf{x_3}$	**1**
	1	**−1/2**	0	1/2	0	x_1	1
	0	**1**	0	0	1	x_5	2
$c_j - z_j$:	0	3/2	0	−1/2	0		

When the simplex criterion is applied to each nonbasic variable in the first tableau, x_2 will enter the basis next. The departing variable fractions are $r_1 = 1/(1/2) = 2$, $r_3 = 2/1 = 2$; r_2 is not computed since $a_{22} = -1/2 \leq 0$. After use of the tie-breaking rule, x_3 (the first-row variable corresponds to the fraction r_1) is selected to leave the basis. The second tableau is given in Table 2.21. The solution in the second tableau ($x_1^* = x_2^* = 2$, $x_3^* = x_4^* = x_5^* = 0$) is the terminal and optimal solution; it is also degenerate.

Table 2.21. Second Tableau for Example 2.14

Cost coefficient:	1	1	0	0	0	Basic	
Variable:	x_1	x_2	x_3	x_4	x_5	variable	Index
	0	1	2	1	0	x_2	2
	1	0	1	1	0	x_1	2
	0	0	−2	−1	1	x_5	0
$c_j - z_j$:	0	0	−3	−2	0		

In this problem, degenerate solutions are encountered in the initial and second (and terminal) tableaus, but the simplex method properly identifies the basic feasible optimal solution. The tip-off that a degenerate solution will be encountered in the next tableau is a tie occurring among the fractions r_i used to select the *departing variables* in the present tableau. In the above example, in moving from the first to the second tableau, $r_1 = r_3 = 2$ and either x_3 or x_5 could have been chosen as the departing variable. The fact that a degenerate solution has been encountered in a simplex iteration does not guarantee that cycling will occur, as Example 2.14 demonstrates.

The degeneracy encountered in this problem results from a redundant constraint, as Figure 2.12 illustrates; the constraint $x_2 \leq 2$ could have been eliminated from the original formulation of the problem.

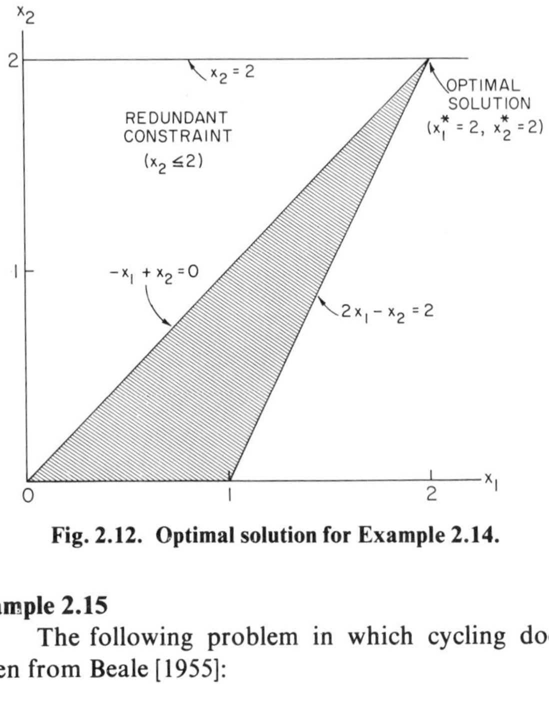

Fig. 2.12. Optimal solution for Example 2.14.

Example 2.15

The following problem in which cycling does occur is taken from Beale [1955]:

$$\text{maximize} \quad z = 0.75x_1 - 150x_2 + 0.02x_3 - 6x_4$$
subject to

$$\begin{aligned} 0.25x_1 - 60x_2 - 0.04x_3 + 9x_4 + x_5 \qquad &= 0 \\ 0.50x_1 - 90x_2 - 0.02x_3 + 3x_4 \qquad + x_6 \qquad &= 0 \\ x_3 \qquad\qquad + x_7 &= 1 \\ x_i \geq 0 \qquad i = 1, 2, \ldots, 7 & \end{aligned}$$

The initial simplex tableau is given in Table 2.22. The initial solution is $(x_1 = x_2 = x_3 = x_4 = x_5 = x_6 = 0,\ x_7 = 1)$, which is clearly degenerate. On the basis of the simplex criterion, the nonbasic variable selected to enter the basis is x_1. There is a tie for the departing variable since $r_1 = 0/0.25 = 0 = r_2 = 0/0.50$. After use of the tie-breaking rule, x_5 is selected to leave the basis. Using the simplex algorithm and the tie-breaking rule when needed, the next six tableaus are given in Tables 2.23–2.28.

Table 2.22. Initial Tableau for Example 2.15

Cost coefficient:	**0.75**	−150	0.02	−6	0	0	0		
Variable:	$\mathbf{x_1}$	x_2	x_3	x_4	x_5	x_6	x_7	Basic variable	Index
	[0.25]	**−60**	**−0.04**	**9**	**1**	**0**	**0**	$\mathbf{x_5}$	**0**
	0.50	−90	−0.02	3	0	1	0	x_6	0
	0	0	1	0	0	0	1	x_7	1

Table 2.23. First Tableau for Example 2.15

Cost coefficient:	0.75	**−150**	0.02	−6	0	0	0		
Variable:	x_1	$\mathbf{x_2}$	x_3	x_4	x_5	x_6	x_7	Basic variable	Index
	1	**−240**	−0.16	36	4	0	0	x_1	0
	0	**[30]**	**0.06**	**−15**	**−2**	**1**	**0**	$\mathbf{x_6}$	**0**
	0	**0**	1	0	0	0	1	x_7	1

Table 2.24. Second Tableau for Example 2.15

Cost coefficient:	0.75	−150	**0.02**	−6	0	0	0		
Variable:	x_1	x_2	$\mathbf{x_3}$	x_4	x_5	x_6	x_7	Basic variable	Index
	1	**0**	**[0.32]**	**−84**	**−12**	**8**	**0**	$\mathbf{x_1}$	**0**
	0	1	**0.002**	−0.50	−1/15	1/30	0	x_2	0
	0	0	**1**	0	0	0	1	x_7	1

Table 2.25. Third Tableau for Example 2.15

Cost coefficient:	0.75	−150	0.02	**−6**	0	0	0		
Variable:	x_1	x_2	x_3	$\mathbf{x_4}$	x_5	x_6	x_7	Basic variable	Index
	25/8	0	.1	**−262.5**	−37.5	25	0	x_3	0
	−1/160	**1**	**0**	**[0.025]**	**1/120**	**−1/60**	**0**	$\mathbf{x_2}$	**0**
	−25/8	0	0	**262.5**	37.5	−25	1	x_7	1

Table 2.26. Fourth Tableau for Example 2.15

Cost coefficient:	0.75	−150	0.02	−6	**0**	0	0	Basic	
Variable:	x_1	x_2	x_3	x_4	$\mathbf{x_5}$	x_6	x_7	variable	Index
	−62.5	**10,500**	**1**	**0**	[50]	**−150**	**0**	$\mathbf{x_3}$	**0**
	−0.25	40	0	1	**1/3**	−2/3	0	x_4	0
	62.5	−10,500	0	0	**−50**	150	1	x_7	1

Table 2.27. Fifth Tableau for Example 2.15

Cost coefficient:	0.75	−150	0.02	−6	0	**0**	0	Basic	
Variable:	x_1	x_2	x_3	x_4	x_5	$\mathbf{x_6}$	x_7	variable	Index
	−1.25	210	0.02	0	1	**−3**	0	x_5	0
	1/6	**−30**	**−1/150**	**1**	**0**	[1/3]	**0**	$\mathbf{x_4}$	**0**
	0	0	1	0	0	**0**	1	x_7	1

Table 2.28. Sixth Tableau for Example 2.15

Cost coefficient:	0.75	−150	0.02	−6	0	0	0	Basic	
Variable:	x_1	x_2	x_3	x_4	x_5	x_6	x_7	variable	Index
	0.25	−60	−0.04	9	1	0	0	x_5	0
	0.50	−90	−0.02	3	0	1	0	x_6	0
	0	0	1	0	0	0	1	x_7	1

The sixth tableau is identical to the initial tableau (compare Tables 2.22 and 2.28). The seventh tableau will therefore be identical to the first tableau and so on. A cycle has been encountered, and the simplex method will not identify the basic feasible optimal solution ($x_1^* = 0.04$, $x_3^* = 1$, $x_5^* = 0.03$, $x_2^* = x_4^* = x_6^* = x_7^* = 0$) in this case.

Methods have been devised to break the cycle when a cycling sequence of basic feasible solutions is encountered in the presence of degeneracy. For a description of one such method (the Charnes perturbation method), the reader is directed to Hadley [1962, pp. 174–96].

Fortunately, the problem of cycling rarely if ever occurs in practice when solving real-world problems. As a precaution, however, most good linear programming computer codes check for degeneracy and cycling and have provisions for applying corrective measures, such as the perturbation method, if the cycling problem does arise.

Nonunique Solution

To illustrate the behavior of the simplex algorithm when applied to a problem that does not have a unique basic feasible optimal solution, consider the following example.

Example 2.16

$$\begin{aligned} \text{maximize} \quad & z = x_1 + 1.5x_2 \\ \text{subject to} \quad & 2x_1 + 3x_2 \le 6 \\ & 2x_1 + x_2 \le 4 \\ & x_1, x_2 \ge 0 \end{aligned}$$

After addition of two slack variables x_3 and x_4 the initial tableau appears as given in Table 2.29. The proper pivot row and column for forming the next tableau are designated in Table 2.29; the first tableau is given in Table 2.30.

Table 2.29. Initial Tableau for Example 2.16

Cost coefficient:	1	**1.5**	0	0	Basic	
Variable:	x_1	$\mathbf{x_2}$	x_3	x_4	variable	Index
	2	[3]	**1**	**0**	$\mathbf{x_3}$	**6**
	2	**1**	0	1	x_4	4
$c_j - z_j$:	1	1.5	0	0		

Table 2.30. First Tableau for Example 2.16

Cost coefficient:	**1**	1.5	0	0	Basic	
Variable:	$\mathbf{x_1}$	x_2	x_3	x_4	variable	Index
	2/3	1	1/3	0	x_2	2
	[4/3]	**0**	**−1/3**	**1**	$\mathbf{x_4}$	**2**
$c_j - z_j$:	0	0	−1/2	0		

The simplex criteria for the nonbasic variables are

$$x_1\colon c_1 - z_1 = 1 - [1.5, 0]\begin{bmatrix} 2/3 \\ 4/3 \end{bmatrix} = 0$$

$$x_3\colon c_3 - z_3 = 0 - [1.5, 0]\begin{bmatrix} 1/3 \\ -1/3 \end{bmatrix} = -1/2$$

Since the criterion for x_1 is not negative, the process may continue. The second tableau, given in Table 2.31, contains the basic feasible optimal solution ($x_1^* = 1, x_2^* = 1.5, x_3 = x_4 = 0$).

Table 2.31. Second Tableau for Example 2.16

Cost coefficient:	1	1.5	0	0	Basic	
Variable:	x_1	x_2	x_3	x_4	variable	Index
	0	1	0.50	−0.50	x_2	1
	1	0	−0.25	0.75	x_1	1.5
$c_j - z_j$:	0	0	−0.5	0		

From Figure 2.13 it is clear that the optimizing point ($x_1^* = 1.5, x_2^* = 1$) is not unique, however. Indeed, the solution in the first tableau ($x_1 = 0, x_2 = 2$) yields the same functional value [$f(1.5, 1) = 3 = f(0, 2)$]. The fact that the simplex criterion yielded a zero value for the nonbasic variable x_1 in proceeding to the second tableau is a tip-off that a unique basic feasible optimal solution does not exist. Any point lying on the line $2x_1 + 3x_2 = 6$ for $0 \leq x_1 \leq 1.5$ will yield a maximum objective function value of $z^* = 3$.

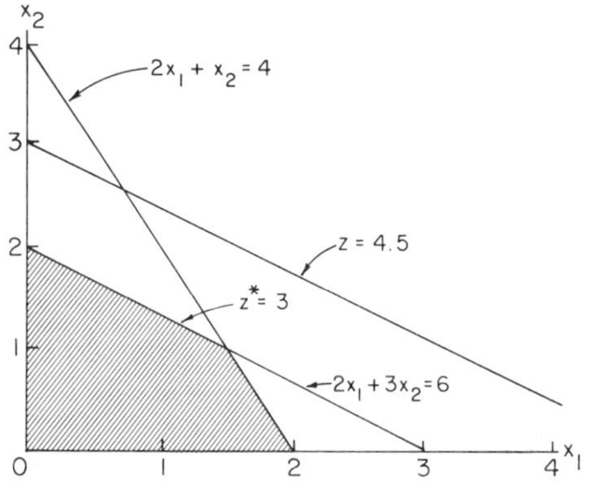

Fig. 2.13. Graphic solution of Example 2.16.

No Feasible Solution

How does the simplex algorithm respond when applied to a problem for which the set of feasible solutions is null? Consider the following example.

Example 2.17

$$\begin{aligned} \text{maximize} \quad & z = x_1 + 2x_2 \\ \text{subject to} \quad & x_1 + x_2 \geq 4 \\ & 2x_1 + 3x_2 \leq 6 \\ & x_1, x_2 \geq 0 \end{aligned}$$

From the graphic presentation of this problem in Figure 2.14, it is apparent that no feasible solution exists.

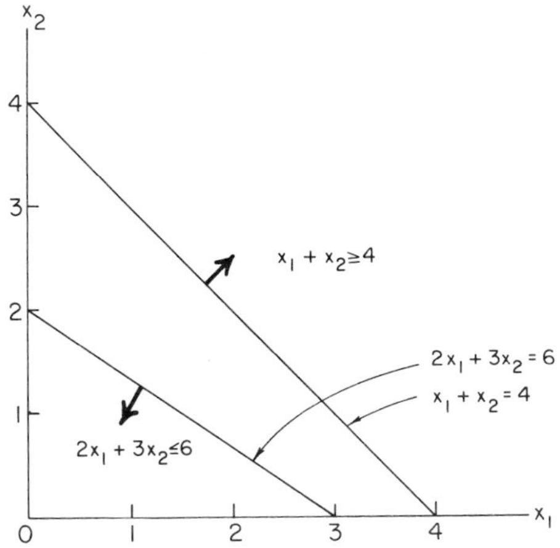

Fig. 2.14. Graphic solution of Example 2.17.

The problem may be converted to the equality constraint form by adding a surplus variable x_3, an artificial variable x_4, and a slack variable x_5.

$$\begin{aligned} \text{maximize} \quad & z = x_1 + 2x_2 + (0)x_3 - Mx_4 + (0)x_5 \\ \text{subject to} \quad & x_1 + x_2 - x_3 + x_4 = 1 \\ & 2x_1 + 3x_2 + x_5 = 1 \\ & x_i \geq 0 \quad i = 1, 2, \ldots, 5 \end{aligned}$$

Rather than writing in a number for the cost coefficient associated with the artificial variable x_4, the large constant M will be used. The initial tableau is given in Table 2.32.

Table 2.32. Initial Tableau for Example 2.17

Cost coefficient:	1	2	0	$-M$	0	Basic	
Variable:	x_1	x_2	x_3	x_4	x_5	variable	Index
	1	1	-1	1	0	x_4	4
	2	[3]	0	0	1	x_5	6
$c_j - z_j$:	$1 + M$	$2 + M$	$-M$	0	0		

The simplex criteria for the nonbasic variables are

$$x_1: c_1 - z_1 = 1 - [-M, 0]\begin{bmatrix}1\\2\end{bmatrix} = 1 + M$$

$$x_2: c_2 - z_2 = 2 - [-M, 0]\begin{bmatrix}1\\3\end{bmatrix} = 2 + M$$

$$x_3: c_3 - z_3 = 0 - [-M, 0]\begin{bmatrix}-1\\0\end{bmatrix} = -M$$

It should now be clear why the value of M need not be specified when the big M method is used. The letter M can be carried along through the iterations and be made large enough when necessary to insure that artificial variables drop out of the basis. Based on the simplex criterion, x_2 is selected as the entering variable; x_5 departs. The first tableau is given in Table 2.33.

Table 2.33. First Tableau for Example 2.17

Cost coefficient:	1	2	0	$-M$	0	Basic	
Variable:	x_1	x_2	x_3	x_4	x_5	variable	Index
	1/3	0	-1	1	$-1/3$	x_4	2
	[2/3]	1	0	0	1/3	x_2	2
$c_j - z_j$:	$(1/3)M - 1/3$	0	$-M$	0	$-(1/3)M - 2/3$		

Table 2.34. Second Tableau for Example 2.17

Cost coefficient:	1	2	0	$-M$	0	Basic	
Variable:	x_1	x_2	x_3	x_4	x_5	variable	Index
	0	$-1/2$	-1	1	$-1/2$	x_4	1
	1	3/2	0	0	1/2	x_1	3
$c_j - z_j$:	0	$1/2 - (1/2)M$	$-M$	0	$-1/2 - (1/2)M$		

The appropriate pivot column and row for forming the second tableau are designated in Table 2.33; the second tableau is displayed in Table 2.34.

The simplex criteria for the nonbasic variables in the second tableau are

$$x_2: 2 - [-M, 1] \begin{bmatrix} -1/2 \\ 3/2 \end{bmatrix} = 1/2 - (1/2)M$$

$$x_3: 0 - [-M, 1] \begin{bmatrix} -1 \\ 0 \end{bmatrix} = -M$$

$$x_5: 0 - [-M, 1] \begin{bmatrix} -1/2 \\ 1/2 \end{bmatrix} = -1/2 - (1/2)M$$

Since for any $M > 0$ all simplex criteria are negative, the terminal tableau has been encountered; but the artificial variable x_4 is still in the basis. Indeed, no matter how large the constant M is made, the simplex algorithm will terminate in the second tableau with the artificial variable in the basis.

Thus if the simplex algorithm terminates with an artificial variable in the basis, the implication is that no feasible solution exists, and the formulation of the model should be carefully checked.

Unboundedness

Occasionally, models have been improperly specified so that the maximum of the objective function is not bounded. The simplex algorithm will identify this condition in another nonnormal termination as Example 2.18 illustrates.

Example 2.18

$$\begin{aligned} \text{maximize} \quad & z = 2x_1 + 3x_2 \\ \text{subject to} \quad & x_1 \leq 2 \\ & x_1 - x_2 \leq 0 \\ & x_1, x_2 \geq 0 \end{aligned}$$

The initial tableau is given in Table 2.35 where slack variables x_3 and x_4 have been added as indicated.

Table 2.35. Initial Tableau for Example 2.18

Cost coefficient:	**2**	3	0	0	Basic	
Variable:	$\mathbf{x_1}$	x_2	x_3	x_4	variable	Index
	1	0	1	0	x_3	2
	[1]	**−1**	**0**	**1**	$\mathbf{x_4}$	**0**
$c_j - z_j$:	2	3	0	0		

The appropriate pivot row and column used in forming the first tableau are designated in Table 2.35; the first tableau is given in Table 2.36. The second tableau is given in Table 2.37.

Table 2.36. First Tableau for Example 2.18

Cost coefficient:	2	**3**	0	0	Basic	
Variable:	x_1	$\mathbf{x_2}$	x_3	x_4	variable	Index
	0	[1]	**1**	**−1**	$\mathbf{x_3}$	**2**
	1	**−1**	0	1	x_1	0
$c_j - z_j$:	0	**5**	0	−2		

Table 2.37. Second Tableau for Example 2.18

Cost coefficient:	2	3	0	**0**	Basic	
Variable:	x_1	x_2	x_3	$\mathbf{x_4}$	variable	Index
	0	1	1	**−1**	x_2	2
	1	0	1	**0**	x_1	2
$c_j - z_j$:	0	0	−5	**3**		

The simplex criteria for the nonbasic variables in the second tableau are

$$x_3: 0 - [3, 2] \begin{bmatrix} 1 \\ 1 \end{bmatrix} = -5$$

$$x_4: 0 - [3, 2] \begin{bmatrix} -1 \\ 0 \end{bmatrix} = 3$$

Since the slack variable x_4 has a nonnegative simplex criterion, the process must continue. However, it is immediately apparent that no basic variable can leave the second tableau since a_{14} =

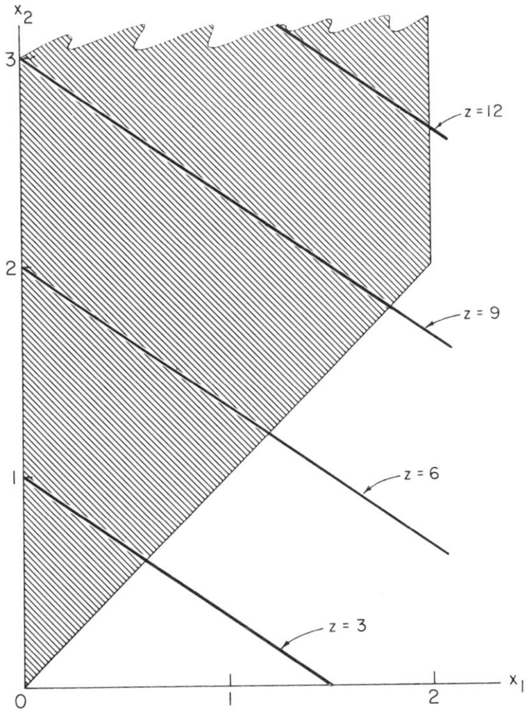

Fig. 2.15. Graphic solution of Example 2.18.

$-1 \leq 0$ and $a_{24} = 0 \leq 0$. Thus the variable x_4 should go back into the basis, but no present basic variable is willing to depart.

Figure 2.15 illustrates the reason for this dilemma. The objective function $z = 2x_1 + 3x_2$ can be made large without bound by increasing x_2 without bound when $x_1 = 0$, for example. Thus a problem with an unbounded solution is identified by the simplex algorithm in the following way. Suppose at the tth iteration the kth nonbasic variable is selected to enter the basis. The departing variable is determined by identifying the pivot row associated with the minimum nonnegative r_i, where

$$r_i = b_i^t / a_{ik}^t \qquad a_{ik}^t > 0; i = 1, 2, \ldots, m$$

If all the coefficients in the pivot column (a_{ik}^t, $i = 1, 2, \ldots, m$) are less than or equal to zero, an unbounded problem has been encountered.

2.6 SIMPLEX THEORY[2]

Notation

Assume that the linear programming problem has been converted to the equality-constraint form in (2.10)–(2.12), by the addition of s slack and/or surplus variables; the problem is

$$\begin{array}{ll} \text{maximize} & z = f(\bar{\mathbf{x}}) = \bar{\mathbf{c}}'\bar{\mathbf{x}} \\ \text{subject to} & \bar{\mathbf{A}}\mathbf{x} = \bar{\mathbf{b}} \qquad \bar{\mathbf{x}} \geq \mathbf{0} \end{array}$$

where $\bar{\mathbf{c}}$, $\bar{\mathbf{x}}$, and $\bar{\mathbf{A}}$ are given by (2.13)–(2.15) respectively. Denote by $\bar{\mathbf{a}}_j$ the jth column of $\bar{\mathbf{A}}$, $j = 1, 2, \ldots, n + s$, and by $\mathbf{B}$ the $m \times m$ matrix whose columns are any set of m linearly independent column vectors in $\bar{\mathbf{A}}$. The columns of $\mathbf{B}$ form a vector basis for the $n + s$ column vectors in $\bar{\mathbf{A}}$, and thus the matrix $\mathbf{B}$ is referred to as the *basis matrix*. The columns of $\mathbf{B}$ are denoted by $\mathbf{b}_1, \mathbf{b}_2, \ldots, \mathbf{b}_m$. The ith column vector $\mathbf{b}_i$ may be associated with any column vector in $\bar{\mathbf{A}}$ as long as the set of vectors in $\bar{\mathbf{A}}$ corresponding to $\mathbf{b}_1, \mathbf{b}_2, \ldots, \mathbf{b}_m$ constitutes a linearly independent set.

Since the columns of $\mathbf{B}$ form a vector basis in the m-dimensional space, any column vector in $\bar{\mathbf{A}}$ can be expressed as a linear combination of the column vectors in $\mathbf{B}$; scalars v_{ij} exist so that for $\bar{\mathbf{a}}_j$ not in $\mathbf{B}$,

$$\bar{\mathbf{a}}_j = v_{1j}\mathbf{b}_1 + v_{2j}\mathbf{b}_2 + \cdots + v_{mj}\mathbf{b}_m = \sum_{i=1}^{m} v_{ij}\mathbf{b}_i = \mathbf{B}\mathbf{v}_j \tag{2.43}$$

where

$$\mathbf{v}_j' = [v_{1j}, v_{2j}, \ldots, v_{mj}] \tag{2.44}$$

Assume for the moment that the first m columns of $\bar{\mathbf{A}}$ form $\mathbf{B}$ so that $\bar{\mathbf{A}}$ may be partitioned in the form,

$$\bar{\mathbf{A}} = [\mathbf{B}, \bar{\mathbf{B}}] \tag{2.45}$$

where $\bar{\mathbf{B}}$ is an $m \times (n + s - m)$ matrix composed of the column vectors in $\bar{\mathbf{A}}$ but not in $\mathbf{B}$. Then

---

2. This section provides the theoretical bases for the simplex method and can be skipped without loss of continuity.

$$\bar{\mathbf{A}}\bar{\mathbf{x}} = [\mathbf{B}, \bar{\mathbf{B}}]\begin{bmatrix}\bar{\mathbf{x}}_B \\ \bar{\mathbf{x}}_{\bar{B}}\end{bmatrix} = \mathbf{b}$$

where $\bar{\mathbf{x}}$ has been partitioned into $\bar{\mathbf{x}}_B$ and $\bar{\mathbf{x}}_{\bar{B}}$, corresponding to the partitioning on $\bar{\mathbf{A}}$. Now $\mathbf{B}\bar{\mathbf{x}}_B + \bar{\mathbf{B}}\bar{\mathbf{x}}_{\bar{B}} = \mathbf{b}$, but from the definition of a basic solution, $\bar{\mathbf{x}}_{\bar{B}} = \mathbf{0}$. Thus $\mathbf{B}\bar{\mathbf{x}}_B = \mathbf{b}$ and

$$\bar{\mathbf{x}}_B = \mathbf{B}^{-1}\mathbf{b} \tag{2.46}$$

It is possible to write (2.46), since $\mathbf{B}$ is an $m \times m$ nonsingular matrix by construction. The solution $(\bar{\mathbf{x}}_B, \bar{\mathbf{x}}_{\bar{B}} = \mathbf{0})$ constitutes a basic solution, and the m-component vector $\bar{\mathbf{x}}_B$ contains the basic variables.

The vector of costs $\bar{\mathbf{c}}$ can be partitioned in a similar manner: $\bar{\mathbf{c}}' = [\bar{\mathbf{c}}_B', \bar{\mathbf{c}}_{\bar{B}}']$ where

$$\bar{\mathbf{c}}_B' = [c_{B1}, c_{B2}, \ldots, c_{Bm}] \tag{2.47}$$

and c_{Bi} is the cost coefficient corresponding to the ith basic column vector $\mathbf{b}_i$. Now

$$z = f(\bar{\mathbf{x}}) = \bar{\mathbf{c}}'\bar{\mathbf{x}} = [\bar{\mathbf{c}}_B', \bar{\mathbf{c}}_{\bar{B}}']\begin{bmatrix}\bar{\mathbf{x}}_B \\ \bar{\mathbf{x}}_{\bar{B}}\end{bmatrix} = \bar{\mathbf{c}}_B'\bar{\mathbf{x}}_B$$

since $\bar{\mathbf{x}}_{\bar{B}} = \mathbf{0}$.

Notationally, the dichotomization between the basic variables $\bar{\mathbf{x}}_B$ and the nonbasic variables $\bar{\mathbf{x}}_{\bar{B}}$ has been set out. Notice that the basis matrix $\mathbf{B}$ need not be the first m columns of $\bar{\mathbf{A}}$ as assumed in the partitioning of $\bar{\mathbf{A}}$ in (2.45). The columns of $\mathbf{B}$ may correspond to any set of linearly independent columns in $\bar{\mathbf{A}}$; the components in $\bar{\mathbf{x}}_B$ and $\bar{\mathbf{c}}_B$ are then the components in $\bar{\mathbf{x}}$ and $\bar{\mathbf{c}}$ respectively, corresponding to the selected m linearly independent columns of $\bar{\mathbf{A}}$.

Define a new scalar quantity z_j

$$z_j = v_{1j}c_{B1} + v_{2j}c_{B2} + \cdots + v_{mj}c_{Bm} = \sum_{i=1}^{m} c_{Bi}v_{ij} = \bar{\mathbf{c}}_B'\mathbf{v}_j \tag{2.48}$$

where $\bar{\mathbf{c}}_B'$ and $\mathbf{v}_j'$ are given in (2.47) and (2.44) respectively. The form of z_j should be familiar; it is a part of the simplex criterion $c_j - z_j$ for selecting a nonbasic variable to enter the basis in the next simplex iteration. It will now be shown how the simplex algorithm mathematically selects a new set of basic variables, while guaranteeing that the objective function will be improved.

Improvement of a Basic Feasible Solution

Assume that the basic solution $\bar{\mathbf{x}}_B$ in (2.46) is also a basic feasible solution. The simplex algorithm generates a new basic feasible solution by removing a column vector from $\mathbf{B}$ and replacing it with a column vector from $\bar{\mathbf{B}}$. If the column vector $\bar{\mathbf{a}}_j$ is not in $\mathbf{B}$,

$$\bar{\mathbf{a}}_j = \sum_{i=1}^{m} v_{ij}\mathbf{b}_i \tag{2.49}$$

From linear algebra it is known that the column vector $\bar{\mathbf{a}}_j$ may replace any vector in $\mathbf{B}$, say $\mathbf{b}_k$, for which $v_{kj} \neq 0$; and the new set of column vectors in $\mathbf{B}$ will still form a basis in the m-dimensional space. Suppose that a column vector $\bar{\mathbf{a}}_j$ in $\bar{\mathbf{B}}$ is now selected with the condition that at least one v_{ij} in (2.49) is greater than zero. If it is decided to drop the kth column vector $\mathbf{b}_k$ from the present basis $\mathbf{B}$ where $v_{kj} \neq 0$ and replace it by $\bar{\mathbf{a}}_j$, it follows from (2.49) that

$$\mathbf{b}_k = (1/v_{kj})\bar{\mathbf{a}}_j - \sum_{\substack{i=1 \\ i \neq k}}^{m} (v_{ij}/v_{kj})\mathbf{b}_i \tag{2.50}$$

The present basic feasible solution can be written as $\mathbf{B}\bar{\mathbf{x}}_B = \mathbf{b}$ or

$$\sum_{i=1}^{m} x_{Bi}\mathbf{b}_i = x_{B1}\mathbf{b}_1 + \cdots + x_{B(k-1)}\mathbf{b}_{(k-1)} + x_{Bk}\mathbf{b}_k + x_{B(k+1)}\mathbf{b}_{(k+1)} + \cdots + x_{Bm}\mathbf{b}_m = \mathbf{b} \tag{2.51}$$

where x_{Bi} is the ith component of the basic variable vector $\mathbf{x}_B$. By substituting the right-hand side of (2.50) for $\mathbf{b}_k$ in (2.51), it follows that the new basic solution is specified by

$$\sum_{\substack{i=1 \\ i \neq k}}^{m} [x_{Bi} - x_{Bk}(v_{ij}/v_{kj})]\mathbf{b}_i + (x_{Bk}/v_{kj})\bar{\mathbf{a}}_j = \mathbf{b}$$

The new basic variable vector, designated by $\hat{\mathbf{x}}_B$, contains the following components:

$$\hat{x}_{Bi} = x_{Bi} - x_{Bk}(v_{ij}/v_{kj}) \qquad i = 1, 2, \ldots, m;\ i \neq k \tag{2.52}$$

$$\hat{x}_{Bk} = x_{Bk}/v_{kj} \tag{2.53}$$

If the new basic solution $\hat{\mathbf{x}}_B$ is to be feasible, it follows from the nonnegativity restriction on $\bar{\mathbf{x}}$ that

$$x_{Bi} - x_{Bk}(v_{ij}/v_{kj}) \geq 0 \qquad i = 1, 2, \ldots, m; i \neq k \tag{2.54}$$

$$x_{Bk}/v_{kj} \geq 0 \tag{2.55}$$

In constructing the new basic vector $\hat{\mathbf{x}}_B$, $\mathbf{b}_k$ is chosen corresponding to a coefficient v_{kj} that cannot be zero. Thus from (2.55) it is seen that if $x_{Bk} > 0$, then v_{kj} must be greater than zero. If, in addition, $v_{ij} \leq 0$ for $i \neq k$, the conditions (2.54) and (2.55) will be satisfied. If at least one $v_{ij} > 0$ $(i \neq k)$, condition (2.54) may not hold. In this case, whether (2.54) holds depends on which column vector $\mathbf{b}_k$ has been removed from the present basis matrix $\mathbf{B}$. Suppose that $v_{ij} > 0$ for some particular i so that the nonnegativity of $\hat{x}_{Bi}$ is in doubt. Then for this subscript value i, the condition in (2.54) can be divided by v_{ij} without altering the sense of the inequality:

$$x_{Bi}/v_{ij} - x_{Bk}/v_{kj} \geq 0 \qquad i \neq k \tag{2.56}$$

To insure that condition (2.56) holds for any i such that $v_{ij} > 0$, choose the column k to be removed from the basis by determining

$$\alpha = x_{Bk}/v_{kj} = \min_i \, (x_{Bi}/v_{ij} = r_i, v_{ij} > 0) \tag{2.57}$$

If the column index k is chosen by employing (2.57), then (2.56) will be satisfied for any i where $v_{ij} > 0$. Therefore, by employing the simplex rule (2.57) for selecting the departing basic variable (and column from $\mathbf{B}$), the new basic solution $\hat{\mathbf{x}}_B$ will be feasible, since the rule will insure that conditions (2.54) and (2.55) are satisfied.

Recall that if a tie for the minimum r_i occurs among two or more columns when the departing column rule in (2.57) is employed, the new basic feasible solution will be degenerate. The reason for this is evident from considering (2.52). Suppose that the tth column in $\mathbf{B}$ takes on the same minimum in (2.57) as the column selected for departure (k). Then $r_t = x_{Bt}/v_{tj} = x_{Bk}/v_{kj} = r_k$. From (2.52) $\hat{x}_{Bt}$ may be written as $\hat{x}_{Bt} = x_{Bt} - x_{Bk}(v_{tj}/v_{kj})$. But from (2.54) $x_{Bt} = v_{tj}x_{Bk}/v_{kj}$. Thus $\hat{x}_{Bt} = 0$. Therefore, if there is a tie for the departing variable in the new basis $\mathbf{B}$, all new basic variables will be zero, corresponding to those columns in $\mathbf{B}$ that have the same value in (2.57) as the eliminated column.

Suppose that α in (2.57) is zero. This can only happen if x_{Bk} is zero from observing (2.57), which means, of course, that the present basis is degenerate. From (2.52) it is seen that if $x_{Bk} = 0$, then $\hat{x}_{Bi} = x_{Bi}$, $i = 1, 2, \ldots, m; i \neq k$ and that $\hat{x}_{Bk} = 0$. Hence the new basic feasible solution is also degenerate. It does not follow that the new basic feasible solution will necessarily be degenerate if the present one is, however. If $v_{ij} < 0$ for every x_{Bi} that is zero in the present basic solution, none of these variables will enter into the computation in (2.57) for α, and α will therefore be positive. As an illustration of this fact, refer to Example 2.14. In this example problem, the initial and terminal basic feasible solutions are degenerate, while the intermediate solution is not. In Example 2.15, however, α is zero in each and every iteration, and the simplex algorithm generates a cycling set of degenerate solutions.

Summarizing to this point, if the rule in (2.57) is used to determine which column in $\mathbf{B}$ to drop out of the basis, the new solution $\hat{\mathbf{x}}_B = \hat{\mathbf{B}}^{-1}\mathbf{b}$ will be basic and feasible.

It is now necessary to show that the simplex algorithm will generate the new solution $\hat{\mathbf{x}}_B$ in such a way that it is an improvement over the present solution $\bar{\mathbf{x}}_B$. The present value of the objective function is given by

$$z = f(\bar{\mathbf{x}}) = \bar{\mathbf{c}}_B'\bar{\mathbf{x}}_B = \sum_{i=1}^{m} c_{Bi}x_{Bi}$$

and the new value, denoted by $\hat{z}$, is given by

$$\hat{z} = \hat{\mathbf{c}}_B'\hat{\mathbf{x}}_B = \sum_{i=1}^{m} \hat{c}_{Bi}\hat{x}_{Bi}$$

where

$$\hat{c}_{Bi} = c_{Bi} \qquad i \neq k \qquad \hat{c}_{Bk} = c_j$$

and c_j is the cost corresponding to the variable entering the new basis. From (2.52) and (2.53), it follows that

$$\hat{z} = \sum_{\substack{i=1 \\ i \neq k}}^{m} c_{Bi}[x_{Bi} - x_{Bk}(v_{ij}/v_{kj})] + (x_{Bk}/v_{kj})c_j \tag{2.58}$$

Since when $i = k$,

$$c_{Bk}[x_{Bk} - x_{Bk}(v_{kj}/v_{kj})] = 0$$

the restriction on the summation variable in (2.58) may be dropped, and $\hat{z}$ may be written as

$$\hat{z} = \sum_{i=1}^{m} c_{Bi}[x_{Bi} - x_{Bk}(v_{ij}/v_{kj})] + (x_{Bk}/v_{kj})c_j$$

or

$$\hat{z} = \sum_{i=1}^{m} c_{Bi}x_{Bi} - (x_{Bk}/v_{kj})\sum_{i=1}^{m} c_{Bi}v_{ij} + (x_{Bk}/v_{kj})c_j$$

However, from (2.48) $\sum_{i=1}^{m} c_{Bi}v_{ij} = z_j$ so that $\hat{z}$ reduces to

$$\hat{z} = z + (x_{Bk}/v_{kj})(c_j - z_j) = z + \alpha(c_j - z_j)$$

Recall that the simplex criterion for selecting the nonbasic variable in the present tableau to become a basic variable in the new tableau is to select the variable corresponding to the maximum positive $c_j - z_j$. By doing so, the value of the objective function will increase by $\alpha(c_j - z_j)$ units.

Thus if a vector $\bar{\mathbf{a}}_j$ in $\bar{\mathbf{B}}$ is chosen to enter the basis for which $c_j - z_j > 0$ and at least one $v_{ij} > 0$ in (2.43), then $\hat{z} \geq z$; if $\alpha > 0$, then $\hat{z} > z$. That is, an improvement may not result in the objective function by applying the simplex criterion if the present solution is degenerate (in which case $\alpha = 0$).

The gain in the objective function by adding the variable x_j to the basis and dropping the variable corresponding to the kth column of $\mathbf{B}$ is αc_j. In the process of doing this, the remaining $m - 1$ basic variables will normally change their values, causing a change of $-\alpha z_j$ in z. The intuitive reasoning presented for using the simplex criterion to select the variable to add to the basis is now substantiated: $c_j - z_j$ represents the net change in the objective function value per unit increase in the value of the added variable when all other nonbasic variables remain at the zero level.

Termination of the Simplex Process

Assume for the moment that degeneracy is not present in the linear programming problem. On the basis of the above information, the simplex algorithm will in this case move from one basic feasible solution to another so that the new value of the objective function is at least as large as the previous value. Since there are only a finite number of basic feasible solutions [bounded above by $(n + s)!/(n + s - m)!\, m!)$] the simplex process must eventually terminate. It can do so in only one of two ways:

1. One or more $c_j - z_j > 0$ and for each $c_j - z_j > 0$, $v_{ij} < 0$ for $i = 1, 2, \ldots, m$.
2. All $z_j - c_j \leq 0$ for the columns of $\bar{\mathbf{A}}$ not in $\mathbf{B}$.

If (1) occurs, an unbounded solution has been encountered. Suppose that a nonbasic variable x_j is chosen to enter the basis (i.e., $c_j - z_j > 0$) and it corresponds to a column vector $\bar{\mathbf{a}}_j$ in $\bar{\mathbf{B}}$ for which $v_{ij} \leq 0$, $i = 1, 2, \ldots, m$. The criterion for selecting the vector to depart is

$$\alpha = x_{Bk}/v_{kj} = \min_i \, (x_{Bi}/v_{ij} = r_i, v_{ij} > 0)$$

If all $v_{ij} \leq 0$, the basis must contain $m + 1$ variables if x_j is allowed to enter. If $\bar{\mathbf{x}}_B$ is the present basic feasible solution,

$$\sum_{i=1}^{m} x_{Bi}\mathbf{b}_i = \mathbf{b} \tag{2.59}$$

A solution can now be formed with $m + 1$ variables different from zero (x_j and the $\bar{\mathbf{x}}_B$ components) by adding and subtracting $\lambda\bar{\mathbf{a}}_j$ from (2.59) where λ is any scalar:

$$\sum_{i=1}^{m} x_{Bi}\mathbf{b}_i - \lambda\bar{\mathbf{a}}_j + \lambda\bar{\mathbf{a}}_j = \mathbf{b} \tag{2.60}$$

but

$$\bar{\mathbf{a}}_j = \sum_{i=1}^{m} v_{ij}\mathbf{b}_i \tag{2.61}$$

and by substituting the right-hand side of (2.61) for $\bar{\mathbf{a}}_j$ in (2.60), the following expression is obtained:

$$\sum_{i=1}^{m} (x_{Bi} - \lambda v_{ij})\mathbf{b}_i + \lambda\bar{\mathbf{a}}_j = \mathbf{b} \tag{2.62}$$

If $\lambda > 0$, then $x_{Bi} - \lambda v_{ij} \geq 0$ since $v_{ij} \leq 0$ for $i = 1, 2, \ldots, m$. Thus (2.62) is a feasible solution in which $m + 1$ variables are different from zero. Since it has $m + 1$ nonzero variables, it is not a basic solution. It will be a feasible solution, however, regardless of how large λ is. The new feasible solution is:

$$\begin{aligned} \hat{x}_{Bi} &= x_{Bi} - \lambda v_{ij} \qquad i = 1, 2, \ldots, m \\ \hat{x}_{Bi} &= \lambda \qquad i = m + 1 \end{aligned} \tag{2.63}$$

The objective function value $\hat{z}$ corresponding to this feasible solution is

$$\hat{z} = \sum_{i=1}^{n+s} c_i \hat{x}_i = \sum_{i=1}^{m} c_{Bi}(x_{Bi} - \lambda v_{ij}) + c_j \lambda = z + \lambda(c_j - z_j)$$

Thus by making λ in (2.63) arbitrarily large, $\hat{z}$ can be made as large as desired since $c_j - z_j > 0$. Hence the simplex algorithm will identify an unbounded problem by terminating with one or more $c_j - z_j < 0$, where for each $c_j - z_j > 0$, $v_{ij} \leq 0$ for all $i = 1, 2, \ldots, m$.

If the process terminates in the second way (2), the optimal basic feasible solution has been determined. This fact may be demonstrated as follows. Suppose that $\bar{\mathbf{x}}_B^*$ is the present basic feasible solution where $\bar{\mathbf{x}}_B^* = \mathbf{B}^{-1}\mathbf{b}$ and that $c_j - z_j \leq 0$ for every column $\bar{\mathbf{a}}_j$ in $\bar{\mathbf{B}}$. Further, denote by z^* the objective function value corresponding to the solution $\bar{\mathbf{x}}_B^*$.

Let $\bar{\mathbf{x}}^{\cdot}$ be any feasible solution to $\bar{\mathbf{A}}\bar{\mathbf{x}} = \mathbf{b}$; the corresponding value of the objective function, denoted by $z^{\cdot}$, is

$$z^{\cdot} = \sum_{i=1}^{n+s} x_i^{\cdot} c_i \tag{2.64}$$

Since $\bar{\mathbf{x}}^{\cdot}$ is feasible,

$$x_1^{\cdot}\bar{\mathbf{a}}_1 + x_2^{\cdot}\bar{\mathbf{a}}_2 + \cdots + x_{n+s}^{\cdot}\bar{\mathbf{a}}_{n+s} = \mathbf{b} \tag{2.65}$$

For any column vector $\bar{\mathbf{a}}_j$ in $\bar{\mathbf{A}}$,

$$\bar{\mathbf{a}}_j = \sum_{i=1}^{m} v_{ij}\mathbf{b}_i \tag{2.66}$$

where $\mathbf{b}_i$, $i = 1, 2, \ldots, m$, are the linearly independent vectors in $\bar{\mathbf{A}}$ forming the basis. By substituting the right-hand side of (2.66) for each $\bar{\mathbf{a}}_j$ in (2.65), the following expression is obtained:

$$x_1^{\cdot}\sum_{i=1}^{m} v_{i1}\mathbf{b}_i + x_2^{\cdot}\sum_{i=1}^{m} v_{i2}\mathbf{b}_i + \cdots + x_{n+s}^{\cdot}\sum_{i=1}^{m} v_{i,n+s}\mathbf{b}_i = \mathbf{b}$$

or

$$\left(\sum_{j=1}^{n+s} x_j^{\cdot} v_{1j}\right)\mathbf{b}_1 + \left(\sum_{j=1}^{n+s} x_j^{\cdot} v_{2j}\right)\mathbf{b}_2 + \cdots + \left(\sum_{j=1}^{n+s} x_j^{\cdot} v_{mj}\right)\mathbf{b}_m = \mathbf{b} \qquad (2.67)$$

The present basic feasible solution $\mathbf{x}_B^*$ must satisfy the equation

$$x_{B1}^*\mathbf{b}_1 + x_{B2}^*\mathbf{b}_2 + \cdots + x_{Bm}^*\mathbf{b}_m = \mathbf{b} \qquad (2.68)$$

Since the expression of any vector (say $\mathbf{b}$) in terms of the basis vectors is unique, from (2.67) and (2.68) it follows that

$$x_{Bi}^* = \sum_{j=1}^{n+s} x_j^{\cdot} v_{ij} \qquad i = 1, 2, \ldots, m \qquad (2.69)$$

Now it must be shown that $z^* \geq z^{\cdot}$. For every column vector in $\bar{\mathbf{A}}$ that is not in $\mathbf{B}$, $c_j - z_j \leq 0$ by assumption. For the column vectors of $\bar{\mathbf{A}}$ that are in $\mathbf{B}$, $\mathbf{v}_j = \mathbf{B}^{-1}\bar{\mathbf{a}}_j = \mathbf{B}^{-1}\mathbf{b}_i = \mathbf{e}_i$ if $\bar{\mathbf{a}}_j$ is the ith column vector in $\mathbf{B}$ and $\mathbf{e}_i$ is an $m \times 1$ column vector with a one in the ith row. Thus $z_j = \mathbf{c}_B'\mathbf{v}_j = \mathbf{c}_B'\mathbf{e}_i = c_{Bi} = c_j$. Therefore, for the columns of $\bar{\mathbf{A}}$ in the basis, $c_j - z_j \equiv 0$. Thus, for *every* column in $\bar{\mathbf{A}}$, $c_j - z_j \leq 0$.

Now, using the fact that $z_j \geq c_j$, $j = 1, 2, \ldots, n + s$, from (2.64) it follows that

$$z_1 x_1^{\cdot} + z_2 x_2^{\cdot} + \cdots + z_{n+s} x_{n+s}^{\cdot} \geq z^{\cdot} \qquad (2.70)$$

since $x_j^{\cdot} \geq 0$, $j = 1, 2, \ldots, n + s$. However, by definition,

$$z_j = \sum_{i=1}^{m} c_{Bi} v_{ij} \qquad (2.71)$$

By substituting the right-hand side of (2.71) for z_j in (2.70), it follows that

$$\left(\sum_{i=1}^{m} c_{Bi} v_{i1}\right) x_1^{\cdot} + \left(\sum_{i=1}^{m} c_{Bi} v_{i2}\right) x_2^{\cdot} + \cdots + \left(\sum_{i=1}^{m} c_{Bi} v_{i,n+s}\right) x_{n+s}^{\cdot} \geq z^{\cdot}$$

or

$$\left(\sum_{j=1}^{n+s} x_j^{\cdot} v_{ij}\right) c_{B1} + \left(\sum_{j=1}^{n+s} x_j^{\cdot} v_{2j}\right) c_{B2} + \cdots + \left(\sum_{j=1}^{n+s} x_j^{\cdot} v_{mj}\right) c_{Bm} \geq z^{\cdot}$$

But from (2.69) it follows that

$$x_{B1}^{*} c_{B1} + x_{B2}^{*} c_{B2} + \cdots + x_{Bm}^{*} c_{Bm} \geq z^{\cdot} \qquad (2.72)$$

However, the left-hand side of (2.72) is z^*, the objective function value corresponding to the present basic feasible solution $\bar{\mathbf{x}}_B^*$. Thus, for *any* feasible solution of $\bar{\mathbf{x}}^{\cdot}$, $z^* \geq z^{\cdot}$.

In summary, if $c_j - z_j \leq 0$ for every $\bar{\mathbf{a}}_j$ in $\bar{\mathbf{A}}$, the basic feasible solution $\bar{\mathbf{x}}_B$ corresponding to the basis with this property is optimal. In the absence of degeneracy there will be an optimal basic feasible solution with $c_j - z_j \leq 0$ for $j = 1, 2, \ldots, n + s$. Notice that the simplex process will correctly terminate at a degenerate optimal solution, since (if all $c_j - z_j \leq 0$) the above development will clearly hold if one or more components of $\bar{\mathbf{x}}_B$ is zero. The development therefore implies a very important result: *the optimal solution to a linear programming problem will never contain more than m nonzero variables.*

The reader is referred to Example 2.14 for an illustration of a problem in which the simplex process terminates at an optimal basic feasible but degenerate solution.

Degeneracy

If degeneracy occurs in solving a linear programming problem by using the simplex method, the objective function value in a particular iteration may not be greater than the one encountered in the previous iteration. Indeed, as Example 2.15 illustrates, it is possible to cycle through a sequence of solutions so that the terminal conditions (1) and (2) will not be reached in a finite number of steps. Thus if degeneracy appears, the simplex method does not insure that there is an optimal basic feasible solution with all $c_j - z_j \leq 0$. Furthermore, the optimal solution may not be basic in the presence of degeneracy.

Fortunately, degeneracy occurs infrequently in the resolution of real-world linear programming problems. The reader is again referred to Hadley [1962] for a thorough treatment of the degeneracy problem.

Nonunique Solution

The optimal value of the objective function of a linear programming problem is unique, but the solution point may not be. Indeed, if there are two or more basic feasible optimal solutions, it is easy to show that there will be an infinite number of optimal solutions.

Let $\bar{\mathbf{x}}_1^*, \bar{\mathbf{x}}_2^*, \ldots \bar{\mathbf{x}}_r^*$ represent r basic feasible optimal solutions. Thus $\bar{\mathbf{A}}\bar{\mathbf{x}}_i^* = \mathbf{b}$, $i = 1, 2, \ldots, r$, and $z^* = f^*(\bar{\mathbf{x}}) = f(\bar{\mathbf{x}}_i^*)$, $i = 1, 2, \ldots, r$.

Now consider any convex combination of these vectors, $\bar{\mathbf{x}} = \sum_{i=1}^r \lambda_i \bar{\mathbf{x}}_i^*$, where $\lambda_i \geq 0$, $i = 1, 2, \ldots, r$ and $\sum_{i=1}^r \lambda_i = 1$. The vector $\bar{\mathbf{x}}$ is feasible, since $\bar{\mathbf{x}} \geq 0$ (this follows since $\bar{\mathbf{x}}_i^* \geq 0$ and $\lambda_i \geq 0$ for $i = 1, 2, \ldots, r$) and $\bar{\mathbf{A}}\bar{\mathbf{x}} = \mathbf{b}$ (this follows since $\bar{\mathbf{A}}\bar{\mathbf{x}}_i^* = \mathbf{b}$, $i = 1, 2, \ldots, r$). Since $z^* = \bar{\mathbf{c}}'\bar{\mathbf{x}}_i^*$, $i = 1, 2, \ldots, r$, the value of the objective function for $\bar{\mathbf{x}}$ is

$$\bar{\mathbf{c}}'\bar{\mathbf{x}} = \bar{\mathbf{c}}' \sum_{i=1}^{r} \lambda_i \bar{\mathbf{x}}_i^* = \sum_{i=1}^{r} \lambda_i \bar{\mathbf{c}}'\bar{\mathbf{x}}_i^* = \sum_{i=1}^{r} \lambda_i z^* = z^* \sum_{i=1}^{r} \lambda_i = z^*$$

Therefore, $\bar{\mathbf{x}}$ is an optimal solution (although it may be nonbasic).

Initial Basic Feasible Solution

In analyzing the way the simplex method moves from solution to solution while improving the objective function, it is assumed from the onset that $\mathbf{x}_B$ is a basic feasible solution. If the statement of the problem does not lend itself to structuring a basic feasible initial solution, one must be synthesized because the simplex algorithm works only on such solutions. In Section 2.5 the use of artificial variables is introduced as one means of synthesizing an initial basic feasible solution. Since the artificial variables have no meaning relative to the interpretation of the problem to which they are applied, they must be forced out of the basis in the first few tableaus to insure that they do not appear in the terminal tableau. The big M method for driving the artificial variables out of the basis has been presented and applied to an example problem in Section 2.5.

Another method for generating an initial basic feasible solution is called the two-phase method. It has been developed by Dantzig, Orden, and others at the RAND Corporation for the purpose of efficiently solving linear programming problems that involve the use of artificial variables when a computer is used. The big M method is not well suited for computer applications since the computer cannot carry the letter M in storage; a numerical quantity must be assigned to it. A large number must be chosen to insure that all artificial variables are driven out of the basis; but the magnitude of M can cause round-off problems to arise, which may seriously affect the precision of the solution. Although the two-phase method is designed for computer adaptation, it is usually no more difficult to apply by hand than the big M method. In subsequent chapters (particularly Chapter 5) the two-phase method will be used exclusively to solve linear programming problems in which artificial variables have been used to synthesize an initial basic feasible solution.

As the name of the method implies, the process used to solve a problem is dichotomized into two phases. In the first, simplex-type iterations are employed to drive all artificial variables out of the basis. The second phase commences when all artificial variables have departed from the basis or one or more artificial variables remain at the zero level. The second-phase iterations employ essentially the standard simplex process.

PHASE I

Suppose that s slack and/or surplus variables have been adjoined to the original set of n variables in a linear programming problem but that h artificial variables are required to identify an initial basic feasible solution. Consider the following function:

$$Y = \sum_{j=1}^{n} c_j x_j + \sum_{j=n+1}^{n+s} c_j x_j + \sum_{j=n+s+1}^{n+s+h} c_j x_j \tag{2.73}$$

The constants $c_j, j = 1, 2, \ldots, n$, denote the cost coefficients in the objective function associated with the original n variables $c_j = 0$, $j = n + 1, \ldots, n + s$, since $x_j, j = n + 1, \ldots, n + s$, is a slack or surplus variable. The Phase I process applies the standard simplex process to the initial tableau (containing the initial basic feasible solution) with one exception: instead of maximizing the objective function $z = \bar{\mathbf{c}}'\bar{\mathbf{x}}$, the function Y in (2.73) is maximized where $c_j = 0, j = 1, 2, \ldots, n + s$, and $c_j = -1, j = n + s + 1, \ldots, n + s + h$. Then zero costs are assigned to the set of original, slack, and surplus variables, and a cost of -1 is assigned to each artificial variable. Thus the object of the Phase I iterations is to maximize

$$Y = -\sum_{j=n+s+1}^{n+s+h} x_j \tag{2.74}$$

Since $x_j \geq 0, j = 1, 2, \ldots, n + s + h$, the maximum value of Y in (2.74) is zero; and this value of Y will be attained when either the basis contains no artificial variables or those that do remain in the basis are at the zero level. If, on the other hand, $Y < 0$ at any Phase I iteration, at least one artificial variable remains at a positive level.

To see why the Phase I process will tend to remove the artificial variables from the basis, suppose that in a Phase I iteration one artificial variable is basic and assume it corresponds to the rth vector among the m basic vectors. Then, for a nonbasic vector $\bar{\mathbf{a}}_j$ that is not an artificial vector, from (2.48)

$$z_j = \sum_{i=1}^{m} c_{Bi} v_{ij} = \sum_{\substack{i=1 \\ i \neq r}}^{m} (0) v_{ij} + (-1) v_{rj} = -v_{rj}$$

and $c_j - z_j = 0 - (-v_{rj}) = v_{rj}$.

Thus the value of $c_j - z_j$ will be positive if $v_{rj} > 0$, which implies that the value of Y will be larger than the value at the previous iteration if the artificial vector $\mathbf{b}_r$ is replaced by $\bar{\mathbf{a}}_j$. The above argument can be generalized to show that the Phase I process will force artificial variables out of the basis if more than one is in the present basis.

Once an artificial variable has been eliminated from the basis in the Phase I process, it can be entirely dropped from the problem, since the objective function Y and the properties of the simplex algorithm will insure that the removed artificial variable can never reenter the basis.

The Phase I process can be stopped as soon as Y becomes zero; it is not necessary to wait until the usual simplex completion criterion (all $c_j - z_j \leq 0$) occurs. At the termination of the Phase I process, one of three conditions will exist:

1. The Phase I process terminates with all $c_j - z_j \leq 0$, but $Y^* < 0$; i.e., one or more artificial variables remain in the terminal basis at a positive level. This indicates (as it did in applying the big M method) that the problem has no feasible solution.
2. $Y^* = 0$, and no artificial variables appear in the basis. A basic feasible solution to the original problem has been identified.
3. $Y^* = 0$, and one or more artificial variables remain in the basis but at the zero level. A feasible solution to the original problem has been identified.

PHASE II

If Phase I terminates in (2) or (3), Phase II commences. In Phase II the terminal tableau in Phase I is used as the initial tableau except that the objective function now becomes $z = \bar{\mathbf{c}}'\bar{\mathbf{x}}$ instead of Y and any artificial vectors in the basis are assigned *zero* costs.

If (2) occurs, the initial tableau in Phase II will contain $n + s$ columns, since the h columns corresponding to the artificial variables will have been eliminated in Phase I. The initial Phase II tableau will therefore contain a basic feasible solution to the original problem, and the usual simplex process may be applied to it to determine the optimal basic feasible solution.

If (3) occurs, special care must be taken to insure that the artificial variables contained in the Phase I basis at termination will not appear in a later basis in Phase II at positive levels. The simplex process will not guarantee this alone since, initially, zero costs are assigned to the artificial variables in Phase II.

Suppose that (3) arises and nonbasic vector $\bar{\mathbf{a}}_r$ is selected to enter the basis at some iteration in the Phase II process. With regard to the possibility of an artificial vector presently in the basis (say $\mathbf{b}_k$) being replaced by $\bar{\mathbf{a}}_j$ and hence departing the basis, four possible cases may occur:

1. If $v_{ir} > 0$ for one or more indices i corresponding to the columns in the basis containing artificial vectors at the zero level, the simplex criterion for selecting a departing vector provides the opportunity for an artificial vector to depart. If an artificial vector does depart, since it did so at the zero level (the value of the variable corresponding to the artificial vector is zero in the present basis), $\bar{\mathbf{a}}_r$ will enter at the zero level (the value of the variable corresponding to $\bar{\mathbf{a}}_r$ in the new basis will be zero). This implies that all other basic variables will remain unchanged in value so that any remaining artificial vectors will continue in the basis at the zero level.
2. If $v_{ir} = 0$ for all indices i corresponding to the columns in the basis containing the artificial vectors at the zero level, no artificial vector will be removed from the present basis. Suppose that the nonartificial vector $\mathbf{b}_h$ is removed. Then the values of the artificial variables in the new basis are $\hat{x}_{Bi} = x_{Bi} - (v_{ir}/v_{hr})x_{Bh} = 0$, since $x_{Bi} = 0$ and $v_{ir} = 0$.
3. If $v_{ij} = 0$ for an index i corresponding to a column in the basis containing an artificial vector at the zero level and for all indices j associated with nonartificial variables, the artificial variable corresponding to this column (i) will *never* leave the basis. This can be seen by considering v_{ij} in the new basis, denoted by $\hat{v}_{ij}$:

$$\hat{v}_{ij} = v_{ij} - (v_{ir}/v_{hr})v_{hj} = 0 \tag{2.75}$$

The scalar $\hat{v}_{ij} = 0$ for any j corresponding to a nonartificial variable, since in (2.75) $v_{ij} = v_{ir} = 0$.

In this case the row containing zeroes associated with nonartificial variables and the ith column can be dropped from the tableau; a redundancy in the constraints is the cause of this situation, and the optimal basic feasible solution will be identified by applying the Phase II process to the smaller basis and tableau. Case (2) provides the guarantee that if an artificial variable is deleted from the basis in this manner, any remaining artificial variables will continue to be at the zero level.

4. If $v_{ir} \leq 0$ for all indices i corresponding to the columns in the basis containing artificial vectors, and $v_{ir} < 0$ for at least one i, the usual simplex criterion for selecting a vector to leave the basis will not remove an artificial vector that may result in the appearance of an artificial variable in the new basis at a positive level. To see this, suppose that the nonartificial vector $\mathbf{b}_h$ is removed and $x_{Bh} > 0$. Then the values of the artificial variables in the new basis with $v_{ir} < 0$ are

$$\hat{x}_{Bi} = x_{Bi} - (v_{ir}/v_{hr})x_{Bh} = -(v_{ir}/v_{hr})x_{Bh} > 0$$

Thus the only case in which an artificial variable may appear in a new basis at a positive level is (4). To insure that this possibility does not occur, instead of removing a nonartificial variable in case (4), remove an artificial one with $v_{ir} < 0$. Since $v_{ir} \neq 0$, the new solution will be basic. It will also be feasible since $\bar{\mathbf{a}}_r$ will enter at the zero level. In this instance, the objective function value associated with the new basis will be equal to the objective function value associated with the present basis.

Thus the only potential difficulty in applying the Phase I–Phase II process occurs if, at the termination of Phase I, one or more artificial variables remain in the basis at the zero level and case (4) arises. However, this difficulty is easily resolved by modifying the simplex criterion for selecting a departing vector as described above.

Three problems are presented in Examples 2.19, 2.20, and 2.21 and solved by using the two-phase method.

Example 2.19

$$\begin{aligned}
\text{maximize} \quad & z = -3x_1 - x_2 - 2x_3 \\
\text{subject to} \quad & 8x_1 + 2x_2 + 3x_3 \geq 5 \\
& 4x_1 + 5x_2 + 6x_3 \geq 5 \\
& x_1, x_2, x_3 \geq 0
\end{aligned}$$

By subtracting the surplus variables x_4 and x_5 and adding the artificial variables x_6 and x_7, the constraint set becomes

$$\begin{aligned} 8x_1 + 2x_2 + 3x_3 - x_4 \qquad\quad + x_6 \qquad &= 5 \\ 4x_1 + 5x_2 + 6x_3 \qquad - x_5 \qquad + x_7 &= 5 \\ x_i \geq 0 \qquad i = 1, 2, \ldots, 7 & \end{aligned}$$

PHASE I

$$\begin{aligned} \text{maximize} \quad & Y = -x_6 - x_7 \\ \text{subject to} \quad & 8x_1 + 2x_2 + 3x_3 - x_4 \qquad\quad + x_6 \qquad = 5 \\ & 4x_1 + 5x_2 + 6x_3 \qquad - x_5 \qquad + x_7 = 5 \\ & x_i \geq 0 \qquad i = 1, 2, \ldots, 7 \end{aligned}$$

The Phase I tableaus are illustrated in Tables 2.38–2.40. The standard simplex process is used to move from tableau to tableau; the pivot elements are bracketed. The first iteration removes artificial variable x_6 and the second iteration removes x_7.

Table 2.38. Initial Phase I Tableau

Cost coefficient:	0	0	0	0	0	−1	−1	Basic	
Variable:	x_1	x_2	x_3	x_4	x_5	x_6	x_7	variable	Index
	[8]	2	3	−1	0	1	0	x_6	5
	4	5	6	0	−1	0	1	x_7	5

$Y = -(5) - (5) = -10$

Table 2.39. First Phase I Tableau

Cost coefficient:	0	0	0	0	0	−1	Basic	
Variable:	x_1	x_2	x_3	x_4	x_5	x_7	variable	Index
	1	1/4	3/8	−1/8	0	0	x_1	5/8
	0	4	[9/2]	1/2	−1	1	x_7	5/2

$Y = -(0) - (5/2) = -5/2$

Table 2.40. Second and Terminal Phase I Tableau

Cost coefficient:	0	0	0	0	0	Basic	
Variable:	x_1	x_2	x_3	x_4	x_5	variable	Index
	1	−1/12	0	−1/6	1/12	x_1	5/12
	0	8/9	1	1/9	−2/9	x_3	5/9

$Y^* = -(0) - (0) = 0$

The terminal tableau in Table 2.40 contains a basic feasible solution to the original problem, and all artificial variables have been removed from the basis.

PHASE II

The objective function to be maximized now becomes $z = -3x_1 - x_2 - 2x_3 + (0)x_4 + (0)x_5$.

The initial tableau in Phase II is the Phase I terminal tableau in Table 2.40 with the Phase I costs replaced by the costs for the original variables; the initial tableau is illustrated in Table 2.41. Applying the simplex process to the initial Phase II tableau results in the optimal basic feasible solution being identified in one iteration. The first (and terminal) Phase II tableau is illustrated in Table 2.42. The optimal solution is $x_1^* = 15/32$, $x_2^* = 5/8$, and $x_3^* = 0$ with $z^* = -65/32$.

Table 2.41. Initial Phase II Tableau

Cost coefficient:	-3	-1	-2	0	0	Basic	
Variable:	x_1	x_2	x_3	x_4	x_5	variable	Index
	1	$-1/12$	0	$-1/6$	$1/12$	x_1	$5/12$
	0	[8/9]	1	$1/9$	$-2/9$	x_3	$5/9$

$z = -85/36$

Table 2.42. First and Terminal Phase II Tableau

Cost coefficient:	-3	-1	-2	0	0	Basic	
Variable:	x_1	x_2	x_3	x_4	x_5	variable	Index
	1	0	$3/32$	$-5/32$	$1/16$	x_1	$15/32$
	0	1	$9/8$	$1/8$	$-1/4$	x_2	$5/8$

The reader should refer to Example 2.12, where this problem is solved by using the big M method. Notice that the first three tableaus in Tables 2.14–2.16 correspond to the Phase I tableaus in Tables 2.38–2.40. With the big M, as with the two-phase method, once artificial variables leave the basis, they may be dropped from the tableaus. Thus the column corresponding to x_6 in Table 2.15 and the columns corresponding to x_6 and x_7 in Table 2.16 could have been dropped from the tableaus as in Tables 2.39 and 2.40 for the Phase I process. The tableaus in Tables 2.16 and 2.17 correspond to the Phase II tableaus in Tables 2.41 and 2.42.

Example 2.20

$$\begin{aligned} \text{maximize} \quad & z = 2x_1 + 3x_2 \\ \text{subject to} \quad & 3x_1 + 2x_2 \le 6 \\ & x_1 - x_2 = 1 \\ & 2x_1 - 2x_2 = 2 \\ & x_1, x_2 \ge 0 \end{aligned}$$

The constraints $x_1 - x_2 = 1$ and $2x_1 - 2x_2 = 2$ are redundant. The application of the two-phase method to the problem in this form will illustrate the retention of an artificial variable at the zero level at the termination of the Phase I process. By adding the slack variable x_3 and two artificial variables x_4 and x_5, the constraint set becomes

$$\begin{aligned} 3x_1 + 2x_2 + x_3 \qquad\qquad &= 6 \\ x_1 - x_2 \qquad + x_4 \qquad &= 1 \\ 2x_1 - 2x_2 \qquad\qquad + x_5 &= 2 \\ x_i \ge 0 \qquad i = 1, 2, \ldots, 5 & \end{aligned}$$

Table 2.43. Initial Phase I Tableau

Cost coefficient:	0	0	0	−1	−1	Basic	
Variable:	x_1	x_2	x_3	x_4	x_5	variable	Index
	3	2	1	0	0	x_3	6
	[1]	−1	0	1	0	x_4	1
	2	−2	0	0	1	x_5	2
$c_j - z_j$:	3	−3	0	0	0		

$Y = -1 - 2 = -3$

In Phase I, the objective function to be maximized is $Y = -x_4 - x_5$. The initial Phase I tableau is illustrated in Table 2.43. The simplex criterion values for selecting the variable to enter the basis are given in the last row of Table 2.43. Since $c_1 - z_1 = 3$ is the maximum, nonbasic variable value, x_1 is selected to enter the next basis. There is a tie for the departing variable, since $r_2 = 1/1 = r_3 = 2/2 = \alpha$. It must follow that the next basis will contain a degenerate basic feasible solution. Using the tie-breaking convention of selecting the departing variable (among the tied variables) with the smallest subscript, x_4 is chosen to leave the basis. The first Phase I tableau is illustrated in Table 2.44.

Table 2.44. First and Terminal Phase I Tableau

Cost coefficient:	0	0	0	−1	Basic	
Variable:	x_1	x_2	x_3	x_5	variable	Index
	0	5	1	0	x_3	3
	1	−1	0	0	x_1	1
	0	0	0	1	x_5	0
$c_j - z_j$:	0	0	0	0		

$Y^* = -0 - 0 = 0$

The solution in the tableau in Table 2.44 ($x_1 = 1$, $x_3 = 3$, $x_2 = x_4 = x_5 = 0$) is clearly degenerate since fewer than $m = 3$ variables are positive. Further, this tableau is the terminal Phase I tableau since $Y = -x_4 - x_5 = -(0) - (0) = 0$. Thus termination occurs with a degenerate solution caused by an artificial variable appearing in the basis at the zero level. Since the row corresponding to the artificial variable in the basis contains zeroes in the columns for the nonartificial variables x_1, x_2, and x_3, case (3) has occurred, the artificial variable x_5 will never leave the basis in Phase II. The case (3) remedy is to cross out of the tableau the row and column corresponding to this variable. The initial Phase II tableau appears in Table 2.45.

Table 2.45. Initial Phase II Tableau

Cost coefficient:	2	3	0	Basic	
Variable:	x_1	x_2	x_3	variable	Index
	0	[5]	1	x_3	3
	1	−1	0	x_1	1
$c_j - z_j$:	0	5	0		

$z = 2$

By applying the simplex process to the tableau in Table 2.45, the terminal Phase II tableau is encountered in one iteration and is illustrated in Table 2.46. The optimal basic feasible solution is ($x_1^* = 8/5$, $x_2^* = 3/5$), with $z^* = 5$.

Table 2.46. First and Terminal Phase II Tableau

Cost coefficient:	2	3	0	Basic	
Variable:	x_1	x_2	x_3	variable	Index
	0	1	1/5	x_2	3/5
	1	0	1/5	x_1	8/5

Example 2.21

$$\begin{aligned} \text{maximize} \quad & z = x_1 + x_2 + 2x_3 \\ \text{subject to} \quad & x_1 \qquad + x_3 = 5 \\ & x_1 + x_2 + x_3 = 5 \\ & 2x_1 \qquad + x_3 = 10 \\ & x_1, x_2, x_3 \geq 0 \end{aligned}$$

By adding artificial variables x_4 and x_5, the constraint set becomes

$$\begin{aligned} x_1 \qquad + x_3 + x_4 \qquad &= 5 \\ x_1 + x_2 + x_3 \qquad &= 5 \\ 2x_1 \qquad + x_3 \qquad + x_5 &= 10 \\ x_i \geq 0 \quad i = 1, 2, \cdots, 5 & \end{aligned}$$

The initial Phase I tableau for maximizing $Y = -x_4 - x_5$ appears in Table 2.47. The variable x_1 is selected to enter the next basis, and by using the tie-breaking convention, x_4 is selected to depart. Since $Y = 0$ in the next tableau, the Phase I process terminates with a degenerate solution containing an artificial variable at the zero level, as illustrated in Table 2.48. The initial Phase II tableau appears in Table 2.49.

Table 2.47. Initial Phase I Tableau

Cost coefficient:	0	0	0	−1	−1	Basic	
Variable:	x_1	x_2	x_3	x_4	x_5	variable	Index
	[1]	0	1	1	0	x_4	5
	1	1	1	0	0	x_2	5
	2	0	1	0	1	x_5	10
$c_j - z_j$:	3	0	2	0	0		

$Y = -5 - 10 = -15$

Table 2.48. First and Terminal Phase I Tableau

Cost coefficient:	0	0	0	−1	Basic	
Variable:	x_1	x_2	x_3	x_5	variable	Index
	1	0	1	0	x_1	5
	0	1	0	0	x_2	0
	0	0	−1	1	x_5	0

$Y^* = -0 - 0 = 0$

Table 2.49. Initial Phase II Tableau

			Pivot column			
Cost coefficient:	1	1	2	0	Basic	
Variable:	x_1	x_2	x_3	x_5	variable	Index
	1	0	(1)	0	x_1	5
	0	1	0	0	x_2	0
	0	0	[−1]	1	x_5	0
$c_j - z_j$:	0	0	1	0		

$z = 5$

Since $c_3 - z_3 = 1$, x_3 is chosen to enter the next basis. Based upon the simplex criterion for selecting the departing variable

$$\alpha = \min_i \left(x_{Bi}/v_{ij}, v_{ij} > 0\right)$$

the artificial variable x_5 cannot leave the basis since $v_{33} = -1 < 0$ (the bracketed element in Table 2.49). In fact, only x_1 can depart. If x_1 is allowed to leave the basis, the next tableau (pivoting on the element in parentheses in Table 2.49) is given in Table 2.50. Notice that the artificial variable now appears at a positive level ($x_5 = 5$); case (4) described above has occurred in the first Phase II tableau (Table 2.49) that led to this problem.

Table 2.50. Tableau Resulting from Selecting x_1 to Leave the Basis in the First Phase II Tableau

Cost coefficient:	1	1	2	0	Basic	
Variable:	x_1	x_2	x_3	x_5	variable	Index
	1	0	1	0	x_3	5
	0	1	0	0	x_2	0
	1	0	0	1	$\mathbf{x_5}$	**5**

The resolution to the case (4) difficulty is to remove the artificial variable for which $v_{ir} < 0$ for $\bar{\mathbf{a}}_r$ entering the basis. Thus x_5 is removed from the basis in the tableau in Table 2.49 instead of x_1. The resulting tableau is illustrated in Table 2.51. Since no variables remain that can enter the basis, the optimal solution has been reached ($x_1^* = 5$, $x_2^* = x_3^* = 0$), with $z^* = 5$. Indeed, from observing Example 2.21, it is seen that the optimal solution is the only feasible solution.

Table 2.51. Proper First Phase II Tableau

Cost coefficient:	1	1	2	Basic	
Variable:	x_1	x_2	x_3	variable	Index
	1	0	0	x_1	5
	0	1	0	x_2	0
	0	0	1	x_3	0
$c_j - z_j$:	0	0	0		

2.7 DUALITY

Corresponding to the linear programming problem,

$$\text{maximize} \quad z = \mathbf{c}'\mathbf{x} \tag{2.76}$$
$$\text{subject to} \quad \mathbf{Ax} \leq \mathbf{b} \tag{2.77}$$
$$\mathbf{x} \geq \mathbf{0} \tag{2.78}$$

is another linear programming problem called the *dual* problem, whose solution is related to that for (2.76)–(2.78), which is called the standard form *primal* problem.

The dual linear programming problem is constructed from the standard form primal problem by using the following rules:

1. A new set of variables is employed: $\lambda_1, \lambda_2, \cdots, \lambda_m$. There will be as many new variables as there are constraints in the primal problem.
2. The direction of the constraint inequalities in the dual problem are the reverse of those in the primal problem.
3. The coefficient matrix of the constraint set for the dual problem is the transpose of the coefficient matrix of the constraints in the primal problem.
4. The right-hand side constants in the constraint set of the dual problem are the coefficients of the objective function from the primal problem.
5. The coefficients of the objective function of the dual problem are the constants from the right-hand side of the constraint set of the primal problem.
6. The objective function of the dual problem is to be minimized when the primal problem is a maximization problem.

The statement of the dual problem with variables $\lambda_1, \lambda_2, \cdots,$

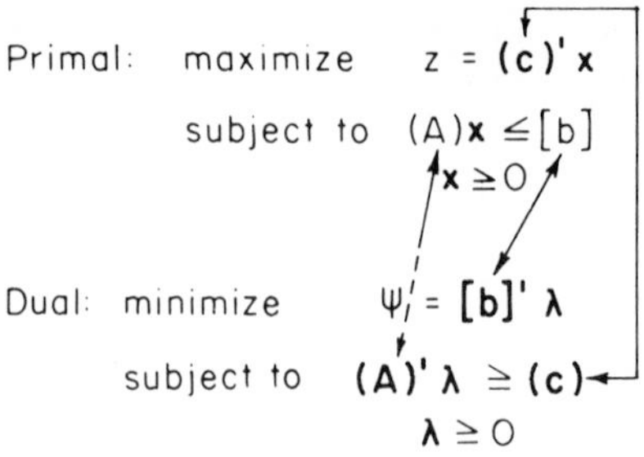

Fig. 2.16. Relationship between the primal problem and its dual linear programming problem.

λ_m and the relationship between it and the primal problem are illustrated in Figure 2.16.

Example 2.22

Consider the linear programming problem

$$\begin{aligned}\text{maximize} \quad & z = 10x_1 + 20x_2 \\ \text{subject to} \quad & 2x_1 + 3x_2 \leq 18 \\ & 2x_1 + x_2 \leq 10 \\ & x_1, x_2 \geq 0\end{aligned}$$

In matrix form, the statement of this problem is

$$\begin{aligned}\text{maximize} \quad & z = \mathbf{c}'\mathbf{x} \\ \text{subject to} \quad & \mathbf{Ax} \leq \mathbf{b} \qquad \mathbf{x} \geq \mathbf{0}\end{aligned}$$

where

$$\mathbf{c}' = [10, 20] \qquad \mathbf{x}' = [x_1, x_2] \qquad \mathbf{b}' = [18, 10]$$

$$\mathbf{A} = \begin{bmatrix} 2 & 3 \\ 2 & 1 \end{bmatrix}$$

By use of the rules for writing the dual of a linear programming problem, the dual of the above problem is

$$\begin{aligned}\text{minimize} \quad & \psi = 18\,\lambda_1 + 10\,\lambda_2 \\ \text{subject to} \quad & 2\,\lambda_1 + 2\,\lambda_2 \geq 10 \\ & 3\,\lambda_1 + \lambda_2 \geq 20 \\ & \lambda_1, \lambda_2 \geq 0\end{aligned}$$

or in matrix form,

$$\begin{aligned} &\text{minimize} \quad \psi = \mathbf{b}'\boldsymbol{\lambda} \\ &\text{subject to} \quad \mathbf{A}'\boldsymbol{\lambda} \geq \mathbf{c} \qquad \boldsymbol{\lambda} \geq \mathbf{0} \end{aligned}$$

where $\boldsymbol{\lambda}' = [\lambda_1, \lambda_2]$ and $\mathbf{b}'$, $\mathbf{c}'$, and $\mathbf{A}$ are given above.

If the primal problem is not in standard form, the rules for constructing the dual may be used after converting the primal problem to standard form. For example, suppose that the primal is a minimization problem (minimize z). The primal objective can be converted to maximization by maximizing the negative of the objective function (maximize $-z$). If the primal problem contains a "greater than or equal to" constraint, the sense of the inequality can be put into standard form ($\leq$) by multiplying both sides of the inequality by -1. A primal equality constraint does not require adjustment, but does modify the dual: the dual variable corresponding to the primal equality constraint is unrestricted in sign. Thus the primal problem

$$\begin{aligned} \text{minimize } z &= 2x_1 + 4x_2 \\ \text{subject to } \quad & 6x_1 + 10x_2 \leq 60 \\ & 2x_1 + x_2 \geq 10 \\ & 2x_1 + 3x_2 = 18 \\ & x_1, x_2 \geq 0 \end{aligned}$$

placed in standard form becomes

$$\begin{aligned} \text{maximize } z^0 &= -2x_1 - 4x_2 \\ \text{subject to } \quad & 6x_1 + 10x_2 \leq 60 \\ & -2x_1 - x_2 \leq -10 \\ & 2x_1 + 3x_2 = 18 \\ & x_1, x_2 \geq 0 \end{aligned}$$

The dual problem corresponding to this standard form primal problem is

$$\begin{aligned} \text{minimize } \psi &= 60\lambda_1 - 10\lambda_2 + 18\lambda_3 \\ \text{subject to } \quad & 6\lambda_1 - 2\lambda_2 + 2\lambda_3 \geq -2 \\ & 10\lambda_1 - \lambda_2 + 3\lambda_3 \geq -4 \\ & \lambda_1, \lambda_2 \geq 0; \ \lambda_3 \text{ unrestricted in sign} \end{aligned}$$

The importance of the dual problem in linear programming is based on Theorem 2.5 and following theorems. Furthermore, the optimal dual solution $\boldsymbol{\lambda}^*$ is related to the optimal primal solution $\boldsymbol{x}^*$ in such a way that the optimal tableau corresponding to either problem provides the necessary information to write the optimal solution to its dual. This fact can be used to efficiently solve a linear programming problem if the dual problem is easier to solve computationally than the primal problem, for example. Additionally, the dual variables λ_i, $i = 1, 2, \ldots, m$, have an important economic interpretation in many linear programming problems.

Theorem 2.5. The Duality Theorem. In a linear programming problem the dual to a maximization (minimization) problem is a minimization (maximization) problem. The optimal values of the objective function for the original (primal) problem and its dual are equal; i.e., $z^* = \psi^*$.

PROOF

Consider the linear programming problem

$$\text{maximize} \quad z = \textstyle\sum_{j=1}^{n} c_j x_j \tag{2.79}$$
$$\text{subject to} \quad \textstyle\sum_{j=1}^{n} a_{ij} x_j \leq b_i \qquad i = 1, 2, \ldots, m \tag{2.80}$$
$$x_j \geq 0 \qquad j = 1, 2, \ldots, n \tag{2.81}$$

In Section 2.4 the possibility of solving a linear programming problem by the method of Lagrangian multipliers has been discussed. This idea will now be pursued, for it will be seen that the dual of the problem in (2.79)–(2.81) and a proof of the duality theorem can be developed from the Lagrangian multiplier method for solving linear programming problems.

For the moment disregard the nonnegativity restrictions on the variables in (2.81) and consider the Lagrangian function:

$$g(\mathbf{x}, \boldsymbol{\lambda}) = \sum_{j=1}^{n} c_j x_j + \sum_{i=1}^{m} \lambda_i \left(b_i - \sum_{j=1}^{n} a_{ij} x_j \right) \tag{2.82}$$

where $\lambda_1, \lambda_2, \ldots, \lambda_m$ are the Lagrangian multipliers required to move the constraints in (2.80) into the objective function. Denote by $\boldsymbol{x}^{*\prime} = [x_1^*, x_2^*, \ldots, x_n^*]$ and $\boldsymbol{\lambda}^{*\prime} = [\lambda_1^*, \lambda_2^*, \ldots, \lambda_m^*]$ the vectors containing the values of the variables for which $g(\mathbf{x}, \boldsymbol{\lambda})$ is a maximum. If proper conditions are placed on the x_j and λ_i, the values of the variables contained in $\mathbf{x}^*$ for maximizing $g(\mathbf{x}, \boldsymbol{\lambda})$ in (2.82) will also maximize z in (2.79). Clearly, one of these conditions is that $x_j \geq 0$, $j = 1, 2, \ldots, n$. Furthermore, from (2.80)

$$\sum_{j=1}^{n} a_{ij}x_j \le b_i \qquad i = 1, 2, \ldots, m \tag{2.83}$$

To insure that (2.83) is satisfied in maximizing $g(\mathbf{x}, \boldsymbol{\lambda})$, it is necessary to require that

$$\lambda_i \left(b_i - \sum_{j=1}^{n} a_{ij}x_j \right) \ge 0 \qquad i = 1, 2, \ldots, m$$

This requirement can be satisfied by insuring that $\lambda_i \ge 0$, $i = 1, 2, \ldots, m$. Thus if the process of maximizing $g(\mathbf{x}, \boldsymbol{\lambda})$ is going to produce the point $\mathbf{x}^*$ at which z in (2.79) achieves its maximum, two necessary conditions are (1) $x_j \ge 0$, $j = 1, 2, \ldots, n$, and (2) $\lambda_i \ge 0$, $i = 1, 2, \ldots, m$.

The set of necessary conditions for the existence of an *unconstrained* maximum to $g(\mathbf{x}, \boldsymbol{\lambda})$ is

$$\frac{\partial g(\mathbf{x}, \boldsymbol{\lambda})}{\partial x_j} = c_j - \sum_{i=1}^{m} \lambda_i a_{ij} = 0 \qquad j = 1, 2, \ldots, n$$

$$\frac{\partial g(\mathbf{x}, \boldsymbol{\lambda})}{\partial \lambda_i} = b_i - \sum_{j=1}^{n} a_{ij}x_j = 0 \qquad i = 1, 2, \ldots, m$$

It is now necessary to modify this set of necessary conditions to insure that $g(\mathbf{x}, \boldsymbol{\lambda})$ is maximized so that the nonnegativity restrictions $\mathbf{x} \ge 0$ and $\boldsymbol{\lambda} \ge \mathbf{0}$ are satisfied. These revised conditions are now presented.

The point $\mathbf{x}^*$ will be a maximum to the linear programming problem in (2.79)–(2.81) and to $g(\mathbf{x}, \boldsymbol{\lambda})$ only if a vector $\boldsymbol{\lambda}^*$ exists such that if

$$x_j^* > 0 \qquad \frac{\partial g(\mathbf{x}^*, \boldsymbol{\lambda}^*)}{\partial x_j} = c_j - \sum_{i=1}^{m} \lambda_i^* a_{ij} = 0 \tag{2.84}$$

$$x_j^* = 0 \qquad \frac{\partial g(\mathbf{x}^*, \boldsymbol{\lambda}^*)}{\partial x_j} = c_j - \sum_{i=1}^{m} \lambda_i^* a_{ij} \le 0 \tag{2.85}$$

$$\lambda_i^* > 0 \qquad \frac{\partial g(\mathbf{x}^*, \boldsymbol{\lambda}^*)}{\partial \lambda_i} = b_i - \sum_{j=1}^{n} a_{ij}x_j^* = 0 \tag{2.86}$$

$$\lambda_i^* = 0 \qquad \frac{g(\mathbf{x}^*, \boldsymbol{\lambda}^*)}{\partial \lambda_i} = b_i - \sum_{j=1}^{n} a_{ij}x_j^* \le 0 \tag{2.87}$$

The conditions, 2.84 and 2.86 follow since if x_j^* and λ_i^* are not zero, the constraints $\mathbf{x}^* \geq \mathbf{0}$, $\boldsymbol{\lambda}^* \geq \mathbf{0}$ have no effect on the maximization of $g(\mathbf{x}, \boldsymbol{\lambda})$. Suppose, however, that $x_j^* = 0$. Then the maximum of $g(\mathbf{x}, \boldsymbol{\lambda})$ must occur at the boundary of the jth dimension. From observing Figure 2.17, if this is the case, it must follow that $\partial g(\mathbf{x}^*, \boldsymbol{\lambda}^*)/\partial x_j \leq 0$.

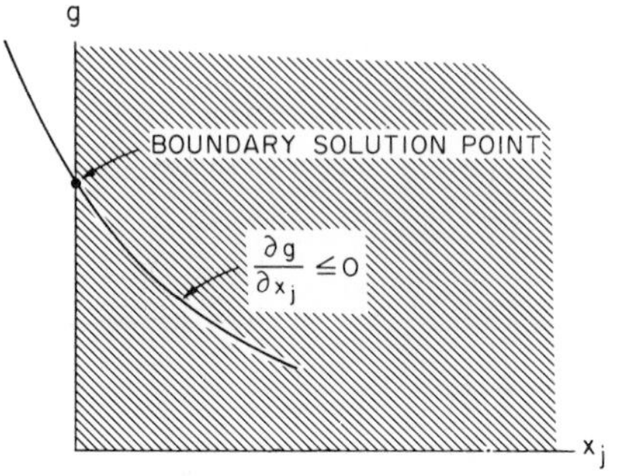

Fig. 2.17. Demonstration of condition (2.85).

Thus if $\mathbf{x}^*$ jointly maximizes $g(\mathbf{x}, \boldsymbol{\lambda})$ and z in (2.79), a vector $\boldsymbol{\lambda}^*$ must exist such that

$$c_j - \sum_{i=1}^{m} \lambda_i^* a_{ij} \leq 0 \qquad j = 1, 2, \ldots, n \tag{2.88}$$

$$b_i - \sum_{j=1}^{n} a_{ij} x_j^* \leq 0 \qquad i = 1, 2, \ldots, m \tag{2.89}$$

Furthermore, it follows from (2.84) and (2.85) that

$$\sum_{j=1}^{n} x_j^* \left(c_j - \sum_{i=1}^{m} \lambda_i^* a_{ij} \right) = 0 \tag{2.90}$$

and from (2.86) and (2.87) that

$$\sum_{i=1}^{m} \lambda_i^* \left(b_i - \sum_{j=1}^{n} a_{ij} x_j^* \right) = 0 \tag{2.91}$$

The conditions in (2.88) and (2.89) represent the Kuhn-Tucker conditions for a linear programming problem. In Chapter 3, more will be said about the Kuhn-Tucker conditions in relation to the general mathematical programming problem.

The condition in (2.90) may be rewritten as

$$\sum_{j=1}^{n} x_j^* c_j = \sum_{i=1}^{m} \lambda_i^* \left(\sum_{j=1}^{n} a_{ij} x_j^* \right) \tag{2.92}$$

However, if the ith constraint is inactive $(b_i - \sum_{j=1}^{n} a_{ij} x_j^* < 0)$, from (2.87) $\lambda_i^* = 0$. Thus the right-hand side of (2.92) can be replaced by the equivalent expression $\sum_{i=1}^{m} \lambda_i^* (b_i)$.

Now (2.92) becomes

$$\sum_{j=1}^{n} x_j^* c_j = \sum_{i=1}^{m} \lambda_i^* b_i$$

or in matrix notation,

$$\mathbf{c}'\mathbf{x}^* = \mathbf{b}'\boldsymbol{\lambda}^* \tag{2.93}$$

Notice that (2.93) is precisely the relationship that the duality theorem claims will hold for the objective functions of the primal problem and its dual linear programming program. It now must be shown that the Lagrangian variables $\lambda_1, \lambda_2, \ldots, \lambda_m$ are indeed the dual variables in the primal-dual relationship.

Since the condition (2.80) must be satisfied for any $\mathbf{x} \geq \mathbf{0}$, it follows for $\boldsymbol{\lambda} \geq \mathbf{0}$ that

$$\lambda_i b_i - \lambda_i \sum_{j=1}^{n} a_{ij} x_j \leq 0 \qquad i = 1, 2, \ldots, m$$

or

$$\lambda_i \sum_{j=1}^{n} a_{ij} x_j \leq \lambda_i b_i \qquad i = 1, 2, \ldots, m$$

Thus

$$\sum_{i=1}^{m} \lambda_i b_i \geq \sum_{i=1}^{m} \lambda_i \left(\sum_{j=1}^{n} a_{ij} x_j \right) = \sum_{j=1}^{n} \left(\sum_{i=1}^{m} \lambda_i a_{ij} \right) x_j \geq \sum_{j=1}^{n} c_j x_j \tag{2.94}$$

Therefore, since (2.94) must also hold at the optimizing point $\mathbf{x}^*$,

$$\sum_{i=1}^{m} \lambda_i b_i \geq \sum_{j=1}^{n} c_j x_j^* \tag{2.95}$$

The condition (2.95) taken together with the Kuhn-Tucker conditions (2.88)–(2.91) imply that $\boldsymbol{\lambda}^*$ must be the solution to the following linear programming problem:

$$\begin{aligned} &\text{minimize} \quad \psi = \textstyle\sum_{i=1}^{m} \lambda_i b_i && && (2.96)\\ &\text{subject to} \quad \textstyle\sum_{i=1}^{m} \lambda_i a_{ij} \geq c_j && j = 1, 2, \ldots, n && (2.97)\\ &\qquad\qquad\qquad \lambda_i \geq 0 && i = 1, 2, \ldots, m && (2.98) \end{aligned}$$

or in matrix notation

$$\begin{aligned} &\text{minimize} \quad \psi = \mathbf{b}'\boldsymbol{\lambda}\\ &\text{subject to} \quad \mathbf{A}'\boldsymbol{\lambda} \geq \mathbf{c} \qquad \boldsymbol{\lambda} \geq \mathbf{0} \end{aligned}$$

It is seen that (2.96)–(2.98) is precisely the dual of the linear programming problem in (2.79)–(2.81). Thus the application of the Lagrangian multiplier method to the primal linear programming problem results in the generation of the dual problem and the fact that the optimal values of the objective functions in the primal and dual problem are equivalent. Moreover, if a solution exists for the primal problem, the Lagrangian derivation insures that a solution exists for the dual problem.

The development of duality theory by using the Lagrangian multiplier method additionally illustrates some important properties of the dual variables. The ith constraint $\sum_{j=1}^{n} a_{ij} x_j \leq b_i$ is called *inactive* if $\sum_{j=1}^{n} a_{ij} x_j^* < b_i$. At the optimal solution $\mathbf{x}^*$ the ith constraint is slack; not all available units b_i of the ith resource have been utilized. In this case λ_i^* will be zero. The ith constraint is called *active* if $\sum_{j=1}^{n} a_{ij} x_j^* = b_i$; all available units of the ith resource have been used at the optimal solution $\mathbf{x}^*$. In this case λ_i^* will be greater than zero. The fact that $\lambda_i^* = 0$ or $\lambda_i^* > 0$ therefore indicates the nature of the utilization of the available units b_i of the ith resource.

The value of λ_i^* when $\lambda_i^* > 0$ also is of importance. To develop the economic interpretation of λ_i^*, consider the equation

$$\mathbf{B}\mathbf{w}_1 = \mathbf{e}_1 \tag{2.99}$$

where $\mathbf{B}$ is the $m \times m$ basis matrix, $\mathbf{w}_1' = [w_{11}, w_{12}, \ldots, w_{1m}]$, and $\mathbf{e}_1' = [1, 0, 0, \ldots, 0]$. In terms of the columns of the basis matrix $\mathbf{B}$, denoted here by $\mathbf{B}_i$, (2.99) may be expressed as

$$\sum_{i=1}^{m} w_{1i}\mathbf{B}_i = \mathbf{e}_1 \qquad (2.100)$$

The vector $\mathbf{e}_1$ in (2.100) may be interpreted as an indicator vector representing the addition of one unit of the first resource and zero units of all other resources ($i = 1, 2, \ldots, m$). The set of scalar values $w_{11}, w_{12}, \ldots, w_{1m}$ represents the amount of change in $x_{B1}, x_{B2}, \ldots, x_{Bm}$ that will result by adding one unit of the first resource while the number of units of all other resources remains the same.

To illustrate this notion, suppose that the basis matrix $\mathbf{B}$ and the resource vector $\mathbf{b}$ assume the values below when $m = 2$:

$$\mathbf{B} = \begin{bmatrix} 2 & 1 \\ 1 & 1 \end{bmatrix} \qquad \mathbf{b} = \begin{bmatrix} 50 \\ 30 \end{bmatrix}$$

Then $x_{B1} = 20$ and $x_{B2} = 10$, since

$$x_{B1}\mathbf{B}_1 + x_{B2}\mathbf{B}_2 = 20\begin{bmatrix} 2 \\ 1 \end{bmatrix} + 10\begin{bmatrix} 1 \\ 1 \end{bmatrix} = \begin{bmatrix} 50 \\ 30 \end{bmatrix} = \mathbf{b}$$

For the matrix $\mathbf{B}$ and $\mathbf{e}_1' = [1, 0]$, it follows that $w_{11} = 1$ and $w_{12} = -1$ in (2.99) since

$$w_{11}\begin{bmatrix} 2 \\ 1 \end{bmatrix} + w_{12}\begin{bmatrix} 1 \\ 1 \end{bmatrix} = 1\begin{bmatrix} 2 \\ 1 \end{bmatrix} + (-1)\begin{bmatrix} 1 \\ 1 \end{bmatrix} = \begin{bmatrix} 1 \\ 0 \end{bmatrix} = \mathbf{e}_1$$

Now if the first resource is increased by one unit so that $\mathbf{b}' = [51, 30]$,

$$\dot{x}_{B1}\mathbf{B}_1 + \dot{x}_{B2}\mathbf{B}_2 = 21\begin{bmatrix} 2 \\ 1 \end{bmatrix} + 9\begin{bmatrix} 1 \\ 1 \end{bmatrix} = \begin{bmatrix} 51 \\ 30 \end{bmatrix} = \mathbf{b} + \mathbf{e}$$

where $\dot{x}_{Bi}$ represents the new solution for x_{Bi} when $\mathbf{e}$ is adjoined to $\mathbf{b}$. Notice that $\dot{x}_{Bi} = x_{Bi} + w_{1i}$. The set of values $w_{11}, w_{12}, \ldots, w_{1m}$ therefore gives the changes in x_{Bi} incurred by adding one unit of the first resource while all other resources remain at the same levels.

The quantity $P_1 = \sum_{i=1}^{m} c_{Bi}w_{1i}$ is the revenue that can be generated by the marginal unit increment of the first resource. It turns out that P_1 is identically λ_1. This may be shown as follows. Since $\mathbf{B}$ in (2.99) is nonsingular,

$$\mathbf{w}_1 = \mathbf{B}^{-1}\mathbf{e}_1 \tag{2.101}$$

Denote by b_{ij}^{-1} the inverse elements of $\mathbf{B}$. Then from (2.101) $w_{1i} = b_{1i}^{-1}$, $i = 1, 2, \ldots, m$; but for the basic feasible solution $\bar{\mathbf{x}}_B' = [x_{B1}, x_{B2}, \ldots, x_{Bm}]$, $\bar{\mathbf{x}}_B = \mathbf{B}^{-1}\mathbf{b}$ so that $z = \bar{\mathbf{c}}_B'\bar{\mathbf{x}}_B = (\bar{\mathbf{c}}_B'\mathbf{B}^{-1})\mathbf{b} = \boldsymbol{\lambda}'\mathbf{b}$ where $\boldsymbol{\lambda}' = \bar{\mathbf{c}}_B'\mathbf{B}^{-1}$. Thus,

$$\lambda_k = \sum_{i=1}^{m} c_{Bi} b_{ik}^{-1}$$

$$\lambda_1 = \sum_{i=1}^{m} c_{Bi} b_{i1}^{-1} = \sum_{i=1}^{m} c_{Bi} w_{1i} = P_1$$

In general, for λ_k let $\mathbf{B}\mathbf{w}_k = \mathbf{e}_k$ where

$$\mathbf{w}_k' = [w_{k1}, w_{k2}, \ldots, w_{km}]$$

$$\mathbf{e}_k' = [0, 0, \ldots, 0, 1, 0, \ldots, 0] \qquad k\text{th component} = 1$$

Then $P_k = \sum_{i=1}^{m} c_{Bi} w_{ki} = \lambda_k$ represents the revenue generated by the marginal unit of the kth resource. The dual variables $\lambda_1, \lambda_2, \ldots, \lambda_m$ are frequently referred to as *shadow prices*, since they are the marginal returns to increments of the available resources. One should be willing to pay up to but no more than the marginal returns for the additional resource unit.

Care must be exercised in the interpretation of the shadow prices λ_i, $i = 1, 2, \ldots, m$. At the optimal solution, λ_k^* gives the revenue resulting from a *one-unit* increase in the kth resource, while all remaining resources are fixed at their original levels. It does not necessarily follow, therefore, that an increase of h units of the kth resource will result in an increase in revenue of $h\lambda_k^*$ units. As the number of units of the kth resource is increased beyond one unit, the optimal basis of the problem may change. The variables comprising the basis may fundamentally change so that the present basis can no longer accurately reflect the increase in revenue to be expected by increasing the kth resource by h units.

Much of the economic literature on applications of linear programming has been concentrated on the interpretations of the variables in dual programming models. Economic interpretations can usually be provided for every dual problem whose corresponding primal problem has economic meaning.

For the diet problem described in Example 2.1, the primal objective function to be minimized is the cost of a satisfactory diet, and the dual objective function to be maximized is the imputed value of a satisfactory diet. Since the primal and dual objective functions have the same value and the primal objective function is evaluated in dollars, the dual objective function must be valued in dollars. The activity levels in the primal objective function are the proportions of three elements used in producing a single pound of food, and the constants in this objective function are costs in dollars; the objective function is $z = \$3x_1 + \$1x_2 + \$2x_3$ for the primal model. Since the constants of the primal right-hand side become the constants in the dual objective function (2000 calories of energy and 1000 units of vitamins) the variables in the dual objective function must be prices or net returns from satisfying the constraints. The dual objective function will be $\psi = 2000y_1 + 1000y_2$, where y_1 and y_2 are these prices.

The constraints of the dual model are

$$\begin{aligned} 3200y_1 + 800y_2 &\leq \$3 \\ 1600y_1 + 1000y_2 &\leq \$1 \\ 2400y_1 + 1200y_2 &\leq \$2 \\ y_1, y_2 &\geq 0 \end{aligned}$$

The variables in the dual model are prices, as delineated above where the primal and dual objective functions and their values are discussed. Therefore the three constraints of the form $a_{1i}y_1 + a_{2i}y_2 \leq \text{cost}_i$ are budget-type constraints, since the right-hand values are the cost coefficients of \$3, \$1, and \$2 from the primal objective function. The a_{1i}'s and a_{2i}'s are the amounts of the first or second nutrient contained in the ith product. The prices y_1 and y_2 are of an unusual type; they are not selling prices found in a supermarket. The values of y_1 and y_2 are shadow prices or values that we should be willing to incur to maximize the value of our diet subject to budget constraints. The three budget constraints in this dual model are not all income constraints; one of these additive linear constraints may be an income constraint and another may be a time constraint where the total available time is evaluated in dollar terms of foregone earnings.

Example 2.23

The solution of the dual problem in Example 2.22 will be determined. The linear programming problem is

$$\begin{aligned} \text{minimize} \quad & \psi = 18\lambda_1 + 10\lambda_2 \\ \text{subject to} \quad & 2\lambda_1 + 2\lambda_2 \geq 10 \\ & 3\lambda_1 + \lambda_2 \geq 20 \\ & \lambda_1, \lambda_2 \geq 0 \end{aligned}$$

Upon adding the surplus variables λ_3 and λ_4 and two artificial variables, the constraint set becomes

$$\begin{aligned} 2\lambda_1 + 2\lambda_2 - \lambda_3 \quad\quad + \lambda_5 \quad\quad &= 10 \\ 3\lambda_1 + \lambda_2 \quad\quad - \lambda_4 \quad\quad + \lambda_6 &= 20 \\ \lambda_i \geq 0 \quad i = 1, 2, \ldots, 6 & \end{aligned}$$

The dual problem is solved using the two-phase method introduced in Section 2.6. The appropriate tableaus for maximizing $Y = -\lambda_5 - \lambda_6$ in Phase I are given in Tables 2.52–2.54.

Table 2.52. Initial Phase I Tableau

Cost coefficient:	0	0	0	0	−1	−1	Basic	
Variable:	λ_1	λ_2	λ_3	λ_4	λ_5	λ_6	variable	Index
	[2]	2	−1	0	1	0	λ_5	10
	3	1	0	−1	0	1	λ_6	20
$c_j - z_j$:	5	3	−1	−1	0	0		

$Y = -10 - 20 = -30.$

Table 2.53. First Phase I Tableau

Cost coefficient:	0	0	0	0	−1	Basic	
Variable:	λ_1	λ_2	λ_3	λ_4	λ_6	variable	Index
	1	1	−1/2	0	0	λ_1	5
	0	−2	[3/2]	−1	1	λ_6	5
$c_j - z_j$:	0	−2	3/2	−1	0		

$Y = -0 - 5 = -5$

Table 2.54. Terminal Phase I Tableau

Cost coefficient:	0	0	0	0	Basic	
Variable:	λ_1	λ_2	λ_3	λ_4	variable	Index
	1	1/3	0	−1/3	λ_1	20/3
	0	−4/3	1	−2/3	λ_3	10/3

$Y^* = -0 - 0 = 0$

The initial Phase II tableau for maximizing $-18\lambda_1 - 10\lambda_2$ is terminal (Table 2.55). The optimal solution is ($\lambda_1^* = 20/3$, $\lambda_2^* = 0$, $\lambda_3^* = 10/3$, $\lambda_4^* = 0$).

Table 2.55. Initial (and Terminal) Phase II Tableau

Cost coefficient:	-18	-10	0	0	Basic	
Variable:	λ_1	λ_2	λ_3	λ_4	variable	Index
	1	1/3	0	$-1/3$	λ_1	20/3
	0	$-4/3$	1	$-2/3$	λ_3	10/3
$c_j - z_j$:	0	-4	0	-6		

$\psi = 120$

The primal problem corresponding to this dual problem is

$$\begin{aligned} \text{maximize} \quad & z = 10x_1 + 20x_2 \\ \text{subject to} \quad & 2x_1 + 3x_2 \leq 18 \\ & 2x_1 + x_2 \leq 10 \\ & x_1, x_2 \geq 0 \end{aligned}$$

The optimal solution to the primal problem is ($x_1^* = 0$, $x_2^* = 6$, $x_3^* = 0$, $x_4^* = 4$), where x_3 and x_4 are the appropriate slack variables. The second constraint ($2x_1 + x_2 \leq 10$) is inactive for $x_1^* = 0$, $x_2^* = 6$, $\lambda_2^* = 0$, as determined above. The revenue z at the optimal solution is $10(0) + 20(6) = \$120$. Since $\lambda_1^* = 20/3$, the *increase* in the revenue by increasing the first resource by one unit (from 18 to 19) is \$20/3. Recall that this problem is introduced as a manufacturing problem in Section 2.1, where x_i represents the number of units to produce of the ith product, $i = 1, 2$. The first product requires 2 hours per unit on the pressing machine and 2 hours on the stamping machine, the second product requires 3 hours on the pressing machine and 1 hour on the stamping machine, and 18 and 10 hours per day are available on the pressing and stamping machines respectively. The marginal revenue incurred by increasing the availability of the pressing machine is therefore \$20/3. The cost of the unit increase on the pressing unit should be less than \$20/3 if it is desired to increase the *net* revenue.

The terminal simplex tableau for the primal problem is shown in Table 2.56. Particular attention should be given to the $c_j - z_j$ row in this table. Notice that the negatives of the numbers corresponding to the optimal dual variables appear in this row: $\lambda_1^* = 20/3$ (x_3 column), $\lambda_2^* = 0$ (x_4 column), $\lambda_3^* = 10/3$ (x_1 column), $\lambda_4^* = 0$ (x_2 column).

Table 2.56. Terminal Primal Problem Tableau

Cost coefficient:	10	20	0	0	Basic	
Variable:	x_1	x_2	x_3	x_4	variable	Index
	2/3	1	1/3	0	x_2	6
	4/3	0	−1/3	1	x_4	4
$c_j - z_j$:	−10/3	0	−20/3	0		

Conversely, the terminal dual tableau in Table 2.55 contains the negative numbers corresponding to the optimal primal variables in its $c_j - z_j$ row: $x_1^* = 0$ (λ_3 column), $x_2^* = 6$ (λ_4 column), $x_3^* = 0$ (λ_1 column), $x_4^* = 4$ (λ_2 column).

The relationship between the primal and dual variables may be stated in general. Let the primal problem be stated in standard form so that there are n variables $x_1, x_2, \ldots, x_n, m$ constraints, and m slack variables $x_{n+1}, \ldots, x_{n+m}$ added to the "less than or equal to" primal inequalities to convert them to equality restrictions. The dual problem will contain m variables $\lambda_1, \lambda_2, \ldots, \lambda_m$, n surplus variables $\lambda_{m+1}, \ldots, \lambda_{m+n}$, and n artificial variables $\lambda_{m+n+1}, \ldots, \lambda_{m+2n}$. From the terminal primal tableau, the optimal values of $\lambda_1, \lambda_2, \ldots, \lambda_m$ are equal to the negatives of the $c_j - z_j$ values in the columns corresponding to $x_{n+1}, \ldots, x_{n+m}$. The optimal values of the dual surplus variables $\lambda_{m+1}, \ldots, \lambda_{m+n}$ are equal to the negatives of the $c_j - z_j$ values in the columns corresponding to $x_1, x_2, \ldots, x_n$. In the dual, $\lambda_{m+n+1}^*, \ldots, \lambda_{m+2n}^*$ each equals zero, since artificial variables must be zero in the terminal solution. Conversely, from the terminal dual tableau, the optimal values of $x_1, x_2, \ldots, x_n$ are equal to the negatives of the $c_j - z_j$ values in the columns corresponding to $\lambda_{m+1}, \ldots, \lambda_{m+n}$. The optimal values of the primal slack variables $x_{n+1}, \ldots, x_{n+m}$ are equal to the negatives of the $c_j - z_j$ values in the columns corresponding to $\lambda_1, \lambda_2, \ldots, \lambda_m$.

Thus from Table 2.55, $x_1^* = -(c_3 - z_3) = 0$, $x_2^* = -(c_4 - z_4) = 6$, $x_3^* = -(c_1 - z_1) = 0$, $x_4^* = -(c_2 - z_2) = 4$. From Table 2.56, $\lambda_1^* = -(c_3 - z_3) = 20/3$, $\lambda_2^* = -(c_4 - z_4) = 0$, $\lambda_3^* = -(c_1 - z_1) = 10/3$, $\lambda_4^* = -(c_2 - z_2) = 0$.

Further Aspects of Duality

As we might expect, additional relationships exist between the primal statement of a linear programming problem and its dual. Two relationships are presented in the following theorems.

Theorem 2.6 The dual of the dual is the primal.

PROOF

Suppose that the primal problem is

$$\text{maximize} \quad z = \mathbf{c}'\mathbf{x} \tag{2.102}$$
$$\text{subject to} \quad \mathbf{A}\mathbf{x} \leq \mathbf{b} \tag{2.103}$$
$$\mathbf{x} \geq \mathbf{0} \tag{2.104}$$

The dual is

$$\text{minimize} \quad \psi = \mathbf{b}'\boldsymbol{\lambda}$$
$$\text{subject to} \quad \mathbf{A}'\boldsymbol{\lambda} \geq \mathbf{c} \qquad \boldsymbol{\lambda} \geq \mathbf{0}$$

The dual problem may be rewritten as

$$\text{maximize} \quad (-\psi) = (-\mathbf{b}')\boldsymbol{\lambda} \tag{2.105}$$
$$\text{subject to} \quad (-\mathbf{A}')\boldsymbol{\lambda} \leq (-\mathbf{c}) \tag{2.106}$$
$$\boldsymbol{\lambda} \geq \mathbf{0} \tag{2.107}$$

The dual problem is stated in (2.105)–(2.107) in the form of a primal problem. Apply the rules for writing the dual of a primal to the following:

$$\text{minimize} \quad \psi = -\mathbf{c}'\mathbf{x}$$
$$\text{subject to} \quad (-\mathbf{A})\mathbf{x} \geq -\mathbf{b} \qquad \mathbf{x} \geq \mathbf{0}$$

which is equivalent to

$$\text{maximize } \mathbf{z} = \mathbf{c}'\mathbf{x}$$
$$\text{subject to} \quad \mathbf{A}\mathbf{x} \leq \mathbf{b} \qquad \mathbf{x} \geq \mathbf{0}$$

which is the primal problem in (2.102)–(2.104).

Thus it really does not matter which statement of a linear programming problem is considered to be the primal and the dual.

Theorem 2.7 If the primal problem has an unbounded solution, then the dual problem has no feasible solution.

PROOF

From (2.95) for any $\boldsymbol{\lambda}$, $\mathbf{c}'\mathbf{x}^* \leq \mathbf{b}'\boldsymbol{\lambda}$. But if the primal problem is unbounded, $\mathbf{c}'\mathbf{x}^* = \infty$. Thus for any $\boldsymbol{\lambda}$,

$$\infty \leq \mathbf{b}'\boldsymbol{\lambda} \tag{2.108}$$

But if $\boldsymbol{\lambda}$ is to be feasible, then from (2.108), the components of $\boldsymbol{\lambda}$ must be at least infinitely large. Hence there is no feasible $\boldsymbol{\lambda}$ for which all components are finite.

The relationship between the primal problem and its dual can be used to achieve a considerable computational advantage in solving some linear programming problems. Suppose, for example, that the primal problem contains 100 variables and 1000 constraints. The basis matrix **B**, therefore, will be a 1000 × 1000 matrix that must be carried from simplex iteration to iteration. The dual problem contains 1000 variables and 100 constraints so that its basis matrix is 100 × 100. Since many computer codes written to solve linear programming problems require the repeated inversion of the basis matrix **B**, the dual problem in this case would be much easier to solve than the primal problem. The dual solution could then be converted by the rules provided above to secure the primal solution.

2.8 SENSITIVITY ANALYSIS

When an optimal solution to a linear programming problem has been determined, it is often desirable to investigate how sensitive the optimal solution is to changes in the given values of c_j, b_i, and a_{ij} in the problem. For example, suppose the resource corresponding to the kth constraint b_k is increased by two units. Will the optimal solution change? It is necessary to be precise in what is meant by optimal solution in this context. In the previous section, it has been shown that if b_k is increased by one unit and the kth constraint is active in the optimal solution, the value of the objective function increases by the value of the dual variable λ_k corresponding to the kth constraint in a maximizing problem. This change in the value of the objective function will be effected by an appropriate change in the values of the optimal solution basic variables. This is a *numerical* change in the optimal solution; i.e., the optimal solution basic variables remain basic but assume different values.

By adding additional units to b_k (say h units) when the kth constraint is active, the objective function will not necessarily increase by $h\lambda_k$. As b_k is increased, a basic variable in the optimal solution may be forced out of the basis, resulting in a *structural* change in the optimal solution, i.e., a new basis.

The second type of change (structural) will be investigated in this section. A structural change generally involves a basic change in policy in applications of linear programming to business problems. For example, an optimal solution to a production problem involving two products may indicate that 10 units of the first product and no units of the second product should be produced. However, if three additional hours of time are added on a machine required to make the two products, the optimal solution may change to one that requires the production of one or more units of both products. Upper level management would most likely want to fully study the implications of shifting the production mix to include at least one unit of the second product.

Over what ranges of c_j and b_i values will the structure of the optimal solution be maintained? The determination of these ranges is important in analyzing the solution of a linear programming problem, since management is often interested in the optimal solutions for different sets of c_j and b_i values and the c_j and b_i values are commonly estimates of the true but unknown values. The ranges will give some insight into the effects of making errors in estimating the true values.

The process of determining these ranges is called *sensitivity analysis*. The analysis may be applied to changes in the constraint resources b_i, the cost coefficients c_j, or the technical or input-output coefficients a_{ij}. The analysis does have one drawback. It looks at the effect on the optimal solution structure by changing one variable at a time with all other variables held constant. For instance, the analysis could be used to determine the range of values for the kth constraint resource b_k, over which the optimal solution structure does not change while fixing the values of $b_1, b_2, \ldots, b_{k-1}, b_{k+1}, \ldots, b_m$. If two or more resources are varied at the same time, sensitivity analysis may not produce the proper joint region of resource values over which the optimal solution structure is maintained. To analyze the effect of jointly changing two or more resources, costs, or input-output coefficients, *parametric programming* must be used. At the end of this section the basic methods of parametric programming will be described. The emphasis here will be on the sensitivity analysis of the constraint resources b_i and the cost coefficients c_j.

Change in the Constraint Resources b_i

If the kth constraint is active in the optimal solution, a unit increase in b_k will produce an increase of λ_k units in the objective function value in a maximizing problem. Over what range of the resource b_k is the shadow price λ_k valid? If the kth constraint is inactive in the optimal solution, then $\lambda_k = 0$, indicating that a zero increase in the value of the objective function will result by increasing b_k by one unit in a maximizing problem. Over what range of the resource b_k will the optimal solution structure be maintained? Both questions can be answered from the tableau corresponding to the optimal solution.

Consider the maximizing form of the linear programming problem

$$\begin{array}{ll} \text{maximize} & z = \mathbf{c}'\mathbf{x} \\ \text{subject to} & \mathbf{A}\mathbf{x} \le \mathbf{b} \qquad \mathbf{x} \ge \mathbf{0} \end{array}$$

Suppose the kth resource level b_k is changed to b_k^0 and define $\mathbf{b}^{0\prime} = [b_1, b_2, \ldots, b_{k-1}, b_k^0, b_{k+1}, \ldots, b_m]$. By changing the kth element of $\mathbf{b}$ from b_k to b_k^0, the optimality criteria in the final tableau are not changed. To see this, write

$$c_j - z_j = c_j - \bar{\mathbf{c}}_B'\mathbf{v}_j = c_j - \bar{\mathbf{c}}_B'\mathbf{B}^{-1}\mathbf{a}_j \tag{2.109}$$

where $\bar{\mathbf{c}}_B$ is the vector containing the cost coefficients of the basic variables, $\mathbf{v}_j$ is the jth column vector in the terminal tableau, $\mathbf{B}$ is the basis matrix and $\mathbf{a}_j$ is the jth column vector in the initial tableau. It is evident from (2.109) that the replacement of $\mathbf{b}$ by $\mathbf{b}^0$ does not affect the optimality criteria. However, this replacement will affect the values of the basic variables. Recall that the basic variable solution is given by $\bar{\mathbf{x}}_B = \mathbf{B}^{-1}\mathbf{b}$. Let $\bar{\mathbf{x}}_B^0 = \mathbf{B}^{-1}\mathbf{b}^0$. Then the ith components of $\bar{\mathbf{x}}_B$ and $\bar{\mathbf{x}}_B^0$ are given by

$$x_{B_i} = \sum_{l=1}^{m} b^{il} b_l \qquad i = 1, 2, \ldots, m$$

$$x_{B_i}^0 = \sum_{\substack{l=1 \\ l \neq k}}^{m} b^{il} b_l + b^{ik} b_k^0 \qquad i = 1, 2, \ldots, m \tag{2.110}$$

respectively, where b^{ij} is the i, jth component of $\mathbf{B}^{-1}$. Equation (2.110) can be rewritten as

$$x^0_{B_i} = \sum_{l=1}^{m} b^{il} b_l + b^{ik}(b^0_k - b_k) = x_{B_i} + b^{ik}(b^0_k - b_k) \qquad i = 1, 2, \ldots, m$$

If $b^{ik}(b^0_k - b_k) \geq 0$, $x^0_{B_i}$ will be greater than zero since $x_{B_i} > 0$. But, if $b^{ik}(b^0_k - b_k) < 0$, then $x^0_{B_i}$ may be driven to zero, causing a change in the basis structure. Thus the present basis will be maintained if

$$x_{B_i} + b^{ik}(b^0_k - b_k) > 0 \qquad i = 1, 2, \ldots, m$$

or if $b^0_k > b_k - [x_{B_i}/b^{ik}]$.

From this inequality it is possible to determine the minimum value of b^0_k, denoted by b^{0-}_k, and the maximum value of b^0_k, denoted by b^{0+}_k, that define the range over which the optimal solution structure will be maintained.

$$b^{0+}_k = \min_i \left[b_k - (x_{B_i}/b^{ik})\right] \qquad \text{for } b^{ik} < 0 \tag{2.111}$$

$$b^{0-}_k = \max_i \left[b_k - (x_{B_i}/b^{ik})\right] \qquad \text{for } b^{ik} > 0 \tag{2.112}$$

Example 2.24

To illustrate the use of (2.111) and (2.112), consider the production problem introduced in section 2.1 and used throughout for illustrative purposes [(2.1)–(2.4)]. From Section 2.6 the optimal solution is ($x_1^* = 0$, $x_2^* = 6$, $x_3^* = 0$, $x_4^* = 4$), with $z^* = 120$, where x_3 and x_4 are slack variables for the first and second constraint respectively. The final tableau is shown in Table 2.57.

Table 2.57. Final Tableau

Cost coefficient:	10	20	0	0	Basic	
Variable:	x_1	x_2	x_3	x_4	variable	Index
	2/3	1	1/3	0	x_2	6
	4/3	0	−1/3	1	x_4	4
$c_j - z_j$:	−10/3	0	−20/3	0		

In this problem, $\mathbf{b}' = [18, 10]$ and the basis inverse matrix is

$$\mathbf{B}^{-1} = \begin{bmatrix} 1/3 & 0 \\ -1/3 & 1 \end{bmatrix}$$

and

$$\overline{\mathbf{x}}_B = \mathbf{B}^{-1}\mathbf{b} = \begin{bmatrix} 1/3 & 0 \\ -1/3 & 1 \end{bmatrix} \begin{bmatrix} 18 \\ 10 \end{bmatrix} = \begin{bmatrix} 6 \\ 4 \end{bmatrix} = \begin{bmatrix} x_2^* \\ x_4^* \end{bmatrix}$$

In the optimal solution the first constraint is active since $x_3^* = 0$, and the second constraint is inactive since $x_4^* = 4$. The limits

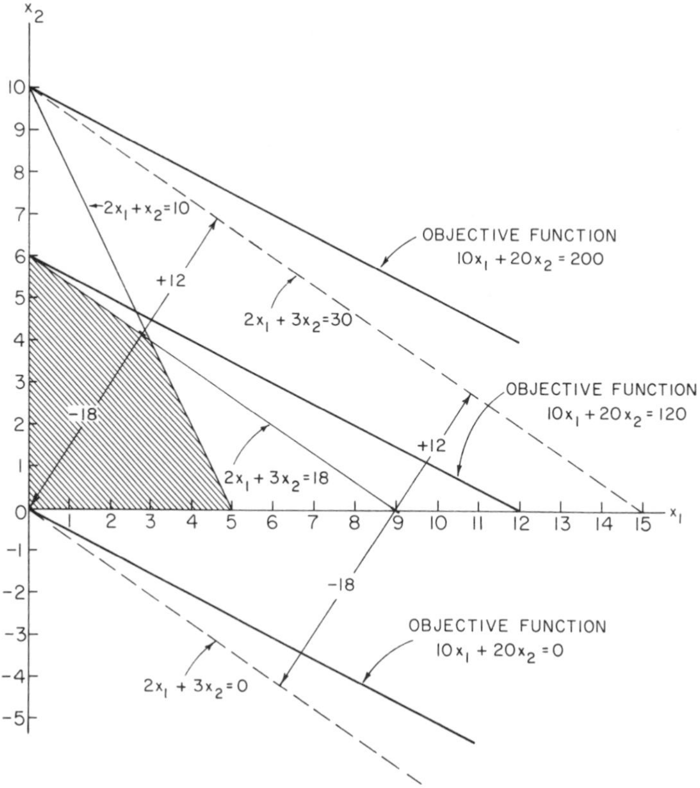

Fig. 2.18. Variation in the first constraint resource.

on b_1^0 are given by

$$b_1^{0+} = 18 - \frac{4}{-1/3} = 30 \qquad b_1^{0-} = 18 - \frac{6}{1/3} = 0$$

In the determination of b_1^{0+} and b_1^{0-} the optimizing process is simplified, since only one term is optimized due to the restrictions $b^{ik} < 0$ and $b^{ik} > 0$ in each case. Thus the range over which b_k may vary without causing a structural change in the basis is 0 to 30. In Figure 2.18 the effect of varying b_1 from 0 to 30 is illustrated. When b_1 becomes 0, the basic variable x_2 will drop out of the optimal solution and will be replaced by x_3. When b_1 becomes 30, the slack variable in the second inequality x_4 becomes 0 and drops out of the basis. If b_1 assumes a value between 0 and 30, x_2 and x_4 will remain basic, as they are in the optimal solution when $b_1 = 18$.

The limits on b_2^0 are given by:

$$b_2^{0+} = \infty \qquad b_2^{0-} = 10 - (4/1) = 6$$

The upper limit on b_2^0 is unbounded, since the condition $b^{12} < 0$ is not satisfied by either element in the second column of $\mathbf{B}^{-1}$.

The ranges on b_1 and b_2 are summarized in Table 2.58. These limits can be determined in a direct manner by examining each constraint and utilizing the information in the final tableau.

Table 2.58. Sensitivity Ranges on b_1 and b_2

Constraint	b_i	Upper limit	Lower limit
1	18	30	0
2	10	∞	6

Consider the second constraint, $2x_1 + x_2 + x_4 = 10$. In the optimal solution, $x_4^* = 4$ so that this constraint is inactive. The resource $b_2 = 10$ can be made as large as desired without affecting the optimal solution structure, since not all this resource is being presently used. However, b_2 can be decreased by no more than $x_4^* = 4$ units without changing the basic structure. When $b_2 = 10 - 4 = 6$, $x_4^* = 0$ and x_4 departs from the basis. Thus the range on b_2 is 6 to ∞.

The determination of the range on the resource corresponding to an inactive constraint is quite easy, as the above analysis illustrates. If the constraint is active, a bit more work is required; consider the first constraint, $2x_1 + 3x_2 + x_3 = 18$. In the optimal solution, $x_3^* = 0$, hence this constraint is active. The range on b_1 can be determined by considering the variable x_3 as the pivot column variable in an additional simplex iteration. If x_3 were allowed to enter the basis from the final tableau shown in Table 2.57, then x_2 would depart and $6/(1/3) = 18$ units of x_3 could be entered in the new basis. By bringing 18 units of x_3 into the basis, the resource $b_2 = 18$ would be expended; this sets the lower range on b_2 at 0. To determine the upper range on b_2, consider bringing a *negative* number of units of x_3 into the basis so that the third column of the final tableau in Table 2.57 becomes

0
$(-1)x_3$
$-1/3$
$+1/3$

The departing variable now becomes x_4 and $4/(1/3) = 12$ units of $(-1)x_3$ can be brought into the new basis. By bringing a negative 12 units of x_3 into the basis, the resource $b_2 = 18$ would be augmented by 12 units—this sets the upper range on b_2 of 30. The reader should relate this method of determining the ranges on b_1 and b_2 to the use of the formulas for the upper and lower bounds defining the ranges given by (2.111) and (2.112).

It should be remembered that this analysis applies only to one constraint resource at a time, with all others held constant. If both b_1 and b_2 are changed *simultaneously* within the ranges given in Table 2.58, the structure of the optimal solution basis *may* change.

Changes in the Cost Coefficient c_j

The effect of changing a cost coefficient can be determined in a similar manner to the method used to perform sensitivity analysis on the constraint resources, since in the dual problem the cost coefficients become the resources:

Primal:

maximize $z = \mathbf{c}'\mathbf{x}$
subject to $\mathbf{A}\mathbf{x} \le \mathbf{b} \quad \mathbf{x} \ge \mathbf{0}$

Dual:

minimize $\psi = \mathbf{b}'\boldsymbol{\lambda}$
subject to $\mathbf{A}'\boldsymbol{\lambda} \ge \mathbf{c} \quad \boldsymbol{\lambda} \ge \mathbf{0}$

By using the dual relationship and the formulas given for the limits defining the ranges on the constraint resources, it is possible to construct the upper and lower limits defining the range on c_k over which the optimal solution structure will not change. Let c_k^+ and c_k^- set the upper and lower limits, respectively, on the range over which c_k can vary without changing the structure of the optimal solution basis. By noticing that **c** is the resource vector in the dual problem, (2.111) and (2.112) may be used to establish the limits c_k^+ and c_k^- from the dual problem:

$$c_k^+ = \min_i [c_k - (\lambda_{B_i}/d^{ik})] \qquad d^{ik} < 0$$

$$c_k^- = \max_i [c_k - (\lambda_{B_i}/d^{ik})] \qquad d^{ik} > 0$$

where c_k is the present value of the kth cost coefficient, d^{ik} is the ikth element of the basis matrix inverse $\mathbf{D}^{-1}$ in the optimal *dual* tableau and λ_{B_i} is the ith basic dual variable in the final tableau.

Example 2.25

Consider again the problem given in (2.1)–(2.4). In this problem the dual is

$$\begin{aligned} \text{minimize} \quad & \psi = 18\lambda_1 + 10\lambda_2 + M\lambda_5 + M\lambda_6 \\ \text{subject to} \quad & 2\lambda_1 + 2\lambda_2 - \lambda_3 + \lambda_5 = 10 \\ & 3\lambda_1 + \lambda_2 - \lambda_4 + \lambda_6 = 20 \\ & \lambda_1, \lambda_2, \ldots, \lambda_6 \geq 0 \end{aligned}$$

where λ_3 and λ_4 are slack variables, λ_5 and λ_6 are artificial variables, and M is an arbitrarily large number. This dual problem was solved by the two-phase algorithm in Section 2.7, and the final dual tableau is given in Table 2.59.

Table 2.59. Final Dual Tableau

Cost coefficient:	-18	-10	0	0	$-M$	$-M$	Basic	
Variable:	λ_1	λ_2	λ_3	λ_4	λ_5	λ_6	variable	Index
	1	1/3	0	$-1/3$	0	1/3	λ_1	20/3
	0	$-4/3$	1	$-2/3$	-1	2/3	λ_3	10/3
$c_j - z_j$:	0	-4	0	-6	$-M$	$-M+6$		

Thus $\mathbf{c}' = [10, 20]$, $\boldsymbol{\lambda}^{*\prime} = [20/3, 10/3] = [\lambda_1^*, \lambda_3^*]$ and

$$\mathbf{D}^{-1} = \begin{bmatrix} 0 & 1/3 \\ -1 & 2/3 \end{bmatrix}$$

The range on c_1 is:

$$c_1^+ = \min_i \, [c_1 - (\lambda_{B_i}/d^{il})] = 10 - \frac{10/3}{-1} = 40/3$$

$$c_1^- = \max_i \, (\cdot) = -\infty$$

Thus c_1 can range from $-\infty$ to 40/3 without causing the optimal basis structure to change. This is evident from studying the optimal primal tableau given in Table 2.57. The nonbasic variable x_1 will enter the basis if c_1 is made sufficiently large such that $c_1 - z_1 > 0$. In the optimal primal tableau, $c_1 - z_1 = -10/3$; if c_1 were increased by more than 40/3, x_1 would enter the basis. In Figure 2.19 the effect of varying c_1 within the range

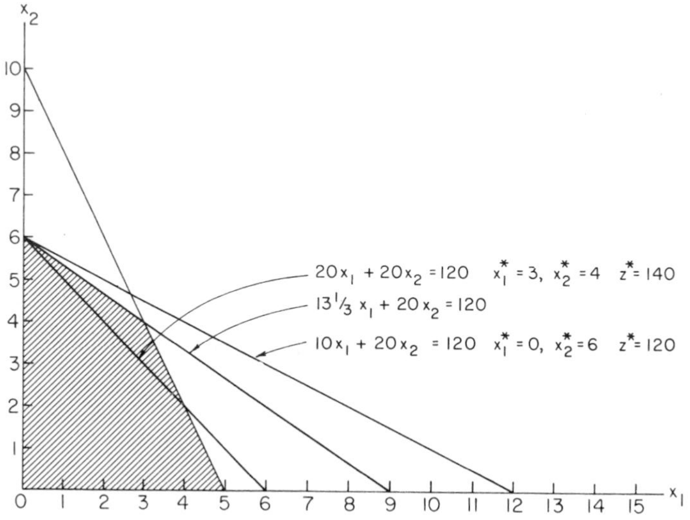

Fig. 2.19. Variation in the cost coefficient c_1.

of $-\infty$ to 40/3 is illustrated. From Figure 2.19 it is apparent that as c_1 increases beyond 40/3, the solution shifts from $(x_1 = 0, x_2 = 6)$ to $(x_1 = 3, x_2 = 4)$, resulting in a structural change in the optimal solution basis.

The range on c_2 is:

$$c_2^+ = \min_i [c_2 - (\lambda_{B_i}/d^{i2})] = \infty$$

$$c_2^- = \max_i [c_2 - (\lambda_{B_i}/d^{i2})] = \max\left\{20 - \frac{20/3}{1/3}, 20 - \frac{10/3}{2/3}\right\} = 15$$

Thus the range on c_2 is 15 to $+\infty$. The ranges on the cost coefficients are summarized in Table 2.60. The range on c_2 can be determined by considering the second dual constraint

$$3\lambda_1 + \lambda_2 - \lambda_4 + \lambda_6 = 20 \qquad (2.113)$$

Table 2.60. Sensitivity Ranges on c_1 and c_2

Cost coefficient	c_j	Upper limit	Lower limit
1	10	40/3	$-\infty$
2	20	$+\infty$	15

In the final dual tableau, $\lambda_6^* = \lambda_4^* = 0$ and the constraint is active. As in the sensitivity analysis on b_i, consider bringing into the basis the nonbasic variable λ_4. From Table 2.59 it is clear that no variable can depart if λ_4 is brought in, since the column vector corresponding to λ_4 contains all negative elements. This implies that as much λ_4 may be added as desired. From (2.113) it is evident that if $\lambda_4 = +\infty$, the resource $c_2 = 20$ can be increased without bound. This sets the upper limit of $+\infty$ on c_2. Now consider bringing in *negative* units of λ_4. The appropriate column vector in the final tableau in Table 2.59 becomes

$$\begin{array}{c} 0 \\ (-1)\lambda_4 \\ \hline 1/3 \\ 2/3 \\ \hline \end{array}$$

At most, 5 units of negative λ_4 may be brought in, and λ_4 will replace λ_3 in the basis. From (2.113) if a negative 5 units of λ_4 were added, the level of c_2 would be *decreased* by 5 units; this sets the lower bound of 15 on c_2.

These methods can be extended to perform sensitivity analysis on the input-output coefficients a_{ij}. The reader is directed to any of the books listed in the Bibliography on linear programming for a development of these methods and a complete treatment of sensitivity analysis.

It should be remembered that sensitivity analysis applies to only one constant at a time, with all other constants remaining fixed. Parametric programming methods may be used to investigate the effect of jointly changing a set of constants. To illustrate the utility of parametric programming, suppose it is desired to determine the effect on optimality of allowing both cost coefficients in Example 2.24 to vary. Let δ_1 and δ_2 be parameters reflecting the change in the constants c_1 and c_2 and

$$\begin{aligned} \text{maximize} \quad & z = (c_1 + \delta_1)x_1 + (c_2 + \delta_2)x_2 \\ \text{subject to} \quad & 2x_1 + 3x_2 \le 18 \\ & 2x_1 + x_2 \le 10 \\ & x_1, x_2 \ge 0 \end{aligned}$$

It may be of interest to determine over what *region* of δ_1 and δ_2 values the current optimal solution remains optimal *in value;* i.e., the solution ($x_1^* = 0$, $x_2^* = 6$, $x_3^* = 0$, $x_4^* = 4$) is retained. The region in the δ_1, δ_2 space can be determined by investigating the tableaus given in Tables 2.61 and 2.62. Table 2.61 gives the initial tableau for the problem, where the cost coefficients corresponding to x_1 and x_2 are functions of δ_1 and δ_2 respectively. If the second column in this tableau is to continue as the pivot column, $20 + \delta_2 > 10 + \delta_1$, or $\delta_2 > \delta_1 - 10$.

Table 2.61. Initial Primal Tableau

Cost coefficient:	$10 + \delta_1$	$20 + \delta_2$	0	0	Basic	
Variable:	x_1	x_2	x_3	x_4	variable	Index
	2	3	1	0	x_3	18
	2	1	0	1	x_4	10
$c_j - z_j$:	$10 + \delta_1$	$20 + \delta_2$	0	0		

Table 2.62. Final Primal Tableau

Cost coefficient:	$10 + \delta_1$	$20 + \delta_2$	0	0	Basic	
Variable:	x_1	x_2	x_3	x_4	variable	Index
	2/3	1	1/3	0	x_2	6
	4/3	0	−1/3	1	x_4	4
$c_j - z_j$:	$(10 + \delta_1) - (2/3)(20 + \delta_2)$	0	$(-1/3)(20 + \delta_2)$	0		

In Table 2.62, optimality is retained if all $c_j - z_j \leq 0$. This implies that

$$(10 + \delta_1) - (2/3)(20 + \delta_2) \leq 0 \qquad (-1/3)(20 + \delta_2) \leq 0$$

Thus a sufficient set of conditions for the current solution to remain optimal in value is:

$$\delta_2 \geq (3/2)\delta_1 - 5 \qquad \delta_2 \geq -20$$

This region is illustrated in Figure 2.20.

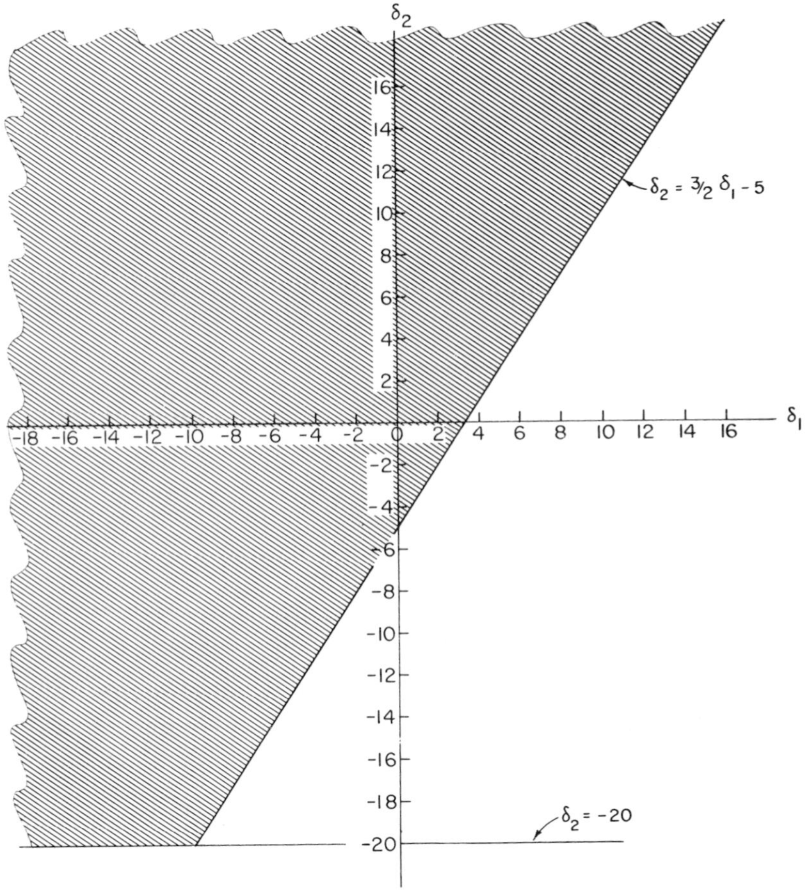

Fig. 2.20. The δ_1, δ_2 region.

It is possible to construct a region for which optimality in value is maintained in terms of the cost coefficients directly. Let $c_1^* = c_1 + \delta_1$ and $c_2^* = c_2 + \delta_2$. If it is further assumed that $c_1^* > 0$, $c_2^* > 0$ (as would be appropriate in most business applications) the (c_1^*, c_2^*) region is as illustrated in Figure 2.21. As long as $c_2^* > (3/2)c_1^*$, the optimal solution in value will be maintained.

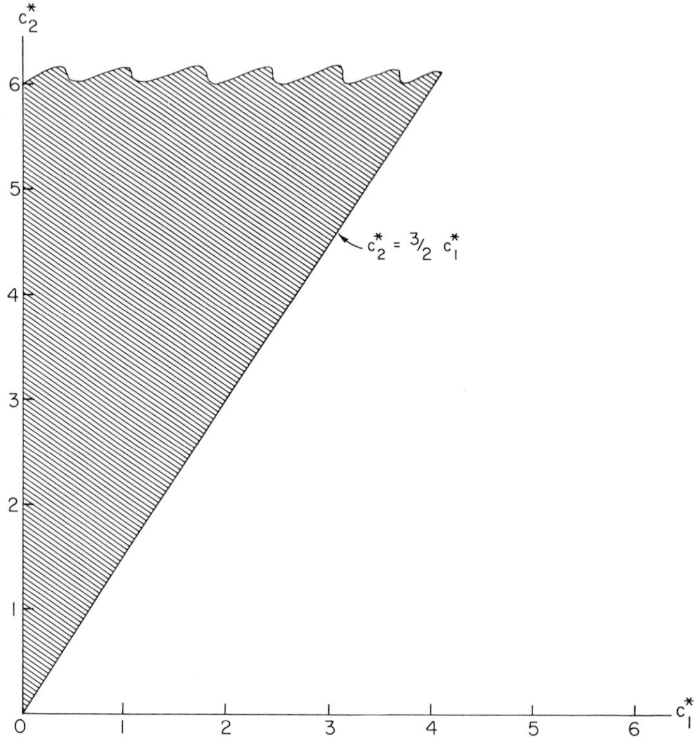

Fig. 2.21. The c_1^*, c_2^* region.

Parametric programming may be used to determine the region over which the values of the constants may change as a function of one or more parameters (δ_1 and δ_2 in the above example). It can also be used to construct a new basis when the constants vary outside the region for which the current optimal solution is maintained. Like sensitivity analysis, it may be applied to the resource constants b_i and the input-output coefficients a_{ij} as well. Again the reader is referred to any of the books on linear programming listed in the Bibliography for a thorough treatment of parametric programming.

Both sensitivity analysis and parametric programming are indispensable tools in analyzing the solution of a linear programming problem. Many linear programming computer codes automatically perform these analyses and print out the results together with the solution to the linear programming problem.

2.9 APPLICATIONS OF LINEAR PROGRAMMING

Even a simple list of the important applications of linear programming would require several pages, so it is not possible to summarize a vast number of cases in this short section. Vadja [1958] and Charnes and Cooper [1961] have written texts in which applications are described exclusively. Gass [1969, Ch. 1] mentions more than twenty cases; specific examples are provided throughout a text by Dorfman et al. [1958]. The theory of the firm in a linear framework is discussed in numerous books and articles; two of these are by Boulding and Spivey [1960] and Dorfman [1951]. A few applications are described in this section. Another application is the use of linear programming to solve game theory models, discussed in Section 2.10.

The applications described here are to indicate the variety of possibilities and to emphasize the role linearity plays in economic theory. For each case the variables and coefficients are interpreted, and numerical examples are provided.

Linear Models of the Firm

One of the most important economic applications of linear programming is to linear models of the firm. Some of this presentation is based on the model of the firm described by Dorfman et al. [1958, pp. 173–78]. Suppose that the original problem is to maximize the firm's net revenue (profit) subject to restrictions on the allocation of the resources available to the firm.

Let p_j = net revenue from producing and selling product j
b_i = amount of resource i available
a_{ij} = amount of resource i required to produce one unit of product j
x_j = amount of product j produced and sold
$i = 1, \ldots, n; j = 1, \ldots, m$

Suppose that the firm chooses to maximize its net revenue z. The restrictions and decisions involve the allocation of the quantities of the n resources b_i, $i = 1, \ldots, n$, among the m production processes and products. The problem is

$$\begin{aligned} \text{maximize} \quad & z = \sum_{j=1}^{m} p_j x_j + \sum_{i=1}^{n} 0 v_i \\ \text{subject to} \quad & \sum_{j=1}^{m} a_{ij} x_j + v_i = b_i \qquad i = 1, \ldots, n \\ & x_j \geq 0 \qquad j = 1, \ldots, m \\ & v_i \geq 0 \qquad i = 1, \ldots, n \end{aligned}$$

The slack variables v_i, $i = 1, \ldots, n$, have been included. The values of the p_j's, a_{ij}'s, and b_i's are presumed to be known or estimated.

Example 2.26

As an example consider the case of a firm that manufactures four products. The firm has limitations on its machine time, warehouse space, and labor available. There are 180 hours of machine time open, of which a unit of product 1 will take 2 hours, a unit of product 2 will take 6.3 hours, a unit of product 3 will take 1.8 hours, and a unit of product 4 will take 6 hours. A unit of each of the four products will use 1.5, 2, 4, and 5 square feet of warehouse space respectively. There are 148 square feet available. Of the 40 man-hours of labor available, a unit of each product will require 0.8, 0.6, 0.9, and 0.4 man-hours respectively. It is assumed that the products will be sold at constant prices. For each of the four products, these are \$3, \$5, \$4, and \$4.50 respectively.

Let x_i = amount of the ith product produced $i = 1, 2, 3, 4$
z = the total revenue gained from the four products

The linear programming problem is

$$\begin{aligned} \text{maximize} \quad & z = 3x_1 + 5x_2 + 4x_3 + 4.5x_4 \\ \text{subject to} \quad & 2.0x_1 + 6.3x_2 + 1.8x_3 + 6.0x_4 \leq 180 \\ & 1.5x_1 + 2.0x_2 + 4.0x_3 + 5.0x_4 \leq 148 \\ & 0.8x_1 + 0.6x_2 + 0.9x_3 + 0.4x_4 \leq 40 \\ & x_1, x_2, x_3, x_4 \geq 0 \end{aligned}$$

where the first three constraints represent the limitations on machine time, warehouse space, and labor respectively. Solving this by means of linear programming yields a maximum profit of \$217.3703, where

$x_1^* = 8.3393 \qquad x_2^* = 18.9538 \qquad x_3^* = 24.3959 \qquad x_4^* = 0.0$

The dual to the above problem is

$$\begin{aligned} \text{minimize} \quad & \psi = \textstyle\sum_{i=1}^{n} b_i \lambda_i \\ \text{subject to} \quad & \textstyle\sum_{i=1}^{n} a_{ij}\lambda_i - \theta_j = p_j \qquad j = 1, \ldots, m \\ & \lambda_i \geq 0 \qquad i = 1, \ldots, n \\ & \theta_j \geq 0 \qquad j = 1, \ldots, m \end{aligned}$$

Here the firm's total value of the resources is minimized, where λ_i is the value of the ith resource and the per unit cost the firm would be willing to pay for a unit of the ith resource. Then $\sum_{j=1}^{n} a_{ij}\lambda_i$ is the firm's imputed cost for the jth production process. This imputed value must be at least as great as the price of the jth product p_j. From the Lagrangian multiplier approach to linear programming and the duality theorem, two results that are important in economic theory can be developed. First, the maximum net revenue equals the minimum total resource value. Second, $x_j = 0$ or $\theta_j = 0$, $j = 1, \ldots, m$; and $v_i = 0$ or $\lambda_i = 0$, $i = 1, \ldots, n$. If $v_i = 0$, the ith resource is fully utilized, since $\sum_{j=1}^{m} a_{ij}x_j = b_i$. Then $\lambda_i > 0$, and the ith resource has a positive value; the firm should be willing to pay a positive price to acquire more of this resource. If $v_i > 0$, the ith resource is not fully employed since $\sum_{j=1}^{m} a_{ij}x_j < b_i$. Then $\lambda_i = 0$, and the firm has no reason to offer anything but a zero price for an additional amount of the ith resource. Any additional quantity of the ith resource that could be acquired would not be utilized and has no value to the firm.

Analogously, either $x_j = 0$ or $\theta_j = 0$. If $x_j = 0$, then $\theta_j > 0$, and $\sum_{i=1}^{n} a_{ij}\lambda_i > p_j$. This means that the imputed value of the jth production process to the firm is greater than the net revenue the firm could earn from producing a unit of product j. The result is the firm's decision to set $x_j = 0$ and not to produce any of the jth product. If $x_i > 0$ and $\theta_j = 0$, then $\sum_{i=1}^{n} a_{ij}\lambda_i = p_j$. Thus the imputed value for production process j to the firm equals the net revenue the firm earns by producing product j, and product j is produced ($x_j > 0$).

Example 2.27

The dual of Example 2.26 is:

$$\begin{aligned} \text{minimize} \quad & \psi = 180\lambda_1 + 148\lambda_2 + 40\lambda_3 \\ \text{subject to} \quad & 2.0\lambda_1 + 1.5\lambda_2 + 0.8\lambda_3 \geq 3.0 \\ & 6.3\lambda_1 + 2.0\lambda_2 + 0.6\lambda_3 \geq 5.0 \end{aligned}$$

$$1.8\lambda_1 + 4.0\lambda_2 + 0.9\lambda_3 \geq 4.0$$
$$6.0\lambda_1 + 5.0\lambda_2 + 0.4\lambda_3 \geq 4.5$$
$$\lambda_i \geq 0 \qquad i = 1, 2, 3$$

The optimal solution to the dual problem is

$$\psi^* = 217.3703 \qquad \lambda_1^* = 0.5057$$
$$\lambda_2^* = 0.3687 \qquad \lambda_3^* = 1.7944$$

The reader should reconsider the interpretation of the simplex tableau for a model of the firm. For each tableau the values in the $c_j - z_j$ row are shadow prices. Recall that the values in this row are called opportunity costs when the simplex method is explained in Section 2.6. For a linear programming model of the firm the opportunity costs are shadow prices.

The values that appear within the tableau have been identified as substitution coefficients. These indicate the effect on each constraint from removing a variable from the solution and introducing another. For a maximization model of the firm, these substitution coefficients determine the amount of the resource that must be employed to produce one unit of an item not currently being produced (changing from $x_j = 0$ to $x_j > 0$).

It is often said that linear programming models of the firm presume perfect competition in the input and output markets. Some models do assume this. However, in the real world, most consumer purchases are at fixed prices that are marked on the items by the sellers. The firm's payment for nearly every resource is settled by contract and may be quoted at an average price per hour, per acre, or per unit purchased.

Table 2.63. Leontief's Input-Output Model Matrix

Sales \ Purchase	To industry: 1	2	$\cdots$ N	Unsold inventory	Government demand	Total
From industry:						
1	x_{11}	x_{12}	$\cdots$ x_{1N}	V_1	G_1	x_1
2	x_{21}	x_{22}	$\cdots$ x_{2N}	V_2	G_2	x_2
$\vdots$	$\vdots$	$\vdots$	$\vdots$	$\vdots$	$\vdots$	$\vdots$
N	x_{N1}	x_{N2}	$\cdots$ x_{NN}	V_N	G_N	x_N
Labor:	x_{01}	x_{02}	$\cdots$ x_{0N}	U_0		L_0

Input-Output Model

Leontief's input-output model of an economy can be employed as an appropriate set of constraints, subject to which a nation's gross national product is maximized. An input-output table is shown in Table 2.63.

Here the quantity of labor employed by industry j is x_{0j}, and the total labor force is L_0. The level of unemployment is the value of U_0. The quantity held by industry i in the form of inventories is V_i. The other variables in the table are:

let x_{ij} = the quantity of goods sold by industry i to industry j
G_i = the quantity of goods sold by industry i to the government
x_i = the total output of industry i

Thus

$$L_0 = \sum_{j=1}^{N} x_{0j} + U_0 \tag{2.114}$$

and

$$x_i = V_i + G_i + \sum_{j=1}^{N} x_{ij} \qquad i = 1, \ldots, N \tag{2.115}$$

Let $a_{ij} = x_{ij}/x_j$, $j = 1, \ldots, N$, where a_{ij} (an input-output coefficient) is the amount of good i needed to produce a unit of good j. Also, let $a_{0j} = x_{0j}/x_j, j = 1, \ldots, N$. Here a_{0j} is the amount of labor needed to produce a unit of good j. The input-output table is converted to Table 2.64.

Table 2.64. Input-Output Model Matrix Expressed in Terms of a_{ij}

Input / Output	To industry: 1	2	⋯ N	Unsold inventory	Government demand	Total
From industry:						
1	a_{11}	a_{12}	$\cdots a_{1N}$	V_1	G_1	x_1
2	a_{21}	a_{22}	$\cdots a_{2N}$	V_2	G_2	x_2
⋮	⋮	⋮	⋮	⋮	⋮	⋮
N	a_{N1}	a_{N2}	$\cdots a_{NN}$	V_N	G_N	x_N
Labor:	a_{01}	a_{02}	$\cdots a_{0N}$	U_0		L_0

This table can be described in matrix terms as

$$\mathbf{Ax} + \mathbf{V} + \mathbf{G} = \mathbf{x} \tag{2.116}$$

and

$$\mathbf{A}_0'\mathbf{x} + U_0 = L_0 \tag{2.117}$$

where

$$\mathbf{A} = \begin{bmatrix} a_{11} & a_{12} & \cdots & a_{1N} \\ a_{21} & a_{22} & \cdots & a_{2N} \\ \vdots & & & \vdots \\ a_{N1} & a_{N2} & \cdots & a_{NN} \end{bmatrix} \quad \mathbf{x} = \begin{bmatrix} x_1 \\ x_2 \\ \vdots \\ x_N \end{bmatrix} \quad \mathbf{G} = \begin{bmatrix} G_1 \\ G_2 \\ \vdots \\ G_N \end{bmatrix} \quad \mathbf{V} = \begin{bmatrix} V_1 \\ V_2 \\ \vdots \\ V_N \end{bmatrix}$$

$$\mathbf{A}_0' = [a_{01}, a_{02}, \ldots, a_{0N}]$$

and U_0 and L_0 are scalars.

Equations (2.116) and (2.117) are equivalent to Table 2.64. The ith row of the product $\mathbf{Ax}$ is $\sum_{j=1}^{N} a_{ij}x_j = \sum_{j=1}^{N} x_{ij}$, since $a_{ij} = x_{ij}/x_j$. Also $\sum_{j=1}^{N} a_{0j}x_j = \sum_{j=1}^{N} x_{0j}$, since $a_{0j} = x_{0j}/x_j$. Thus (2.116) and (2.117) are the matrix equivalents of (2.114) and (2.115) respectively. The vector $\mathbf{x}$ is equivalent to $\mathbf{Ix}$ where $\mathbf{I}$ is the $n \times n$ identity matrix.

The constraints of the input-output table (Table 2.64), can now be written as

$$\mathbf{A}_0'\mathbf{x} + U_0 = L_0 \tag{2.118}$$

and

$$(\mathbf{I} - \mathbf{A})\mathbf{x} = \mathbf{V} + \mathbf{G} \tag{2.119}$$

Allowing for the possibilities that inventories may be held and that there may be unemployment, (2.118) and (2.119) become

$$\mathbf{A}_0'\mathbf{x} \leq L_0 \tag{2.120}$$

and

$$(\mathbf{I} - \mathbf{A})\mathbf{x} \geq \mathbf{G} \tag{2.121}$$

These inequalities will be the constraints of the linear programming model. To convert these constraints from inequalities to equations, there will be one slack variable added to the left-hand side of (2.120) and N slack and artificial variables included in the left-hand side of (2.121).

The objective function is $Y = \mathbf{P}'\mathbf{x}$, where $\mathbf{P}'$ is a row vector of N average prices for a unit of output in industries $1, 2, \ldots, N$. Thus Y is the nation's gross national product in monetary terms. The optimization problem is

$$\begin{aligned} \text{maximize} \quad & Y = \mathbf{P}'\mathbf{X} \\ \text{subject to} \quad & \mathbf{A}_0'\mathbf{x} \leq L_0 \\ & (\mathbf{I} - \mathbf{A})\mathbf{x} \geq \mathbf{G} \\ & \mathbf{x} \geq \mathbf{0} \end{aligned}$$

The elements of $\mathbf{P}$, $\mathbf{A}_0$, $\mathbf{I}$, $\mathbf{A}$, and $\mathbf{0}$ are known. The scalar value L_0 is also presumed.

Example 2.28

For a previous period, suppose that a simple economy is described by the data in Table 2.65. The economy's interindustry purchases, sales, and labor use are converted into the input-output coefficients presented in Table 2.66.

Table 2.65. Leontief Model Example

Sales \ Purchases	To industry: 1	2	3	Unsold inventory	Government demand	Total
From industry:						
1	320	150	210	90	30	800
2	80	60	140	5	15	300
3	160	210	280	35	15	700
Labor:	400	240	280	60		980

Table 2.66. Conversion of Table 2.65 to Input-Output Coefficients

Sales \ Purchases	To industry: 1	2	3	Unsold inventory	Government demand	Total
From industry:						
1	0.4	0.5	0.3	90	30	800
2	0.1	0.2	0.2	5	15	300
3	0.2	0.7	0.4	35	15	700
Labor:	0.5	0.8	0.4	60		980

If the government has specified the following demand for the coming period,

$$\mathbf{G} = \begin{bmatrix} 105 \\ 40 \\ 75 \end{bmatrix}$$

and the size of the labor force for the coming period will be 1600 (that is, $L_0 = 1600$), the constraints of (2.121) and (2.120) are specified as

$$\left\{ \begin{bmatrix} 1 & 0 & 0 \\ 0 & 1 & 0 \\ 0 & 0 & 1 \end{bmatrix} - \begin{bmatrix} 0.4 & 0.5 & 0.3 \\ 0.1 & 0.2 & 0.2 \\ 0.2 & 0.7 & 0.4 \end{bmatrix} \right\} \begin{bmatrix} x_1 \\ x_2 \\ x_3 \end{bmatrix} \geq \begin{bmatrix} 105 \\ 40 \\ 75 \end{bmatrix}$$

or

$$\begin{bmatrix} 0.6 & -0.5 & -0.3 \\ -0.1 & 0.8 & -0.2 \\ -0.2 & -0.7 & 0.6 \end{bmatrix} \begin{bmatrix} x_1 \\ x_2 \\ x_3 \end{bmatrix} \geq \begin{bmatrix} 105 \\ 40 \\ 75 \end{bmatrix} \qquad (2.122)$$

and

$$[0.5, 0.8, 0.4] \begin{bmatrix} x_1 \\ x_2 \\ x_3 \end{bmatrix} \leq 1600 \qquad (2.123)$$

If the unit prices for the three products are \$15, \$20, and \$35 respectively, the objective becomes

$$\text{maximize} \quad Y = [15, 20, 35] \begin{bmatrix} x_1 \\ x_2 \\ x_3 \end{bmatrix} \qquad (2.124)$$

The optimization problem is to maximize (2.124) subject to the constraints of (2.122) and (2.123). The solution to this problem is:

$$Y^* = \$76{,}129.8701 \qquad \mathbf{x}^* = \begin{bmatrix} 1283.1169 \\ 539.6104 \\ 1316.8831 \end{bmatrix}$$

Another application of linear programming appears in Section 2.10, where linear programming is applied to solve a matrix game.

2.10 GAME THEORY AND LINEAR PROGRAMMING

Expositions of game theory and the solutions to game problems have been presented in mathematical as well as popular editions. The earliest analyst of game theory was von Neumann. His first papers, cited in *Theory of Games and Economic Behavior* [1953], were in German. Williams [1954] has contributed an entertaining elementary presentation of game theory. Much of this section is based on Chapters 15 and 16 of Dorfman et al. [1958]. The relationship between game theory and linear programming will be described in this section. The presentation does not involve the esoteric mathematical analysis required to solve some of the advanced problems in game theory.

A simple game is a representation of a conflict situation between two opponents. The opponents may be persons, countries, organizations, or nature versus a single opponent. The opponents are presumed not to collude. The games considered here are all two-person, zero-sum games. This means that the gain of one player equals the loss of the second player. The combined gain and loss for the two players is zero. The final payoff to each player is the value of the game.

Each player knows the alternatives from which he can choose. Suppose that player I chooses from among $A_1, A_2, \ldots, A_n$, and player II chooses from among $B_1, B_2, \ldots, B_m$. Player I sees the game as in Table 2.67, where a_{ij} is the payoff to player I if he chooses A_i when player II selects B_j.

Table 2.67. Payoff Matrix

		II			
		B_1	B_2	$\cdots$	B_m
I	A_1	a_{11}	a_{12}	$\cdots$	a_{1m}
	A_2	a_{21}	a_{22}	$\cdots$	a_{2m}
	$\vdots$	$\vdots$	$\vdots$		$\vdots$
	A_n	a_{n1}	a_{n2}	$\cdots$	a_{nm}

The payoff matrix for player II is the same as in Table 2.67, but the entries are player II's losses to player I. Player I's matrix provides the gains he can achieve, but entries that are negative are the cases where player I pays player II.

Each player conceives a strategy. A player employs a pure strategy if he selects a single alternative and plays that alternative for certain. If a player decides to mix his choices among several alternatives, the player employs a mixed strategy. Usually a mixed strategy will consist of several alternatives to be played probabilistically.

Example 2.29

Consider the case in which the payoff matrix in dollars for player I is given by Table 2.68. If player I selects A_3, he would win \$4 or \$5. The best decision for player II is B_1. Upon seeing II select B_1, the largest return I can obtain occurs from selecting A_1. Now the game will settle at the pair of strategies (A_1, B_1). Any other pair of strategies will enable one of the players to improve his situation by changing his strategy. If I replaces A_1 by A_2 or A_3, II can select B_1 or B_2 so that less than \$6 are paid to I. *Eventually,* player I will decide on A_1, and player II will settle on B_1.

Table 2.68. Simple Game Payoff

		II	
		B_1	B_2
I	A_1	[6]	8
	A_2	3	9
	A_3	4	5

If there is a strategy for each player at which the game will settle, the game has a saddle point. For the case presented in the previous paragraph, (A_1, B_1) is a saddle point. If a game has a saddle point that occurs when players select pure strategies, each player will select a pure strategy.

The value of the game is the gain of I and the loss of II. If a game has a saddle point, the value of the game is determined by the minimax criterion, which can be illustrated using player I's payoff matrix again.

Let α_i = the minimum value across row i
β_j = the maximum value down column j

Then the payoff matrix can be pictured as in Table 2.69.

Table 2.69. Minimax Payoff Table

		II	min row
		B_1 B_2 $\cdots$ B_m	
I	A_1	a_{11} a_{12} $\cdots$ a_{1m}	α_1
	A_2	a_{21} a_{22} $\cdots$ a_{2m}	α_2
	$\vdots$	$\vdots$ $\vdots$ $\vdots$	$\vdots$
	A_n	a_{n1} a_{n2} $\cdots$ a_{nm}	α_n
	max col	β_1 β_2 $\cdots$ β_m	

$\alpha_i = \min_j a_{ij} \quad i = 1, \ldots, n$

$\beta_j = \max_i a_{ij} \quad j = 1, \ldots, m$

Denote the maximum value among $\alpha_1, \alpha_2, \ldots, \alpha_n$ as α^*, and let β^* be the minimum value among $\beta_1, \beta_2, \ldots, \beta_m$. Then

$$\alpha^* = \max_i \alpha_i = \max_i (\min_j a_{ij})$$

$$\beta^* = \min_j \beta_j = \min_j (\max_i a_{ij})$$

If v is the value of the game,

$$\alpha^* = \max_i (\min_j a_{ij}) \leq v \leq \min_j (\max_i a_{ij}) = \beta^* \tag{2.125}$$

This holds because $\mathbf{min}_j\, a_{ij}$ is a vector of the smallest values and $\max_i(\min_j a_{ij})$ is only the largest of these small values. On the other hand, $\mathbf{max}_i\, \mathbf{a}_{ij}$ is a vector of the large values, of which the smallest is $\min_j(\max_i a_{ij})$. The inequalities of (2.125) express the fundamental theorem of games. A rigorous proof of this theorem is offered by Kuhn [1957].

If it so happens that $\alpha^* = \beta^*$,

$$\max_i (\min_j a_{ij}) = v = \min_j (\max_i a_{ij}) \tag{2.126}$$

the game has a saddle point, and the value of the game is v. In this case, each player will select a pure strategy. Player I chooses A_i, and player II picks B_j. Each time that either changes his selection to try to improve his payoff, the opponent can change his alternative to nullify the first variant. After a series of moves and countermoves, each player will learn that his optimal strategy is to select such alternatives that the payoff is v in (2.126). This solution is obtained by the minimax criterion, which is the most famous criterion for solving a simple game when there are pure strategies.

The example that has been considered before has a saddle point at (A_1, B_1), and the value is \$6 (the payoff to Player I). The values for the column labeled α_i are the row minima. The largest value of each column is β_j. The largest α_i equals the smallest β_j, which is 6. See Table 2.70.

Table 2.70. Illustration of a Saddle Point

		II		
		B_1	B_2	min row $= \alpha_i$
I	A_1	[6]	8	[6]
	A_2	3	9	3
	A_3	4	5	4
	max col $= \beta_j$	[6]	9	$v = 6$

min(max col) = v = max(min row)

Example 2.30

Suppose players I and II are duopolists processing a single product. Each is presumed to play one of four alternatives, and the selection of a strategy determines how many units of output a competitor will produce. The payoff matrix for player I is given in Table 2.71.

Table 2.71. Payoff Matrix

		II				
		B_1	B_2	B_3	B_4	min row $= \alpha_i$
I	A_1	2	8	10	3	2
	A_2	6	8	9	[5]	[5]
	A_3	3	−9	−12	−2	−12
	A_4	7	−4	−1	4	−4
	max col $= \beta_j$	7	8	10	[5]	$v = 5$

max(min row) = min(max col) = v = 5

The choices (A_2, B_4) provide a saddle point with a value of 5. Any other set of strategies will allow one of the other players to improve his position by selecting a new alternative. At

(A_2, B_4) any variation by one player can be counteracted with a change by the opponent.

Games with Mixed Optimal Strategies

A payoff matrix that happens to have a saddle point and $\alpha^* = v = \beta^*$ is the simplest possible case. Presuming that a duopolist or any competitor can always play a single strategy is not very realistic. A more likely incidence is where the players apply mixed decision strategies.

Suppose that player I selects strategies $A_1, A_2, \ldots, A_n$ with probabilities $x_1, x_2, \ldots, x_n$ respectively, and player II chooses strategies $B_1, B_2, \ldots, B_m$ with probabilities $y_1, y_2, \ldots, y_m$ respectively.

$$0 \leq x_i \leq 1 \qquad i = 1, \ldots, n \qquad \sum_{i=1}^{n} x_i = 1$$

$$0 \leq y_j \leq 1 \qquad j = 1, \ldots, m \qquad \sum_{j=1}^{m} y_j = 1$$

These strategies are denoted by

$$\mathbf{S}_{\mathrm{I}} = \begin{bmatrix} A_1 & A_2 & \cdots & A_n \\ x_1 & x_2 & \cdots & x_n \end{bmatrix} \qquad \mathbf{S}_{\mathrm{II}} = \begin{bmatrix} B_1 & B_2 & \cdots & B_m \\ y_1 & y_2 & \cdots & y_m \end{bmatrix}$$

The value $\sum_{i=1}^{n} a_{ij} x_i$ is the expected return to player I if player II selects strategy j. Player I will want to select a mix of the alternatives (values for $x_1, x_2, \ldots, x_n$) so that he obtains the largest possible expected return when player II plays his own best strategy. Let

$$E_j = \sum_{i=1}^{n} a_{ij} x_i \qquad j = 1, \ldots, m$$

and suppose that $V = \min(E_1, E_2, \ldots, E_m)$. Then

$$\sum_{i=1}^{n} a_{ij} x_i \geq V \qquad j = 1, \ldots, m$$

and $\sum_{i=1}^{n} x_i = 1$ are the $m + 1$ constraints that player I conceives. Player I will want to make his expected gain V as large as possible. Thus his mixed strategy problem is the linear programming problem:

$$\begin{aligned}
&\text{maximize} \quad V && && (2.127)\\
&\text{subject to} \quad \textstyle\sum_{i=1}^{n} a_{ij}x_i - V \geq 0 && j = 1, \ldots, m && (2.128)\\
&\qquad\qquad \textstyle\sum_{i=1}^{n} x_i = 1 && && (2.129)\\
&\qquad\qquad x_i \geq 0 && i = 1, \ldots, n && (2.130)
\end{aligned}$$

Player II's problem is to select the y's. Let

$$F_i = \sum_{j=1}^{m} a_{ij}y_j \qquad i = 1, \ldots, n$$

The F_i's are player II's expected losses from the mixed strategy $\mathbf{S}_{\text{II}}$ if player I selects A_j. If $W = \max(F_1, F_2, \ldots, F_n)$ player II will want

$$\sum_{j=1}^{m} \alpha_{ij} y_j \leq W \qquad i = 1, \ldots, n$$

and $\sum_{j=1}^{m} y_j = 1$.

Player II's goal is to minimize his expected loss, and his programming problem is:

$$\begin{aligned}
&\text{minimize} \quad W && && (2.131)\\
&\text{subject to} \quad \textstyle\sum_{j=1}^{m} a_{ij}y_i - W \leq 0 && i = 1, \ldots, n && (2.132)\\
&\qquad\qquad \textstyle\sum_{j=1}^{m} y_j = 1 && && (2.133)\\
&\qquad\qquad y_j \geq 0 && j = 1, \ldots, m && (2.134)
\end{aligned}$$

These two players' problems are the duals of each other. The optimal solution is that the gain of player I equals the loss of player II, and the game is zero-sum. This relates to the duality theorem of linear programming and the fundamental theorem of games (2.125). In the notation of these dual programming problems, $V \leq v \leq W$, and $\max V = v = \min W$. The optimal solution $\max V = \min W$ is the value of the game.

Example 2.31

Suppose that the duopolists whose game matrix has been presented in Table 2.71 learn that their payoff matrix is slightly revised. The revised payoff matrix for player I is given by Table 2.72. This game does not have a saddle point for the case when each player selects a single or pure strategy.

Table 2.72. Revised Payoff Matrix

		II				
		B_1	B_2	B_3	B_4	min row
I	A_1	2	6	10	3	2
	A_2	6	8	9	8	6
	A_3	3	−9	−12	−2	−12
	A_4	7	−4	−1	4	−4
	max col	7	8	10	8	

The optimal strategies for each player and the value of the game are determined by solving two dual linear programming problems. For player I the problem is

$$\begin{aligned}
&\text{maximize} && V\\
&\text{subject to} && 2x_1 + 6x_2 + 3x_3 + 7x_4 - V \geq 0\\
& && 6x_1 + 8x_2 - 9x_3 - 4x_4 - V \geq 0\\
& && 10x_1 + 9x_2 - 12x_3 - x_4 - V \geq 0\\
& && 3x_1 + 8x_2 - 2x_3 + 4x_4 - V \geq 0\\
& && x_1 + x_2 + x_3 + x_4 = 1\\
& && x_j \geq 0 \quad j = 1, 2, 3, 4
\end{aligned}$$

This model is simply that of (2.127)–(2.130) using the data in Table 2.72. The optimal solution is

$$V^* = 6.15385 \qquad x_1^* = 0 \qquad x_2^* = 0.84615$$
$$x_3^* = 0 \qquad x_4^* = 0.15385$$

This means that player I should never employ strategies 1 or 3. He should vary his choices between strategies 2 and 4. About 85% of the time he will select strategy 2, and strategy 4 will be chosen the remaining 15% of the time.

The problem facing player II is:

$$\begin{aligned}
&\text{minimize} && W\\
&\text{subject to} && 2y_1 + 6y_2 + 10y_3 + 3y_4 - W \leq 0\\
& && 6y_1 + 8y_2 + 9y_3 + 8y_4 - W \leq 0\\
& && 3y_1 - 9y_2 - 12y_3 - 2y_4 - W \leq 0\\
& && 7y_1 - 4y_2 - y_3 + 4y_4 - W \leq 0\\
& && y_1 + y_2 + y_3 + y_4 = 1\\
& && y_i \geq 0 \quad i = 1, 2, 3, 4
\end{aligned}$$

This is the application of the data in Table 2.72 for player II. His linear programming problem has been given by (2.131)–(2.134). The optimal results are:

$$W^* = 6.15385 \qquad y_1^* = 0.92308 \qquad y_2^* = 0.07692$$
$$y_3^* = 0 \qquad y_4^* = 0$$

Player II should select strategies 1 and 2 in the proportions of 0.92 and 0.08. His optimal strategy is never to select alternative 3 or 4. Since $V^* = W^* = 6.15385$, this is the value of the game to the two players. Player I gains \$6.15, and this is the loss incurred by player II.

PROBLEMS

2.1. Using the graphic technique, find the values of the x's that solve each of the following problems.

(a) maximize $z = x_1 + 2x_2$
subject to
$$x_1 + x_2 \le 4$$
$$0 \le x_1 \le 3$$
$$0 \le x_2 \le 3$$

(b) maximize $z = 3x_1 + 4x_2$
subject to
$$x_1 + x_2 \le 4$$
$$x_1 - x_2 \ge 2$$
$$x_1, x_2 \ge 0$$

(c) minimize $z = 3x_1 + 5x_2$
subject to
$$x_1 + 2x_2 \ge 4$$
$$2x_1 + x_2 \le 6$$
$$x_1, x_2 \ge 0$$

2.2. Determine which of the following constraints if any define a convex set. Use graphs to your best advantage.

(a) $-4x_1 + x_2^2 \le 4$

(b) $x_1^2 + x_2^2 \ge 1$

(c)
$$x_1 - 5x_2^2 \ge 5$$
$$2x_1 + 5x_2 \ge 10$$
$$3x_1 + 2x_2 \ge 12$$

(d)
$$x_1 - 5x_2^2 \le 5$$
$$2x_1 + 5x_2 \le 10$$
$$3x_1 + 2x_2 \ge 17$$

2.3. Find the basic feasible solutions of the following linear programming problem:

$$\begin{aligned} \text{maximize} \quad & z = 7x_1 + 3x_2 \\ \text{subject to} \quad & 4x_1 + 2x_2 \le 15 \\ & 3x_1 + 8x_2 \le 25 \\ & x_1, x_2 \ge 0 \end{aligned}$$

2.4. Use the simplex method for this problem.

(a) Find the values of x_1, x_2, x_3, and z that solve the following:

$$\begin{aligned} \text{maximize} \quad & z = 3x_1 + 4x_2 + 3.5x_3 \\ \text{subject to} \quad & 10x_1 + 2x_2 + x_3 \le 60 \\ & 8x_1 + 4x_2 + 4x_3 \le 48 \\ & x_1, x_2, x_3 \ge 0 \end{aligned}$$

(b) Give the interpretation of each element of the final simplex tableau.

(c) Formulate the dual to this problem and solve it using the simplex method.

(d) From the final tableau of the maximization problem, give the optimal solution to the dual problem. What is the optimal level of each dual variable?

2.5. Consider the following linear programming problem:

$$\begin{aligned} \text{maximize} \quad & z = 3x_1 + 4x_2 + 3.5x_3 \\ \text{subject to} \quad & 9x_1 + 5x_2 + 3x_3 \le 30 \\ & x_1, x_2, x_3 \ge 0 \end{aligned}$$

(a) Find the optimal values for x_1, x_2, x_3, and z using the simplex method.

(b) Find the optimal values for x_1, x_2, x_3, and z using the complete description method.

(c) Find the optimal values for x_1, x_2, x_3, and z using the Lagrangian multiplier method.

(d) Set up and solve the dual of the programming problem given here. Use any solution technique you like.

2.6. Using the simplex method, find x_1, x_2, and z:

$$\begin{aligned} \text{maximize} \quad & z = 3x_1 + 5x_2 \\ \text{subject to} \quad & x_1 \le 4 \qquad x_2 \le 6 \\ & 3x_1 + 2x_2 \le 18 \\ & x_1, x_2 \ge 0 \end{aligned}$$

2.7. Consider the firm that employs two resources in the process of producing three products (the levels are in thousands of dozens of units). Twenty acres of land and 160 hours of labor are available. To produce one unit of product 1, two acres of land and 30 hours of labor are required. To produce one unit of product 2, four acres of land and 10 hours of labor are required. To produce one unit of product 3, two acres of land and 20 hours of labor are required. The selling prices of the three products are \$5, \$12, and \$9 respectively.

(a) Set up and solve the linear programming problem to maximize the firm's sales revenue subject to the resource restrictions. Use the simplex method.

(b) Set up and solve the dual to this problem. Use the simplex method. Give the economic interpretation of the variables and results. Compare the results with the solutions to the dual that appear in part (a) of this problem.

2.8. A firm has two plants in which no finished goods may be stored. The output per day in plant 1 never exceeds 100 units. In plant 2 the maximum output is 150 units. The firm has three warehouses. In warehouse A the inventory must be exactly 80 units; in warehouse B, exactly 75 units; in warehouse C, exactly 60 units. The costs of shipping each unit from a plant to a warehouse are given below. Set up this problem in a linear programming framework.

	To *A*	To *B*	To *C*
From plant 1	5	6	8
From plant 2	6	7	4

2.9. Consider the simple economy that has the following input-output table.

	To plant 1	To plant 2	Demand
From plant 1	0.7	0.4	50
From plant 2	0.3	0.6	80
Labor	0.45	0.55	

Total labor force = 60

The average prices of the output in the two industries are \$3 and \$4 respectively. Set up the linear programming model whose optimal solution will provide the nation's maximum gross national product.

2.10. Each of the following game matrices has a saddle point that can be achieved if each player selects a pure strategy.

(a) Find the strategy for each player that provides the saddle point solution.

(b) What is the value of each game.

Each matrix is the payoff table for player I, who is the maximizer.

Game A

		II	
		B_1	B_2
I	A_1	3	7
	A_2	2	4

Game B

		II			
		B_1	B_2	B_3	B_4
I	A_1	100	50	75	100
	A_2	60	40	90	200

Game C

		II		
		B_1	B_2	B_3
	A_1	12	8	6
I	A_2	2	5	3
	A_3	14	9	7

2.11. The following games have saddle points, but the saddle point is not determined if the players select pure strategies. Check this.

(a) Set up the linear programming problem to determine the value of each game for each player and each player's optimal strategy.

(b) Give the interpretation of the variables in the linear programming model.

Player I is the maximizing player.

Game A

		II			
		B_1	B_2	B_3	B_4
I	A_1	100	50	75	100
	A_2	60	80	90	20

Game B

		II		
		B_1	B_2	B_3
	A_1	12	8	6
I	A_2	2	5	3
	A_3	4	9	7

3 NONLINEAR PROGRAMMING

3.1 INTRODUCTION

In Chapter 1, the general mathematical programming problem has been delineated as the determination of values for the n variables $x_1, x_2 \ldots, x_n$ that optimize a function (the objective function)

$$z = f(\mathbf{x}) = f(x_1, x_2, \ldots, x_n) \qquad (3.1)$$

subject to m constraints of the form

$$g_i(\mathbf{x}) = g_i(x_1, x_2, \ldots, x_n)\{\leq \;=\; \geq\}\; b_i \qquad i = 1, 2, \ldots, m \qquad (3.2)$$

where it is usually assumed additionally that

$$x_i \geq 0 \qquad i = 1, 2, \ldots, n \qquad (3.3)$$

Throughout this chapter, it is assumed that the function f in (3.1) and the functions $g_1, g_2, \ldots, g_m$ in (3.2) are differentiable functions of $x_1, x_2, \ldots, x_n$.

The linear programming problem in Chapter 2 assumes that the function f in (3.1) and the functions $g_i(\mathbf{x})$, $i = 1, 2, \ldots, m$, in (3.2) are linear in the variables $x_1, x_2, \ldots, x_n$. In the nonlinear programming model, the linear assumption is relaxed; one or more functions in the set of functions $f, g_1, g_2, \ldots, g_m$ is allowed to be nonlinear in $x_1, x_2, \ldots, x_n$. The introduction of nonlinear functions in the mathematical programming problem usually insures more difficulty in solving the problem than if all functions are linear. The primary difficulty introduced by the nonlinear functions is the potential existence of relative or local minima or maxima. The existence of local optima can arise due to the nonlinearity of the objective function $f(\mathbf{x})$, the nonlinearity of one or more constraint functions $g_i(\mathbf{x})$, or a combination effect of the nonlinearity in $f(\mathbf{x})$ and in one or more of the constraint functions.

Suppose, for example, that the objective function $f(\mathbf{x})$ is nonlinear in one variable. The existence of a local maximum is illustrated in Figure 3.1. There are two maximizing points x^0 and x^* in the interval (x_l, x_u), and $f(x^*) > f(x^0)$. At the point x^0 a relative or local maximum occurs, while the absolute or global maximum occurs at x^*.

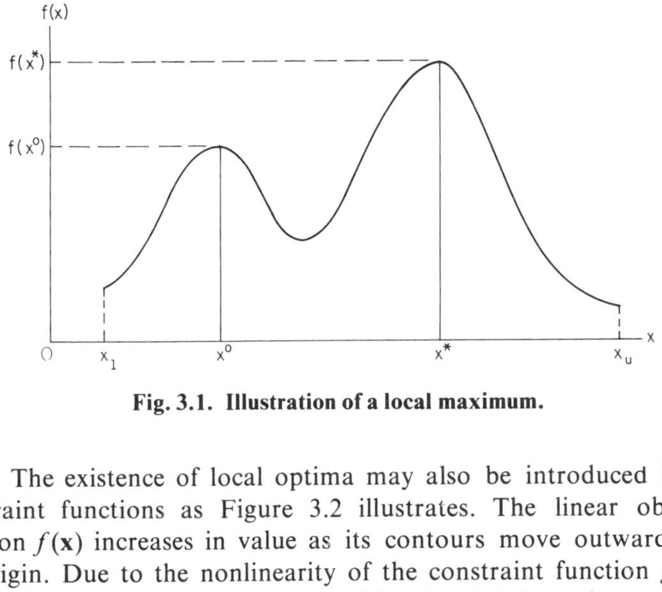

Fig. 3.1. Illustration of a local maximum.

The existence of local optima may also be introduced by the constraint functions as Figure 3.2 illustrates. The linear objective function $f(\mathbf{x})$ increases in value as its contours move outward from the origin. Due to the nonlinearity of the constraint function g_2, the point (x_1^0, x_2^0) is locally maximizing while the global maximum occurs at the point (x_1^*, x_2^*).

In many nonlinear programming problems, numerous local optima may exist that considerably complicate the search for the global optimum. Unfortunately, there is no algorithm that will guarantee locating the global optimum for the general nonlinear programming problem and that is computationally feasible to apply for all such problems. This is in marked contrast to the linear programming problem where the simplex algorithm can solve almost all linear problems with considerable computational efficiency.

Although many economics and business constrained optimiza-

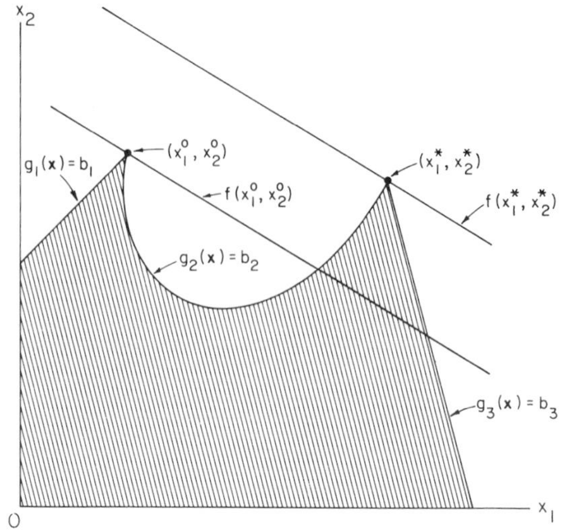

Fig. 3.2. Illustration of local optima.

tion problems have been successfully modeled and solved using linear programming, an increasing number of problems require nonlinear analysis. Therefore, the interest in applying nonlinear programming methods to economic and business problems has increased dramatically. Many algorithms have been devised that will guarantee the determination of the global optimum for certain types of the general nonlinear programming problem. These algorithms have already been applied successfully to a variety of nonlinear business optimization problems that concern such topics as optimal cash flow in a banking environment, portfolio selection for the investor, and resource allocation.

It is the purpose of this chapter to develop the basic theory upon which methods devised to solve the nonlinear programming problem are typically based. Among the topics considered are the definitions of local and global optima, the necessary and sufficient conditions for identifying an optimizing point, and the mathematical evils introduced into this identification process by nonlinearity. The final section contains some applications of this material to nonlinear programming example problems.

3.2 PRELIMINARIES

The concepts of "local" and "global" optima play an extremely important role in nonlinear programming. In this section, the theory of maximizing an unconstrained function $f(\mathbf{x})$ is reviewed. This theory can be conveniently applied to the constrained maximization problem, as will be seen in Section 3.3.

Definition 3.1 Global Maximum (unconstrained problem). The unconstrained function $f(\mathbf{x})$ is said to take on its global maximum at the point $\mathbf{x}^*$ if $f(\mathbf{x}) \leq f(\mathbf{x}^*)$ for all $\mathbf{x}$ over which the function $f(\mathbf{x})$ is defined.

Definition 3.2 Local Maximum (unconstrained problem). The unconstrained function $f(\mathbf{x})$ is said to take on a local maximum at the point $\mathbf{x}^0$ if constants ϵ and δ, $0 < \epsilon < \delta$, exist such that for all $\mathbf{x}$ satisfying $0 < | \mathbf{x} - \mathbf{x}^0 | < \epsilon$, $f(\mathbf{x}) \leq f(\mathbf{x}^0)$, where $f(\mathbf{x})$ is defined for all points in some δ-neighborhood of $\mathbf{x}^0$.

Figure 3.1 illustrates a local and global maximum for a univariate function. Notice from Definitions 3.1 and 3.2 that a global maximum is also a local maximum. A familiar theorem from differential calculus is now introduced, which states the necessary conditions for a point $\mathbf{x}^0$ to be a local (or global) maximum.

Theorem 3.1 If $f(\mathbf{x})$ assumes a relative (local) maximum at $\mathbf{x}^0$, then $\mathbf{x}^0$ must be a solution to the set of n equations

$$\frac{\partial f(\mathbf{x})}{\partial x_j} = 0 \qquad j = 1, 2, \ldots, n$$

PROOF

Suppose that $f(\mathbf{x})$ assumes a local maximum at $\mathbf{x}^0$. Then from the definition of a local maximum, an $\epsilon > 0$ must exist such that for all points $\mathbf{x}$ in a δ-neighborhood of $\mathbf{x}^0$, $f(\mathbf{x}) \leq f(\mathbf{x}^0)$. In particular, consider a point in the δ-neighborhood of $\mathbf{x}^0$ of the form $\mathbf{x} = \mathbf{x}^0 + h\mathbf{e}_j$ where $\mathbf{e}_j' = [0, 0, \ldots, 0, 1, 0 \cdots 0]$ with the 1 placed in the jth position of $\mathbf{e}_j$ and $0 < | h | < \epsilon$. Then

$$f(\mathbf{x}^0 + h\mathbf{e}_j) \leq f(\mathbf{x}^0) \qquad j = 1, 2, \ldots, n \tag{3.4}$$

for all h, $0 < | h | < \epsilon$. Dividing (3.4) by h results in the expressions

$$\frac{f(\mathbf{x}^0 + h\mathbf{e}_j) - f(\mathbf{x}^0)}{h} \leq 0 \text{ if } h > 0 \qquad j = 1, 2, \ldots, n \tag{3.5}$$

$$\frac{f(\mathbf{x}^0 + h\mathbf{e}_j) - f(\mathbf{x}^0)}{h} \geq 0 \text{ if } h < 0 \qquad j = 1, 2, \ldots, n \tag{3.6}$$

On taking the limits of (3.5) and (3.6) as $h \to 0$, it follows from the definition of a partial derivative that

$$\frac{\partial f(\mathbf{x}^0)}{\partial x_j} \leq 0 \text{ for } h \to 0 \qquad h > 0$$

$$\frac{\partial f(\mathbf{x}^0)}{\partial x_j} \geq 0 \text{ for } h \to 0 \qquad h < 0$$

Thus

$$\frac{\partial f(\mathbf{x}^0)}{\partial x_j} = 0 \qquad j = 1, 2, \ldots, n \tag{3.7}$$

The condition in (3.7) can be conveniently displayed in vector notation in terms of the *gradient vector* of $f(\mathbf{x})$.

Definition 3.3 The Gradient Vector. The gradient vector of $f(\mathbf{x}) = f(x_1, x_2, \ldots, x_n)$, denoted by $\nabla f(\mathbf{x})$, is the $n \times 1$ column vector whose components are the first-order partial derivatives of $f(\mathbf{x})$:

$$\nabla f(\mathbf{x}) = \begin{bmatrix} \dfrac{\partial f(\mathbf{x})}{\partial x_1} \\ \dfrac{\partial f(\mathbf{x})}{\partial x_2} \\ \vdots \\ \dfrac{\partial f(\mathbf{x})}{\partial x_n} \end{bmatrix}$$

The condition in (3.7) stated in vector form is $\nabla f(\mathbf{x}^0) = \mathbf{0}$.

If a point $\mathbf{x}^0$ satisfies (3.7), it might not be a maximizing point. Theorem 3.1 provides only the necessary condition for $\mathbf{x}^0$ to be a maximizing point. In the univariate case (3.7) may be satisfied at a minimizing point, a maximizing point, or a point of inflection as illustrated in Figure 3.3. In the n-dimensional case where $\mathbf{x}' = [x_1, x_2, \ldots, x_n]$, the analogy to the univariate case is a minimizing point, maximizing point, or saddle point. A saddle point is the multidimensional analogy to the inflection point in the univariate case. A saddle point for the bivariate case $[\mathbf{x}' = (x_1, x_2)]$ is illustrated in Figure 3.4.

The sufficient condition for $\mathbf{x}^0$ to be a maximizing point can be expressed as a property of the *Hessian matrix* of $f(\mathbf{x})$.

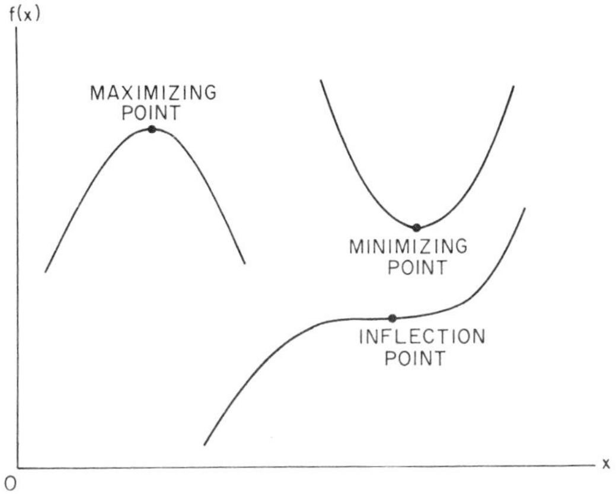

Fig. 3.3. Possible solution points to $df(\mathbf{x})/dx_j = 0$.

Definition 3.4 The Hessian Matrix. The Hessian matrix of $f(\mathbf{x}) = f(x_1, x_2, \ldots, x_n)$, denoted by $H(\mathbf{x})$, is the $n \times n$ matrix whose elements are the second-order partial derivatives of $f(\mathbf{x})$:

$$H(\mathbf{x}) = \begin{bmatrix} \dfrac{\partial^2 f(\mathbf{x})}{\partial x_1^2} & \dfrac{\partial^2 f(\mathbf{x})}{\partial x_1 \partial x_2} & \cdots & \dfrac{\partial^2 f(\mathbf{x})}{\partial x_1 \partial x_n} \\ \dfrac{\partial^2 f(\mathbf{x})}{\partial x_2 \partial x_1} & \dfrac{\partial^2 f(\mathbf{x})}{\partial x_2^2} & \cdots & \dfrac{\partial^2 f(\mathbf{x})}{\partial x_2 \partial x_n} \\ \vdots & & & \vdots \\ \dfrac{\partial^2 f(\mathbf{x})}{\partial x_n \partial x_1} & \dfrac{\partial^2 f(\mathbf{x})}{\partial x_n \partial x_2} & \cdots & \dfrac{\partial^2 f(\mathbf{x})}{\partial x_n^2} \end{bmatrix}$$

Theorem 3.2 A sufficient condition for $f(\mathbf{x}) = f(x_1, x_2, \ldots, x_n)$ to have a local maximum at the point $\mathbf{x}^0$ where $\nabla f(\mathbf{x}^0) = \mathbf{0}$ is that the Hessian matrix $H(\mathbf{x})$ be negative definite; i.e., for any $\mathbf{y}' = [y_1, y_2, \ldots, y_n]$, except $\mathbf{y} \equiv \mathbf{0}$, $\mathbf{y}'H(\mathbf{x})\mathbf{y} < 0$.

PROOF

This theorem can be proved by applying Taylor's theorem to the function $f(\mathbf{x})$. Taylor's theorem states that for any two points $\mathbf{x}_1$ and $\mathbf{x}_2 = \mathbf{x}_1 + \mathbf{h}$, there exists a scalar θ, $0 \leq \theta \leq 1$, such that

$$f(\mathbf{x}_2) = f(\mathbf{x}_1) + \nabla f'(\mathbf{x}_1)\mathbf{h} + 0.5\mathbf{h}'H[\theta\mathbf{x}_1 + (1 - \theta)\mathbf{x}_2]\mathbf{h} \quad (3.8)$$

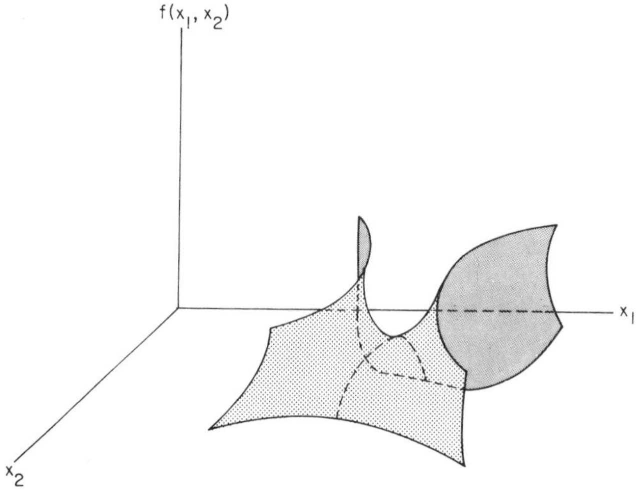

Fig. 3.4. Two-dimensional [$\mathbf{x}' = (x_1, x_2)$] saddle point.

For a proof of Taylor's theorem the reader is referred to Apostol [1957].

Applying (3.8) to $f(\mathbf{x})$, where $\mathbf{x}_1 = \mathbf{x}^0$ and $\mathbf{x}_2 = \mathbf{x}^0 + \mathbf{h}$, produces the expression

$$f(\mathbf{x}^0 + \mathbf{h}) = f(\mathbf{x}^0) + \nabla f'(\mathbf{x}^0)\mathbf{h} + 0.5\mathbf{h}'H[\theta\mathbf{x}^0 + (1 - \theta)(\mathbf{x}^0 + \mathbf{h})]\mathbf{h}$$

Since $\nabla f(\mathbf{x}^0) = \mathbf{0}$,

$$f(\mathbf{x}^0 + \mathbf{h}) = f(\mathbf{x}^0) + 0.5\mathbf{h}'H[\theta\mathbf{x}^0 + (1 - \theta)(\mathbf{x}^0 + \mathbf{h})]\mathbf{h}$$

or

$$f(\mathbf{x}^0 + \mathbf{h}) - f(\mathbf{x}^0) = 0.5\mathbf{h}'H[\theta\mathbf{x}^0 + (1 - \theta)(\mathbf{x}^0 + \mathbf{h})]\mathbf{h} \qquad (3.9)$$

If the right-hand side of (3.9) is negative for all $\mathbf{h}$ in a δ-neighborhood of $\mathbf{x}^0$, by Definition 3.2, $\mathbf{x}^0$ must be a local maximum, since $f(\mathbf{x}^0 + \mathbf{h}) - f(\mathbf{x}^0) \leq \mathbf{0}$ if this is the case. The second partial derivatives $\partial^2 f(\mathbf{x}^0)/\partial x_i \partial x_j$ will have the same sign as $\partial^2 f[\theta\mathbf{x}^0 + (1 - \theta)(\mathbf{x}^0 + \mathbf{h})]/\partial x_i \partial x_j$ provided that the point $\theta\mathbf{x}^0 + (1 - \theta)(\mathbf{x}^0 + \mathbf{h})$ is in a suitable δ-neighborhood of $\mathbf{x}^0$. Thus the right-hand side of (3.9) is negative only if $\mathbf{h}'H(\mathbf{x}^0)\mathbf{h} < 0$; i.e., the Hessian matrix evaluated at $\mathbf{x}^0$, $H(\mathbf{x}^0)$, must be negative definite to insure that $\mathbf{x}^0$ is a maximizing point.

Example 3.1

Determine the maximum of

$$f(\mathbf{x}) = f(x_1, x_2, x_3) = 16x_1 + 24x_2 - 4x_1^2 - 3x_2^2 - x_3^2$$

$$\nabla f(\mathbf{x}) = \begin{bmatrix} \dfrac{\partial f(\mathbf{x})}{\partial x_1} \\ \dfrac{\partial f(\mathbf{x})}{\partial x_2} \\ \dfrac{\partial f(\mathbf{x})}{\partial x_3} \end{bmatrix} = \begin{bmatrix} -8x_1 + 16 \\ -6x_2 + 24 \\ -2x_3 \end{bmatrix} \overset{\text{set}}{=} \begin{bmatrix} 0 \\ 0 \\ 0 \end{bmatrix}$$

The condition $\nabla f(\mathbf{x}) = \mathbf{0}$ generates a system of three linear equations in three unknowns. The solution to this system is $\mathbf{x}' = [x_1, x_2, x_3] = [2, 4, 0]$. The Hessian matrix $H(\mathbf{x})$ is now determined:

$$\frac{\partial^2 f(\mathbf{x})}{\partial x_1^2} = -8 \qquad \frac{\partial^2 f(\mathbf{x})}{\partial x_1 \partial x_2} = 0 \qquad \frac{\partial^2 f(\mathbf{x})}{\partial x_1 \partial x_3} = 0$$

$$\frac{\partial^2 f(\mathbf{x})}{\partial x_2^2} = -6 \qquad \frac{\partial^2 f(\mathbf{x})}{\partial x_2 \partial x_3} = 0 \qquad \frac{\partial^2 f(\mathbf{x})}{\partial x_3^2} = -2$$

Thus the Hessian matrix evaluated at $\mathbf{x}^{0'} = [2, 4, 0]$ is

$$H(\mathbf{x}^0) = \begin{bmatrix} -8 & 0 & 0 \\ 0 & -6 & 0 \\ 0 & 0 & -2 \end{bmatrix}$$

The scalar quantity $\mathbf{y}'H(\mathbf{x}^0)\mathbf{y}$ is

$$[y_1, y_2, y_3] \begin{bmatrix} -8 & 0 & 0 \\ 0 & -6 & 0 \\ 0 & 0 & -2 \end{bmatrix} \begin{bmatrix} y_1 \\ y_2 \\ y_3 \end{bmatrix} = -8y_1^2 - 6y_2^2 - 2y_3^2$$

which is clearly less than zero for any $\mathbf{y}' = [y_1, y_2, y_3]$; $\mathbf{y} \neq \mathbf{0}$. Thus $\mathbf{x}^{0'} = [2, 4, 0]$ is a maximizing point.

One additional theorem is required before proceeding to the development of the theory for the constrained optimization problem.

The theorem is stated without proof, but the proof may be found in Apostol [1957].

Theorem 3.3 The Implicit Function Theorem. Consider a set of m equations in n variables ($m < n$) of the form

$$g_i(\mathbf{x}) = 0 \qquad i = 1, 2, \ldots, m$$

and define the $m \times m$ matrix $\mathbf{G}$ by

$$\mathbf{G} = \begin{bmatrix} \dfrac{\partial g_1(\mathbf{x})}{\partial x_1} & \dfrac{\partial g_1(\mathbf{x})}{\partial x_2} & \cdots & \dfrac{\partial g_1(\mathbf{x})}{\partial x_m} \\ \dfrac{\partial g_2(\mathbf{x})}{\partial x_1} & \dfrac{\partial g_2(\mathbf{x})}{\partial x_2} & \cdots & \dfrac{\partial g_2(\mathbf{x})}{\partial x_m} \\ \vdots & & & \vdots \\ \dfrac{\partial g_m(\mathbf{x})}{\partial x_1} & \dfrac{\partial g_m(\mathbf{x})}{\partial x_2} & \cdots & \dfrac{\partial g_m(\mathbf{x})}{\partial x_m} \end{bmatrix} \tag{3.10}$$

If the rank of $\mathbf{G}$ at $\mathbf{x}^0$ is m, then there exist m functions ϕ_i, $i = 1, 2, \ldots, m$, such that $x_i = \phi_i(x_{m+1}, x_{m+2}, \ldots, x_n)$, $i = 1, 2, \ldots, m$; and these functions are unique, continuous, and differentiable in some neighborhood of $\mathbf{x}^0$.

To appreciate the relevance of this theorem in the context of mathematical programming, consider the following example.

Example 3.2

Suppose in a linear programming problem, the constraint set is specified by

$$3x_1 + 2x_2 \leq 6$$
$$x_1 + 4x_2 \leq 4$$
$$x_1, x_2 \geq 0$$

By adding the slack variables x_3 and x_4, the constraint set becomes

$$3x_1 + 2x_2 + x_3 = 6$$
$$x_1 + 4x_2 + x_4 = 4$$
$$x_1, x_2, x_3, x_4 \geq 0$$

In this system of equations, $m = 2$ and $n = 4$. Recall from Chapter 2 that if m variables are selected, corresponding to independent column vectors in $\bar{\mathbf{A}}$, the selected column vectors form the basis matrix $\mathbf{B}$ and the remaining $n - m$ variables may be expressed in terms of these m variables. In this problem

$$\bar{\mathbf{A}} = \begin{bmatrix} 3 & 2 & 1 & 0 \\ 1 & 4 & 0 & 1 \end{bmatrix}$$

and if x_3 and x_4 are chosen to form the basis matrix $\mathbf{B}_1$,

$$\mathbf{B}_1 = \begin{bmatrix} 1 & 0 \\ 0 & 1 \end{bmatrix}$$

and

$$x_3 = 6 - 3x_1 - 2x_2 = \phi_1(x_1, x_2)$$
$$x_4 = 4 - x_1 - 4x_2 = \phi_2(x_1, x_2)$$

Thus the remaining $n - m = 4 - 2 = 2$ variables (x_1 and x_2) can be written in terms of the $m = 2$ variables (x_3 and x_4). Notice that x_1 and x_2 could have been chosen to form the basis as well. In this case,

$$\mathbf{B}_2 = \begin{bmatrix} 3 & 2 \\ 1 & 4 \end{bmatrix}$$

and

$$x_1 = (1/5)(8 - 2x_3 + x_4) \qquad x_2 = (1/10)(6 + x_3 - 3x_4)$$

Indeed, any one of the $\binom{n}{m} = n!/(m!)(n - m!)$ set of m variables can be used to form $\phi_1, \phi_2, \ldots, \phi_m$.

The implicit function theorem extends this result for linear systems to any set of m equations in n variables when the conditions of Theorem 3.3 are satisfied. For example, in the linear system in Example 3.2, the fact that x_3 and x_4 can be expressed in terms of x_1 and x_2 is guaranteed by this theorem since

$$\mathbf{G} = \mathbf{B}_1 = \begin{bmatrix} 1 & 0 \\ 0 & 1 \end{bmatrix}$$

and the rank of $\mathbf{B}_1$ is clearly $2 = m$.

3.3 LAGRANGIAN MULTIPLIERS AND EQUALITY-CONSTRAINED PROBLEMS

Before investigating the general nonlinear programming problem given in (3.1)–(3.3), it is necessary to first introduce the method of Lagrangian multipliers for solving the equality-constrained mathematical programming problem. The problem is specified as

$$\text{maximize} \quad z = f(\mathbf{x}) = f(x_1, x_2, \ldots, x_n) \tag{3.11}$$

$$\text{subject to} \quad g_i(\mathbf{x}) = g_i(x_1, x_2, \ldots, x_n) = b_i \tag{3.12}$$
$$i = 1, 2, \ldots, m$$

The method of Lagrangian multipliers has been introduced in Chapter 2.

The Lagrangian function corresponding to (3.11) and (3.12) is

$$F(\mathbf{x}, \boldsymbol{\lambda}) = f(\mathbf{x}) + \sum_{i=1}^{m} \lambda_i [b_i - g_i(\mathbf{x})] \tag{3.13}$$

The necessary conditions for a point $[\mathbf{x}^*, \boldsymbol{\lambda}^*]' = [x_1^*, x_2^*, \ldots, x_n^*, \lambda_1^*, \lambda_2^*, \ldots \lambda_m^*]$ to maximize $F(\mathbf{x}, \boldsymbol{\lambda})$ are, from Theorem 3.1,

$$\frac{\partial F(\mathbf{x}, \boldsymbol{\lambda})}{\partial x_j} = \frac{\partial f(\mathbf{x})}{\partial x_j} - \sum_{i=1}^{m} \lambda_i \frac{\partial g_i(\mathbf{x})}{\partial x_j} = 0 \qquad j = 1, 2, \ldots, n \tag{3.14}$$

$$\frac{\partial F(\mathbf{x}, \boldsymbol{\lambda})}{\partial \lambda_i} = b_i - g_i(\mathbf{x}) = 0 \qquad i = 1, 2, \ldots, m \tag{3.15}$$

In the case where $f(\mathbf{x}), g_1(\mathbf{x}), \ldots, g_m(\mathbf{x})$ are linear in $\mathbf{x}$, it has been mentioned in Chapter 2 that if a point $\mathbf{x}^*$ satisfies (3.14) and (3.15), it is the maximizing point to the corresponding linear programming problem (3.11) and (3.12), where $f(\mathbf{x}), g_1(\mathbf{x}), \ldots, g_m(\mathbf{x})$ are linear functions. For the general nonlinear programming problem, it will now be shown why a solution $\mathbf{x}^*$ to (3.14) and (3.15), which is a local maximum of $F(\mathbf{x}, \boldsymbol{\lambda})$ in (3.13), is also a local maximum for (3.11) and (3.12).

To demonstrate this result, first assume that $n = 2$ and $m = 1$ so that the (3.11) and (3.12) problem is

$$\text{maximize} \quad z = f(x_1, x_2) \tag{3.16}$$

$$\text{subject to} \quad g_1(x_1, x_2) = b_1 \tag{3.17}$$

If the conditions of the implicit function theorem (Theorem 3.3) are satisfied, it must be possible to write x_2 in terms of x_1 so that $x_2 =$

$\phi_1(x_1)$. The theorem then guarantees that $\phi_1(x_1)$ is differentiable. The objective function can be written using $\phi_1(\mathbf{x})$ as a univariate function in x_1 and the (3.16) and (3.17) problem is equivalent to the unconstrained problem

$$\text{maximize} \quad z = f[x_1, \phi_1(x_1)]$$

The necessary condition for x_1^0 to be a local optimum of $f[x_1, \phi_1(x_1)]$ is

$$\frac{df[x_1, \phi_1(x_1)]}{dx_1} = 0$$

But recall from differential calculus that the total derivative d/dx_1 of $f(x_1, x_2)$ can be written as

$$\frac{df(x_1, x_2)}{dx_1} = \frac{\partial f(x_1, x_2)}{\partial x_1} + \frac{\partial f(x_1, x_2)}{\partial x_2} \frac{dx_2}{dx_1} \tag{3.18}$$

But $x_2 = \phi_1(x_1)$. If $\phi_1(x_1)$ is substituted for x_2 in (3.18) and the total derivative df/dx_1 is evaluated at (x_1^0, x_2^0),

$$\frac{d}{dx_1} f(x_1^0, x_2^0) = \frac{\partial f(x_1^0, x_2^0)}{\partial x_1} + \frac{\partial f(x_1^0, x_2^0)}{\partial x_2} \frac{d\phi_1(x_1^0)}{dx_1} = 0 \tag{3.19}$$

Since $g_1(x_1, x_2) = b_1$,

$$\frac{dg_1(x_1, x_2)}{dx_1} = \frac{\partial g_1(x_1, x_2)}{\partial x_1} + \frac{\partial g_1(x_1, x_2)}{\partial x_2} \frac{d\phi_1(x_1)}{dx_1} = 0 \tag{3.20}$$

where $\phi_1(x_1)$ has been substituted for x_2 in the last term. From (3.20)

$$\frac{d\phi_1(x_1)}{dx_1} = -\frac{\partial g_1(x_1, x_2)}{\partial x_1} \bigg/ \frac{\partial g_1(x_1, x_2)}{\partial x_2} \tag{3.21}$$

Now substitute the right-hand side of (3.21) for $d\phi_1(x_1)/dx_1$ in (3.19) where $d\phi_1(x_1)/dx_1$ is evaluated at (x_1^0, x_2^0). Then

$$\frac{\partial f(x_1^0, x_2^0)}{\partial x_1} - \frac{\partial f(x_1^0, x_2^0)}{\partial x_2} \left[\frac{\partial g_1(x_1^0, x_2^0)}{\partial x_1} \bigg/ \frac{\partial g_1(x_1^0, x_2^0)}{\partial x_2} \right] = 0 \tag{3.22}$$

and define λ_1 as

$$\lambda_1 = \frac{\partial f(x_1^0, x_2^0)}{\partial x_2} \bigg/ \frac{\partial g_1(x_1^0, x_2^0)}{\partial x_2}$$

Then (3.22) can be written as

$$\frac{\partial f(x_1^0, x_2^0)}{\partial x_1} - \lambda_1 \frac{\partial g_1(x_1^0, x_2^0)}{\partial x_1} = 0 \qquad (3.23)$$

Directly from the definition of λ_1 it follows that

$$\frac{\partial f(x_1^0, x_2^0)}{\partial x_2} - \lambda_1 \frac{\partial g_1(x_1^0, x_2^0)}{\partial x_2} = 0 \qquad (3.24)$$

Additionally, (x_1^0, x_2^0) must satisfy

$$g_1(x_1^0, x_2^0) = b_1 \qquad (3.25)$$

Therefore, by using the implicit function theorem, it is possible to write the necessary conditions for determining a local maximum to (3.16) and (3.17) in the form (3.23)–(3.25).

Now consider the Lagrangian function corresponding to (3.16) and (3.17):

$$F(\mathbf{x}, \boldsymbol{\lambda}) = F(x_1, x_2, \lambda_1) = f(x_1, x_2) + \lambda_1[b_1 - g_1(x_1, x_2)]$$

The necessary conditions for maximizing $F(\mathbf{x}, \boldsymbol{\lambda})$ are, from Theorem 3.1,

$$\frac{\partial F(\mathbf{x}, \boldsymbol{\lambda})}{\partial x_1} = \frac{\partial f(x_1, x_2)}{\partial x_1} - \lambda_1 \frac{\partial g_1(x_1, x_2)}{\partial x_1} = 0 \qquad (3.26)$$

$$\frac{\partial F(\mathbf{x}, \boldsymbol{\lambda})}{\partial x_2} = \frac{\partial f(x_1, x_2)}{\partial x_2} - \lambda_1 \frac{\partial g_1(x_1, x_2)}{\partial x_2} = 0 \qquad (3.27)$$

$$\frac{\partial F(\mathbf{x}, \boldsymbol{\lambda})}{\partial \lambda_1} = b_1 - g_1(x_1, x_2) = 0 \qquad (3.28)$$

The necessary conditions for a point $\mathbf{x}^0$ to maximize $F(\mathbf{x}, \boldsymbol{\lambda})$, given by (3.26)–(3.28), are identical to the necessary conditions for the equality-constrained problem in (3.16) and (3.17).

It is possible to extend the above argument from the $n = 2$ case to the general n-variate case to show that the necessary conditions to maximize the Lagrangian function $F(\mathbf{x}, \boldsymbol{\lambda})$ in (3.13) are equivalent to the necessary conditions to maximize $f(\mathbf{x})$ in the equality-constrained problem (3.11) and (3.12). Before doing this, it is first necessary to modify the definitions of a local and a global maximum in the presence of constraints.

Definition 3.5 Global Maximum (constrained problem). The function $f(\mathbf{x})$ is said to take on its global maximum at the point $\mathbf{x}^*$ if $f(\mathbf{x}) \leq f(\mathbf{x}^*)$ for all $\mathbf{x}$ (including $\mathbf{x}^*$) that belong to the feasible set of points $\mathbf{X}$, where the set $\mathbf{X}$ represents the constraint region.

In the equality-constrained problem, for example, $\mathbf{x}$ belongs to $\mathbf{X}$ if $\mathbf{x}$ satisfies $g_i(\mathbf{x}) = b_i$, $i = 1, 2, \ldots, m$.

Definition 3.6 Local Maximum (constrained problem). The function $f(\mathbf{x})$ is said to take on a local maximum at $\mathbf{x}^0$ if $\mathbf{x}^0$ belongs to $\mathbf{X}$ and there exists an $\epsilon > 0$ such that for every $\mathbf{x} \neq \mathbf{x}^0$ that belongs to $\mathbf{X}$ and is in an ϵ-neighborhood of $\mathbf{x}^0$, $f(\mathbf{x}) \leq f(\mathbf{x}^0)$.

Now, suppose that $f(\mathbf{x})$ takes on a local maximum for the equality-constrained set of feasible points, $\mathbf{X}$, at $\mathbf{x}^0$. Furthermore, assume that at $\mathbf{x}^0$ the conditions of the implicit function theorem are satisfied so that the rank of $\mathbf{G}$, denoted by $r(\mathbf{G})$, is m. Then for $\hat{\mathbf{x}}^{0\prime} = [x^0_{m+1}, x^0_{m+2}, \ldots, x^0_n]$, there exist m functions $\phi_i(\hat{\mathbf{x}}^0)$, such that

$$x_i = \phi_i(\hat{\mathbf{x}}^0) \qquad i = 1, 2, \ldots, m \tag{3.29}$$

Now consider the total differentials of $f(\mathbf{x})$ and $g_i(\mathbf{x})$:

$$df(\mathbf{x}) = \sum_{j=1}^{n} \frac{\partial f(\mathbf{x})}{\partial x_j}\, dx_j \tag{3.30}$$

$$dg_i(\mathbf{x}) = \sum_{j=1}^{n} \frac{\partial g_i(\mathbf{x})}{\partial x_j}\, dx_j \qquad i = 1, 2, \ldots, m \tag{3.31}$$

Since $\partial f(\mathbf{x}^0)/\partial x_j = 0$, $j = 1, 2, \ldots, n$, if $\mathbf{x}^0$ is a stationary point, (3.30) may be written as

$$\sum_{j=1}^{n} \frac{\partial f(\mathbf{x}^0)}{\partial x_j} dx_j = 0 \tag{3.32}$$

Additionally, since $g_i(\mathbf{x}) = b_1, i = 1, 2, \ldots, m$, (3.31) may be written as

$$\sum_{j=1}^{n} \frac{\partial g_i(\mathbf{x})}{\partial x_j} dx_j = 0 \qquad i = 1, 2, \ldots, m \tag{3.33}$$

From the rules for differentiating compound functions,

$$\frac{\partial h(\mathbf{x})}{\partial x_j} = \sum_{i=1}^{m} \frac{\partial f(\mathbf{x})}{\partial x_i} \frac{\partial \phi_i(\mathbf{x})}{\partial x_j} + \frac{\partial f(\mathbf{x})}{\partial x_j} \qquad j = m+1, m+2, \ldots, n \tag{3.34}$$

where $h(\mathbf{x}) = f(\phi_1(\hat{\mathbf{x}}^0), \ldots, \phi_m(\hat{\mathbf{x}}^0), \hat{\mathbf{x}}^0)$. It is now possible to proceed as has been done for the two-dimensional case. Identify an expression for $\partial \phi_i(\hat{\mathbf{x}})/\partial x_j$ in terms of the partial derivatives $\partial g_i(\mathbf{x})/\partial x_j$ and substitute this expression for $\partial \phi_i(\hat{\mathbf{x}})/\partial x_j$ in (3.34). However, it is not possible to arrive at the desired result by solving for $\partial \phi_i(\hat{\mathbf{x}})/\partial x_j$ directly. Introduce the Lagrangian multiplier λ_i and write

$$df(\mathbf{x}) - \sum_{i=1}^{m} \lambda_i \, dg_i(\mathbf{x}) \tag{3.35}$$

At the point $\mathbf{x}^0$, it follows from (3.32) and (3.33) that (3.35) may be written as

$$\sum_{j=1}^{n} \left[\frac{\partial f(\mathbf{x}^0)}{\partial x_j} - \sum_{i=1}^{m} \lambda_i \frac{\partial g_i(\mathbf{x}^0)}{\partial x_j} \right] dx_j = 0 \tag{3.36}$$

Since the first m variables can be expressed in terms of the remaining $n - m$ by (3.29), the set of $n - m$ variables may be thought of as independent variables in (3.29). Thus dx_j, $j = m+1, m+2, \ldots, n$, may be considered as independent variables; and if (3.36) is satisfied, it must follow that

$$\frac{\partial f(\mathbf{x}^0)}{\partial x_j} - \sum_{i=1}^{m} \lambda_i \frac{\partial g_i(\mathbf{x}^0)}{\partial x_j} \equiv 0 \qquad j = m+1, m+2, \ldots, n \tag{3.37}$$

Now (3.36) can be rewritten excluding the components in the sum for

$j = m + 1, m + 2, \ldots, n$, since by (3.37) they are zero:

$$\sum_{j=1}^{m} \left[\frac{\partial f(\mathbf{x}^0)}{\partial x_j} - \sum_{i=1}^{m} \lambda_i \frac{\partial g_i(\mathbf{x}^0)}{\partial x_j} \right] dx_j = 0 \tag{3.38}$$

Since the $dx_j, j = 1, 2, \ldots, m$, are the dependent variables determined uniquely by $dx_j, j = m + 1, m + 2, \ldots, n$, in (3.29), the coefficients of $dx_j, j = 1, 2, \ldots, m$, in (3.38) must be identically zero. Thus

$$\frac{\partial f(\mathbf{x}^0)}{\partial x_j} - \sum_{i=1}^{m} \lambda_i \frac{\partial g_i(\mathbf{x}^0)}{\partial x_j} \equiv 0 \qquad j = 1, 2, \ldots, m \tag{3.39}$$

Combining (3.37) and (3.39), it is seen that the following condition must be satisfied for $\mathbf{x}^0$:

$$\frac{\partial f(\mathbf{x}^0)}{\partial x_j} - \sum_{i=1}^{m} \lambda_i \frac{\partial g_i(\mathbf{x}^0)}{\partial x_j} = 0 \qquad j = 1, 2, \ldots, n \tag{3.40}$$

Additionally,

$$g_i(\mathbf{x}^0) - b_i = 0 \qquad i = 1, 2, \ldots, m \tag{3.41}$$

It is seen that these conditions in (3.40) and (3.41) are identical to the necessary conditions for maximizing the Lagrangian function $F(\mathbf{x}, \boldsymbol{\lambda})$ given by (3.14) and (3.15).

In summary, the necessary conditions for $\mathbf{x}^0$ to be a local maximum to the equality-constrained problem

$$\text{maximize} \quad z = f(\mathbf{x}) = f(x_1, x_2, \ldots, x_n) \tag{3.42}$$

$$\text{subject to} \quad g_i(\mathbf{x}) = b_i \qquad i = 1, 2, \ldots, m \tag{3.43}$$

can be generated by defining the Lagrangian function

$$F(\mathbf{x}, \boldsymbol{\lambda}) = f(\mathbf{x}) + \sum_{i=1}^{m} \lambda_i [b_i - g_i(\mathbf{x})] \tag{3.44}$$

These conditions for $\mathbf{x}^0$ to be a maximizing point to (3.44) are

$$\frac{\partial F(\mathbf{x}^0, \boldsymbol{\lambda}^0)}{\partial x_j} = \frac{\partial f(\mathbf{x}^0)}{\partial x_j} - \sum_{i=1}^{m} \lambda_i^0 \frac{\partial g_i(\mathbf{x}^0)}{\partial x_j} = 0 \qquad j = 1, 2, \ldots, n \tag{3.45}$$

$$\frac{\partial F(\mathbf{x}^0, \boldsymbol{\lambda}^0)}{\partial \lambda_i} = g_i(\mathbf{x}^0) - b_i = 0 \qquad i = 1, 2, \ldots, m \tag{3.46}$$

If $r(\mathbf{G}) = m$ at $\mathbf{x}^0$, then (3.45) and (3.46) will also be the necessary conditions for $\mathbf{x}^0$ to be a local maximum to (3.42) and (3.43).

Example 3.3 illustrates the use of (3.44)–(3.46) in solving a problem of the type given by (3.42) and (3.43).

Example 3.3

The problem is

$$\begin{aligned} &\text{maximize} \quad z = f(\mathbf{x}) = f(x_1, x_2, x_3) = x_1 x_2 x_3 \\ &\text{subject to} \quad x_1^2 + x_2^2 + x_3^2 = 27 \end{aligned} \tag{3.47}$$

The Lagrangian function is

$$F(\mathbf{x}, \boldsymbol{\lambda}) = x_1 x_2 x_3 + \lambda_1 (27 - x_1^2 - x_2^2 - x_3^2)$$

and the four equations resulting from (3.14) and (3.15) are

$$\begin{aligned} x_2 x_3 - 2\lambda_1 x_1 &= 0 \\ x_1 x_3 - 2\lambda_1 x_2 &= 0 \\ x_1 x_2 - 2\lambda_1 x_3 &= 0 \\ 27 - x_1^2 - x_2^2 - x_3^2 &= 0 \end{aligned} \tag{3.48}$$

There are eight possible solutions to this set of equations, which are the combinations of $x_1 = x_2 = x_3 = \pm 3$ with $\lambda_1 = 3/2$ and $\lambda_1 = -3/2$ in four solutions each.

The maximum of $f(\mathbf{x})$ is not unique; four points in the variable space of x_1, x_2, and x_3(3, 3, 3), (3, −3, −3), (−3, 3, −3), and (−3, −3, 3) will give the maximum value of $f(\mathbf{x})$, which is 27. The other four points resulting from (3.48) give the minimum of $f(\mathbf{x})$ on the shell of the sphere delineated by (3.47).

Unfortunately, if $f(\mathbf{x})$ takes on a local maximum at $\mathbf{x}^*$, then $\mathbf{x}^*$ may not always satisfy (3.45) and (3.46). To determine when this pathological case may arise, it is necessary to investigate the behavior of the functions $f, g_1 \ldots, g_m$ at the critical point $\mathbf{x}^*$ in a more general way than has been presented above. Denote by ∇f and ∇g_i the column gradient vectors associated with the functions $f(\mathbf{x})$ and $g_i(\mathbf{x})$, where

$$\nabla f' = \left[\frac{\partial f}{\partial x_1}, \frac{\partial f}{\partial x_2}, \ldots, \frac{\partial f}{\partial x_n}\right]$$

and

$$\nabla g_i' = \left[\frac{\partial g_i}{\partial x_1}, \frac{\partial g_i}{\partial x_2}, \ldots, \frac{\partial g_i}{\partial x_n}\right]$$

and define the $(m + 1) \times n$ matrix $\mathbf{G}^0$ and the $m \times n$ matrix $\mathbf{G}$ by

$$\mathbf{G}^0 = \begin{bmatrix} \nabla g_1' \\ \vdots \\ \nabla g_m' \\ \nabla f' \end{bmatrix} \qquad \mathbf{G} = \begin{bmatrix} \nabla g_1' \\ \vdots \\ \nabla g_m' \end{bmatrix} \tag{3.49}$$

respectively.

Then the Lagrangian function can be written in a more general form:

$$F^0(\mathbf{x}, \boldsymbol{\lambda}) = \lambda_0 f(\mathbf{x}) + \sum_{i=1}^{m} \lambda_i [b_i - g_i(\mathbf{x})]$$

where λ_0 is either 0 or 1. If $f(\mathbf{x})$ takes on a local maximum (or minimum) at $\mathbf{x}^*$, then $\mathbf{x}^*$ must satisfy

$$\frac{\partial F^0(\mathbf{x}, \boldsymbol{\lambda})}{\partial x_j} = \lambda_0 \frac{\partial f(\mathbf{x})}{\partial x_j} - \sum_{i=1}^{m} \lambda_i \frac{\partial g_i(\mathbf{x})}{\partial x_j} = 0 \qquad j = 1, 2, \ldots, n \tag{3.50}$$

and

$$\frac{\partial F^0(\mathbf{x}, \boldsymbol{\lambda})}{\partial \lambda_i} = b_i - g_i(\mathbf{x}) = 0 \qquad i = 1, 2, \ldots, m \tag{3.51}$$

where $\lambda_0 = 0$ or 1.

Notice that if $\lambda_0 = 1$, the results are the usual Lagrangian necessary conditions given in (3.45) and (3.46). In the special cases where $\mathbf{x}^*$ will not satisfy (3.45) and (3.46), $\mathbf{x}^*$ will satisfy (3.50) and (3.51) when

$\lambda_0 = 0$. The problem may now be broken down in the following cases:

CASE 1: $r(\mathbf{G}^0) = m + 1$ at $\mathbf{x}^*$

Notationally, $r(\mathbf{G}^0)$ denotes the rank of the matrix $\mathbf{G}^0$ in (3.49). In this case, $f(\mathbf{x})$ does not take on a local maximum at $\mathbf{x}^*$. This can be demonstrated by assuming that $z = f(\mathbf{x}^*)$ is a local maximum or minimum, for if it is,

$$\frac{\partial F(\mathbf{x}^*,\boldsymbol{\lambda}^*)}{\partial x_j} = \frac{\partial f(\mathbf{x}^*)}{\partial x_j} - \sum_{i=1}^{m} \lambda_i^* \frac{\partial g_i(\mathbf{x}^*)}{\partial x_j} = 0 \qquad j = 1, 2, \ldots, n \tag{3.52}$$

which implies that

$$\frac{\partial f(\mathbf{x}^*)}{\partial x_j} = \sum_{i=1}^{m} \lambda_i^* \frac{\partial g_i(\mathbf{x}^*)}{\partial x_j} \qquad j = 1, 2, \ldots, n$$

or in vector notation

$$\nabla f(\mathbf{x}^*) = \sum_{i=1}^{m} \lambda_i^* \nabla g_i(\mathbf{x}^*) \tag{3.53}$$

But if $r(\mathbf{G}^0) = m + 1$ at $\mathbf{x}^*$, the rows of $\mathbf{G}^0$ which are $\nabla g_1', \nabla g_2', \ldots, \nabla g_m', \nabla f'$ are linearly independent, which means that ∇f cannot be expressed as a linear combination of ∇g_i, $i = 1, 2, \ldots, m$, as in (3.53). Thus by contradiction, $f(\mathbf{x}^*)$ cannot be a relative minimum or maximum; i.e., $\mathbf{x}^*$ will not satisfy (3.52). In this case $\mathbf{x}^*$ will satisfy (3.50) and (3.51) if $\lambda_0 = 0$, however. Now consider a very simple example of this case.

Example 3.4

The problem is

$$\begin{aligned} &\text{maximize} \quad z = f(x_1, x_2) = x_1 + 2x_2 \\ &\text{subject to} \quad g(x_1, x_2) = x_1 + 3x_2 = 3 \end{aligned}$$

For this problem $n = 2$, $m = 1$, and

$$\mathbf{G}^0 = \begin{bmatrix} 1 & 3 \\ 1 & 2 \end{bmatrix}$$

so that $r(\mathbf{G}^0) = 2 > 1 = m$. The Lagrangian function is

$$F^0(\mathbf{x}, \boldsymbol{\lambda}) = (x_1 + 2x_2)\lambda_0 + \lambda_1(3 - x_1 - 3x_2)$$

and

$$\frac{\partial F^0(\mathbf{x},\boldsymbol{\lambda})}{\partial x_1} = \lambda_0 - \lambda_1 = 0$$

$$\frac{\partial F^0(\mathbf{x},\boldsymbol{\lambda})}{\partial x_2} = 2\lambda_0 - 3\lambda_1 = 0$$

$$\frac{\partial F^0(\mathbf{x},\boldsymbol{\lambda})}{\partial \lambda_1} = 3 - x_1 - 3x_2 = 0$$

A quick look at these equations indicates that they are inconsistent if $\lambda_0 = 1$. The solution with $\lambda_0 = 0$ is: $\lambda_1 = 0$ and any (x_1, x_2) satisfying $3 - x_1 - x_2 = 0$. The problem is represented graphically in Figure 3.5. From Figure 3.5 it is obvious that the maximum of $z = f(x_1, x_2)$ is not finite, and hence a local maximum (or minimum) of $f(x_1, x_2)$ does not exist.

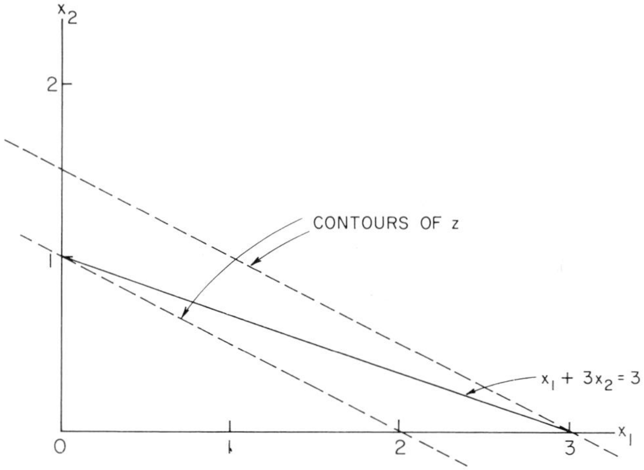

Fig. 3.5. Graphic solution to Example 3.4.

CASE 2: $r(\mathbf{G}^0) < m + 1$

This case may be further broken down into four subcases.

Subcase 2.1: $r(\mathbf{G}^0) = r(\mathbf{G}) = m$ at $\mathbf{x}^*$

In this subcase, (3.50) will be a consistent set of equations at $\mathbf{x}^*$ when $\lambda_0 = 1$.

Example 3.5

The problem is

$$\begin{aligned} &\text{maximize} \quad z = f(x_1, x_2) = x_1 + 2x_2 \\ &\text{subject to} \quad g(x_1, x_2) = 4x_1 + 8x_2 = 8 \end{aligned} \tag{3.54}$$

The appropriate Lagrangian function is

$$F^0(\mathbf{x}, \boldsymbol{\lambda}) = \lambda_0(x_1 + 2x_2) + \lambda_1(8 - 4x_1 - 8x_2)$$

The equations using (3.50) with $\lambda_0 = 1$ are

$$\frac{\partial F^0(\mathbf{x}, \boldsymbol{\lambda})}{\partial x_1} = 1 - 4\lambda_1 = 0 \qquad \frac{\partial F^0(\mathbf{x}, \boldsymbol{\lambda})}{\partial x_2} = 2 - 8\lambda_1 = 0$$

$$\frac{\partial F^0(\mathbf{x}, \boldsymbol{\lambda})}{\partial \lambda_1} = 8 - 4x_1 - 8x_2$$

The solution to the above set of equations is $\lambda_1 = 1/4$ and any (x_1, x_2) satisfying (3.54). In this problem, the maximum of $f(x_1, x_2)$ is 2 and is not unique, for the slope of $f(x_1, x_2)$ is identical to that of g.

Subcase 2.2: $r(\mathbf{G}^0) = m$ and $r(\mathbf{G}) < m$
In this subcase, a solution to (3.50) will exist only if $\lambda_0 = 0$.

Subcase 2.3: $r(\mathbf{G}^0) = r(\mathbf{G}) = p < m$
A solution to (3.50) will exist with $\lambda_0 = 1$.

Subcase 2.4: $r(\mathbf{G}^0) = p > r(\mathbf{G})$ with $p < m$
This subcase is similar to Case 1. It is not possible to write ∇f as a linear combination of ∇g_i, $i = 1, 2, \ldots, m$; therefore a solution to (3.50) will exist only if $\lambda_0 = 0$.

To further illustrate situations that may arise in Cases 1 and 2, three additional examples will be presented. In each example, it is assumed that there are only two constraints ($m = 2$) and two variables ($n = 2$).

Example 3.6

Figure 3.6 depicts an example for which $r(\mathbf{G}^0) = r(\mathbf{G}) = m = 2$. The gradient vector ∇f may be written as a linear combination of ∇g_1 and ∇g_2; i.e., there exist two scalars c_1

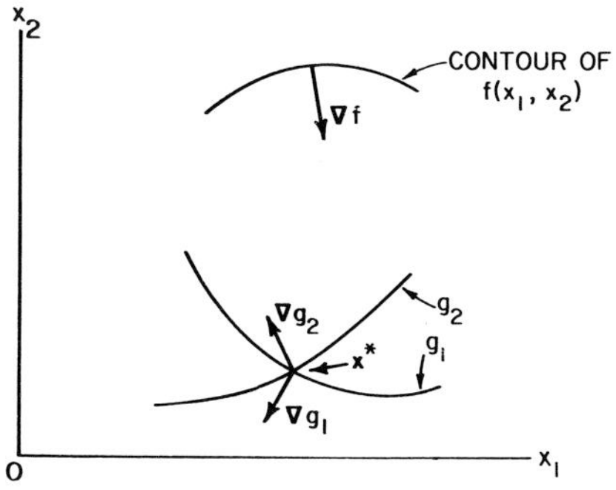

Fig. 3.6. Illustration of Subcase 2.1.

and c_2 such that $\nabla f = c_1 \nabla g_1 + c_2 \nabla g_2$. Thus at $\mathbf{x}^* = [x_1^*, x_2^*]$ a solution to (3.50) will exist when $\lambda_0 = 1$.

Example 3.7

Figure 3.7 illustrates an example in which g_1 is tangent to g_2 at $\mathbf{x}^*$ and $f(x_1, x_2)$ is tangent to both g_1 and g_2 at this point. In this instance at $\mathbf{x}^*$, $r(\mathbf{G}^0) = r(\mathbf{G}) = 1$ and $m = 2$. Hence a solution to (3.50) exists with $\lambda_0 = 1$, but ∇f cannot be written as a unique linear combination of ∇g_1 and ∇g_2; i.e., λ_1 and λ_2 will not be unique.

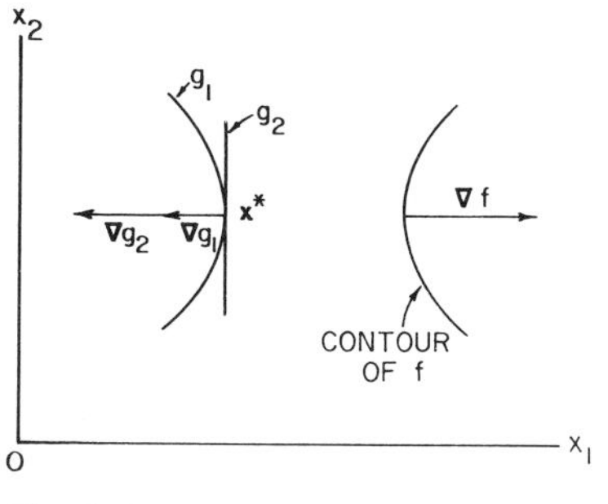

Fig. 3.7. Illustration of Subcase 2.3.

Example 3.8

In this example, $r(\mathbf{G}^0) = 2 > r(\mathbf{G}) = 1$. It is obvious from Figure 3.8 that ∇f cannot be written as a linear combination of ∇g_1 and ∇g_2; therefore, a solution to (3.50) will exist at $\mathbf{x}^*$ only if $\lambda_0 = 0$.

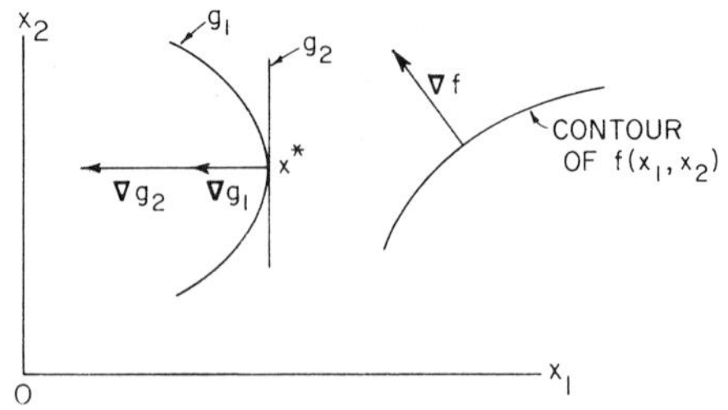

Fig. 3.8. Illustration of Subcase 2.4.

In summary, if $f(\mathbf{x})$ is a local minimum or maximum at $\mathbf{x}^*$, a general set of necessary conditions that must be satisfied at $\mathbf{x}^*$ is

$$\lambda_0 \nabla f(\mathbf{x}) = \sum_{i=1}^{m} \lambda_i \nabla g_i(\mathbf{x})$$

and

$$g_i(\mathbf{x}) = b_i \qquad i = 1, 2, \ldots, m$$

where $\lambda_0 = 0$ or 1.

If $f, g_1, \ldots, g_m$ are well-behaving functions at $\mathbf{x}^*$, then λ_0 may be set to one and the usual Lagrangian necessary conditions are satisfied. Most often this will be the case. However, if no solutions to (3.50) and (3.51) can be found with $\lambda_0 = 1$, then set $\lambda_0 = 0$ and solve.

If there are nonnegative restrictions on the n variables $x_1, x_2, \ldots, x_n$, they should be directly augmented to (3.12) by introducing n new variables w_i^2, $i = 1, 2, \ldots, n$, such that

$$w_i^2 - x_i = 0 \qquad i = 1, 2, \ldots, n \tag{3.55}$$

In this case the Lagrangian function will now depend on $2n + 2m$ variables and is given by

$$F(\mathbf{x}, \mathbf{w}, \lambda) = f(\mathbf{x}) + \sum_{i=1}^{m} \lambda_i [b_i - g_i(\mathbf{x})] + \sum_{i=m+1}^{m+n} \lambda_i (x_{i-m} - w_{i-m}^2)$$

This method of dealing with nonnegative restrictions on the $\mathbf{x}$ variables has been introduced in Chapter 2 and will be used in Chapter 5. Notice that the inclusion of (3.55) increases the size of the problem. If it can be determined that the solution will not occur on a boundary where at least one $x_i = 0$, the exclusion of (3.55) in the Lagrangian function will ease the computations considerably.

In the next section the Lagrangian procedure will be applied to the more general problem given in (3.1) and (3.2) where inequality constraints exist in order to generate a set of necessary conditions for an optimizing point $\mathbf{x}^*$. It will be assumed in the next section that no unusual behavior of $f, g_1, \ldots, g_m$ occurs at $\mathbf{x}^*$, as illustrated above. A discussion of difficulties of this nature, which may arise in the inequality-constrained problem, will be delayed until Section 3.5.

3.4 KUHN-TUCKER CONDITIONS

In 1951 Kuhn and Tucker [1951] developed a set of necessary and sufficient conditions for optimal solutions to the general mathematical programming problem. The Kuhn-Tucker conditions provided a framework from which numerous computational methods have been developed for solving certain types of the nonlinear programming problem. A thorough treatment of the conditions is therefore essential before discussing these computational methods. Furthermore, the Kuhn-Tucker conditions provide considerable insight into the nature of the nonlinear problem.

In developing the necessary conditions, the nonlinear programming problem in (3.1) and (3.2) is recast in the form

$$\text{maximize} \quad z = f(\mathbf{x}) \tag{3.56}$$

$$\begin{aligned} \text{subject to} \quad & g_i(\mathbf{x}) \le b_i && i = 1, 2, \ldots, r \\ & g_i(\mathbf{x}) \ge b_i && i = r + 1, \ldots, s \\ & g_i(\mathbf{x}) = b_i && i = s + 1, \ldots, m \\ & \mathbf{x} \ge \mathbf{0} \end{aligned} \tag{3.57}$$

Normally, equality constraints will not often occur in a mathematical programming problem. The inclusion of equality constraints does perhaps surprisingly introduce further difficulties in the solving of this problem; these will be discussed and dealt with later. By adding r slack variables x_{ti}, $i = 1, 2, \ldots, r$, and s surplus variables x_{ti}, $i = r + 1, \ldots, s$, the above problem may be rewritten as

$$\begin{aligned} \text{maximize} \quad & z = f(\mathbf{x}) \\ \text{subject to} \quad & g_i(\mathbf{x}) + x_{ti} = b_i && i = 1, 2, \ldots, r \\ & g_i(\mathbf{x}) - x_{ti} = b_i && i = r + 1, \ldots, s \\ & g_i(\mathbf{x}) \phantom{+ x_{ti}} = b_i && i = s + 1, \ldots, m \\ & [\mathbf{x}, \mathbf{x}_t] \ge \mathbf{0} \end{aligned} \tag{3.58}$$

where $[\mathbf{x}, \mathbf{x}_t]' = [x_1, x_2, \ldots, x_n; x_{t1}, x_{t2}, \ldots, x_{ts}]$

In this form the problem is now amenable to treatment by the Lagrangian function technique presented in the last section. For this problem the Lagrangian function is

$$F(\mathbf{x},\boldsymbol{\lambda}) = f(\mathbf{x}) + \sum_{i=1}^{r} \lambda_i[b_i - g_i(\mathbf{x}) - x_{ti}] + \sum_{i=r+1}^{s} \lambda_i[b_i - g_i(\mathbf{x}) + x_{ti}]$$
$$+ \sum_{i=s+1}^{m} \lambda_i[b_i - g_i(\mathbf{x})] \qquad (3.59)$$

where $\lambda_i, i = 1,2,\ldots,m$, are the Lagrangian multipliers associated with the constraints in (3.58).

By differentiating $F(\mathbf{x},\boldsymbol{\lambda})$ with respect to $\boldsymbol{\lambda}$ and $\mathbf{x}$ (including x_{ti}, $i = 1,2,\ldots,s$) and setting these differentials equal to zero, a set of necessary conditions for optimizing the unconstrained function $F(\mathbf{x},\boldsymbol{\lambda})$ results; they are

$$\frac{\partial F(\mathbf{x},\boldsymbol{\lambda})}{\partial x_j} = \frac{\partial f(\mathbf{x})}{\partial x_j} - \sum_{i=1}^{m} \lambda_i \frac{\partial g_i(\mathbf{x})}{\partial x_j} = 0 \qquad j = 1,2,\ldots,n \qquad (3.60)$$

$$\frac{\partial F(\mathbf{x},\boldsymbol{\lambda})}{\partial \lambda_i} = b_i - g_i(\mathbf{x}) - x_{ti} = 0 \qquad i = 1,2,\ldots,r \qquad (3.61)$$

$$\frac{\partial F(\mathbf{x},\boldsymbol{\lambda})}{\partial \lambda_i} = b_i - g_i(\mathbf{x}) + x_{ti} = 0 \qquad i = r+1,\ldots,s \qquad (3.62)$$

$$\frac{\partial F(\mathbf{x},\boldsymbol{\lambda})}{\partial \lambda_i} = b_i - g_i(\mathbf{x}) = 0 \qquad i = s+1,\ldots,m \qquad (3.63)$$

$$\frac{\partial F(\mathbf{x},\boldsymbol{\lambda})}{\partial x_{ti}} = -\lambda_i = 0 \qquad i = 1,2,\ldots r \qquad (3.64)$$

$$\frac{\partial F(\mathbf{x},\boldsymbol{\lambda})}{\partial x_{ti}} = +\lambda_i = 0 \qquad i = r+1,\ldots,s \qquad (3.65)$$

Thus if $x_{ti} > 0$ for $i = 1,2,\ldots,s$, then $\lambda_i = 0, i = 1,2,\ldots,s$, by virtue of (3.64) and (3.65). The implication of this result is important; if $x_{ti} > 0$ for $i = 1,2,\ldots,s$, the s inequality constraints may be ignored since they all will be inactive at the optimizing point $[\mathbf{x}^*, \mathbf{x}_t^*, \boldsymbol{\lambda}^*]' = [x_1^*, x_2^*, \ldots, x_n^*;\ x_{t1}^*, x_{t2}^*, \ldots, x_{ts}^*;\ \lambda_1^*, \lambda_2^*, \ldots, \lambda_m^*]$, satisfying (3.60)–(3.63). It also follows that at the optimizing point, $x_{ti}^*\lambda_i^* = 0$, since if $x_{ti}^* > 0$, then $\lambda_i^* = 0$.

A second observation regarding the solution of (3.56) and (3.57) concerns the interpretation of the Lagrangian multipliers. The values of $\mathbf{x}^*$ and $\boldsymbol{\lambda}^*$ depend on the values of b_i, $i = 1, 2, \ldots, m$. Assume, for the moment, that all m constraints are equalities and consider the rate of change of $z^* = f(\mathbf{x}^*) = f(x_1^*, x_2^*, \ldots, x_n^*)$ with respect to b_i, $i = 1, 2, \ldots, m$. From the chain rule it follows that

$$\frac{\partial f^*}{\partial b_i} = \sum_{j=1}^{n} \frac{\partial f}{\partial x_j^*} \frac{\partial x_j^*}{\partial b_i} \tag{3.66}$$

and

$$\frac{\partial g_k}{\partial b_i} = \sum_{j=1}^{n} \frac{\partial g_k}{\partial x_j^*} \frac{\partial x_j^*}{\partial b_i} \qquad k = 1, 2, \ldots, m$$

Since for the kth equality constraint, $g_k(\mathbf{x}^*) = b_k$,

$$\frac{\partial g_k}{\partial b_i} = \sum_{j=1}^{n} \frac{\partial g_k}{\partial x_j^*} \frac{\partial x_j^*}{\partial b_i} = \delta_{ik} \begin{cases} = 1 \text{ if } i = k \\ = 0 \text{ if } i \neq k \end{cases}$$

Therefore,

$$\delta_{ik} - \sum_{j=1}^{n} \frac{\partial g_k}{\partial x_j^*} \frac{\partial x_j^*}{\partial b_i} = 0 \tag{3.67}$$

and by multiplying both sides of (3.67) by λ_k^*,

$$\lambda_k^* \delta_{ik} - \sum_{j=1}^{n} \lambda_k^* \frac{\partial g_k}{\partial x_j^*} \frac{\partial x_j^*}{\partial b_i} = 0 \tag{3.68}$$

If (3.68) is now summed over k and adjoined to (3.66),

$$\frac{\partial f^*}{\partial b_i} = \sum_{k=1}^{m} \lambda_k^* \delta_{ik} + \sum_{j=1}^{n} \left(\frac{\partial f}{\partial x_j^*} - \sum_{k=1}^{m} \lambda_k^* \frac{\partial g_k}{\partial x_j^*} \right) \frac{\partial x_j^*}{\partial b_i}$$

But since $\mathbf{x}^*$ and $\boldsymbol{\lambda}^*$ must satisfy (3.60), it follows that

$$\frac{\partial f^*}{\partial b_i} = \lambda_i^* \tag{3.69}$$

Thus if $f(\mathbf{x})$ is a cost function, for example, and b_i is the number of physical units of some resource, λ_i^* gives an approximation of how much the minimum cost will be changed if the quantity of resource i is increased by one unit.

Suppose now that among the m constraints are s inequality constraints. If the ith inequality constraint is inactive at $\mathbf{x}^*$, $\partial f^*/\partial b_i = 0$, since $x_{ti} > 0$ and $\lambda_i^* = 0$. If the ith inequality constraint is active, $\partial f^*/\partial b_i$ may be computed as above, treating it as an equality constraint resulting in (3.69).

Now return to investigating the optimization of the Lagrangian function in (3.59). Assume that at $\mathbf{x}^*$, $f(\mathbf{x})$ takes on its global maximum. Let J be the subset of indices $j = 1, 2, \ldots, n$ containing the indices j for which $x_j^* > 0$, and let $\bar{J}$ be the subset containing indices for which $x_j^* = 0$. Let I be the subset of indices $i = 1, 2, \ldots, m$ containing the indices i for which the ith constraint is active at $\mathbf{x}^*$, and let $\bar{I}$ be the subset of indices for which the indices i associate with inactive constraints at $\mathbf{x}^*$. From the above, $\mathbf{x}^*$ and $\boldsymbol{\lambda}^*$ will satisfy

$$\frac{\partial f(\mathbf{x}^*)}{\partial x_j} - \sum_{i=1}^{m} \lambda_i^* \frac{\partial g_i(\mathbf{x}^*)}{\partial x_j} = 0 \text{ for } j \in J \tag{3.70}$$

Also, it has been shown that $\lambda_i^* = 0$ for $i \in \bar{I}$.

It is now natural to inquire what happens to (3.70) when $j \in \bar{J}$. To answer this question, the original problem given in (3.56) and (3.57) is recast by adding the lower bound constraints on $\mathbf{x}$ directly to the m existing constraints. Thus n additional constraints are augmented to the m constraints in (3.58) of the form $x_j - x_{t,s+j} = 0$, $j = 1, 2, \ldots, n$, where $x_{t,s+j}$, $j = 1, 2, \ldots, n$, are n additional nonnegative surplus variables. The original problem now becomes

$$\begin{array}{lrll}
\text{maximize} & z = f(\mathbf{x}) & & \\
\text{subject to} & g_i(\mathbf{x}) + x_{ti} = b_i & i = 1, 2, \ldots, r & \\
& g_i(\mathbf{x}) - x_{ti} = b_i & i = r + 1, \ldots, s & \\
& g_i(\mathbf{x}) = b_i & i = s + 1, \ldots, m & (3.71) \\
& x_{i-m} - x_{t,s+i-m} = 0 & i = m + 1, \ldots, m + n & \\
& x_{ti} \geq 0 & i = 1, 2, \ldots, n + s &
\end{array}$$

Thus the problem now has $2n + s$ variables and $m + n$ equality constraints. The new Lagrangian function is

$$\begin{aligned}
F(\mathbf{x}, \boldsymbol{\lambda}) = f(\mathbf{x}) &+ \sum_{i=1}^{r} \lambda_i[b_i - g_i(\mathbf{x}) - x_{ti}] \\
&+ \sum_{i=r+1}^{s} \lambda_i[b_i - g_i(\mathbf{x}) + x_{ti}] + \sum_{i=s+1}^{m} \lambda_i[b_i - g_i(\mathbf{x})] \\
&+ \sum_{i=m+1}^{m+n} \lambda_i(-x_{i-m} + x_{t,s+i-m})
\end{aligned}$$

The differential equation $\partial F(\mathbf{x}, \boldsymbol{\lambda})/\partial \mathbf{x}_j = 0$ evaluated at $[\mathbf{x}^* \boldsymbol{\lambda}^*]'$ becomes

$$\frac{\partial f(\mathbf{x}^*)}{\partial x_j} - \sum_{i=1}^{m} \lambda_i^* \frac{\partial g_i(\mathbf{x}^*)}{\partial x_j} - \lambda_{m+j}^* = 0 \qquad j = 1, 2, \ldots, n \tag{3.72}$$

where

$$\lambda_i^* = 0 \qquad i \in \bar{I} \text{ and } \lambda_{m+j}^* = 0 \qquad j \in J \tag{3.73}$$

Notice that since $\lambda_{m+j}^* = 0$ for $j \in J$, (3.72) is equivalent to (3.70). Now what happens to (3.73) when $j \in \bar{J}$? From (3.69) it follows that

$$\frac{\partial f^*}{\partial b_{m+j}} = \lambda_{m+j}^* \qquad j = 1, 2, \ldots, n$$

The lower bound constraints on $x_j, j = 1, 2, \ldots, n$, require that

$$x_j \leq b_{m+j} = 0 \qquad j = 1, 2, \ldots, n \tag{3.74}$$

Suppose that b_{m+j} is decreased, then z^* can only increase or remain unchanged, since any feasible solution to $x_j \leq b_{m+j} < 0, j = 1, 2, \ldots, n$, also satisfies (3.74). Therefore, $\partial f^*/\partial b_{m+j} \leq 0$, which implies that

$$\lambda_{m+j}^* \leq 0 \tag{3.75}$$

Now

$$\frac{\partial f(\mathbf{x}^*)}{\partial x_j} - \sum_{i=1}^{m} \lambda_i^* \frac{\partial g_i(\mathbf{x}^*)}{\partial x_j} \leq 0 \text{ for } j \in \bar{J}$$

due to (3.75), and the question is answered. Furthermore, if $b_i, i = 1, 2, \ldots, r$, is increased, z^* can only increase or remain unchanged. Hence $\partial f^*/\partial b_i \geq 0$, which implies that $\lambda_i^* \geq 0$ for $i = 1, 2, \ldots, r$. Also, if $b_i, i = r + 1, \ldots, s$, is decreased, z^* can only decrease or remain unchanged. Hence $\partial f^*/\partial b_i \leq 0$, which implies that $\lambda_i^* \leq 0$ for $i = r + 1, \ldots, s$. For $i = s + 1, \ldots, m$, λ_i^* may be either positive or negative.

The above necessary conditions may now be summarized and stated in the following form. Given the Lagrangian function

$$F(\mathbf{x}, \boldsymbol{\lambda}) = f(\mathbf{x}) + \sum_{i=1}^{m} \lambda_i [b_i - g_i(\mathbf{x})]$$

where the m constraints take the form

$$\begin{aligned} g_i(\mathbf{x}) &\leq b_i \qquad i = 1, 2, \ldots, r \\ g_i(\mathbf{x}) &\geq b_i \qquad i = r + 1, \ldots, s \\ g_i(\mathbf{x}) &= b_i \qquad i = s + 1, \ldots, m \end{aligned} \tag{3.76}$$

if $f(\mathbf{x}^*)$ is the global maximum of $f(\mathbf{x})$ for $\mathbf{x}$ satisfying (3.76), then a $\boldsymbol{\lambda}^*$ must exist such that

(I) $$\frac{\partial f(\mathbf{x}^*)}{\partial x_j} - \sum_{i=1}^{m} \lambda_i^* \frac{\partial g_i(\mathbf{x}^*)}{\partial x_j} \leq 0 \qquad j = 1, 2, \ldots, n \tag{3.77}$$

with strict equality holding for $j \in J$.

(II) $$\sum_{j=1}^{n} x_j^* \left[\frac{\partial f(\mathbf{x}^*)}{\partial x_j} - \sum_{i=1}^{m} \lambda_i^* \frac{\partial g_i(\mathbf{x}^*)}{\partial x_j} \right] = 0 \tag{3.78}$$

(III) $$\begin{aligned} b_i - g_i(\mathbf{x}^*) &\geq 0 \qquad i = 1, 2, \ldots, r \\ b_i - g_i(\mathbf{x}^*) &\leq 0 \qquad i = r + 1, \ldots, s \\ b_i - g_i(\mathbf{x}^*) &= 0 \qquad i = s + 1, \ldots, m \end{aligned} \tag{3.79}$$

(IV) $$\sum_{i=1}^{m} \lambda_i^* [b_i - g_i(\mathbf{x}^*)] = 0 \tag{3.80}$$

(V) for $i = 1, 2, \ldots, r$, $\lambda_i^* \geq 0$
for $i = r + 1, \ldots, s$, $\lambda_i^* \leq 0$ (3.81)
for $i = s + 1, \ldots, m$, λ_i^* may be either positive or negative

The inequalities and equations (3.77)–(3.81) are the Kuhn-Tucker necessary conditions for a global maximum of $f(\mathbf{x})$. Notice that (3.78) follows from (3.77) when $j \in J$ and the fact that $x_j^* = 0$ for $j \in \bar{J}$. The conditions in (3.79) reflect the requirement that $\mathbf{x}^*$ be feasible. Also, (3.80) follows from (3.79) for $i = s + 1, \ldots, m$ and the fact that if $i \in \bar{I}$, $\lambda_i^* = 0$ and if $i \in I$, $b_i - g_i(\mathbf{x}^*) = 0$.

It is appropriate now to reflect on what (3.77)–(3.81) provide. The global optimizing point $\mathbf{x}^*$ for the problem delineated in (3.56) and (3.57) must satisfy these four necessary conditions. Unfortunately, any optimizing point (local minima or maxima) to (3.56) and (3.57) will also satisfy these conditions. Therefore, there is no guarantee that just because $\mathbf{x}^*$ satisfies (3.77)–(3.81), it will be the global maximum. This brings up the obvious question of sufficiency for a global optimum, and it shall be answered now.

The necessary conditions for $(\mathbf{x}^*, \boldsymbol{\lambda}^*)$ to be an optimizing point also are necessary for the Lagrangian function to have a saddle point at $(\mathbf{x}^*, \boldsymbol{\lambda}^*)$. By a saddle point, it is meant that $F(\mathbf{x}, \boldsymbol{\lambda})$ is a maximum with respect to $\mathbf{x}$ and a minimum with respect to $\boldsymbol{\lambda}$;

$$F(\mathbf{x}, \boldsymbol{\lambda}^*) \leq F(\mathbf{x}^*, \boldsymbol{\lambda}^*) \leq F(\mathbf{x}^*, \boldsymbol{\lambda})$$

and $[\mathbf{x}^*, \boldsymbol{\lambda}^*]$ is a global saddle point if

$$\sup_{\mathbf{x}} F(\mathbf{x}, \boldsymbol{\lambda}^*) = F(\mathbf{x}^*, \boldsymbol{\lambda}^*) = \inf_{\boldsymbol{\lambda}} F(\mathbf{x}^*, \boldsymbol{\lambda})$$

From the theory of saddle points, the set of necessary conditions (3.77)–(3.81) for a global saddle point will also be sufficient if $f(\mathbf{x})$ is a concave function and the constraints $g_i(\mathbf{x}) = b_i$, $i = 1, 2, \ldots, m$, form a convex set. The notion of a saddle point has been introduced in Chapter 2 in connection with game theory. The reader is directed to Kuhn [1957] for a thorough treatment of saddle point theory and to Appendix B for a review of convex functions and sets. When $f(\mathbf{x})$ and the constraint set have these properties, $F(\mathbf{x}, \boldsymbol{\lambda})$ will be a concave function of $\mathbf{x}$ for all nonnegative values of $\mathbf{x}$, and since $F(\mathbf{x}^*, \boldsymbol{\lambda})$ is linear in $\boldsymbol{\lambda}$, it will be a convex function of $\boldsymbol{\lambda}$. Then $F(\mathbf{x}, \boldsymbol{\lambda})$ will have a global saddle point at $(\mathbf{x}^*, \boldsymbol{\lambda}^*)$. It remains to be shown that $f(\mathbf{x})$ will take on its global maximum at $\mathbf{x}^*$ under these conditions.

From condition (3.80) it follows that $F(\mathbf{x}^*, \boldsymbol{\lambda}^*) = f(\mathbf{x}^*)$. Consider any nonnegative $\mathbf{x}$ satisfying the constraints. It must also satisfy

$$F(\mathbf{x}, \boldsymbol{\lambda}^*) = f(\mathbf{x}) + \sum_{i=1}^{m} \lambda_i^* [b_i - g_i(\mathbf{x})]$$

and if it can be shown that

$$\sum_{i=1}^{m} \lambda_i^* [b_i - g_i(\mathbf{x})] \geq 0 \tag{3.82}$$

the global optimality of $f(\mathbf{x})$ at $\mathbf{x}^*$ will be insured. Since $\mathbf{x}$ must satisfy the constraints, $g_i(\mathbf{x}) = b_i$ for $i = s + 1, \ldots, m$, and

$$\lambda_i^* [b_i - g_i(\mathbf{x})] = 0 \tag{3.83}$$

for $i = s + 1, \ldots, m$. For the first r constraints, $g_i(\mathbf{x}) \leq b_i$; but if the constraint is a strict inequality, it has been shown that $\lambda_i^* \geq 0$ so that $\lambda_i^*[b_i - g_i(\mathbf{x})] \geq 0$ for $i = 1, 2, \ldots, r$. If one or more of the first r constraints is active, (3.83) will be satisfied for these active constraints. For the next set of $s - r$ constraints, $g_i(\mathbf{x}) \geq b_i$; but it has been shown that if strict inequality exists, $\lambda_i^* \leq 0$. Therefore, $\lambda_i^*[b_i - g_i(\mathbf{x})] \geq 0$ for $i = r + 1, \ldots, s$. If any of the $s - r$ constraints are active, (3.83) will be satisfied for them. Therefore, (3.82) must hold for any $\mathbf{x}$ satisfying the m constraints so that $f(\mathbf{x}) \leq F(\mathbf{x}, \boldsymbol{\lambda}^*)$. From this fact it follows that $f(\mathbf{x}) \leq f(\mathbf{x}^*)$, and thus $f(\mathbf{x})$ takes on its global maximum at $\mathbf{x}^*$.

In summary, if $f(\mathbf{x})$ is a concave function and $g_i(\mathbf{x}) \{\leq \; = \; \geq\} = b_i$, $i = 1, 2, \ldots, m$, form a convex set, the Kuhn-Tucker conditions (3.77)–(3.81) form a set of necessary and sufficient conditions for a global optimum to the mathematical programming problem delineated in (3.56) and (3.57). Under these circumstances, (3.77)–(3.81) also provide a set of necessary and sufficient conditions for a global saddle point for the Lagrangian function $F(\mathbf{x}, \boldsymbol{\lambda})$.

The connection between global saddle point theory and the resolution of the mathematical programming problem has a most interesting history. Lagrangian functions have been used in the physical sciences for many years, but it was only in the 1950s that the utility of this form of analysis in resolving mathematical programming problems was recognized. Indeed, the Lagrangian procedure outlined in the development of the Kuhn-Tucker conditions may be directly applied in a computational algorithm that will guarantee the global maximum of the nonlinear programming problem. There is a catch, of course, and it will become readily apparent in the following outline of the Lagrangian solution to (3.56) and (3.57).

The Lagrangian solution method involves the following steps:

1. Find the unconstrained maximum of $f(\mathbf{x})$. Frequently, by inspecting the function, it is apparent that the unconstrained maximum will not be feasible, so that this step may be deleted. If this solution is feasible, it will be the global maximum, and there is no need to proceed to the next step.
2. Solve the Lagrangian function based only on the $m - s$ equality constraints $g_i(\mathbf{x}) = b_i$, $i = s + 1, \ldots, m$. If this solution satisfies the remaining constraints, it will be the global maximum of $f(\mathbf{x})$ and the process may be stopped.
3. Add one of the inequality constraints to the Lagrangian function in (2) treating it as if it were active. Solve this Lagrangian system. If the solution satisfies the remaining $s - 1$ constraints, stop. Otherwise, drop the current inequality constraint, add another, and repeat the process. If all s inequality constraints fail to yield a feasible solution when treated individually as equality constraints, proceed to (4).

4. Repeat the process by now adjoining pairs of inequality constraints to the Lagrangian function in (2), treating them as active constraints. Continue until a feasible solution to the $s - 2$ remaining constraints is encountered or all $C_2^s = s!/2!(s - 2)!$ pairs have been exhausted. If the latter occurs, proceed to (5).
5. Continue the process taking all C_a^s combinations for $a = 3, 4, \ldots, s$ until a feasible solution is encountered.

Provided that the sufficiency conditions of the Kuhn-Tucker theorem are satisfied and the problem is well-behaved (the behavior of a nonlinear programming problem will be discussed in Section 3.5), this algorithm will guarantee locating the global maximum. If the sufficiency conditions of the Kuhn-Tucker theorem are not satisfied, it may be necessary to evaluate all the feasible solutions encountered by the algorithm to guarantee the attainment of the global maximum. In this case it may not be possible to stop after the first feasible solution is encountered. In this event, the computational effort in employing this algorithm may be enormous. The feasibility of implementing this algorithm depends on the number of constraints m and the degree of nonlinearity in $F(\mathbf{x}, \boldsymbol{\lambda})$. Additionally, it may not be possible to find the optimal solution from maximizing $F(\mathbf{x}, \boldsymbol{\lambda})$ at any stage, since the differential equations

$$\frac{\partial F(\mathbf{x}, \boldsymbol{\lambda})}{\partial \mathbf{x}} = 0 \qquad \frac{\partial F(\mathbf{x}, \boldsymbol{\lambda})}{\partial \boldsymbol{\lambda}} = 0 \tag{3.84}$$

might form a system of highly nonlinear equations that may resist solution. Naturally, numerical methods can be applied to solving (3.84), but they may not be able to guarantee the best solution. Furthermore, the system of equations in (3.84) must generally be solved many times, increasing the computational labor required. One additional difficulty exists. If it is required that $\mathbf{x}$ be nonnegative, these conditions must be adjoined to the existing m constraints forming $m + n$ constraints in all, as done in (3.71). If the original mathematical programming problem portrayed in (3.56) and (3.57) is especially wicked (viciously nonconvex) to solve, by applying this algorithm to it, a more wicked problem will generally result.

This algorithm does provide, however, a systematic method for generating solutions to the Kuhn-Tucker conditions and thus gives a clue to their practicality. By themselves the Kuhn-Tucker conditions do not provide much help in locating the global maximum, for if $F(\mathbf{x}, \boldsymbol{\lambda})$ is a highly nonlinear function of $\mathbf{x}$, attempting to find a solution by utilizing the Lagrangian procedure upon which the Kuhn-Tucker conditions are based may be extremely difficult, as suggested above. The

Kuhn-Tucker conditions do provide a framework that can suggest solution procedures, and they give a convenient set of necessary conditions for checking purposes. For certain classes of problems, the Kuhn-Tucker results have led to efficient solution procedures, the most notable of which is Wolfe's algorithm for solving the quadratic programming problem for which $f(\mathbf{x})$ is a concave quadratic function and $g_i(\mathbf{x})$, $i = 1, 2, \ldots, m$, are all linear functions. Note in this case that the equations in (3.84) will be linear in both $\mathbf{x}$ and $\boldsymbol{\lambda}$. Wolfe's algorithm will be thoroughly discussed in Chapter 5.

In the Section 3.5, anomalies in the functions $f, g_1, g_2, \ldots, g_m$ are investigated to consider cases where the optimal solution $\mathbf{x}^*$ will not satisfy the Kuhn-Tucker conditions. Furtunately, these cases rarely if ever occur in practice, but it is well to be advised of their existence. Before proceeding, two example problems are presented to illustrate the Lagrangian algorithm and the application of the Kuhn-Tucker conditions.

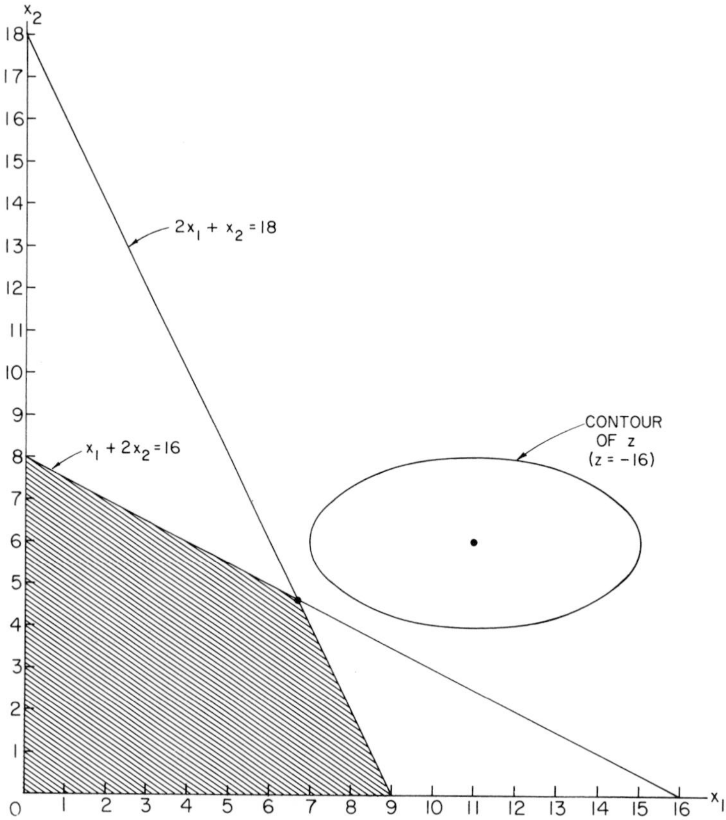

Fig. 3.9. Graphic solution to Example 3.9.

Example 3.9

The problem is

$$\begin{aligned} \text{maximize} \quad & z = f(x_1, x_2) = -(x_1 - 11)^2 - 4(x_2 - 6)^2 \\ \text{subject to} \quad & 2x_1 + x_2 \leq 18 \\ & x_1 + 2x_2 \leq 16 \\ & x_1, x_2 \geq 0 \end{aligned}$$

From the graphic representation of the problem given in Figure 3.9, it is apparent that the solution occurs at the intersection of the two lines $2x_1 + x_2 = 18$ and $x_1 + 2x_2 = 16$. The Lagrangian algorithm will now be applied to verify this conjecture and to illustrate the technique.

STEP 1:

The unconstrained global maximum from inspection of $f(x_1, x_2)$ occurs at (11, 6), which clearly is not feasible.

STEP 2:

Form the Lagrangian function using the constraint $x_1 + 2x_2 = 16$.

$$F_1(\mathbf{x}, \boldsymbol{\lambda}) = -(x_1 - 11)^2 - 4(x_2 - 6)^2 + \lambda_1(16 - x_1 - 2x_2)$$

$$\begin{aligned} \frac{\partial F_1(\mathbf{x}, \boldsymbol{\lambda})}{\partial x_1} &= -2(x_1 - 11) - \lambda_1 = 0 \\ \frac{\partial F_1(\mathbf{x}, \boldsymbol{\lambda})}{\partial x_2} &= -8(x_2 - 6) - 2\lambda_1 = 0 \\ \frac{\partial F_1(\mathbf{x}, \boldsymbol{\lambda})}{\partial \lambda_1} &= 16 - x_1 - 2x_2 = 0 \end{aligned} \tag{3.85}$$

The solution to (3.85) is $x_1 = 7.5$, $x_2 = 4.25$, $\lambda_1 = 7$, and $\lambda_2 = 0$. Since the point (7.5, 4.25) is not feasible, the process continues.

STEP 3:

Form the Lagrangian function using the constraint $2x_1 + x_2 = 18$.

$$F_2(\mathbf{x}, \boldsymbol{\lambda}) = -(x - 11)^2 - 4(x_2 - 6)^2 + \lambda_2(18 - 2x_1 - x_2)$$

$$\begin{aligned} \frac{\partial F_2(\mathbf{x}, \boldsymbol{\lambda})}{\partial x_1} &= -2(x_1 - 11) - 2\lambda_2 = 0 \\ \frac{\partial F_2(\mathbf{x}, \boldsymbol{\lambda})}{\partial x_2} &= -8(x_2 - 6) - \lambda_2 = 0 \end{aligned} \tag{3.86}$$

$$\frac{\partial F_2(\mathbf{x}, \lambda)}{\partial \lambda_2} = 18 - 2x_1 - x_2 = 0$$

The solution to (3.86) is $x_1 = 6.3$, $x_2 = 5.4$, $\lambda_1 = 0$, and $\lambda_2 = 4.8$. Since (6.3, 5.4) also is not feasible, the process continues.

STEP 4:

Form the Lagrangian function using both constraints $x_1 + 2x_2 = 16$ and $2x_1 + x_2 = 18$.

$$\begin{aligned}
F_3(\mathbf{x}, \lambda) &= -(x_1 - 11)^2 - 4(x_2 - 6)^2 + \lambda_1(16 - x_1 - 2x_2) \\
&\quad + \lambda_2(18 - 2x_1 - x_2) \\
\frac{\partial F_3(\mathbf{x}, \lambda)}{\partial x_1} &= -2(x_1 - 11) - \lambda_1 - 2\lambda_2 = 0 \\
\frac{\partial F_3(\mathbf{x}, \lambda)}{\partial x_2} &= -8(x_2 - 6) - 2\lambda_1 - \lambda_2 = 0 \qquad (3.87) \\
\frac{\partial F_3(\mathbf{x}, \lambda)}{\partial \lambda_1} &= 16 - x_1 - 2x_2 = 0 \\
\frac{\partial F_3(\mathbf{x}, \lambda)}{\partial \lambda_2} &= 18 - 2x_1 - x_2 = 0
\end{aligned}$$

The solution to (3.87) is $x_1 = 6.67$, $x_2 = 4.67$, $\lambda_1 = 4.2$, and $\lambda_2 = 2.23$. The point (6.67, 4.67) is the intersection of the two lines and hence is feasible. Therefore, the global maximum occurs at the point $\mathbf{x}^* = (6.67, 4.67)$, and $f(\mathbf{x}^*)$ is -25.82.

Notice that the nonnegative constraints have been ignored, although technically they should have been incorporated in the solution process. However, Figure 3.9 illustrates that they will not participate in the global maximization of $f(\mathbf{x})$. Eliminating them obviously results in a significant savings of computation time and suggests the utility of learning as much as possible about the problem before proceeding with a computational solution method. The cursory analytical and numerical investigation of a mathematical programming problem, before embarking with a solution procedure, can be of considerable value in saving computation time and will be fully discussed in Chapter 4. It is apparent that for bivariate problems a simple graph can be very illuminating. For example, in the above problem if it is desired to minimize $f(x_1, x_2)$ rather than maximize this function, the nonnegative constraints obviously would now play an important role. Indeed, by inspection the minimizing point (x_1, x_2) would be (0, 0).

For Example 3.9 and the optimizing point $[\mathbf{x}^*, \boldsymbol{\lambda}^*]' = [6.67, 4.67, 4.2, 2.23]$, it will now be shown that the Kuhn-Tucker conditions are satisfied at this point. Notice that $m = 2$, $r = 2$, $s = 2$, $J = (1,2)$, $\bar{J} = \varphi$, $I = (1,2)$, and $\bar{I} = \varphi$, where φ is the null set.

(I)
$$\frac{\partial f(\mathbf{x}^*)}{\partial x_1} - \sum_{i=1}^{2} \lambda_i^* \frac{\partial g_i(\mathbf{x}^*)}{\partial x_1} = -2(6.67 - 11) - (4.2) - 2(2.23) = 0$$

$$\frac{\partial f(\mathbf{x}^*)}{\partial x_2} - \sum_{i=1}^{2} \lambda_i^* \frac{\partial g_i(\mathbf{x}^*)}{\partial x_2} = -8(4.67 - 6) - 2(4.2) - (2.23) = 0$$

(II)
$$\sum_{j=1}^{2} x_j^* \left[\frac{\partial f(\mathbf{x}^*)}{\partial x_j} - \sum_{i=1}^{2} \lambda_i^* \frac{\partial g_i(\mathbf{x}^*)}{\partial x_j}\right] = 6.67[-2(6.67 - 11) - (4.2) - 2(2.23)] + 4.67[-8(4.67 - 6) - 2(4.2) - (2.23)] = 0$$

(III)
$$b_1 - g_1(\mathbf{x}^*) = 18 - 2(6.67) - (4.67) = 0$$
$$b_2 - g_2(\mathbf{x}^*) = 16 - (6.67) - 2(4.67) = 0$$

(IV)
$$\sum_{i=1}^{2} \lambda_i^* [b_i - g_i(\mathbf{x}^*)] = (4.2)[18 - 2(6.67) - 4.67] + (2.23)[16 - 6.67 - 2(4.67)] = 0$$

(V)
$$r = 2;\ \lambda_1^* = 4.2 \geq 0 \quad \text{and} \quad \lambda_2^* = 2.23 \geq 0$$

Since the constraint space is convex and $f(x_1, x_2)$ is a concave function, the Kuhn-Tucker conditions are also sufficient so that (6.67, 4.67) is the global maximizing point of $f(x_1, x_2)$.

Example 3.10

The problem is

$$\begin{aligned} \text{maximize} \quad & z = f(x_1, x_2) = -(x_1 - 4)^2 - (x_2 - 4)^2 \\ \text{subject to} \quad & x_1 + x_2 \leq 4 \\ & x_1^2 + x_2^2 = 4 \\ & x_1, x_2 \geq 0 \end{aligned}$$

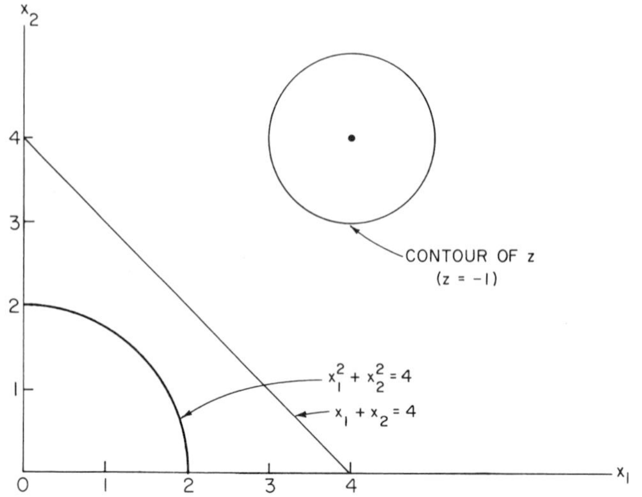

Fig. 3.10. Graphic solution to Example 3.10.

The Lagrangian method will be used to solve this problem also.

STEP 1:

The unconstrained maximum from inspection of $f(x_1, x_2)$ occurs at (4, 4), which is not feasible.

STEP 2:

Form the Lagrangian function using the constraint $x_1 + x_2 = 4$.

$$\begin{aligned}
F_1(\mathbf{x}, \boldsymbol{\lambda}) &= -(x_1 - 4)^2 - (x_2 - 4)^2 + \lambda_1(4 - x_1 - x_2) \\
\frac{\partial F_1(\mathbf{x}, \boldsymbol{\lambda})}{\partial x_1} &= -2(x_1 - 4) - \lambda_1 = 0 \\
\frac{\partial F_1(\mathbf{x}, \boldsymbol{\lambda})}{\partial x_2} &= -2(x_2 - 4) - \lambda_1 = 0 \qquad (3.88) \\
\frac{\partial F_1(\mathbf{x}, \boldsymbol{\lambda})}{\partial \lambda_1} &= 4 - x_1 - x_2 = 0
\end{aligned}$$

The solution to (3.88) is $x_1 = x_2 = 2$, $\lambda_1 = 4$, and $\lambda_2 = 0$. Since (2, 2) is not feasible, the process continues.

STEP 3:

Form the Lagrangian function using the constraint $x_1^2 + x_2^2 = 4$.

$$\begin{aligned} F_2(\mathbf{x}, \boldsymbol{\lambda}) &= -(x_1 - 4)^2 - (x_2 - 4)^2 + \lambda_2(4 - x_1^2 - x_2^2) \\ \frac{\partial F_2(\mathbf{x}, \boldsymbol{\lambda})}{\partial x_1} &= -2(x_1 - 4) - 2x_1\lambda_2 = 0 \\ \frac{\partial F_2(\mathbf{x}, \boldsymbol{\lambda})}{\partial x_2} &= -2(x_2 - 4) - 2x_2\lambda_2 = 0 \\ \frac{\partial F_2(\mathbf{x}, \boldsymbol{\lambda})}{\partial \lambda_2} &= 4 - x_1^2 - x_2^2 = 0 \end{aligned} \tag{3.89}$$

The solution to (3.89) is $x_1 = x_2 = \sqrt{2}$, $\lambda_1 = 0$, and $\lambda_2 = 1.83$. Since $(\sqrt{2}, \sqrt{2})$ is feasible, the maximum is $f(\sqrt{2}, \sqrt{2})$.

The nonnegative constraints on x_1 and x_2 have been ignored since, from Figure 3.10, they clearly will not be involved in the optimal solution. As with Example 3.9, it will now be shown that the optimizing point $[\mathbf{x}^*, \boldsymbol{\lambda}^*]' = [\sqrt{2}, \sqrt{2}, 0, 1.38]$ satisfies the Kuhn-Tucker conditions. Notice that $m = 2$, $r = 1$, $s = 1$, $J = (1, 2)$, $\bar{J} = \varphi$, $I = (2)$, and $\bar{I} = (1)$.

(I) $$\frac{\partial f(\mathbf{x}^*)}{\partial x_1} - \sum_{i=1}^{2} \lambda_i^* \frac{\partial g_i(\mathbf{x}^*)}{\partial x_1} = -2(\sqrt{2} - 4) - 2\sqrt{2}(1.83) = 0$$

$$\frac{\partial f(\mathbf{x}^*)}{\partial x_2} - \sum_{i=1}^{2} \lambda_i^* \frac{\partial g_i(\mathbf{x}^*)}{\partial x_2} = -2(\sqrt{2} - 4) - 2\sqrt{2}(1.83) = 0$$

(II) $$\begin{aligned} \sum_{j=1}^{2} x_j^* \left[\frac{\partial f(\mathbf{x}^*)}{\partial x_j} - \sum_{i=1}^{2} \lambda_i^* \frac{\partial g_i(\mathbf{x}^*)}{\partial x_j}\right] &= \sqrt{2}[-2(\sqrt{2} - 4) - 2\sqrt{2}(1.83)] \\ &\quad + \sqrt{2}[-2(\sqrt{2} - 4) - 2\sqrt{2}(1.83)] = 0 \end{aligned}$$

(III) $$\begin{aligned} b_1 - g_1(\mathbf{x}^*) &= 4 - \sqrt{2} - \sqrt{2} = 1.172 > 0 \\ b_2 - g_2(\mathbf{x}^*) &= 4 - (\sqrt{2})^2 - (\sqrt{2})^2 = 0 \end{aligned}$$

(IV) $$\begin{aligned} \sum_{i=1}^{2} \lambda_i^* [b_i - g_i(\mathbf{x}^*)] &= 0(4 - \sqrt{2} - \sqrt{2}) \\ &\quad + 1.83(4 - 2 - 2) = 0 \end{aligned}$$

(V) $r = 1$; $\lambda_1^* = 0 \geq 0$, and $\lambda_2^* = 1.83$ (unrestricted sign)

Notice that the constraint space is not convex. Therefore, the Kuhn-Tucker conditions are not sufficient for the point $(\sqrt{2}, \sqrt{2}, 0,$

1.83). From the graphic representation of the problem displayed in Figure 3.10, it is clear that this point is the global maximum. In general, it may be extremely difficult, indeed computationally impossible, to demonstrate global optimality when the Kuhn-Tucker conditions are not sufficient. Also notice that the nonconvexity of the constraint space is induced by the nonlinear equality constraint. Equality constraints that are nonlinear can create additional problems by introducing nonconvexity into the constraint region.

3.5 CONSTRAINT QUALIFICATION

In Section 3.3, the mathematical programming problem with only equality constraints has been investigated, and it has been shown that a relative minimum or maximum may not satisfy the necessary condition

$$\nabla f(\mathbf{x}^*) = \sum_{i=1}^{m} \lambda_i \nabla g_i(\mathbf{x}^*)$$

derived from the Lagrangian function

$$F(\mathbf{x}, \boldsymbol{\lambda}) = f(\mathbf{x}) + \sum_{i=1}^{m} \lambda_i [b_i - g_i(\mathbf{x})]$$

Unfortunately, this situation also arises when the Lagrangian analysis is applied to the more general form of the problem where inequality constraints are allowed as delineated by (3.56) and (3.57). In this case, it is possible that a relative minimum or maximum will not satisfy the Kuhn-Tucker necessary conditions given in (3.77)–(3.81). A number of ways of qualifying a problem so that a relative minimum or maximum will satisfy (3.77)–(3.81) have therefore been proposed. It is instructive to study problems in which the above difficulty arises; two examples will be presented in which the optimizing point does not satisfy the Kuhn-Tucker necessary conditions.

Example 3.11

Consider the following problem:

$$\begin{aligned} \text{maximize} \quad & z = f(x_1, x_2) = -(x_1 - 1)^2 - (x_2 - 1)^2 \\ \text{subject to} \quad & (1 - x_1 - x_2)^3 \geq 0 \\ & x_1, x_2 \geq 0 \end{aligned} \tag{3.90}$$

The Lagrangian function is

$$F(\mathbf{x},\boldsymbol{\lambda}) = -(x_1 - 1)^2 - (x_2 - 1)^2 - \lambda_1(1 - x_1 - x_2)^3 \quad (3.91)$$

and the necessary conditions are

$$\frac{\partial F(\mathbf{x},\boldsymbol{\lambda})}{\partial x_1} = -2(x_1 - 1) - 3\lambda_1(1 - x_1 - x_2)^2(-1) = 0$$

$$\frac{\partial F(\mathbf{x},\boldsymbol{\lambda})}{\partial x_2} = -2(x_2 - 1) - 3\lambda_1(1 - x_1 - x_2)^2(-1) = 0 \quad (3.92)$$

$$\frac{\partial F(\mathbf{x},\boldsymbol{\lambda})}{\partial \lambda_1} = (1 - x_1 - x_2)^3 = 0$$

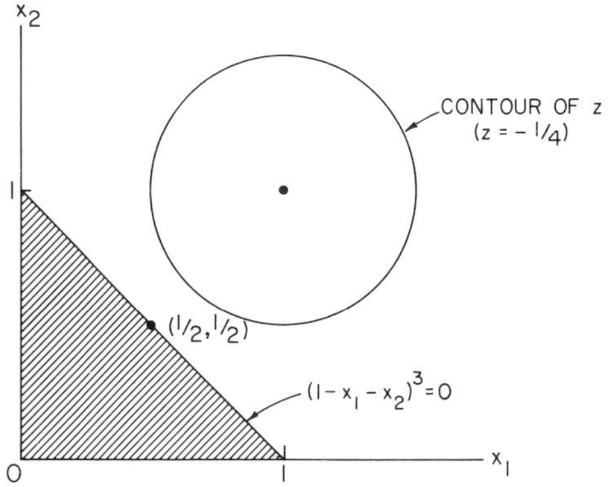

Fig. 3.11. Graphic solution to Example 3.11.

It is apparent that at the maximizing point (1/2, 1/2), deduced from Figure 3.11, the equations in (3.92) are not satisfied. If on the other hand the Lagrangian function had been written as

$$F^0(\mathbf{x},\boldsymbol{\lambda}) = \lambda_0[-(x_1 - 1)^2 - (x_2 - 1)^2] - \lambda_1(1 - x_1 - x_2)^3$$

the differential equation $\lambda_0 \nabla f - \lambda_1 \nabla g = 0$ would be satisfied at $\mathbf{x}^* = (1/2, 1/2)$ if $\lambda_0 = 0$. Also notice that the problem

$$\begin{array}{ll} \text{maximize} & z = f(x_1, x_2) = -(x_1 - 1)^2 - (x_2 - 1)^2 \\ \text{subject to} & x_1 + x_2 \le 1 \qquad x_1, x_2 \ge 0 \end{array}$$

is equivalent to that described in (3.90), and the Kuhn-Tucker conditions will be satisfied at $\mathbf{x}^* = (1/2, 1/2)$.

Example 3.12

The problem is

$$
\begin{aligned}
\text{maximize} \quad & z = f(x_1, x_2) = x_1 \\
\text{subject to} \quad & -(2 - x_1)^3 + (x_2 - 1) \leq 0 \\
& -(2 - x_1)^3 - (x_2 - 1) \leq 0 \\
& x_1, x_2 \geq 0
\end{aligned}
$$

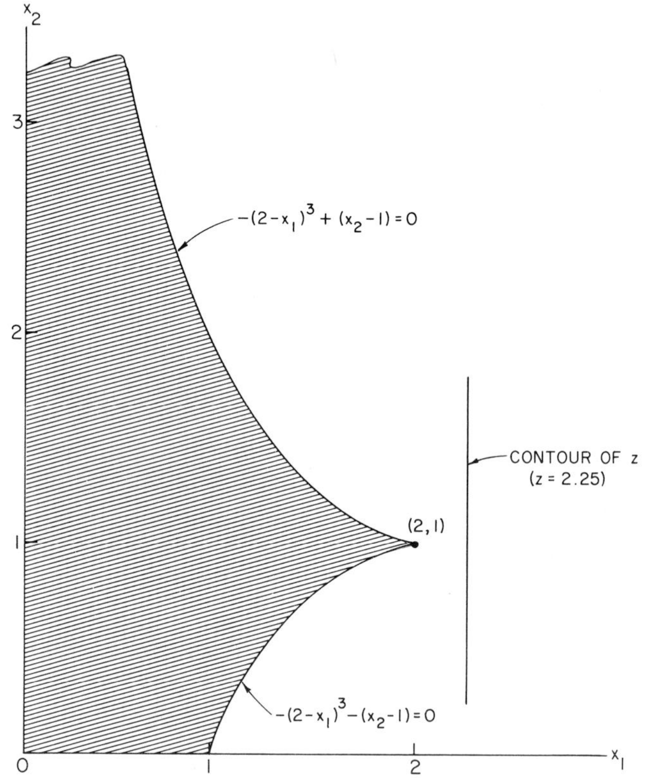

Fig. 3.12. Graphic solution to Example 3.12.

From Figure 3.12 it is apparent that the maximum of $f(\mathbf{x})$ occurs at the point (2, 1). The Kuhn-Tucker necessary conditions for this problem are

(I)
$$
\begin{aligned}
1 - 3\lambda_1(2 - x_1)^2 - 3\lambda_2(2 - x_1)^2 &= 0 \\
0 - \lambda_1 + \lambda_2 &= 0 \qquad (3.93)
\end{aligned}
$$

(II) $x_1[1 - 3\lambda_1(2 - x_1)^2 - 3\lambda_2(2 - x_1)^2]$
$$+ x_2(-\lambda_1 + \lambda_2) = 0$$

(III)
$$-(2 - x_1)^3 + (x_2 - 1) \leq 0$$
$$-(2 - x_1)^3 - (x_2 - 1) \leq 0$$

(IV) $\lambda_1[-(2 - x_1)^3 + (x_2 - 1)]$
$$+ \lambda_2[-(2 - x_1)^3 - (x_2 - 1)] = 0$$

At the point $\mathbf{x}^{*\prime} = [2,1]$ the first equation in (3.93) will not be satisfied regardless of what values λ_1 and λ_2 have.

Notice that once again if the problem had been specified with the general form of the Lagrangian function $F^0(\mathbf{x}, \boldsymbol{\lambda})$, the analog to (3.93) would be satisfied with $\lambda_0 = 0$.

Can situations such as those illustrated in Examples 3.11 and 3.12 be readily identified? The answer to this question is yes, but it may involve considerable computational labor. One approach to this problem is similar to the analysis in Section 3.3 for the equality-constrained problem. In (3.58) the problem involving inequality constraints, (3.56) and (3.57), has been delineated as a strictly equality-constrained problem leading to the Lagrangian function given in (3.59). As suggested in the above examples, it is possible to generalize this function by writing

$$F^0(\mathbf{x}, \mathbf{w}, \boldsymbol{\lambda}) = \lambda_0 f(\mathbf{x}) + \sum_{i=1}^{r} \lambda_i[b_i - g_i(\mathbf{x}) - x_{ti}]$$
$$+ \sum_{i=r+1}^{s} \lambda_i[b_i - g_i(\mathbf{x}) + x_{ti}] + \sum_{i=s+1}^{m} \lambda_i[b_i - g_i(\mathbf{x})]$$
$$+ \sum_{i=m+1}^{m+n} \lambda_i(x_{i-m} - w_{i-m}^2) \tag{3.94}$$

where λ_0 is one or zero and the last term in (3.94) constrains the x's to nonnegative values. It is now possible to classify the problems for which λ_0 should be one or zero at the optimizing point x^* depending upon the rank of $\mathbf{G}^0$ in (3.49) and $\mathbf{G}$ where the order of $\mathbf{G}^0$ is now $(m + n + 1) \times (m + n)$ and the order of $\mathbf{G}$ is $(m + n) \times (m + n)$. However, there is an easier way to investigate the constraints for troublesome behavior at a boundary point $\mathbf{x}^*$. Let K be the set of indices $i = 1, 2, \ldots, m + n$ containing the indices i for which the constraint i is active, and let there be k members in this set. Define the matrix $\mathbf{D}$ with rows composed of the gradient vectors associated with the k active constraints; i.e.,

$$D(\mathbf{x}) = \begin{bmatrix} \nabla g_1'(\mathbf{x}) \\ \nabla g_2'(\mathbf{x}) \\ \cdot \\ \cdot \\ \cdot \\ \nabla g_k'(\mathbf{x}) \end{bmatrix}$$

If the rank of $D(\mathbf{x})$ evaluated at $\mathbf{x}^*$ is k, then the constraint qualification will be satisfied. However, demonstrating that the rank of $\mathbf{D}$ is k may be a difficult computational chore if k is sizable.

In Example 3.11 there is only one constraint, and it is active at $\mathbf{x}^{*\prime} = [1/2, 1/2]$. Thus

$$D(\mathbf{x}^*) = [-3(1 - 1/2 - 1/2), -3(1 - 1/2 - 1/2)]$$

and the rank of $D(\mathbf{x}^*)$ is zero. Thus the constraint qualification is not satisfied and, as has been shown, the point $(1/2, 1/2)$ does not satisfy the Kuhn-Tucker conditions. In Example 3.12 the $\mathbf{D}$ matrix is given by

$$D(\mathbf{x}) = \begin{bmatrix} 3(2 - x_1)^2 & -1 \\ 3(2 - x_1)^2 & +1 \end{bmatrix}$$

and at the optimal point $\mathbf{x}^{*\prime} = [2, 1]$

$$D(\mathbf{x}^*) = \begin{bmatrix} 0 & -1 \\ 0 & +1 \end{bmatrix}$$

The rank of $D(\mathbf{x}^*)$ is one ($k = 2$), and hence the constraint qualification is not satisfied at $\mathbf{x}^*$.

A number of constraint qualifications have been presented in the literature. Each attempts to characterize abnormal behavior of the constraint functions at an optimizing point $\mathbf{x}^*$, which may lead to the failure of $\mathbf{x}^*$ to satisfy the Kuhn-Tucker conditions. A more detailed discussion of particular forms of the constraint qualification is given in Hadley [1964] and Zangwill [1968], and a paper by Arrow et al. [1961] compares and contrasts the various forms that the qualification may take. Additional material on the constraint qualification may be found in Mangasarian [1969].

Since all constraints in the linear and quadratic programming models are linear, the constraint qualification will be satisfied in these cases. In Section 3.6, the convex programming problem will be discussed, and the constraint qualification will also hold for this model. For the general nonlinear problem, care should be exercised in evaluating a contending point $\mathbf{x}^*$ for a minimum or maximum because the failure of $\mathbf{x}^*$ to satisfy the Kuhn-Tucker conditions may be due to the failure of the constraint qualification to hold.

3.6 ROLE OF CONVEXITY

In Sections 3.4 and 3.5, the importance of the convexity of the constraint space and the concavity of the function to be maximized has been emphasized. These results can be restated as follows. If this condition exists, the Kuhn-Tucker conditions are necessary and sufficient for the function $f(\mathbf{x})$ to take on its constrained global maximum. Furthermore, the constraint qualifications will be satisfied in this case so that no unusual behavior will occur at the boundaries, and the optimizing point $\mathbf{x}^*$ will satisfy the Kuhn-Tucker conditions. The mathematical programming problem in this case is referred to as the convex programming model and is described by

$$\text{maximize} \quad z = f(\mathbf{x}) \tag{3.95}$$

where $f(\mathbf{x})$ is a concave function, or by

$$\text{minimize} \quad w = h(\mathbf{x}) \tag{3.96}$$

where $h(\mathbf{x})$ is a convex function,

$$\text{subject to} \quad g_i(\mathbf{x})\{\leq \; = \; \geq\}b_i \qquad i = 1, 2, \ldots, m \tag{3.97}$$

where the constraint space generated by (3.97) forms a convex set of points.

The convexity of sets of points has been discussed in conjunction with the linear programming model presented in Chapter 2, and the methods for demonstrating that functions are either convex or concave are presented in Appendix B. Notice that both the linear and quadratic programming models belong to the general convex class given by (3.95)–(3.97).

There are numerous algorithms that will usually guarantee the optimal solution to the convex programming problem. Since any local optimum will be the global optimum in the convex programming problem (see Hadley [1964, pp. 90–94] for a proof of this fact), the Lagrangian method discussed in this chapter may usually be applied without undue difficulty in solving the problem. In Chapter 4, a number of algorithms will be presented for solving the general mathematical programming problem, and most of these will guarantee the global solution to the convex programming model.

If the problem is nonlinear but not convex, the problem may have many local optima, among which is the global optimum. There is no algorithm computationally feasible to employ that will also guarantee locating the global optimum in this case. However, it may be possible to demonstrate that the global optimum must reside in a partition of the constraint space where the problem is locally convex so that convex programming methods may be applied to secure the solution. Examples of locally convex functions are given in Appendix B. In any event, the more information that can be gained about the nonconvex programming problem before a solution algorithm is applied, the better the chance of guaranteeing global optimality. Methods for gaining information with regard to the convex properties of a mathematical programming problem will be discussed in Chapter 4.

3.7 EXAMPLES

To illustrate some of the basic concepts presented in this chapter, two additional examples will be presented.

Example 3.13

Suppose that two additives *A* and *B* are to be used to enhance the quality of a specific brand of gasoline. In mixing the two additives with the basic gasoline in a tank at the distribution center, the following conditions must be satisfied:

1. The quantity of additive *B* plus twice the quantity of additive *A* must exceed 0.5 pound in the mixing tank.
2. One pound of additive *A* will add 10 octane units per tank, and one pound of additive *B* will add 20 octane units per tank. To insure optimum performance standards of the gasoline, the total number of octane units must not be less than 6.
3. The total of the two additive must not exceed 0.5 pound.

Each pound of additive *A* costs \$153, and each pound of additive *B* costs \$400. What are the optimum amounts of *A* and *B* to add to the gasoline so that the cost is minimized? This is a linear programming model, and if x_1 = amount of additive *A* and x_2 = amount of additive *B*, it may be stated as

$$\begin{aligned} \text{minimize} \quad & z = f(x_1, x_2) = 153x_1 + 400x_2 \\ \text{subject to} \quad & 2x_1 + x_2 \geq 0.5 \\ & 10x_1 + 20x_2 \geq 6 \\ & x_1 + x_2 \leq 0.5 \\ & x_1, x_2 \geq 0 \end{aligned}$$

The problem is represented graphically in Figure 3.13.

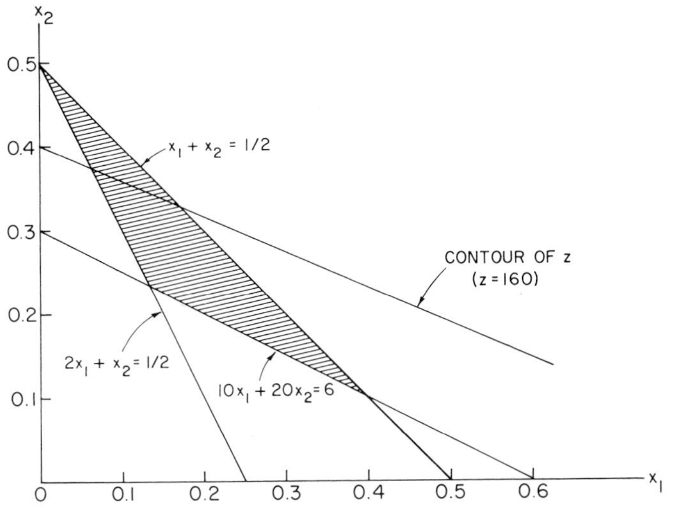

Fig. 3.13. Graphic solution to Example 3.13.

The appropriate Lagrangian function for this problem is

$$\begin{aligned} F(\mathbf{x}, \mathbf{w}, \boldsymbol{\lambda}) = {} & 153x_1 + 400x_2 + \lambda_1(0.5 - 2x_1 - x_2) \\ & + \lambda_2(6 - 10x_1 - 20x_2) \\ & + \lambda_3(0.5 - x_1 - x_2) + \lambda_4(w_1^2 - x_1) \\ & + \lambda_5(w_2^2 - x_2) \end{aligned} \tag{3.98}$$

In applying the Lagrangian method for solving this problem, Step 3 is eventually reached with the following Lagrangian function:

$$\begin{aligned} F_1(\mathbf{x}, \mathbf{w}, \boldsymbol{\lambda}) = {} & 153x_1 + 400x_2 + \lambda_2(6 - 10x_1 - 20x_2) \\ & + \lambda_3(0.5 - x_1 - x_2) \end{aligned} \tag{3.99}$$

The solution to (3.99) is achieved by solving the following set of linear equations:

$$\frac{\partial F_1(\mathbf{x}, \mathbf{w}, \boldsymbol{\lambda})}{\partial x_1} = 153 - \lambda_3 - 10\lambda_2 = 0$$
$$\frac{\partial F_1(\mathbf{x}, \mathbf{w}, \boldsymbol{\lambda})}{\partial x_2} = 400 - \lambda_3 - 20\lambda_2 = 0$$
$$\frac{\partial F_1(\mathbf{x}, \mathbf{w}, \boldsymbol{\lambda})}{\partial \lambda_2} = 6 - 10x_1 - 20x_2 = 0 \qquad (3.100)$$
$$\frac{\partial F_1(\mathbf{x}, \mathbf{w}, \boldsymbol{\lambda})}{\partial \lambda_3} = 0.5 - x_1 - x_2 = 0$$

The solution to (3.100) is $x_1 = 0.4$, $x_2 = 0.1$, $\lambda_2 = 24.7$, and $\lambda_3 = -94$. The point $(0.4, 0.1)$ is the intersection of the constraints $10x_1 + 20x_2 \geq 6$ and $x_1 + x_2 \leq 0.5$.

Since the problem is linear, the Kuhn-Tucker conditions are necessary and sufficient for a global optimum. In forming the conditions from (3.98) and noting that $J = (1, 2)$, $\bar{J} = \varphi$, $I = (2, 3)$, and $\bar{I} = (1, 4, 5)$, it is easy to show that the point $[\mathbf{x}^*, \mathbf{w}^*, \boldsymbol{\lambda}^*] = [0.4, 0.1; \sqrt{0.4}, \sqrt{0.1}; 0, 24.7, -94, 0, 0]$ satisfies the Kuhn-Tucker conditions. At this point, $z = 101.2$.

Also, recall from Section 3.2 the interpretation of the Lagrangian multipliers. If the total number of octane units required is increased by one unit from 6 to 7, the total cost will increase by approximately \$24.70. If the total amount of the two additives A and B is increased one unit from 0.5 pound to 1.5 pounds, the total cost will decrease approximately \$94. The interpretation of λ_2 and λ_3 must be taken independently and ignores the fact that if an attempt were made to choose x_1 and x_2, for example, in a way that a \$94 reduction in cost would result (assuming the third constraint becomes $x_1 + x_2 \leq 1.5$), the variable x_2 would be moved infeasibly to the value -0.9.

This problem could have been solved by using the simplex algorithm. The interpretation of the Lagrangian multipliers would follow from analyzing the dual variable values at the optimal primal solution as pointed out in Section 2.7 of Chapter 2.

Example 3.14

Consider the following:

$$\begin{aligned} \text{minimize} \quad & z = f(x_1, x_2) = (x_1 - 2)^2 + (x_2 - 1)^2 \\ \text{subject to} \quad & g_1(x_1, x_2) = -(x_1^2/4) - x_2^2 + 1 \geq 0 \\ & g_2(x_1, x_2) = x_1 - 2x_2 + 1 \geq 0 \end{aligned}$$

This problem may be restated as

$$\begin{aligned} \text{maximize} \quad & -z = -(x_1 - 2)^2 - (x_2 - 1)^2 \\ \text{subject to} \quad & g_1^0(x_1, x_2) = (x_1^2/4) + x_2^2 - 1 \le 0 \\ & g_2^0(x_1, x_2) = -x_1 + 2x_2 - 1 \le 0 \end{aligned} \qquad (3.101)$$

This is a convex programming problem since $f^0 = -f$, $g_1^0 = -g_1$, and $g_2^0 = -g_2$ are all convex functions. By use of the results in Appendix B, this can be verified as follows:

$$\frac{\partial f^0}{\partial x_1} = -2(x_1 - 2) \qquad \frac{\partial f^0}{\partial x_2} = -2(x_2 - 1) \qquad \frac{\partial^2 f^0}{\partial x_1^2} = -2$$

$$\frac{\partial^2 f^0}{\partial x_1 \partial x_2} = 0 \qquad \frac{\partial^2 f^0}{\partial x_2^2} = -2$$

$$\Delta = (-2)(-2) - (0)^2 = 4 > 0$$

Thus f^0 is a convex function.

$$\frac{\partial g_1^0}{\partial x_1} = \frac{x_1}{2} \qquad \frac{\partial^2 g_1^0}{\partial x_1^2} = \frac{1}{2} \qquad \frac{\partial g_1^0}{\partial x_2} = 2x_2$$

$$\frac{\partial^2 g^0}{\partial x_2^2} = 2 \qquad \frac{\partial^2 g_1^0}{\partial x_1 \partial x_2} = 0$$

$$\Delta = (2)(1/2) - (0)^2 = 1 > 0$$

Thus g_1^0 is convex, and since g_2^0 is linear, it is also convex. To show that the constraint region is a convex set, use the following facts: if $h(\mathbf{x})$ is a convex function, the set $\{\mathbf{x}/h(\mathbf{x}) \le 0\}$ is convex; and the intersection of two convex sets S_1 and S_2 is convex. Since both g_1^0 and g_2^0 are convex functions, it follows that the set of points satisfying (3.101) is convex. The convex constraint region is illustrated in Figure 3.14.

The second step of the Lagrangian method is

$$F_1(\mathbf{x}, \boldsymbol{\lambda}) = (x_1 - 2)^2 + (x_2 - 1)^2 + \lambda_1[1 - (x_1^2/4) - x_2^2]$$

so that

$$\begin{aligned} \frac{\partial F_1(\mathbf{x}, \boldsymbol{\lambda})}{\partial x_1} &= 2(x_1 - 2) - 0.5\lambda_1 x_1 = 0 \\ \frac{\partial F_1(\mathbf{x}, \boldsymbol{\lambda})}{\partial x_2} &= 2(x_2 - 1) - 2x_2\lambda_1 = 0 \\ \frac{\partial F_1(\mathbf{x}, \boldsymbol{\lambda})}{\partial \lambda_1} &= 1 - (x_1^2/4) - x_2^2 = 0 \end{aligned} \qquad (3.102)$$

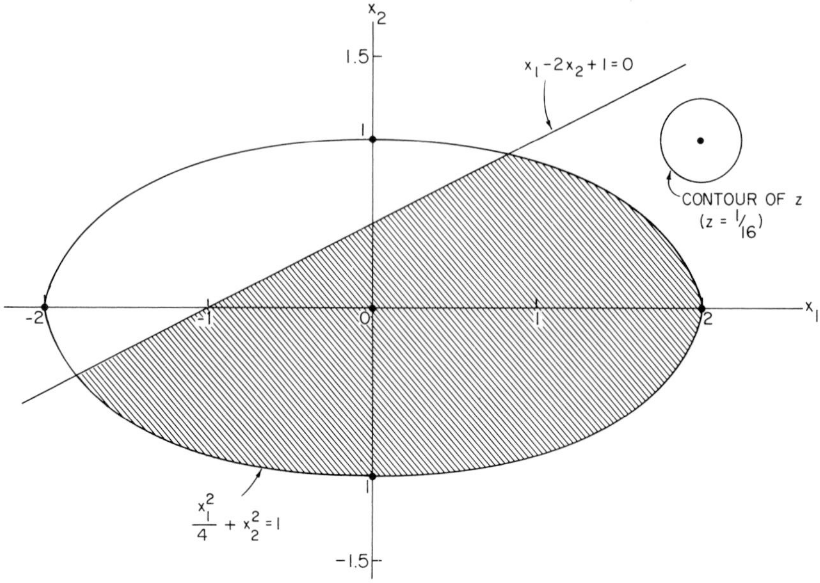

Fig. 3.14. Graphic solution to Example 3.14.

The solution to (3.102) is $x_1 = 1.66$, $x_2 = 0.55$, and $\lambda_1 = -0.818$. Since $(1.66, 0.55)$ satisfies the constraint $x_1 - 2x_2 + 1 \geq 0$, the Lagrangian process would terminate. Furthermore, since this is a convex programming problem, the solution $\mathbf{x}^{*\prime} = [1.66, 0.55]$ is optimal; therefore, the global minimum in the initial problem is $f(\mathbf{x}^*) = 0.311$. The reader should verify that the Kuhn-Tucker conditions are satisfied at the point $[\mathbf{x}^*, \boldsymbol{\lambda}^*]' = [1.66, 0.55, -0.818, 0]$.

PROBLEMS

For Problems 3.1, 3.2, and 3.3 derive the Kuhn-Tucker necessary conditions. Solve the problems graphically and show that the optimal solution point $(x_1^*, x_2^*, \ldots)$ satisfies the Kuhn-Tucker conditions.

3.1. Minimize $z = -3x^3 + 5x^2 + 2x$
subject to $9x \leq 7$
$x \geq 0$

3.2. Maximize $z = 3x_1^2 + 4x_2^3$
subject to
$$\begin{aligned} x_1 + x_2 &\leq 10 \\ 2x_1 + 3x_2 &\geq 15 \\ x_1, x_2 &\geq 0 \end{aligned}$$

3.3. Maximize $z = 12y_1 + 9y_2$
subject to
$$\begin{aligned} 20y_1^2 + 10y_2^2 &\leq 50 \\ y_1 + y_2 &= 1 \end{aligned}$$

3.4. The XYZ Corporation produces and sells two products at daily levels of X_1 and X_2 respectively, while employing 6 hours of labor and 4 hours of machine time each day. Two hours of labor and 1 hour of machine time are required to produce one unit of X_1, and 1 hour of labor and 2 hours of machine time are required to produce one unit of X_2. Furthermore, X_3 is the number of daily hours of unemployed labor and X_4 is the number of daily hours of unutilized machine time. Therefore, the constraints facing XYZ Corporation are

$$\begin{aligned} 2X_1 + X_2 + X_3 &= 6 \\ X_1 + 2X_2 + X_4 &= 4 \\ X_1, X_2, X_3, X_4 &\geq 0 \end{aligned}$$

The corporation wishes to maximize its sales revenue SR from producing X_1 and X_2. The sales revenue from the products is $SR = p_1X_1 + p_2X_2$ where $p_1 = 2 - X_1 + 3X_2$, and $p_2 = 1 - 3X_1$.

The XYZ Corporation's optimization problem is

maximize $SR = 2X_1 - X_1^2 + X_2$
subject to
$$\begin{aligned} 2X_1 + X_2 + X_3 &= 6 \\ X_1 + 2X_2 + X_4 &= 4 \\ X_1, X_2, X_3, X_4 &\geq 0 \end{aligned}$$

Derive the Kuhn-Tucker necessary conditions for the XYZ Corporation. Find the optimal values of X_1, X_2, X_3, X_4; and by using the Lagrangian algorithm, show that the optimal solution point $(X_1^*, X_2^*, X_3^*, X_4^*)$ satisfies the Kuhn-Tucker conditions.

3.5. The PQR Corporation is a middleman, wholesale dealer in widgets. The firm's main concern is inventory control. The corporation's current problem is to determine the desirable inventory levels for 1981, 1982, and 1983, which are defined as *IA*, *IB*, and *IC* respectively. It costs PQR \$1 for each unit of inventory change between 1981 and 1982, \$3 for each unit of inventory change between 1982 and 1983, and \$2 for each unit of inventory held during 1982. PQR's goal is to minimize its inventory costs C where $C = 1(IA - IB) + 3(IC - IB) + 2IB$, or $C = 1IA - 2IB + 3IC$.

The interactions among the three annual inventory levels are

$$-2IA + IB + 3IC = 20$$
$$2IA + 3IB + 4IC = 36$$
$$IA + IB + 2IC = 18$$

For the model,

$$\begin{aligned} \text{minimize} \quad & C = IA - 2IB + 3IC \\ \text{subject to} \quad & -2IA + IB + 3IC = 20 \\ & 2IA + 3IB + 4IC = 36 \\ & IA + IB + 2IC = 18 \end{aligned}$$

derive the Kuhn-Tucker necessary conditions, solve the problem using the Lagrangian algorithm, and show that the optimal solution point $(I1^*, I2^*, I3^*)$ satisfies the Kuhn-Tucker conditions.

3.6. The new government of Sylvania is headed by a graduate of a well-known business school who has studied macroeconomic models. The government has decided that its goal will be to maximize a linear function of national income Y, aggregate consumption C, and aggregate investment I. These three variables are weighted in the national utility function according to their relative importance, with investment being considered by far the most important since it has the largest income multiplier. Sylvania's national goal is

$$\text{maximize } U = 0.30Y + 0.20C + 2.20I$$

where $Y = C + I$. The goal can be restated as

$$\text{maximize} \quad U = 0.30Y + 0.20(Y - I) + 2.20I$$

or

$$\text{maximize} \quad U = 0.50Y + 2.00I$$

As a result of a recent statistical study, it has been learned that for 1973 Sylvania has a national income of \$3.0 million, a total consumption level of \$1.0 million, and a total aggregate private and public investment of \$2.0 million. From these data and the work of government economists it has been decided that reasonable constraints for Sylvania for 1975 are

$$Y + I \le 6 \qquad Y - I \le 1 \qquad 2Y + I \ge 6$$
$$0.5Y + 4 \ge I \qquad Y \ge 1 \qquad I \ge 0$$

Derive the Kuhn-Tucker necessary conditions for 1975, Y^* and I^*, to maximize Sylvania's utility function, $U = 0.50Y + 2.00I$. Solve the problem using the Lagrangian algorithm, and show that the optimal 1975 income and investment levels Y^* and I^* respectively satisfy the Kuhn-Tucker conditions.

3.7. The new Koorsville city government was elected on the strength of a promise to improve the city's return on the costly subway and bus system. The city will operate X_1 subway cars and X_2 buses each year. Because the government believes that subway transportation is 25 times as socially desirable as bus transportation, the city will try to determine the number of subway cars and buses to

$$\text{maximize} \quad z = 25(X_1 - 2)^2 + (X_2 - 2)^2$$

It is essential to have at least 2 public transportation vehicles operating at all times, $X_1 + X_2 \ge 2$.

The operating transportation budget for the city is \$32,000, and the average cost of operating either a bus or subway car is \$4000 per year, so $4000X_1 + 4000X_2 \le 32{,}000$, or $X_1 + X_2 \le 8$ is required.

Furthermore, because of the political considerations of a previous local bond issue and requirements attached to federal highway grants, $X_1 + 3 \ge X_2$ and $X_1 \le 3X_2 + 2$ respectively. Therefore, the Koorsville government would like to know the optimal number of subway cars X_1 and buses X_2 that should be operating. The problem is

$$\begin{aligned} \text{maximize} \quad & z = 25(X_1 - 2)^2 + (X_2 - 2)^2 \\ \text{subject to} \quad & X_1 + X_2 \ge 2 \\ & X_1 + X_2 \le 8 \\ & X_1 + 3 \ge X_2 \\ & X_1 \le 3X_2 + 2 \\ & X_1, X_2 \ge 0 \end{aligned}$$

Solve this problem graphically and by using the Lagrangian algorithm.

3.8. Consider the following problem:

$$\begin{aligned} \text{maximize} \quad & z = x_1 + x_2 \\ \text{subject to} \quad & x_1^2 + x_2^2 \le 32 \\ & 4x_1 + 4x_2 = 32 \\ & x_1, x_2 \ge 0 \end{aligned}$$

Is the constraint qualification satisfied at the optimizing point [determine (x_1^*, x_2^*) graphically]? If the equality constraint is dropped, is the constraint qualification satisfied at (x_1^*, x_2^*)?

3.9 Solve the following problem graphically.

$$\begin{aligned} \text{minimize} \quad & z = x_1^2 + 4x_2^2 \\ \text{subject to} \quad & x_1^3 \ge 0 \\ & 5x_1 + 10x_2 \le 50 \\ & 0 \le x_2 \le 4.5 \end{aligned}$$

Is the constraint qualification satisfied at the optimizing point (x_1^*, x_2^*)? If not, why?

3.10. Show that the following problems either are or are not convex programming problems:

(a) $$\begin{aligned} \text{maximize} \quad & z = -(x_1 - 4)^2 - (x^2 - 4)^2 \\ \text{subject to} \quad & x_1 + x_2 \le 2 \\ & x_1^2 + x_2^2 = 4 \end{aligned}$$

(b) $$\begin{aligned} \text{maximize} \quad & z = -(x_1 - 1)^2 - (x_2 - 1)^2 \\ \text{subject to} \quad & (1 - x_1 - x_2)^3 \ge 0 \\ & x_1, x_2 \ge 0 \end{aligned}$$

(c) $$\begin{aligned} \text{maximize} \quad & z = x_1 \\ \text{subject to} \quad & -(2 - x_1)^3 + (x_2 - 1) \le 0 \\ & -(2 - x_1)^3 - (x_2 - 1) \le 0 \\ & x_1, x_2 \ge 0 \end{aligned}$$

(d) $$\begin{aligned} \text{minimize} \quad & z = x_1^2 + 4x_2^2 \\ \text{subject to} \quad & x_1^3 \le 0 \\ & 5x_1 + 10x_2 \le 30 \\ & x_2 \le 4.5 \\ & x_2 \ge 0 \end{aligned}$$

(e) $$\begin{aligned} \text{minimize} \quad & z = 4x_1^2 + 6x_1 + 10x_2^2 - 5x_2 + 4x_3^2 \\ & \qquad - 2x_3 + x_4^2 \\ \text{subject to} \quad & 4x_1 + 10x_2 + 10x_3 + 2x_4 \le 100 \\ & x_1^2 + x_2^2 \le 81 \\ & 50 - x_4^2 \ge 25 \\ & x_1, x_2, x_3, x_4 \ge 0 \end{aligned}$$

4 NONLINEAR PROGRAMMING ALGORITHMS

4.1 INTRODUCTION

In Chapter 3, the nonlinear programming model was introduced and investigated by the application of the Lagrangian multiplier technique, which led to necessary conditions for a global maximum and suggested a solution procedure based upon the Lagrangian function. However, it is apparent that unless the problem is convex and the number of constraints and variables are small, the Lagrangian procedure will generally be tedious to apply computationally. In fact, if the problem is sizable and severely nonconvex, the application of the Lagrangian method may be infeasible. Fortunately, numerous algorithms have been devised for solving the nonlinear problem. In this chapter, four basic types of algorithms are presented with the purpose of identifying and discussing their fundamental solution principles. Computational procedures utilizing these algorithms typically have been designed for use on computers, and numerous codes have been written for them; many of these are readily available through research groups and university computer centers.

In resolving a nonlinear programming problem, the important decision as to which algorithm to use is presented in Section 4.6. Suggestions are made also with regard to preparatory work which should precede the application of an algorithm, for occasionally global optimality may be insured, although the mathematical programming problem is not convex.

4.2 STEEPEST ASCENT METHODS

The methods of steepest ascent utilize the information provided by the first derivative of a function in seeking out its maximum. As an example, consider the unconstrained maximization of the univariate function

$$z = f(x) = 8 + 4x - x^2 \tag{4.1}$$

The first derivative of z with respect to x is $-2x + 4$ and provides the rate of change of z per unit increase in x. For example, at the point $x = 0$, z is increasing at the rate of 4 units per unit increase in x. In terms of maximizing z, this certainly suggests that x be increased from 0, since z will increase. The question is how much x should be increased, for if x is set equal to 4, for example, than $z = f(x)$ is decreasing at the rate of 4 units per unit increase in x and the maximizing point has been passed. The increment kt that should be added to $x = 0$ to maximize z may be determined by maximizing z at the point $0 + kt$ where k is the gain in z at $x = 0$; i.e.,

$$\max_t f(0 + 4t) = \max_t [8 + 4(4t) - (4t)^2] \tag{4.2}$$

This maximum can be determined by differentiating (4.2) with respect to t, equating to zero, and solving for t:

$$\frac{df(x)}{dt} = 16 - 32t = 0 \qquad t = 1/2$$

The new point is therefore $0 + 4(1/2) = 2$, and at this point z is maximized. The maximizing point $x^* = 2$ could have been determined directly by setting the first derivative of (4.1) equal to zero and solving for x. However, this extremely simple example illustrates the basic idea in most steepest ascent algorithms; the first derivative indicates the direction of steepest ascent (in this case, moving from zero in a positive direction along the x axis) and can be exploited in this manner to determine the maximizing point.

As an additional illustration of the basic method, consider the unconstrained maximization of the bivariate function

$$z = f(x_1, x_2) = 500 - 16(x_1 - 6)^2 - 4(x_2 - 8)^2 \tag{4.3}$$

The function is shown graphically in Figure 4.1 where the third axis, representing the values of the function, is understood to project per-

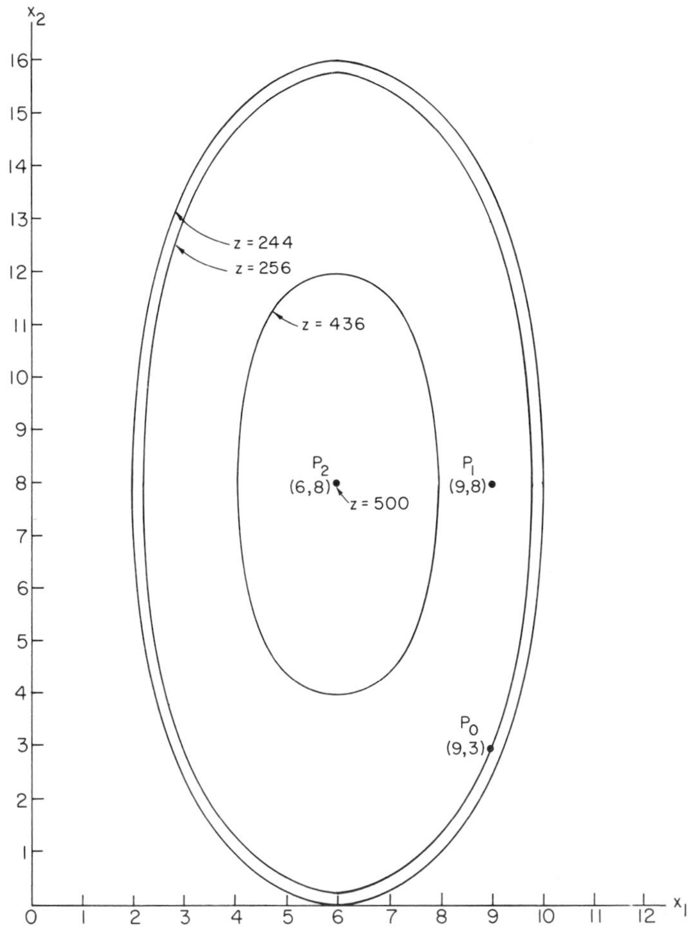

Fig. 4.1. Contours of the function (4.3).

pendicularly out of the page. From (4.3) it is clear that the maximum of the function is 500, and it occurs at the point (6, 8). Three contours of the function are drawn in Figure 4.1 to assist in the geometric interpretation; it is a mountain with its peak of 500 at (6, 8).

The gradient vector ∇f, introduced in the last chapter, gives as its components the first partial derivatives of $f(\mathbf{x})$ with respect to x_i, $i = 1, 2$. More importantly, the vector ∇f points in the direction of steepest ascent. For example, suppose the point $P_0 = (9, 3)$ is selected as the starting point; this is located on the contour $z = 256$ in Figure 4.1. The gradient vector at this point is

$$\nabla f_{P_0} = \begin{bmatrix} -32(x_1 - 6) \\ -8(x_2 - 8) \end{bmatrix}_{P_0} = \begin{bmatrix} -96 \\ 40 \end{bmatrix} \qquad (4.4)$$

Denote by $\mathbf{x}_i$ an $n \times 1$ column vector with a one in the ith component and zeroes elsewhere. Then the vector

$$-96\mathbf{x}_1 + 40\mathbf{x}_2 = -96\begin{bmatrix} 1 \\ 0 \end{bmatrix} + 40\begin{bmatrix} 0 \\ 1 \end{bmatrix} = \begin{bmatrix} -96 \\ 40 \end{bmatrix}$$

points from (9, 3) in the direction of steepest ascent. This vector is normal to the contour $z = 256$ at the point (9, 3) and therefore may not point directly at the peak of the mountain. However, the closer the peak, the more accurate the direction of the gradient will become. From (4.4) it is seen that a unit increase in x_1 will result in a 96-unit decrease in z while a unit increase in x_2 will increase z by 40 units.

In order to increase z, begin by incrementing x_2. To determine the magnitude of the increase, it is necessary to maximize z with respect to t at the point $(9, 3 + 40t)$;

$$\max_t f(9, 3 + 40t) = \max_t (-6400t^2 + 1600t + 256)$$

$$\frac{df}{dt} = -12{,}800t + 1600 = 0$$

$$t = 1/8$$

Therefore, the new point is $[9, 3 + 40(1/8)] = (9, 8) = P_1$. Now

$$\nabla f_{P_1} = \begin{bmatrix} -96 \\ 0 \end{bmatrix}$$

and no further increases in z will result by incrementing x_2.

Now decrease x_1. The magnitude of the decrement must be determined:

$$\max_t f(9 - 96t, 8) = \max_t [356 + 9216t - 16(96)^2t^2]$$

$$\frac{df}{dt} = 9216 - 32(96)^2t = 0$$

$$t = 1/32$$

The new point is $P_2 = [9 - 96(1/32), 8] = (6, 8)$ and

$$\nabla f_{P_2} = \begin{bmatrix} 0 \\ 0 \end{bmatrix}$$

so that no further increase in f is possible. The maximizing point is therefore (6, 8), and $f(\mathbf{x}^*) = 500$. The path generated by this steepest ascent algorithm from the initial point $P_0 = (9, 3)$ to the terminal point $P_2 = (6, 8)$ is illustrated in Figure 4.2. Once again, the problem is very simple, but the procedure used above could be easily programmed for computer resolution to solve larger problems.

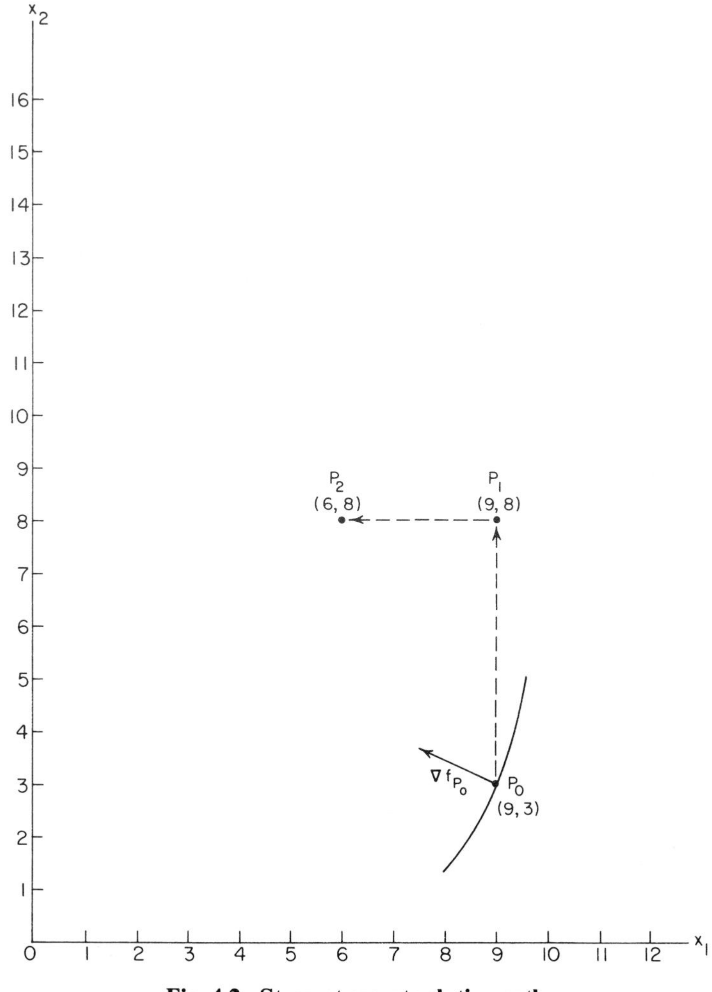

Fig. 4.2. Steepest ascent solution path.

Two observations with regard to the above solution process are in order. First, it is obvious that the move from $P_0 = (9, 3)$ to $(6, 8)$ could be accomplished in one step by selecting s and t so that $f(9 - 96s, 3 + 40t)$ is maximized with respect to s and t. This one-step process more efficiently utilizes the information provided by ∇f and is equivalent in this case to setting $\nabla f = 0$ and solving for x_1 and x_2. Second, what happens if the problem is constrained in some manner? Suppose the original problem in (4.3) is now changed to

$$\begin{aligned} &\text{maximize} \quad f(x_1, x_2) = -16(x_1 - 6)^2 - 4(x_2 - 8)^2 \\ &\text{subject to} \quad x_1 \geq 7 \end{aligned}$$

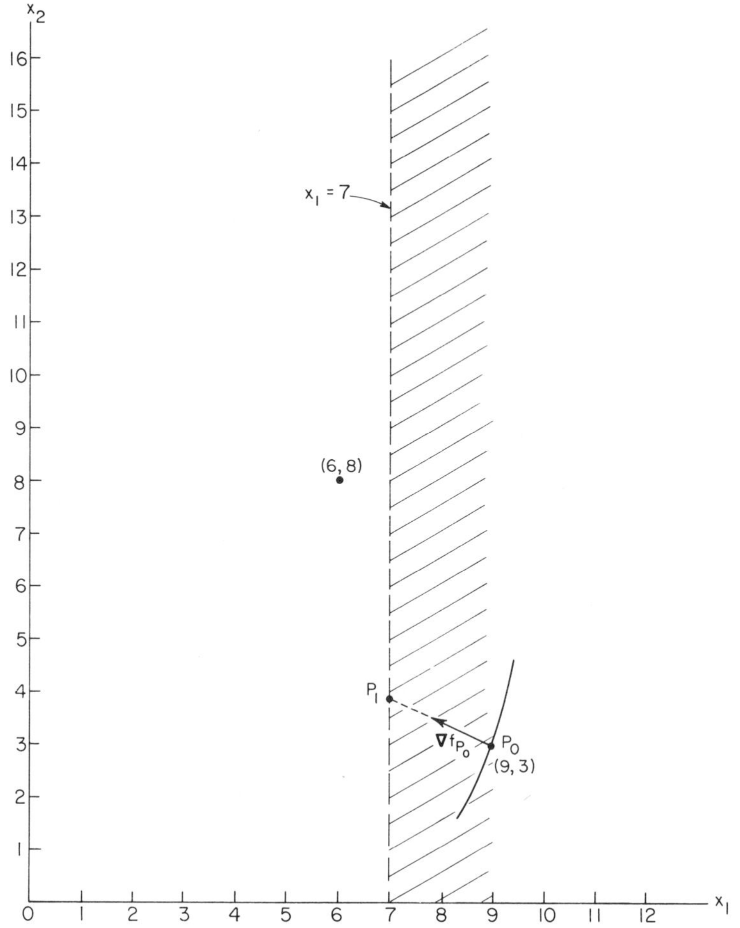

Fig. 4.3. Intersection of the steepest ascent path and the linear constraint $x_1 = 7$.

Starting at the point $P_0 = (9, 3)$ and attempting to move in the direction ∇f_{P_0}, it is clear from Figure 4.3 that the line $x = 7$ is intersected at P_1 before the maximizing point (6, 8) is encountered.

Intuitively, it seems reasonable to move along the line $x_1 = 7$ from P_1 in the positive x_2 direction, for z will certainly increase along this path, at least until $x_2 = 8$. This process may be summarized in three steps:

1. Determine the distance from P_0 to the constraint $x_1 = 7$ in the direction ∇f_{P_0}.
2. Move this distance to P_1 and compute ∇f_{P_1}.
3. Project ∇f_{P_1} onto the constraint $x_1 = 7$ and move in the direction of the projected vector.

The general method for projecting ∇f onto constraints will be delineated momentarily; in this case, the projected vector simply will be $c\vec{\mathbf{x}}_2$ where c is a constant.

The three-step procedure described above is the basis of most steepest ascent algorithms devised for constrained problems, although its application becomes somewhat complicated as the number of constraints and variables increases. As an example, suppose there are three variables and the point P is presently located in such a way that ∇f_P points outside the feasible region. The point P may lie in Euclidean two-space, in E^2 (a face or plane), in E^1 (a line, the intersection of two planes), or in E^0 (a point, the intersection of three planes). In the first case, ∇f_P should be projected onto the plane; in the second, ∇f_P should be projected onto the line; and P must be at a vertex in the third case. In general, the gradient vector is projected onto a manifold (in the above example, plane or line) of the least dimension that contains P. An m-dimensional manifold is a surface in n-dimensional space in which m restrictions have been placed on the function defining the surface. For example, $f(x_1, x_2, \ldots, x_n) = 0$ is an $n - 1$ dimensional manifold (one restriction: f is set to zero), while the intersection of $f_1(x_1, x_2, \ldots, x_n) = 0$ and f is an $n - 2$ dimensional manifold.

An algorithm developed by Rosen [1960], which utilizes these basic ideas for solving a mathematical programming problem with m linear constraints, will now be delineated. Consider the following problem specification:

$$\text{maximize} \quad f(\mathbf{x}) \tag{4.5}$$

$$\text{subject to} \quad g_i(\mathbf{x}) = \mathbf{a}_i'\mathbf{x} \leq 0 \qquad i = 1, 2, \ldots, m \tag{4.6}$$

$$\mathbf{a}_i' = [a_{i1}, a_{i2}, \ldots, a_{in}]$$

where $\mathbf{x}$ is an n-dimensional column vector with components x_j, $j = 1, 2, \ldots, n$, and $\mathbf{a}_i'$ is an n-dimensional row vector with components $a_{i1}, a_{i2}, \ldots, a_{in}$, which are the coefficients associated with the ith linear constraint. If there are any upper or lower bound constraints on $x_j, j = 1, 2, \ldots, n$, they must be included directly in (4.6). Denote by G_i the $n - 1$ dimensional manifold (hyperplane) defined by $g_i(\mathbf{x}) = 0$. Thus $G_i = \{\mathbf{x}/g_i(\mathbf{x}) = 0\}, i = 1, 2, \ldots, m$.

The boundary of the feasible region delineated by (4.6) will consist of all feasible points

$$\{\mathbf{x}/g_i(\mathbf{x}) \leq 0 \text{ for all } i\} \tag{4.7}$$

where

$$g_i(\mathbf{x}) = 0 \text{ for at least one } i \tag{4.8}$$

The vector $\mathbf{a}_i'$ is normal to the hyperplane G_i, and two hyperplanes G_l and G_k are independent if $\mathbf{a}_l$ and $\mathbf{a}_k$ are linearly independent vectors. An $n - l$ dimensional manifold in the n-dimensional space of $\mathbf{x}$ is formed by the intersection of l independent hyperplanes. Suppose now that the requirement is to project a vector ∇f onto an $n - q$ dimensional manifold generated by the intersection of $q < n$ independent hyperplanes $G_1, G_2, \ldots, G_q$. Compute

$$\mathbf{D}_q = \mathbf{I} - \mathbf{A}_q'(\mathbf{A}_q\mathbf{A}_q')^{-1}\mathbf{A}_q \tag{4.9}$$

where $\mathbf{I}$ is an $n \times n$ identity matrix and $\mathbf{A}_q$ is a $q \times n$ matrix with rows $\mathbf{a}_1', \mathbf{a}_2', \ldots, \mathbf{a}_q'$;

$$\mathbf{A}_q = \begin{bmatrix} \mathbf{a}_1' \\ \mathbf{a}_2' \\ \vdots \\ \mathbf{a}_q' \end{bmatrix} \tag{4.10}$$

Let ∇h denote the projected vector. Then

$$\nabla h = \mathbf{D}_q \nabla f \qquad \text{if } 0 < q < n \tag{4.11}$$

$$\nabla h = \nabla f \qquad \text{if } q = 0 \tag{4.12}$$

$$\nabla h = \mathbf{0} \qquad \text{if } q = n \tag{4.13}$$

To illustrate the use of (4.7)–(4.13), suppose $n = 3$ and $q = 2$;

then there are three variables and the projection of ∇f is the $n - q = 3 - 2 = 1$ dimensional manifold (line) generated by the intersection of $q = 2$ linearly independent hyperplanes (planes) G_1 and G_2. In this instance, from (4.10)

$$\mathbf{A}_2 = \begin{bmatrix} a_{11} & a_{12} & a_{13} \\ a_{21} & a_{22} & a_{23} \end{bmatrix}_{2 \times 3}$$

where the a_{ij}'s are the coefficients associated with the constraints $g_1(\mathbf{x}) \leq 0$ and $g_2(\mathbf{x}) \leq 0$. It is assumed that there are $m > 2$ inequality constraints and that they have been reordered so that the first two intersect, forming the manifold. Since $q = 1$, (4.11) would be used to compute ∇h. In the event that $q = 0$, the manifold $n - q = 3 - 0 = 3$ is of full dimension ($n = 3$) so that the present point must be interior to the constraint region. If $q = n$, the manifold is of zero dimension since $n - q = 3 - 3 = 0$, and the point must be at a vertex or corner of the constraint region. It will be useful to let ∇h be the null vector in this case.

Before describing the full algorithm, it is necessary to consider what situations may arise if at the kth step, P_k is a boundary point. Suppose that P_k lies on exactly $q(0 \leq q \leq n)$ independent hyperplanes $G_i, i = 1, 2, \ldots, q$. Assume without loss of generality that it is the first $q, g_i(P_k) = 0, i = 1, 2, \ldots, q$; P_k is a solution to (4.5) and (4.6) if

$$\mathbf{D}_q \nabla f_{P_k} = \mathbf{0} \tag{4.14}$$

and

$$(\mathbf{A}_q \mathbf{A}_q')^{-1} \mathbf{A}_q \nabla f_{P_k} = \boldsymbol{\mu} \geq \mathbf{0} \tag{4.15}$$

The conditions (4.14) and (4.15) insure that ∇f_{P_k} can be written as a linear combination of the normal vectors $\mathbf{a}_1, \mathbf{a}_2, \ldots, \mathbf{a}_q$ to the hyperplanes $G_1, G_2, \ldots, G_k$. In this instance, it will be impossible to move from P_k in any feasible direction and increase $z = f(\mathbf{x})$. If $q = 0$, then (4.14) becomes $\mathbf{D}_0 \nabla f_{P_k} = \nabla f_{P_k} = \mathbf{0}$, which provides a condition for an interior maximizing point. If P_k does not satisfy (4.14) and (4.15), it may be possible to increase z by moving feasibly from P_k to P_{k+1}. There are two cases where this situation may arise.

CASE 1: $\mathbf{D}_q \nabla f_{P_k} \neq \mathbf{0}$

In this case, P_k is not a vertex and the manifold in which it resides must therefore be at least of dimension one (a line). Now project

∇f_{P_k} onto this manifold by using (4.11); i.e., $\nabla h_{P_k} = \mathbf{D}_q \nabla f_{P_k}$. The projection vector ∇h_{P_k} will lie in the manifold and therefore so will $P^* = P_k + \lambda \nabla h_{P_k}$, where λ is a scalar. The problem is to determine how large λ can be and still keep the new point P^* feasible. Since P_k is feasible, $g_i(P_k) < 0$ for $i = q + 1, \ldots, m$.

To determine λ, it is necessary to calculate the distance from P^* to each of the remaining $m - q$ constraints. These distances, denoted by λ_i, are computed from the equation

$$\lambda_i = -g_i(P_k)/\mathbf{a}_i' \nabla h_{P_k} \qquad i = q + 1, \ldots, m \tag{4.16}$$

Furthermore, P^* will become infeasible if λ exceeds λ', where λ' is given by $\lambda' = \min\{\lambda_i/\lambda_i > 0 \text{ for } i = q + 1, \ldots, m\}$. Therefore, set $P^* = P_k + \lambda' \nabla h_{P_k}$. In (4.16) λ_i may be negative or infinite. If λ_i is negative, this implies that the ith constraint is in the opposite direction of ∇h_{P_k} and hence will not be encountered by moving in the direction of ∇h_{P_k}. If λ_i is infinite, ∇h_{P_k} will not initially intersect the ith constraint, since λ_i will not be minimal.

Two special cases of Case 1 must be considered. In moving from P_k to P^*, the largest functional value may be passed. Thus

1. If the maximum value of z on the ray from P_k to P^* occurs at P^*, then set $P_{k+1} = P^*$.
2. If $\nabla h_{P_k}' \nabla f_{P^*} < 0$, then z attains its maximum on the ray between P_k and P^*. The point P_{k+1} should then be chosen such that $\nabla h_{P_k}' \nabla f_{P_{k+1}} = 0$. The point P_{k+1} may be approximately determined in this subcase by linear interpolation: compute

$$\lambda'' = \lambda'(\nabla h_{P_k}' \nabla f_{P_k})/(\nabla h_{P_k}' \nabla f_{P_k} - \nabla h_{P_k}' \nabla f_{P^*}) \tag{4.17}$$

Then $P_{k+1} = P_k + \lambda'' \nabla h_{P_k}$.

CASE 2: $\mathbf{D}_q \nabla f_{P_k} = \mathbf{0}$, but $(\mathbf{A}_q \mathbf{A}_q')^{-1} \mathbf{A}_q \nabla f_{P_k} \ngeq \mathbf{0}$

Since $\mathbf{a}_1, \mathbf{a}_2, \ldots, \mathbf{a}_q$ are linearly independent, it is possible to write ∇f_{P_k} as a linear combination of these vectors; $\nabla f_{P_k} = \sum_{i=1}^{q} \mu_i \mathbf{a}_i = \mathbf{A}_q' \boldsymbol{\mu}$, where $\boldsymbol{\mu}$ is the $q \times 1$ column vector of constants defined in (4.15). Hence Case 2 may be rephrased as $\mathbf{D}_q \nabla f_{P_k} = \mathbf{0}$ and $\boldsymbol{\mu} \ngeq \mathbf{0}$, so that at least one component of $\boldsymbol{\mu}$ is negative. In this case, choose one of the indices for which $\mu_i < 0$ and drop the hyperplane G_i from $\mathbf{A}_q$. Suppose for simplicity one component is negative—the qth; drop $\mathbf{a}_q'$ from $\mathbf{A}_q$

forming $\mathbf{A}_{q-1}$ and proceed as in Case 1, forming $\nabla h_{P_k} = \mathbf{D}_{q-1}\nabla f_{P_k}$, and so on.

Now the complete algorithm can be delineated. It has the following steps:

1. Compute ∇f_{P_k} and form $\mathbf{D}_q$. If P_k is a new boundary point, it may be necessary to add at least one new row to $\mathbf{A}_q$ in forming $\mathbf{D}_q$. Compute $\mathbf{D}_q\nabla f_k$. If $\mathbf{D}_q\nabla f_k = \mathbf{0}$ and $\boldsymbol{\mu} \geq \mathbf{0}$, then P_k is the solution point.
2. If $\mathbf{D}_q\nabla f_{P_k} \neq \mathbf{0}$, compute $\nabla h_{P_k} = \mathbf{D}_q\nabla f_{P_k}$ and go to step (4).
3. If $\mathbf{D}_q\nabla f_{P_k} = \mathbf{0}$ and $\mu_i < 0$ for at least one i, $i = q+1,\ldots,m$, then select a negative μ_i and drop G_i. Compute $\nabla h_{P_k} = \mathbf{D}_{q-1}\nabla f_{P_k}$.
4. Determine λ' using ∇h_{P_k} from either (2) or (3) and set $P^* = P_k + \lambda'\nabla h_{P_k}$.
5. Compute ∇f_{P^*} and $\nabla h'_{P_k}\nabla f_{P^*}$. If $\nabla h'_{P_k}\nabla f_{P^*} \geq 0$, set $P_{k+1} = P^*$ and go to (1).
6. If $\nabla h'_{P_k}\nabla f_{P^*} < 0$, set $P_{k+1} = P_k + \lambda''\nabla h_{P_k}$, where λ'' is given by (4.17) and go to (1).

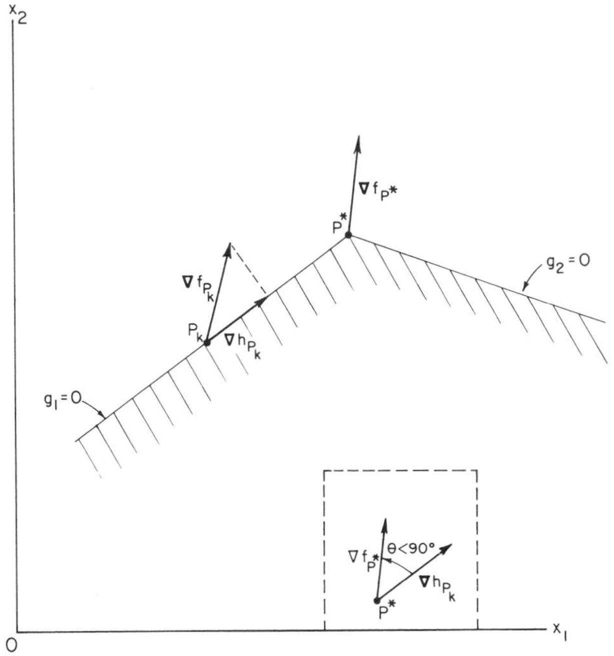

Fig. 4.4. Illustration of the case where $P_{k+1} = P^*$.

It is illuminating to see graphically the situations involved in steps (5) and (6). In Figure 4.4, the angle between ∇h_{P_k} and ∇f_{P*}, shown in the inset, is less than 90° and since $\nabla h' \nabla f = a \cos \theta$ where a is a constant and θ is the angle between ∇h and ∇f, $\nabla h' \nabla f \geq 0$. (From vector analysis, the inner product of two vectors $\mathbf{x}'\mathbf{y}$ is equal to a constant times the cosine of the angle between $\mathbf{x}$ and $\mathbf{y}$.) In this case, $P_{k+1} = P*$

In Figure 4.5, the angle between ∇h_{P_k} and ∇f_{P*} (see inset) will exceed 90°, so that $\nabla h'_{P_k} \nabla f_{P*} < 0$. In this case, $P*$ will be between P_k and P_{k+1} on the ray (line in this case) connecting them.

One additional rule may be added to the above six steps. If $\mathbf{A}_q \nabla f_{P_k} \leq \mathbf{0}$, set $\nabla h_{P_k} = \nabla f_{P_k}$ and proceed to step (4). In this instance, the gradient vector ∇f_{P_k} points to the interior of the constraint region, and there is no need to project ∇f on any constraints.

To further describe the algorithm, it is worthwhile to consider geometric illustrations that depict some of the cases and subcases discussed above. In Figure 4.6 moving in the direction of ∇f_{P_k} violates constraint g_1. This is Case 1 with $n = 2$ and $q = 1$, and ∇f will be projected onto the $n - q$ manifold (line in this instance); λ' would be chosen so that g_2 is not violated.

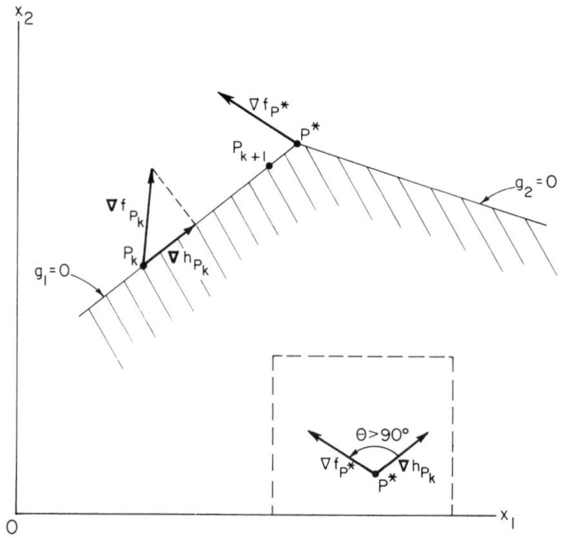

Fig. 4.5. Illustration of the case where P_{k+1} lies between P_k and $P*$.

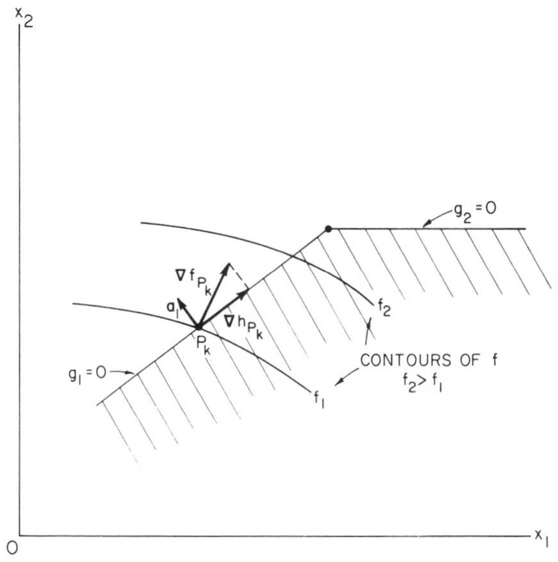

Fig. 4.6. Illustration of Case 1 with $n = 2$ and $q = 1$.

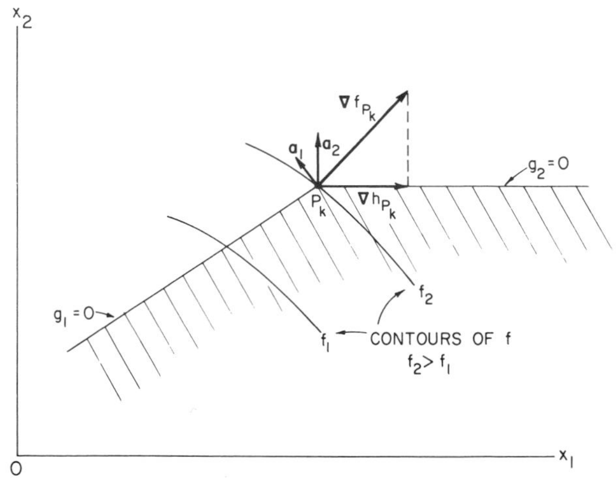

Fig. 4.7. Illustration of Case 2 with $n = 2$ and $q = 2$.

In Figure 4.7 Case 2 is illustrated; P_k is at a vertex with $n = 2$ and $q = 2$ so that $\mathbf{D}_2 \nabla f_{P_k} = \mathbf{0}$, but $\mu_1 < 0$ so that g_1 would be eliminated, q would be reduced to one, and the algorithm would proceed with the third step.

In Figure 4.8 at P_k, $\mathbf{D}_k \nabla f_{P_k} = \mathbf{0}$ and $\mu_1, \mu_2 \geq 0$. Therefore,

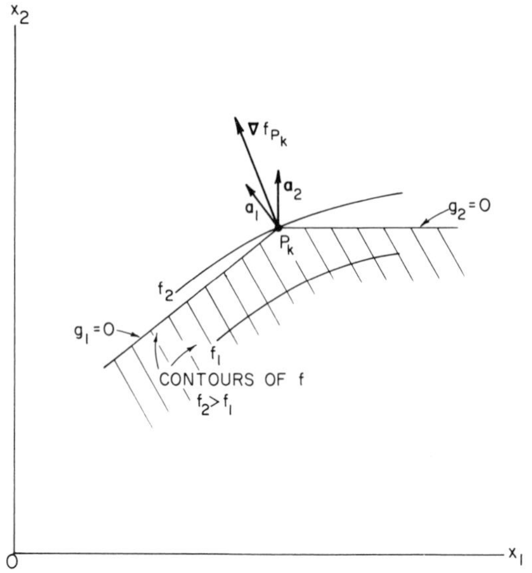

Fig. 4.8. Illustration of a solution point.

P_k is a solution point. Notice that ∇f_{P_k} lies between $\mathbf{a}_1$ and $\mathbf{a}_2$. In general, the condition that $\boldsymbol{\mu} \geq \mathbf{0}$ requires ∇f to lie in a cone spanned by $\mathbf{a}_1, \mathbf{a}_2, \ldots, \mathbf{a}_q$.

Figure 4.9 illustrates the case where the additional rule comes

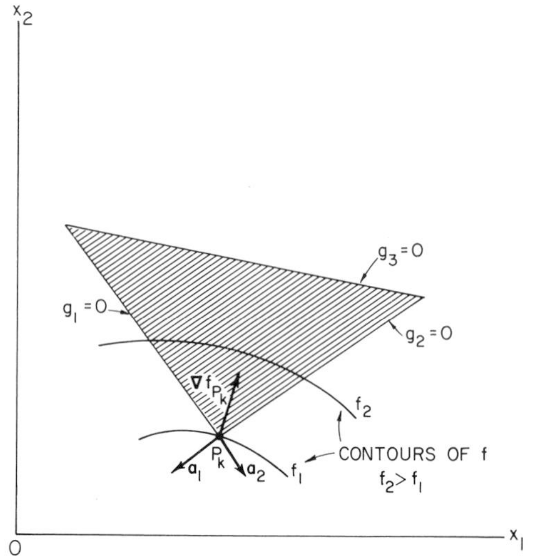

Fig. 4.9. Illustration of a case where the gradient vector points in a feasible direction.

in handy. Here $\mathbf{D}_2 \nabla f_{P_k} = \mathbf{0}$, but μ_1 and μ_2 are both negative. The quickest route to the maximizing point is in the direction of ∇f_{P_k}; therefore, ∇h_{P_k} is set equal to ∇f_{P_k}.

Example 4.1

Consider the following:

$$\begin{aligned}
\text{maximize} \quad & z = f(x_1, x_2) = -6(x_1 - 6)^2 - 4(x_2 - 4)^2 \\
\text{subject to} \quad & x_1 + x_2 \geq 2 \\
& x_1 - x_2 \geq -2 \\
& x_1 + x_2 \leq 6 \\
& x_1 - 3x_2 \leq 2 \\
& x_1, x_2 \geq 0
\end{aligned}$$

The problem must first be put into the form described by (4.5) and (4.6):

$$\begin{aligned}
\text{maximize} \quad & z = f(x_1, x_2) = -6(x_1 - 6)^2 - 4(x_2 - 4)^2 \\
\text{subject to} \quad & -x_1 - x_2 + 2 \leq 0 && (4.18) \\
& -x_1 + x_2 - 2 \leq 0 && (4.19) \\
& x_1 + x_2 - 6 \leq 0 && (4.20) \\
& x_1 - 3x_2 - 2 \leq 0 && (4.21) \\
& -x_1 \leq 0, -x_2 \leq 0 && (4.22)
\end{aligned}$$

The problem is represented graphically in Figure 4.10.

Suppose (0, 2) is chosen as the starting point P_0. Since

$$\nabla f = \begin{bmatrix} -12(x_1 - 6) \\ -8(x_2 - 4) \end{bmatrix}$$

it follows that

$$\nabla f_{P_0} = \begin{bmatrix} 72 \\ 16 \end{bmatrix}$$

At the point P_0, the constraints (4.18) and (4.19) are active. (The nonnegativity constraints (4.22) can be ignored in this problem.) Hence $\mathbf{a}_1' = [-1, -1]$ and $\mathbf{a}_2' = [-1, 1]$ so that

$$\mathbf{A}_2 = \begin{bmatrix} -1 & -1 \\ -1 & 1 \end{bmatrix}$$

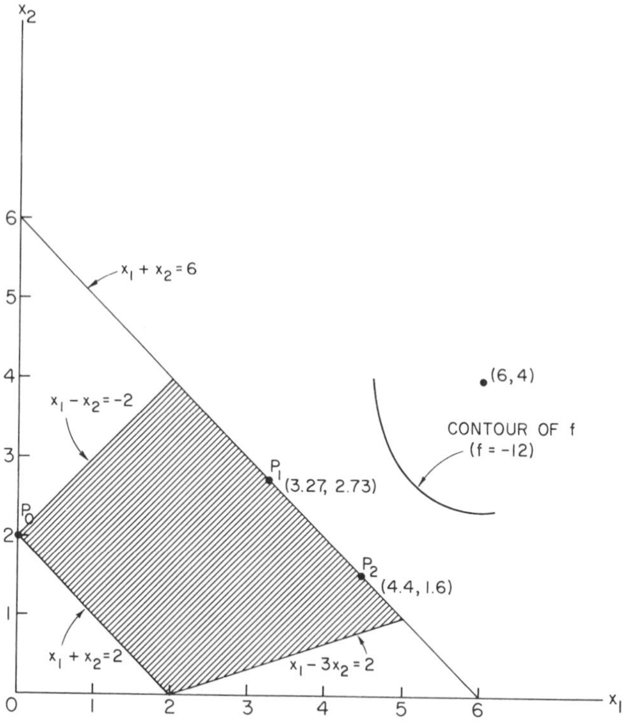

Fig. 4.10. Graphic solution of Example 4.1.

Now

$$\mathbf{A}_2\mathbf{A}_2' = \begin{bmatrix} 2 & 0 \\ 0 & 2 \end{bmatrix}$$

$$(\mathbf{A}_2\mathbf{A}_2')^{-1} = \begin{bmatrix} 1/2 & 0 \\ 0 & 1/2 \end{bmatrix}$$

$$(\mathbf{A}_2\mathbf{A}_2')^{-1}\mathbf{A}_2 = \begin{bmatrix} -1/2 & -1/2 \\ -1/2 & +1/2 \end{bmatrix}$$

From (4.15)

$$\boldsymbol{\mu} = \begin{bmatrix} -1/2 & -1/2 \\ -1/2 & +1/2 \end{bmatrix} \begin{bmatrix} 72 \\ 16 \end{bmatrix} = \begin{bmatrix} -44 \\ -28 \end{bmatrix}$$

Since $q = 2 = n$, $\mathbf{D}_2 \nabla f_{P_0} = \mathbf{0}$; but μ_1 and μ_2 are both negative, so that the process must continue. Now use the additional rule since ∇f_{P_0} points to the interior of the constraint region. Thus

$$\nabla h_{P_0} = \nabla f_{P_0} = \begin{bmatrix} 72 \\ 16 \end{bmatrix}$$

The next step is to determine the distances to the remaining two constraints from P_0 in the direction of ∇h_{P_0}. Although it is not necessary, it is desirable to represent this direction by a unit vector (length one unit). By doing this, the distances λ_3 and λ_4 will be of the proper magnitude. The length of ∇h_{P_0}, denoted by $|\nabla h_{P_0}|$, is computed by taking the square root of the sum of squares of its components. Thus

$$|\nabla h_{P_0}| = [(72)^2 + (16)^2]^{1/2} = 73.756$$

A unit vector in the direction of ∇h_{P_0}, denoted by $\overrightarrow{\nabla h}_{P_0}$ is therefore given by

$$\overrightarrow{\nabla h}_{P_0} = \frac{\nabla h_{P_0}}{|\nabla h_{P_0}|} = \begin{bmatrix} 0.976 \\ 0.217 \end{bmatrix}$$

The distances to (4.20) and (4.21) from P_0 are

$$\lambda_3 = -\quad [1, 1]\begin{bmatrix} 0 \\ 2 \end{bmatrix} - 6 \Big/ [1, 1]\begin{bmatrix} 0.976 \\ 0.217 \end{bmatrix} = 3.35$$

$$\lambda_4 = -\quad [1, -3]\begin{bmatrix} 0 \\ 2 \end{bmatrix} - 2 \Big/ [1, -3]\begin{bmatrix} 0.976 \\ 0.217 \end{bmatrix} = 24.58$$

Since the closest constraint is (4.20),

$$P_1 = \begin{bmatrix} 0 \\ 2 \end{bmatrix} + 3.35 \begin{bmatrix} 0.976 \\ 0.217 \end{bmatrix} = \begin{bmatrix} 3.27 \\ 2.73 \end{bmatrix}$$

The point P_1 lies on the line $x_1 + x_2 = 6$. Now

$$\nabla f_{P_1} = \begin{bmatrix} 32.76 \\ 10.16 \end{bmatrix}$$

and since it points infeasibly, ∇f_{P_1} must be projected onto the line $x_1 + x_2 = 6$.

On this line $q = 1$ and $\mathbf{a}_1' = [1, 1]$ so that $\mathbf{A}_1 = [1, 1]$, and

$$\mathbf{A}_1'(\mathbf{A}_1\mathbf{A}_1')^{-1}\mathbf{A}_1 = \begin{bmatrix} 1/2 & 1/2 \\ 1/2 & 1/2 \end{bmatrix}$$

Therefore, the projection matrix is computed from

$$\mathbf{D}_1 = \begin{bmatrix} 1 & 0 \\ 0 & 1 \end{bmatrix} - \begin{bmatrix} 1/2 & 1/2 \\ 1/2 & 1/2 \end{bmatrix} = \begin{bmatrix} 1/2 & -1/2 \\ -1/2 & 1/2 \end{bmatrix}$$

The projection vector ∇h_{P_1} is given by

$$\nabla h_{P_1} = \mathbf{D}_1 \overrightarrow{\nabla f}_{P_1} = \begin{bmatrix} 1/2 & -1/2 \\ -1/2 & 1/2 \end{bmatrix} \begin{bmatrix} 0.955 \\ 0.297 \end{bmatrix} = \begin{bmatrix} 0.329 \\ -0.329 \end{bmatrix}$$

where

$$\overrightarrow{\nabla f}_{P_1} = \frac{\nabla f_{P_1}}{|\nabla f_P|} = \frac{1}{34.27} \begin{bmatrix} 32.76 \\ 10.16 \end{bmatrix} = \begin{bmatrix} 0.955 \\ 0.297 \end{bmatrix}$$

Now

$$\overrightarrow{\nabla h}_{P_1} = \frac{\nabla h_{P_1}}{|\nabla h_{P_1}|} = \begin{bmatrix} 0.707 \\ -0.707 \end{bmatrix}$$

Since $\nabla h_{P_1}' \nabla f_{P_1} \neq 0$, P_1 is not a terminal point, and z may be increased by moving in the direction of ∇h_{P_1}. The distance from P_1 to the constraint (4.21) is now computed by

$$\lambda_4 = -\left([1, -3]\begin{bmatrix} 3.27 \\ 2.73 \end{bmatrix} - 2\right) \Big/ [1, -3]\begin{bmatrix} 0.707 \\ -0.707 \end{bmatrix} = 2.44$$

Thus

$$P^* = \begin{bmatrix} 3.27 \\ 2.73 \end{bmatrix} + 2.44 \begin{bmatrix} 0.707 \\ -0.707 \end{bmatrix} = \begin{bmatrix} 5 \\ 1 \end{bmatrix}$$

One of the two subcases to Case 1 has occurred; to find out which, it is necessary to compute

$$\overrightarrow{\nabla h'_{P_1}}\,\overrightarrow{\nabla f_{P*}} = [0.707, -0.707]\begin{bmatrix}0.447\\0.894\end{bmatrix} = -0.316$$

where $\overrightarrow{\nabla f_{P*}} = \nabla f_{P*}/|\nabla f_{P*}|$ and

$$\nabla f_{P*} = \begin{bmatrix}12\\24\end{bmatrix}$$

Since $\nabla h'_{P_1}\nabla f_{P*}$ is negative, the maximizing point occurs along the line $x_1 + x_2 = 6$ between P_1 and P^*. To find P_2, solve the following set of equations for x_1 and x_2.

$$\nabla h'_P\ \nabla f = [0.707, -0.707]\begin{bmatrix}-12(x_1 - 6)\\-8(x_2 - 4)\end{bmatrix} = 0$$

$$x_1 + x_2 = 6 \tag{4.23}$$

The solution to (4.23) is $x_1 = 4.4$ and $x_2 = 1.6$. Therefore, $P_2 = (4.4, 1.6)$. To check for the termination of the process, compute

$$\nabla f_{P_2} = \begin{bmatrix}19.2\\19.2\end{bmatrix}$$

$$\mathbf{D}_1\nabla f_{P_2} = \begin{bmatrix}1/2 & -1/2\\-1/2 & 1/2\end{bmatrix}\begin{bmatrix}19.2\\19.2\end{bmatrix} = \begin{bmatrix}0\\0\end{bmatrix}$$

$$(\mathbf{A}_I\mathbf{A}_1')^{-1}\mathbf{A}_1\nabla f_{P_2} = [1/2, 1/2]\begin{bmatrix}19.2\\19.2\end{bmatrix} = 19.2 \geq 0$$

Therefore, $\mathbf{x}_2^{*\prime} = [4.4, 1.6]$ is the constrained maximizing point of z and $f(\mathbf{x}^*) = -38.4$.

Since this is a convex programming problem, the solution in Example 4.1 is the constrained global maximum of $z = f(\mathbf{x})$. In general, unless the programming problem is convex, Rosen's gradient projection algorithm cannot guarantee locating the global optimum. This can very easily be seen from the situation portrayed in Figure 4.11. Here, geometrically, the function has two mountains with peaks at $\mathbf{x}_1^*$ and $\mathbf{x}_2^*$ and a valley between these points. Let L denote the curve that

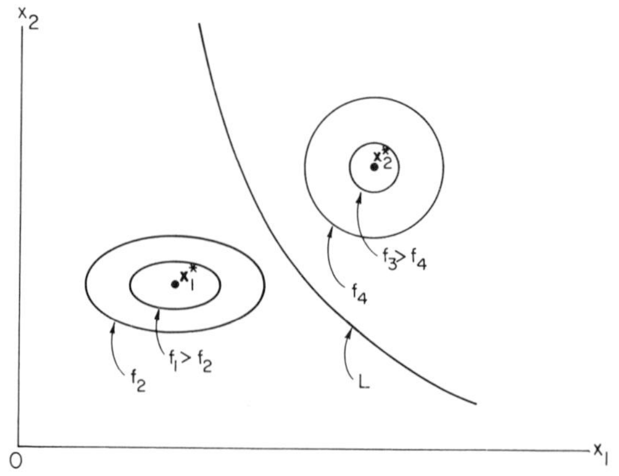

Fig. 4.11. Illustration of a problem with two local maxima.

represents the bottom of this valley. If P_0 is chosen to the left of L, the steepest ascent algorithm will climb the mountain from P_0 to the peak $\mathbf{x}_1^*$, whereas if P_0 is chosen to the right of L, the algorithm will climb the other mountain up to $\mathbf{x}_2^*$. If $f(\mathbf{x}_2^*) < f(\mathbf{x}_1^*)$, then in the latter case above a local maximum is located, and the global maximum, which occurs at $\mathbf{x}_1^*$, is missed.

In summary, if the programming problem is nonconvex, locating the global optimum depends upon the probability of the starting point P_0 being on or at the foot of the highest mountain. To increase the likelihood of starting at a point with this property, a common practice in applying an ascent algorithm to a nonconvex problem is to choose a set of starting points and apply the solution process to each of them. This method will be further discussed in Section 4.6.

There is one additional problem in applying an ascent algorithm to a mathematical programming problem. In (4.5) and (4.6), notice that a requirement on the constraints is that they must all be linear functions of $x_1, x_2, \ldots, x_n$. The reason is readily apparent. How can we project the gradient vector ∇f on a curvilinear surface and how can we computationally move along such a surface? Rosen [1961] provides an answer in a paper on the gradient projection method but it is quite complicated. Basically, the method requires the construction of a tangent hyperplane at the point P_0 on the curvilinear surface. The gradient vector ∇f_{P_0} is projected on this hyperplane, and the movement to a new point is made on it. Figure 4.12 indicates the nature of the process.

Since ∇h_{P_0} on the hyperplane generally will point outside the constraint region, constant correction is necessary to bring the point

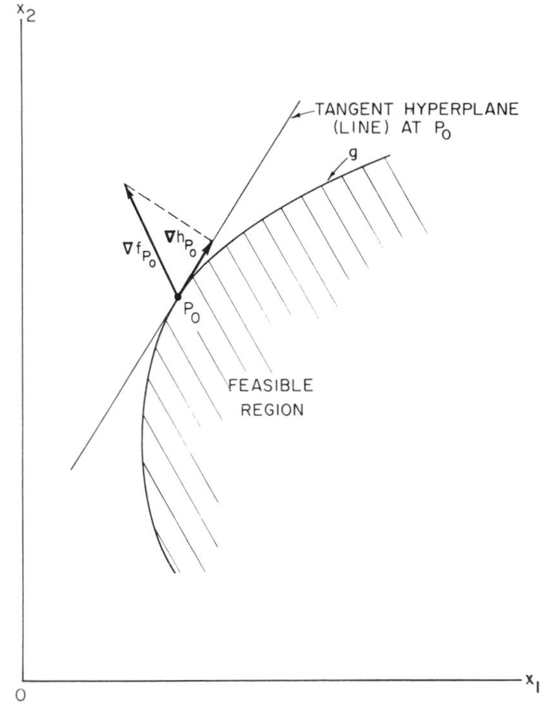

Fig. 4.12. Projection method for a curvilinear constraint.

back on the boundary of the region. This is accomplished by forming a sequence of tangent hyperplanes on g in the direction of steepest ascent. Since this process is quite involved, the addition of nonlinear constraints will usually increase the computation time considerably when the problem is solved by this method. Most computer codes written for the gradient algorithm include provisions for handling nonlinear constraints based upon Rosen's [1961] procedure.

4.3 SEPARABLE PROGRAMMING

In this section an approximate solution method known as separable programming is presented for the problem

$$\text{maximize} \quad z = f(\mathbf{x}) = \textstyle\sum_{j=1}^{n} f_j(x_j) \tag{4.24}$$

$$\text{subject to} \quad g_i = \textstyle\sum_{j=1}^{n} g_{ij}(x_j)\{\leq \; = \; \geq\} b_i \qquad i = 1, 2, \ldots, m \tag{4.25}$$

$$x_j \geq 0 \qquad j = 1, 2, \ldots, n \tag{4.26}$$

Notice that this model is not as general as the one described by (3.1) and (3.4), since it is required that the $m + 1$ functions z, $g_1, g_2, \ldots, g_m$ are separable in the n variables $x_1, x_2, \ldots, x_n$. If the problem initially does not satisfy the separability requirement, it may be possible, through the transformation of variables, to recast the problem so that it does; but this shall be dealt with after discussing the approximate solution to (4.24)–(4.26).

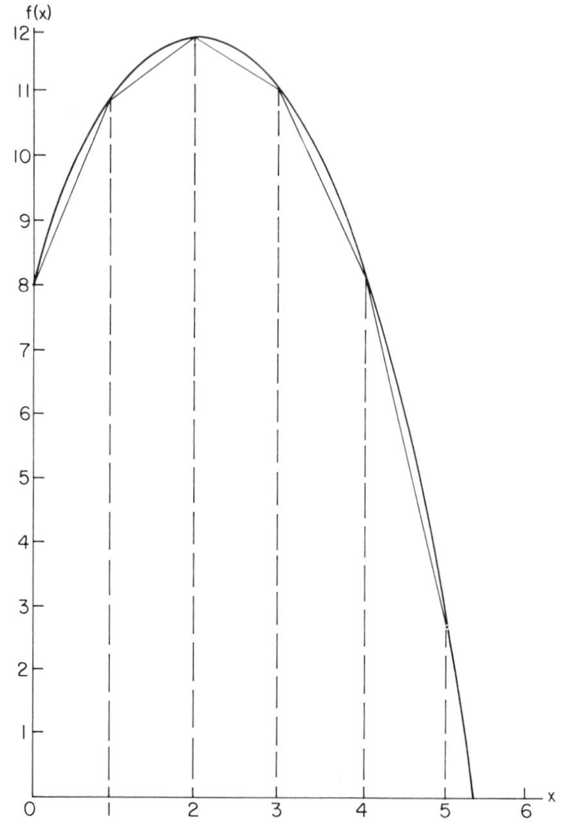

Fig. 4.13. Illustration of the piecewise linear approximation of a curvilinear function.

To introduce the basic mechanism upon which the approximation procedure is based, consider the univariate function $f(x) = 8 + 4x - x^2$, which is plotted in Figure 4.13. By connecting lines at points on z at equal intervals of one unit of length on the x axis beginning at the origin, it is possible to approximate z reasonably well in a piecewise linear fashion. Denote the interval demarking points on the x axis by x_k, $k = 0, 1, \ldots, 5$, and $f(x_k)$ by f_k. It is now possible to write down the approximating piecewise linear function, denoted by $f(\hat{x})$, for each of the five straight lines over the range $0 \leq x \leq 5$ in Figure 4.13:

$$\hat{f}(x) = f_k + \frac{f_{k+1} - f_k}{x_{k+1} - x_k}(x - x_k) \qquad k = 0, 1, \ldots, 4 \tag{4.27}$$

For example, the line segment connecting the point (0, 8) with (1, 11) is given by

$$\hat{f}(x)_1 = 8 + [(11 - 8)/(1 - 0)](x - 0) = 3x + 8$$

It is possible to use $\hat{f}(x)$ to approximate $f(x)$ at some point x, say between the kth and kth + 1 points on the x axis. By defining a constant λ that must satisfy $0 \leq \lambda \leq 1$,

$$x = \lambda x_{k+1} + (1 - \lambda)x_k \tag{4.28}$$

Solving (4.28) for $x - x_k$ gives $x - x_k = \lambda(x_{k+1} - x_k)$; and upon substituting this into (4.27),

$$\hat{f}(x) = f_k + \frac{f_{k+1} - f_k}{x_{k+1} - x_k}\lambda(x_{k+1} - x_k) = f_k + \lambda(f_{k+1} - f_k)$$
$$= \lambda f_{k+1} + (1 - \lambda)f_k$$

In (4.28) let $\lambda = \lambda_{k+1}$ and $(1 - \lambda) = \lambda_k$, thus $x = \lambda_k x_k + \lambda_{k+1}x_{k+1}$.

Hence when x is between x_k and x_{k+1}, it is possible to determine unique values for λ_k and λ_{k+1} such that

$$x = \lambda_k x_k + \lambda_{k+1}x_{k+1} \tag{4.29}$$

$$\hat{f}(x) = \lambda_k f_k + \lambda_{k+1}f_{k+1} \tag{4.30}$$

where

$$\lambda_k + \lambda_{k+1} = 1 \qquad \lambda_k, \lambda_{k+1} \geq 0 \tag{4.31}$$

As an illustration of using the above formulas, suppose that it is desired in the example to provide an approximate value of $f(x)$ when $x = 5/4$. From (4.29) and (4.31) it follows that $\lambda_1 = 3/4$ and $\lambda_2 = 1/4$ since 5/4 lies between $x_1 = 1$ and $x_2 = 2$. Since $f_1 = 11$ and $f_2 = 12$, it follows from (4.30) that $\hat{f}(5/4) = (3/4)(11) + (1/4)(12) = 11.25$ where $f(5/4) = 11.44$.

The above formulas can be generalized by allowing r segments (not necessarily of equal length) over the range of x, specified by $0 \leq x \leq b$ where b is an upper bound on x: for any x, $0 \leq x \leq b$, let

$$x = \sum_{k=0}^{r} \lambda_k x_k \tag{4.32}$$

$$\hat{f}(x) = \sum_{k=0}^{r} \lambda_k f_k \tag{4.33}$$

$$\sum_{k=0}^{r} \lambda_k = 1 \tag{4.34}$$

$$\lambda_k \geq 0 \qquad k = 0, 1, \ldots, r \tag{4.35}$$

with the additional condition that no more than two of the λ_k shall be positive, and if two are positive, they must be adjacent.

The general piecewise linear approximation technique described above for a univariate function may be extended in a straightforward manner to functions of n variables if the functions are separable. For example, suppose

$$f(x_1, x_2) = (x_1 - 2)^2 + (x_2 - 4)^2 \tag{4.36}$$

Since $f(x_1, x_2) = f_1(x_1) + f_2(x_2)$ where $f_1(x_1) = (x_1 - 2)^2$ and $f_2(x_2) = (x_2 - 4)^2$, the results given in (4.32)–(4.35) may be applied to each of the functions $f_1(x_1)$ and $f_2(x_2)$, then

$$f_1(x_1) \doteq \hat{f}_1(x_1) = \sum_{k=0}^{r_1} \lambda_{k1} f_{k1}$$

$$f_2(x_2) \doteq \hat{f}_2(x_2) = \sum_{k=0}^{r_2} \lambda_{k2} f_{k2}$$

where $x_1 = \sum_{k=0}^{r_1} \lambda_{k1} x_{k1}$ and $x_2 = \sum_{k=0}^{r_2} \lambda_{k2} x_{k2}$. The function $f(x_1, x_2)$ in (4.36) may now be approximated by $\hat{f}_1(x_1) + \hat{f}_2(x_2)$. Notice that it is possible to use a different number of intervals on the x_1 and x_2 axes; r_1 might not equal r_2.

The general piecewise linear approximation rules may now be

stated for a separable n-variate function $f(x_1, x_2, \ldots, x_n)$:

$$x_j = \sum_{k=0}^{r_j} \lambda_{kj} x_{kj} \qquad j = 1, 2, \ldots, n \tag{4.37}$$

$$\hat{f}_j(x_j) = \sum_{k=0}^{r_j} \lambda_{kj} f_{kj} \qquad j = 1, 2, \ldots, n \tag{4.38}$$

$$\sum_{k=0}^{r_j} \lambda_{kj} = 1 \qquad j = 1, 2, \ldots, n \tag{4.39}$$

$$\lambda_{kj} \geq 0 \qquad j = 1, 2, \ldots, n; k = 0, 1, \ldots, r_j \tag{4.40}$$

and for any given j no more than two λ_{kj} may be positive, and if exactly two are positive, they must be adjacent. In (4.38) f_{kj} denotes $f_j(x_k)$.

These results may be applied directly to the functions f_j and g_{ij} in (4.24) and (4.25), generating the following approximating functions:

$$\hat{f}_j(x_j) = \sum_{k=0}^{r_j} \lambda_{kj} f_{kj} \qquad j = 1, 2, \ldots, n$$

$$\hat{g}_{ij}(x_j) = \sum_{k=0}^{r_j} \lambda_{kj} g_{kij} \qquad i = 1, 2, \ldots, m; j = 1, 2, \ldots, n$$

where (4.39), (4.40), and the adjacency condition are satisfied. The original problem in (4.24)–(4.26) may now be replaced with the following approximating problem:

$$\text{maximize} \quad z = \hat{f}(\mathbf{x}) = \sum_{j=1}^{n} \sum_{k=0}^{r_j} f_{kj} \lambda_{kj} \tag{4.41}$$

$$\text{subject to} \quad \sum_{j=1}^{n} \sum_{k=0}^{r_j} g_{kij} \lambda_{kj} \{\leq \; = \; \geq\} b_i \qquad i = 1, 2, \ldots, m \tag{4.42}$$

$$\sum_{k=0}^{r_j} \lambda_{kj} = 1 \qquad j = 1, 2, \ldots, n \tag{4.43}$$

$$\lambda_{kj} \geq 0 \qquad j = 1, 2, \ldots, n; \quad k = 0, 1, \ldots, r_j \tag{4.44}$$

and the condition that for any given j, no more than two λ_{kj} may be positive; if exactly two are positive, they must be adjacent.

If it were not for this last condition, (4.41)–(4.44) would delineate a linear programming problem in the λ_{kj}. However, the simplex algorithm may be applied if the basis entry is restricted so that no more than two positive λ_{kj} are allowed for any j; if exactly two λ_{kj} occur for some j, they must be adjacent. Once a solution in the λ_{kj} has been determined in this fashion, (4.37) may be used to convert them to x_j values.

Before proceeding with an example problem, a few observations are in order. The primary requirement of this approximating procedure is separability of the $m + 1$ functions $z = f(\mathbf{x}), g_1, g_2, \ldots, g_m$. It is frequently possible to satisfy this condition through transformation if the original problem does not have this property. Another requirement is that $0 \leq x_j \leq b_j$, $j = 1, 2, \ldots, n$; i.e., the n variables have upper bounds b_j and the same lower bound of zero. If the lower bound requirement is not satisfied, it is usually possible to translate the x_j so that the new variables have zero as a lower bound. Notice also that it is possible to select a different number of grid points for the n variables, and for any particular variable the line segments need not be of equal length. This provides considerable flexibility in representing the functions by piecewise linear approximation. For example, in Figure 4.13 it would clearly be clever to take a denser grid of points on the x axis in and around $x = 2$ than in the region $3 \leq x \leq 5$. In general, the denser the grids on the $x_1, x_2, \ldots, x_n$ axes, the better the approximation procedure will be. However, by increasing the number of grid points $[n(r_j + 1)]$, the size of the linear programming problem in (4.41)–(4.44) also increases, since r_j dictates the number of λ_{kj} for the jth variable. It is very easy to select so many points that the linear programming problem becomes very unattractive to solve due to its size alone.

Example 4.2

Consider the following:

$$\begin{aligned} \text{maximize} \quad & z = f(x_1, x_2) = 3x_1 + 2x_2 \\ \text{subject to} \quad & g(x_1, x_2) = 4x_1^2 + x_2^2 \leq 16 \\ & x_1, x_2 \geq 0 \end{aligned}$$

The problem is illustrated in Figure 4.14. From this figure, it is apparent that by selecting $0 \leq x_1 \leq 2$ and $0 \leq x_2 \leq 4$, the feasible region will be covered. Suppose now that $r_1 = 2$ and

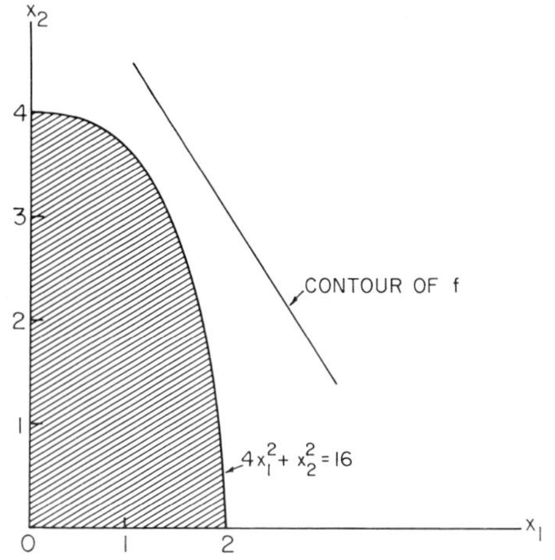

Fig. 4.14. Illustration of Example 4.2.

$r_2 = 4$ are selected. Table 4.1 gives the appropriate values of $x_{k1}, x_{k2}, f_{k1}, f_{k2}, g_{k1}$, and g_{k2}.

Table 4.1. Approximation Values for x_{kj}, f_{kj}, and g_{kj}

k	x_{k1}	x_{k2}	f_{k1}	f_{k2}	g_{k1}	g_{k2}
0	0	0	0	0	0	0
1	1	1	3	2	4	1
2	2	2	6	4	16	4
3		3		6		9
4		4		8		16

The functions f and g may now be approximated by

$$\begin{aligned} \hat{f} &= 3\lambda_{11} + 6\lambda_{21} + 2\lambda_{12} + 4\lambda_{22} + 6\lambda_{32} + 8\lambda_{42} \\ \hat{g} &= 4\lambda_{11} + 16\lambda_{21} + \lambda_{12} + 4\lambda_{22} + 9\lambda_{32} + 16\lambda_{42} \end{aligned} \quad (4.45)$$

The approximating linear programming problem becomes

$$\begin{aligned} \text{maximize} \quad & z = \hat{f} \\ \text{subject to} \quad & \hat{g} \leq 16 \\ & \lambda_{01} + \lambda_{11} + \lambda_{21} = 1 \\ & \lambda_{02} + \lambda_{12} + \lambda_{22} + \lambda_{32} + \lambda_{42} = 1 \end{aligned} \quad (4.46)$$

and therefore depends on three constraints and eight variables.

The simplex algorithm may be applied to solve this problem if the basis entry restriction discussed above is satisfied. Prior to expressing the first tableau (Table 4.2), a slack variable λ must be added to (4.46), giving $\hat{g} + \lambda = 16$.

Table 4.2. First Tableau

Cost coefficient:	0	3	6	0	2	4	6	8	0	Basic	
Variable:	λ_{01}	λ_{11}	λ_{21}	λ_{02}	λ_{12}	λ_{22}	λ_{32}	λ_{42}	λ	variable	Index
	0	4	16	0	1	4	9	16	1	λ	16
	1	1	1	0	0	0	0	0	0	λ_{01}	1
	0	0	0	1	1	1	1	[1]	0	λ_{02}	1
$c_j - z_j$:	0	3	6	0	2	4	6	(8)	0		

From the simplex criteria for basis entry the variable λ_{42} is selected to enter. However, λ_{02} is already in the basis so that λ_{42} cannot enter (two lambdas in the set can be in a basis only if they are adjacent) unless λ_{02} departs. The simplex procedure for basis departure (16/16, 1/0, 1/1) indicates that either λ or λ_{02} can be eliminated. Therefore, λ_{42} is allowed to enter the basis and λ_{02} departs. The pivot element is indicated in Table 4.2 by the bracketed number.

The solution to the first tableau is $\lambda = 16$, $\lambda_{01} = 1$, and $\lambda_{02} = 1$, with λ_{42} entering the new basis and λ_{02} exiting. The second tableau is given in Table 4.3.

Table 4.3. Second Tableau

Cost coefficient:	0	3	6	0	2	4	6	8	0	Basic	
Variable:	λ_{01}	λ_{11}	λ_{21}	λ_{02}	λ_{12}	λ_{22}	λ_{32}	λ_{42}	λ	variable	Index
	0	[4]	16	−16	−15	−12	−7	0	1	λ	0
	1	1	1	0	0	0	0	0	0	λ_{01}	1
	0	0	0	1	1	1	1	1	0	λ_{42}	1
$c_j - z_j$:	0	3	(6)	−8	−6	−4	−2	0	0		

The solution to the second tableau is $\lambda = 0$, $\lambda_{01} = 1$, and $\lambda_{42} = 1$. The simplex basis entry criterion indicates that λ_{21} should enter next, but this is impossible for if it enters, λ would depart, leaving λ_{01} in the basis not adjacent to λ_{21}. The next best contender is λ_{11}. If λ_{11} enters, λ departs, leaving two adjacent lambdas in the first set in the basis, namely, λ_{01} and λ_{11}. Thus λ_{11}

enters the basis, and the slack variable λ departs. The third tableau is given in Table 4.4.

Table 4.4. Third Tableau

Cost coefficient:	0	3	6	0	2	4	6	8	0	Basic	
Variable:	λ_{01}	λ_{11}	λ_{21}	λ_{02}	λ_{12}	λ_{22}	λ_{32}	λ_{42}	λ	variable	Index
	0	1	4	−4	−15/4	−3	−7/4	0	1/4	λ_{11}	0
	1	0	−3	4	15/4	3	[7/4]	0	−1/4	λ_{01}	1
	0	0	0	1	1	1	1	1	0	λ_{42}	1
$c_j - z_j$:	0	0	−6	4	(21/4)	5	13/4	0	−3/4		

The solution to the third tableau is $\lambda_{11} = 0$, $\lambda_{01} = 1$, and $\lambda_{42} = 1$. According to the simplex criterion λ_{12} should enter the basis next; but this is not possible since λ_{01} would depart, leaving two nonadjacent lambdas λ_{12} and λ_{42} in the basis. Therefore, the only lambda that can enter is λ_{32}, with λ_{01} departing. The fourth tableau is given in Table 4.5.

Table 4.5. Fourth Tableau

Cost coefficient:	0	3	6	0	2	4	6	8	0	Basic	
Variable:	λ_{01}	λ_{11}	λ_{21}	λ_{02}	λ_{12}	λ_{22}	λ_{32}	λ_{42}	λ	variable	Index
	1	1	1	0	0	0	0	0	0	λ_{11}	1
	4/7	0	−12/7	16/7	15/7	12/7	1	0	−1/7	λ_{32}	4/7
	−4/7	0	12/7	−9/7	−8/7	−5/7	0	1	1/7	λ_{42}	3/7
$c_j - z_j$:	−13/7	0	−24/7	−24/7	−12/7	−4/7	0	0	−2/7		

The solution to the fourth tableau is $\lambda_{11} = 1$, $\lambda_{32} = 4/7$, and $\lambda_{42} = 3/7$. Upon checking the simplex criterion, it is found that this solution is optimal. From (4.37) it follows that

$$x_1 = (0)\,\lambda_{01} + (1)\,\lambda_{11} + (2)\,\lambda_{21} = 1$$

$$x_2 = (0)\,\lambda_{02} + (1)\,\lambda_{12} + (2)\,\lambda_{22} + (3)\,\lambda_{32} + (4)\,\lambda_{42} = 24/7$$

and from (4.45) $\hat{f} = (3)(1) + (6)(4/7) + (8)(3/7) = 9.86$. The exact answer to this problem is $f^* = 10$ and, although the grids on x_1 and x_2 were very rough, it shows that the solution to the approximating problem is quite good in this case.

It may be possible to separate functions by transformation of variables. As an example, suppose $f(x_1,x_2) = x_1x_2$. Since product terms are the most common causes of nonseparability, two methods will be explained for dealing with this case. First, two new variables could be defined as

$$y_1 = (1/2)(x_1 + x_2) \qquad y_2 = (1/2)(x_1 - x_2) \tag{4.47}$$

so that $x_1x_2 = y_1^2 - y_2^2$. Thus whenever x_1x_2 appears, it can be replaced by $y_1^2 - y_2^2$, and the two equalities in (4.47) would be added to the problem as constraints. Notice that this adds two variables to the original set of two so that two additional sets of lambdas would also have to be added to the approximating problem. A second way to deal with x_1x_2 would be to take its log, forming

$$\log y = \log x_1 + \log x_2 \tag{4.48}$$

where $y = x_1x_2$. Everywhere that x_1x_2 occurs it would be replaced by y, and (4.48) would be added to the constraint set. Although this method adds only one variable, it must be carefully applied, since difficulties arise if either x_1 or x_2 or both can equal zero.

Most functions may be separated by transformation with sufficient cleverness and perseverance, but the cost may be extremely high in terms of enlarging the problem. One of the drawbacks to implementing the separable programming technique is the size of the linear programming problem that results. In Example 4.2 a bivariate and single constraint problem is converted into a problem with three constraints and nine variables (including the slack variable).

There is one final consideration with regard to separable programming. If the problem in (4.24)–(4.26) is convex, the global solution to the approximating problem is obtained. Naturally, as the number of grid points increases, the closer this solution will come to the global solution of the original problem. Hadley [1964] notes and proves a most interesting fact: if the problem is convex and the approximating problem is solved as a straight linear programming problem ignoring the basis restrictions, the global optimum to the approximating problem will be obtained. Therefore, in Example 4.2 the problem could be solved more easily by ignoring the basis restriction; the claim is that the same solution would result. Verification of this is suggested in Problem 4.4.

If the programming problem is not convex, in general the separable programming technique will provide only a local maximum. Additionally, the approximation procedure may introduce extraneous local optima and also may give a solution that is not feasible in the original problem. In the latter case, the solution usually will be just outside the feasible region and can be projected onto it with little difficulty. Although these problems in applying the separable technique to the nonconvex programming problem seem to be severe, experience with this approximating method has been no worse than other methods in this case. Recall that the gradient projection method cannot guarantee global optimality in the nonconvex programming problem either.

4.4 PENALTY FUNCTION METHOD

One of the drawbacks in using the gradient projection method for solving a mathematical programming problem involving nonlinear constraints is the computational difficulty in projecting onto and moving along a curvilinear function. A method that obviates this difficulty has been proposed by Carroll [1961] and implemented computationally by Fiacco and McCormick [1964]. The method represents the original constrained problem by an unconstrained one in such a way that if an attempt is made to move a feasible point outside the constraint region, the move is penalized and the point is forced back to the interior of the feasible region. The manner in which this is accomplished will now be explained.

The nonlinear programming problem in (3.1)–(3.3) is written in the form

$$\begin{aligned} \text{minimize} \quad & h(\mathbf{x}) = -f(\mathbf{x}) \\ \text{subject to} \quad & g_i(\mathbf{x}) \geq 0 \qquad i = 1, 2, \ldots, s & (4.49) \\ & g_i(\mathbf{x}) = 0 \qquad i = s+1, \ldots, m & (4.50) \\ & \mathbf{x} \geq \mathbf{0} & (4.51) \end{aligned}$$

This is equivalent to the form of the problem in (3.1)–(3.3), since minimizing the negative of a function is equivalent to maximizing it, and all inequality constraints may be expressed as in (4.49) with no loss of generality. Consider now the following function:

$$R(\mathbf{x}, r_1) = h(\mathbf{x}) + r_1 \sum_{i=1}^{s} \frac{1}{g_i(\mathbf{x})} + r_1^{-1/2} \sum_{i=s+1}^{m} g_i^2(\mathbf{x}) \qquad (4.52)$$

where $r_1 > 0$. The Fiacco-McCormick algorithm may be stated in the following steps:

1. Select a starting point $\mathbf{x}^0$ such that $g_i(\mathbf{x}^0) > 0$ for $i = 1, 2, \ldots, s$.
2. Move from $\mathbf{x}^0$ to a new point $\mathbf{x}(r_1)$ that approximates the minimum of $R(\mathbf{x}, r_1)$ in (4.52) for $\mathbf{x}$ satisfying (4.50).

3. Form the new function

$$R(\mathbf{x}, r_2) = h(\mathbf{x}) + r_2 \sum_{i=1}^{s} \frac{1}{g_i(\mathbf{x})} + r_2^{-1/2} \sum_{i=s+1}^{m} g_i^2(\mathbf{x})$$

where $r_1 > r_2 > 0$.

4. Starting from the point $\mathbf{x}(r_1)$, approximate the minimum of $R(\mathbf{x}, r_2)$.
5. Continue in this fashion, forming a sequence of points $\{\mathbf{x}(r_j)\}$, $j = 1, 2, \ldots$, minimizing the associated sequence of functions $\{R(\mathbf{x}, r_j)\}$ where the sequence $\{r_j\}$, $j = 1, 2, \ldots$, monotonically decreases to 0.

It is now necessary to show that the above process will converge to the solution of (4.49)–(4.51). The terms of the function $R(\mathbf{x}, r_1)$ in (4.52) merit attention in arguing convergence. The last term is the most intuitive, since as r monotonically decreases to zero, $r^{-1/2}$ is monotonically increasing rapidly; this forces the sum of $g_i^2(\mathbf{x})$, $i = s + 1, \ldots, m$, to zero, satisfying (4.50). The first two terms represent a balancing off of the minimization of $h(\mathbf{x})$ and the avoidance of the boundaries $g_i(\mathbf{x}) = 0$, $i = 1, \ldots, s$. Since $\mathbf{x}^0$ is chosen so that $g_i(\mathbf{x}^0) > 0$, $i = 1, \ldots, s$, and $R(\mathbf{x}, \mathbf{r}_j)$ is minimized for $j = 1, 2, \ldots$, an attempt to approach a boundary will not be tolerated since $g_i(\mathbf{x}) \to 0$ for at least one i in this case, causing $[1/g_i(\mathbf{x})] \to \infty$. Fiacco and McCormick [1968] have applied this algorithm to the problem

$$\begin{aligned} \text{minimize} \quad & z = f(x_1, x_2) = (x_1 - 2)^2 + (x_2 - 1)^2 \\ \text{subject to} \quad & -(x_1^2/4) - x_2^2 + 1 \geq 0 \\ & x_1 - 2x_2 + 1 \geq 0 \end{aligned}$$

The solution produced by the Fiacco-McCormick algorithm is $(x_1^*, x_2^*) = (0.8229, 0.9114)$ and $f(x_1^*, x_2^*) = 1.39$, which is encountered in seven iterations with $r_1 = 1$ and $r_7 = 4.096\,(10)^{-9}$.

This algorithm, referred to as SUMT (sequential unconstrained minimization technique), will determine the global minimum in (4.49)–(4.51) if the problem is convex. No guarantees of global optimality can be made if the problem is nonconvex (see Section 3.5 for the definition of a convex programming problem), but Stong [1965] proved the existence of a sequence of global R function minima converging to the global minimum of the nonconvex programming problem. The difficulty in this case is locating the global minima of the unconstrained R functions, for they will generally be severely nonconvex in $x_1, x_2, \ldots, x_n$.

The use of penalty functions in solving the nonlinear programming problem is most appealing. The difficulty of moving along curvilinear functions is avoided, and a sequence of unconstrained problems is solved rather than a constrained problem. The last point is important since there are numerous algorithms for solving unconstrained problems. Although the same difficulties with nonconvexity arise in the unconstrained problem, at least there are no constraints to worry about. There are problems with applying penalty functions, however. For example, it is clear that (4.52) is not the only unconstrained function that could be formulated for handling the constraints. Another alternative, for example, is

$$Z(\mathbf{x}, r) = h(\mathbf{x}) + r \sum_{i=1}^{s} \frac{1}{g_i^2(\mathbf{x})} + r^{-1/4} \exp \sum_{i=s+1}^{m} g_i^2(\mathbf{x})$$

Depending upon individual creativity, enumerable other functions could be generated to accomplish what $R(\mathbf{x}, r)$ does in (4.52). The additional problem is the selection of constants in the penalty functions (r in this case). Certainly, the selection of r will affect the computational efficiency of the solution, and in a nonconvex problem two different sequences $\{r_j\}$, $\{r_j^*\}$ may lead to different solutions.

4.5 SEARCH PROCEDURES

Since it is generally impossible to guarantee global optimality in the nonconvex programming problem by using the algorithms previously mentioned in this chapter, methods that attempt to avoid being trapped at local optima will now be considered. The procedure begins by searching the feasible region in the vicinity of the starting and optimizing point for better solutions. A variety of search methods have been devised for the nonlinear programming problem, but prior to discussing a few of them, an approximating procedure commonly employed in search algorithms will be considered; this is the Taylor series expansion of a function.

Suppose $f(\mathbf{x})$ has continuous partial derivatives of the second order. Given a point $\mathbf{x}^{(0)}$, $f[\mathbf{x}^{(0)} + \mathbf{h}]$ may be approximated by

$$f[\mathbf{x}^{(0)}] + \mathbf{h}'\nabla f[\mathbf{x}^{(0)}] + 0.5\mathbf{h}'\mathbf{H}_f[\mathbf{x}^{(0)}]\mathbf{h} \tag{4.53}$$

where

$$\mathbf{x}' = [x_1, x_2, \ldots, x_n] \qquad \mathbf{h}' = [h_1, h_2, \ldots, h_n]$$

$$\nabla f(\mathbf{x})' = \left[\frac{\partial f}{\partial x_1}, \frac{\partial f}{\partial x_2}, \ldots, \frac{\partial f}{\partial x_n}\right]$$

$$\mathbf{H}_f(\mathbf{x}) = \begin{bmatrix} \dfrac{\partial^2 f}{\partial x_1^2} & \dfrac{\partial^2 f}{\partial x_1 \partial x_2} & , \ldots , & \dfrac{\partial^2 f}{\partial x_1 \partial x_n} \\ \dfrac{\partial^2 f}{\partial x_2 \partial x_1} & \dfrac{\partial^2 f}{\partial x_2^2} & , \ldots , & \dfrac{\partial^2 f}{\partial x_1 \partial x_n} \\ \vdots & & & \vdots \\ \dfrac{\partial^2 f}{\partial x_n \partial x_1} & \dfrac{\partial^2 f}{\partial x_n \partial x_2} & , \ldots , & \dfrac{\partial^2 f}{\partial x_n^2} \end{bmatrix}$$

Example 4.3

Given the function $f(x_1, x_2) = 2(x_1 - 3)^2 + 5(x_2 - 4)^2$ and the point $\mathbf{x}^{(0)\prime} = [x_1^{(0)}, x_2^{(0)}] = (3, 4)$, approximate $f(x_1, x_2)$ at (5, 5) by using (4.53).

In this example, $\mathbf{h}' = [2, 1]$,

$$\nabla f(\mathbf{x})' = [4(x_1 - 3), 10(x_2 - 4)]$$

$$\mathbf{H}_f(\mathbf{x}) = \begin{bmatrix} 4 & 0 \\ 0 & 10 \end{bmatrix} = \mathbf{H}_f[\mathbf{x}^{(0)}]$$

$$\nabla f[\mathbf{x}^{(0)}]' = [0, 0] \qquad f[\mathbf{x}^{(0)}] = 0$$

Thus

$$f(5, 5) = 0 + [2, 1]\begin{bmatrix} 0 \\ 0 \end{bmatrix} + \frac{1}{2}[2, 1]\begin{bmatrix} 4 & 0 \\ 0 & 10 \end{bmatrix}\begin{bmatrix} 2 \\ 1 \end{bmatrix} = 13$$

In this case the approximate answer is errorless.

If the third term in (4.53) is ignored, the second term represents the approximate increase in the function by moving from $\mathbf{x}^{(0)}$ to $\mathbf{x}^{(0)} + \mathbf{h}$. Notice in the above example the second term is zero, which is a poor approximation of the actual increase of 13. However, in general, $\mathbf{h}'\nabla f[\mathbf{x}^{(0)}]$ will give a reasonably good approximation of $f[\mathbf{x}^{(0)} + \mathbf{h}] - f[\mathbf{x}^{(0)}]$; it is poor in this example due to the fact that $\mathbf{x}^{(0)}$ is the minimum of $f(x_1, x_2)$; $\nabla f[\mathbf{x}^{(0)}]$ is therefore the null vector and little information is provided in $\nabla f[\mathbf{x}^{(0)}]$ as to how rapidly the function is changing at $\mathbf{x}^{(0)} + \mathbf{h}$. This suggests that $\mathbf{h}$ should generally be small so that $\nabla f[\mathbf{x}^{(0)} + \mathbf{h}]$ is not markedly different from $\nabla f[\mathbf{x}^{(0)}]$.

It is possible to construct an algorithm to solve a mathematical programming problem based on the Taylor series expansion of a function. To indicate the nature of such an algorithm, consider the following problem:

$$\begin{aligned} \text{maximize} \quad & f(x_1, x_2) = 2(x_1 - 3)^2 + 5(x_2 - 4)^2 \\ \text{subject to} \quad & 4x_1 + 3x_2 \leq 12 \\ & x_1, x_2 \geq 0 \end{aligned} \tag{4.54}$$

The problem is represented graphically in Figure 4.15.

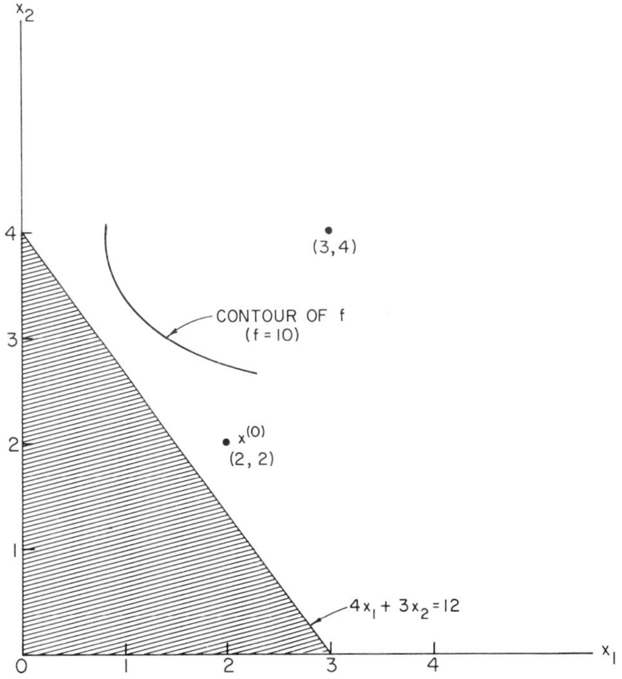

Fig. 4.15. Illustration of Example 4.3.

Begin at the point $\mathbf{x}^{(0)\prime} = (2, 2) = [x_1^{(0)}, x_2^{(0)}]$ as if Rosen's gradient projection method were to be applied. The goal is to move to a new point $\mathbf{x}^{(1)\prime} = [x_1^{(1)}, x_2^{(1)}] = [x_1^{(0)} + h_1, x_2^{(0)} + h_2]$ so that $f(x_1, x_2)$ is increased. The approximate increase of $f(x_1, x_2)$ at $\mathbf{x}^{(1)}$ is given from

(4.53) by ignoring the third term as

$$\mathbf{h}'\nabla f[\mathbf{x}^{(0)}] = [h_1, h_2]\begin{bmatrix}-4\\-20\end{bmatrix} = -4h_1 - 20h_2 \tag{4.55}$$

since

$$\nabla f(\mathbf{x}) = \begin{bmatrix}4(x_1 - 3)\\10(x_2 - 4)\end{bmatrix}$$

Furthermore $x_1^{(1)} = 2 + h_1$ and $x_2^{(1)} = 2 + h_2$ so that the constraints in (4.54) may be expressed in terms of h_1 and h_2 as

$$4(2 + h_1) + 3(2 + h_2) \leq 12$$
$$2 + h_1 \geq 0, \quad 2 + h_2 \geq 0$$

or

$$4h_1 + 3h_2 \leq -2$$
$$h_1 \geq -2, \quad h_2 \geq -2$$

Since (4.55) represents the increase in $f(x_1, x_2)$ due to moving from $\mathbf{x}^{(0)}$ to $\mathbf{x}^{(1)}$, $\mathbf{x}^{(1)}$ may be determined by solving the following linear programming problem in h_1 and h_2:

$$\text{maximize} \quad z = f(h_1, h_2) = -4h_1 - 20h_2 \tag{4.56}$$
$$\text{subject to} \quad 4h_1 + 3h_2 \leq -2 \tag{4.57}$$
$$h_1 \geq -2, \quad h_2 \geq -2 \tag{4.58}$$

The solution to (4.56)–(4.58) must occur at $(-2, -2)$, $(-2, 2)$, or $(1, -2)$; it is apparent that the solution point is $(-2, -2)$. Therefore, $\mathbf{x}^{(1)'} = [2 - 2, 2 - 2] = [0, 0]$ and $f(0, 0) = 98$ which, from Figure 4.15, is clearly the global solution to (4.54).

Fortunately, this problem terminates at the optimizing point in only one move. Normally, this is not the case; furthermore, the solution generated by this method will generally be approximate. For a maximization problem the algorithm may be summarized by the following steps; at the ith point in the process:

1. Calculate $\nabla f[\mathbf{x}^{(i)}]$.
2. Form the function $g_0(\mathbf{h}) = \mathbf{h}'\nabla f[\mathbf{x}^{(i)}]$.
3. Convert the original constraints in $\mathbf{x}$ to $\mathbf{h}$ by using the linear transformation $x_j^{(i)} = x_j^{(i-1)} + h_j, j = 1, 2, \ldots, n$.
4. Solve the following mathematical programming problem for $\mathbf{h}$: maximize $\mathbf{h}'\nabla f[\mathbf{x}^{(i)}]$, subject to the constraints in $\mathbf{h}$ given by (3).

5. Denoting the solution to (4) by $\mathbf{h}^*$, form the new point $\mathbf{x}^{(i+1)\prime} = [x_1^{(i)} + h_1^*, x_2^{(i)} + h_2^*, \ldots, x_n^{(i)} + h_n^*]$.

To initialize the algorithm, $\mathbf{x}^{(0)}$ does not necessarily have to be a feasible point.

The linear programming problem in (4.56)–(4.58) has been solved graphically since there are only three vertex points. Notice that the simplex algorithm could not be applied to this problem due to the possibility of negative h values. However, this is easily remedied by letting $r_1 = h_1 + 2$ and $r_2 = h_2 + 2$ so that (4.56)–(4.58) becomes

$$\text{maximize} \quad z = f(r_1, r_2) = -4r_1 - 20r_2 + 48 \tag{4.59}$$
$$\text{subject to} \quad 4r_1 + 3r_2 \leq 12 \tag{4.60}$$
$$r_1, r_2 \geq 0 \tag{4.61}$$

The solution for h_1 and h_2 may therefore be determined by solving (4.59)–(4.61), which is amenable to simplex resolution. It is immediately apparent that $r_1^* = 0$, $r_2^* = 0$, which is consistent with the solution in (4.56)–(4.58).

To further illustrate the algorithm, it is applied to a problem involving nonlinear constraints in the following example.

Example 4.4

Consider the following:

$$\text{maximize} \quad f(x_1, x_2) = -5(x_1 - 3)^2 - 2(x_2 - 4)^2$$
$$\text{subject to} \quad g_1(x_1, x_2) = x_1^2 + x_2 \leq 3 \tag{4.62}$$
$$g_2(x_1, x_2) = 2x_1 + 3x_2 \leq 6 \tag{4.63}$$

The problem is represented graphically in Figure 4.16.

Let $\mathbf{x}^{(0)\prime} = [1, 1]$, and since $\nabla f' = [-10(x_1 - 3), -4(x_2 - 4)]$,

$$\nabla f[\mathbf{x}^{(0)}] = \begin{bmatrix} 20 \\ 12 \end{bmatrix}$$

In moving to the new point $\mathbf{x}^{(1)\prime} = [x_1^{(0)} + h_1, x_2^{(0)} + h_2]$, the requirement is to maximize $20h_1 + 12h_2$. The nonlinear constraint $g_1 \leq 3$ must be handled in a slightly different way from the linear constraints, since the change in g_1 by moving from $\mathbf{x}^{(0)}$ to $\mathbf{x}^{(1)}$ does not occur at a constant rate. As with $f(x_1, x_2)$

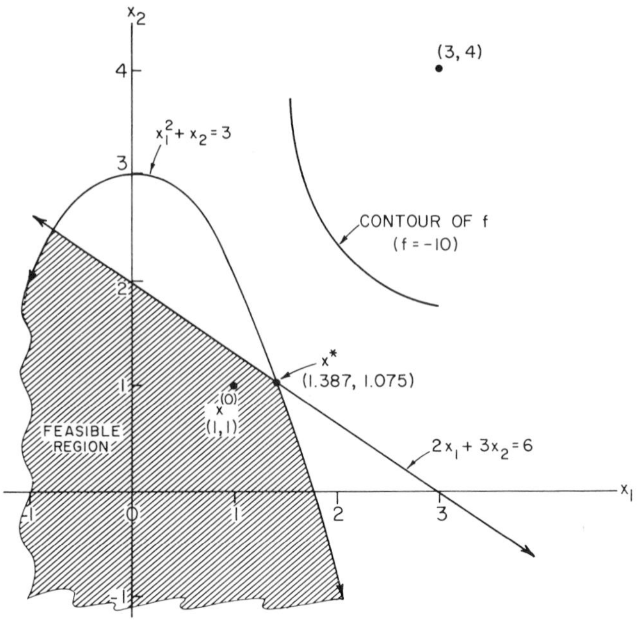

Fig. 4.16. Illustration of Example 4.4.

this change may be approximated by

$$\mathbf{h}' \nabla g_1[\mathbf{x}^{(0)}] \tag{4.64}$$

and since $\nabla g_1' = [2x_1, 1]$, (4.64) becomes

$$2h_1 + h_2 \tag{4.65}$$

Since $g_1(1, 1) = 1^2 + 1 = 2$, the approximated increase in g_1, given by (4.65), due to moving from $\mathbf{x}^{(0)}$ to $\mathbf{x}^{(1)}$ must not exceed $3 - 2 = 1$, or constraint (4.62) would be violated; i.e., $2h_1 + h_2 \leq 1$.

By substituting $x_1^{(1)} = 1 + h_1$ and $x_2^{(1)} = 1 + h_2$ into (4.63) for x_1 and x_2, the constraint in h associated with (4.63) becomes $2h_1 + 3h_2 \leq 1$. The linear programming problem that remains is

$$\text{maximize} \quad 20h_1 + 12h_2 \tag{4.66}$$
$$\text{subject to} \quad 2h_1 + h_2 \leq 1 \tag{4.67}$$
$$2h_1 + 3h_2 \leq 1 \tag{4.68}$$

The solution to (4.66)–(4.68) occurs at the point $h_1^* = 1/2$ and $h_2^* = 0$. Therefore,

$$\mathbf{x}^{(1)} = \begin{bmatrix} x_1^{(0)} + h_1^* \\ x_2^{(0)} + h_2^* \end{bmatrix} = \begin{bmatrix} 1 + 1/2 \\ 1 + 0 \end{bmatrix} = \begin{bmatrix} 3/2 \\ 1 \end{bmatrix}$$

Notice that the point (3/2, 1) is on the line $2x_1 + 3x_2 = 6$ but violates (slightly) the first constraint (4.62). This is due to the fact that (4.62) has been handled only in an approximate way. The process now continues from the point $\mathbf{x}^{(1)\prime} = [3/2, 1]$. Since

$$\nabla f[\mathbf{x}^{(1)}] = \begin{bmatrix} 15 \\ 12 \end{bmatrix}$$

the requirement is to maximize $15h_1 + 12h_2$. At $\mathbf{x}^{(1)}$

$$\nabla g_1[\mathbf{x}^{(1)}] = \begin{bmatrix} 3 \\ 1 \end{bmatrix}$$

so that the approximate increase in moving from $\mathbf{x}^{(1)}$ to $\mathbf{x}^{(2)}$ in g_1 is $3h_1 + h_2$. Since $g_1[\mathbf{x}^{(1)}] = 3.25$, from (4.62) it follows that $3h_1 + h_2 \leq -0.25$. The constraint in (4.63) now becomes $2h_1 + 3h_2 \leq 0$, since $x_1^{(2)} = 3/2 + h_1$ and $x_2^{(2)} = 1 + h_2$.

The linear programming problem in h_1 and h_2 is therefore

$$\begin{aligned} &\text{maximize} && 15h_1 + 12h_2 && (4.69) \\ &\text{subject to} && 3h_1 + h_2 \leq -0.25 && (4.70) \\ & && 2h_1 + 3h_2 \leq 0 && (4.71) \end{aligned}$$

The solution point in (4.70) is $h_1^* = -3/28$ and $h_2^* = 2/28$. Hence, the new point is

$$\mathbf{x}^{(2)} = \begin{bmatrix} x_1^{(1)} + h_1^* \\ x_2^{(1)} + h_2^* \end{bmatrix} = \begin{bmatrix} 3/2 - 3/28 \\ 1 \quad + 2/28 \end{bmatrix} = \begin{bmatrix} 1.393 \\ 1.071 \end{bmatrix}$$

From (4.62) and (4.63), it is seen that $\mathbf{x}^{(2)}$ is not quite feasible. The global maximizing point occurs at the intersection of the line and parabola as indicated in Figure 4.16; $x_1^* = 1.387$, $x_2^* = 1.075$. Thus the global point is rapidly being approached and the process could be continued until the desired degree of accuracy in the solution is achieved.

In approximating the increase of a nonlinear function in moving from one point to another, a more accurate measure will result if the third term in (4.53) is included. However, the programming problem in **h** would no longer be linear. In using only the second term, the increase is approximated with less accuracy, which surely will result in more iterations required for convergence to the optimizing point, but each iteration involves the solution of a linear rather than a nonlinear programming problem.

Notice that this algorithm will produce nonlinear programming problems in **h** if ∇f is not linear. In the above two problems, ∇f is linear and each mathematical programming problem in the sequence is linear. If ∇f is not linear, each problem in the sequence will be nonlinear, requiring nonlinear techniques to determine its solution.

This algorithm is basically an ascent algorithm, for it is attempting to move in a direction that increases the function to be maximized. Since the computation of ∇f is required, if the constraints are all linear, Rosen's procedure (discussed in Section 4.2) certainly should be preferred. If some or all of the constraints are nonlinear, this method becomes more attractive, for it linearizes these constraints. However, the algorithm is rarely used by itself. Most frequently, it is used in conjunction with a search algorithm in the final stages to home in on the optimizing point. The Glass-Cooper [1965] "sequential search" algorithm uses it in this fashion. Specifically, sequential search involves the following steps:

1. An exploratory move. In this step, the gradient vector ∇f is approximated numerically and movement is initialized in this direction.
2. Pattern moves. In this step, movement continues in the direction indicated by step (1) where constant checking for improvement of the objective function and for violation of constraints is made.
3. Alternate moves. The final moves are made by incorporating the Taylor series algorithm described above.

The initial step in sequential search merits further discussion for it obviates computing the gradient vector directly. The way this is accomplished is interesting. Recall that

$$\frac{\partial f}{\partial x_1} = \lim_{\Delta x_1 \to 0} \frac{f(x_1 + \Delta x_1, x_2, \ldots, x_n) - f(x_1, x_2, \ldots, x_n)}{\Delta x_1}$$

Thus the partial derivative $\partial f/\partial x_1$ may be approximated by

$$(1/h_1)\,[f(x_1 + h_1, x_2, \ldots, x_n) - f(x_1, \ldots, x_n)]$$

where h_1 is small. This result may be applied to the gradient vector ∇f giving

$$\nabla f = \begin{bmatrix} (1/h_1)\, f(x_1 + h_1, x_2, \ldots, x_n) - f(x_1, x_2, \ldots, x_n) \\ (1/h_2)\, f(x_1, x_2 + h_2, \ldots, x_n) - f(x_1, x_2 \ldots, x_n) \\ \vdots \\ (1/h_n)\, f(x_1, x_2, \ldots, x_n + h_n) - f(x_1, x_2 \ldots, x_n) \end{bmatrix} \quad (4.72)$$

The approximation of ∇f in (4.72) may be used to gather information with regard to the increase or decrease in $f(\mathbf{x})$ by experimenting with small increments or decrements $h_j, j = 1, 2, \ldots, n$, for x_j, $j = 1, 2, \ldots, n$. It is therefore possible to explore in the $2n$ directions from a point x parallel to the n axes associated with x_j, $j = 1, 2, \ldots, n$, to check for increases in $f(\mathbf{x})$. Additionally, it is not necessary to calculate ∇f, which is advantageous in the sense of writing a general computer code for the algorithm. If the exploratory move is used until step (3) is implemented, it may be possible to avoid being locked into a local optimum, which is a major disadvantage of an algorithm based strictly on the gradient. In Figure 4.17 once a gradient procedure is in the cross-hatched region, it will proceed to P_1, a local optimum in this case. Once in this region, it may still be possible for the sequential search procedure, through its exploratory move, to reach the global optimum P_2.

Very general search procedures are commonly used for nonconvex problems in conjunction with one or more of the algorithms

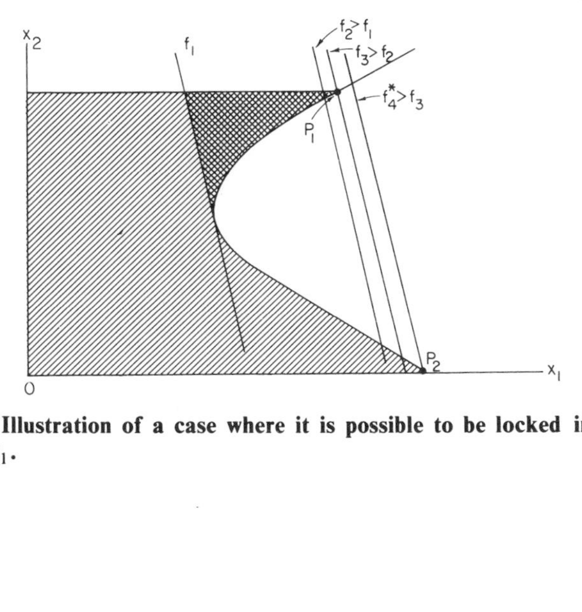

Fig. 4.17. Illustration of a case where it is possible to be locked into a local optimum P_1.

mentioned in this and preceding sections of this chapter. Undoubtedly, the most common is a simple random search of the feasible region. Several feasible points (say N) are randomly selected, and to each is applied an algorithm such as gradient projection, resulting in a set of local optima, the maximum of which is taken as the global maximum. The larger the number of points sampled, the better is the chance of securing a point that will lead to the global optimum when operated upon by an algorithm. The difficulty, referred to by Bellman [1957] as the "curse of dimensionality," is that for practical problems the constraint region will be sizable with its volume increasing exponentially as the number of variables increase. To put it simply, the constraint region is usually a very big place to conduct a random search of points that will lead to the global maximum.

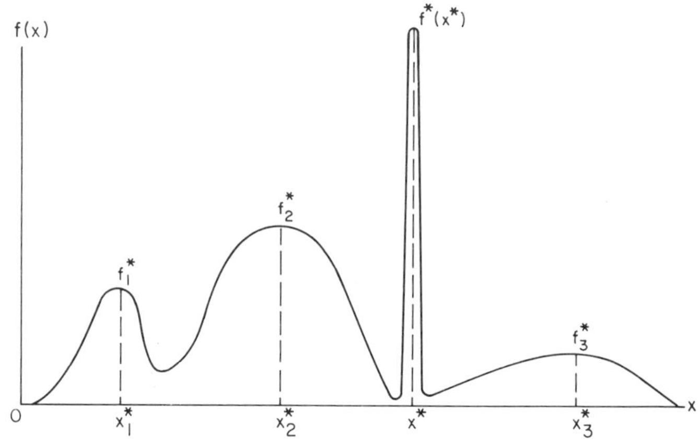

Fig. 4.18. Illustration of a global maximum "spike."

In Figure 4.18 the univariate analog to the multivariate nightmare presented by the worst of all possible nonconvex programming problems is illustrated; namely, a global maximum that has an extremely small set of base points. There is little hope of locating such a maximum by any method without a lot of luck. It has been argued that f_2^* in Figure 4.18, if located, is not a bad alternative to $f(x^*)$ if the problem is one of optimizing output from an elaborate process (e.g., chemical blending), since slight departures of input from x^* could easily plunge the process into the valleys between $f^*(x^*)$, f_2^*, and f_3^*, whereas departures from x_2^* will not affect the output as adversely. Situations such as the one illustrated in Figure 4.18 are not often encountered in practice either because they are rare, or because though not so rare they are seldom exposed. It is hoped the former case is closer to the truth.

Another procedure involves forming grids over the constraint region by segmenting along each variable axis as illustrated in Figure 4.19 for a bivariate case. The function to be maximized is evaluated at each of the grid points, and interesting regions are then explored by applying an algorithm to the points in these regions. Rather than taking grid points, the points at which the function is evaluated may be randomly sampled within each cell or, in general, hypercube. In Figure 4.19 the incomplete cells present no problem, since a randomly selected point in such a cell can be easily checked for feasibility and rejected if it is outside the constraint region.

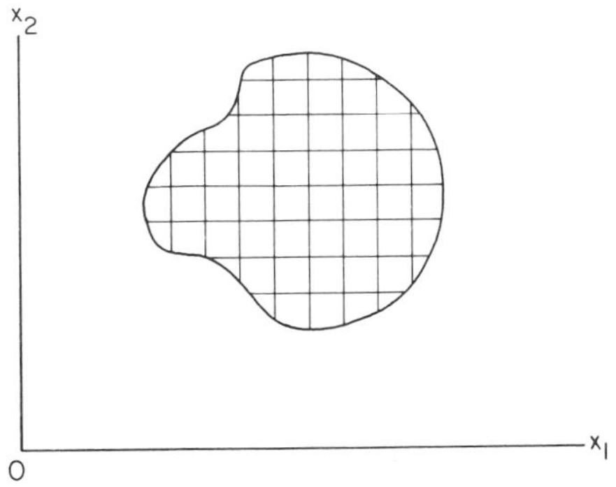

Fig. 4.19. Grid of the feasible region.

Reiter and Rice [1966] have suggested the following search sequence.

1. Randomly select k_1 feasible points.
2. Move these points to local optima by applying an algorithm such as steepest ascent.
3. Determine a subregion of the constraint space which leads to good optima.
4. For the next k_2 points sample the entire constraint region but assign a higher probability of selecting points in the subregion in (3) than outside this region.
5. Continue the process until $N = k_1 + k_2 + \cdots$ points have been sampled.

The last three search procedures are designed for the nonconvex programming problem in which many local optima may occur. There is little alternative to employing a search procedure, since there is no algorithm that will guarantee global optimality in this case and it seems naive to ever expect a feasibly computational one to be developed.

4.6 SELECTION OF AN ALGORITHM

It has been repeatedly pointed out that many algorithms can guarantee (within specified limits of accuracy) the attainment of the global optimum in a convex programming problem. Rosen's gradient projection procedure, separable programming, SUMT, the Hartley-Hocking algorithm [1963], and most search procedures such as the Hooke-Jeeves [1961] and Glass-Cooper algorithms are among the class of algorithms guaranteeing global optimality for the convex problem.

As an obvious preparatory step before selecting an algorithm, as much should be determined analytically about the specific mathematical programming problem to be resolved as possible. If it can be shown that it is convex, global optimality can be insured by employing one of the techniques mentioned above. Commonly, this step may require considerable effort and mathematical skill in assessing convexity, but time spent at this stage is generally well rewarded. For example, by recasting the problem through the addition or deletion of constraints or the replacement of a nonlinear constraint with an approximating linear one, it may be determined that the resulting problem is convex. Additionally, from the knowledge gained by investigating the functions analytically, it may be apparent that the problem is locally convex in the region of the constraint space where the optimum is thought to reside.

After studying the problem analytically, it is usually helpful to acquire a feel for the objective function by forming a very rough grid over the constraint space and evaluating the function at the grid points. This usually requires very little computer time and the resulting map of points may suggest local convexity or at least attractive regions with large functional values that are worth pursuing in greater detail. Admittedly, the evaluation and interpretation of the functional map may be extremely difficult if the number of variables is large, but it is an inexpensive way to gain at least some insight into the behavior of the function.

If the problem is determined to be convex through the analytical or numerical analysis, any one of the algorithms mentioned above may be employed to secure an optimizing point; the one selected generally depends on the availability of computer codes.

If the problem is clearly nonconvex, it is recommended that a number of starting points be chosen over the constraint region and one or more of the above algorithms applied to each of the selected points. If information is available with regard to good regions of the constraint space (good in the sense of having a high probability of leading to large functional values), the starting points should be selected by stratified sampling, giving points in these regions a greater probability of being selected than points outside these regions. Otherwise, the starting points may be selected uniformly. Generally, it is a good idea to apply at least two algorithms, one that strictly depends on a starting point such as gradient projection, SUMT, the Taylor series algorithm, or the Hartley-Hocking algorithm and one that is less dependent on the starting point such as separable programming or, to a lesser degree, the Glass-Cooper or Hooke-Jeeves search algorithms.

The Greenstadt [1972] ricocheting gradient method appears to be quite successful for nonlinear problems where the solution point lies on a curvilinear surface and the objective function is not too severely nonconvex. The penalty function methods seem to experience more difficulty in solving this type of problem than do other algorithms.

If the objective function is highly nonconvex and the constraints are linear, Rosen's [1960] gradient projection method will efficiently locate local optima. The algorithm should be initiated at a number of starting points to increase the likelihood of encountering the global optimum.

If the problem is nonconvex, it may be extremely difficult and computationally expensive to guarantee global optimality. If a better solution can be found than the existing best solution, most mathematical programmers are generally satisfied, although they still may not have encountered $f(\mathbf{x}^*)$, the global solution. If the problem is analogous to the one in Figure 4.18, it is hopeless to secure $f(\mathbf{x}^*)$ with any degree of certainty by any means unless considerable time and money are invested in an exhaustive search of the feasible space.

A number of papers have been published on the comparison of algorithms. The reader is referred to Braitsch [1972], Colville [1968], and Stocker [1969] for discussion of this subject. Concomitant with the interest in computer code and algorithm comparisons is an interest in establishing standards for comparing computer code and algorithmic performance—see Ignizio [1971], for example.

To say which computer code and/or algorithm will work best for a specific problem is difficult. Research efforts on this topic typically lead to general recommendations that depend on the size of the problem (number of constraints, number of variables), the severity of nonconvexity if the problem is not convex, and the existence or nonexistence of boundary restrictions ($\mathbf{x} \geq \mathbf{0}$).

PROBLEMS

4.1. Apply the gradient project method to the following unconstrained problems:

(a) minimize $z = x_1^2 + 6x_1 + 3x_2^2 + 4x_2 + 24$
starting point $(0, 0)$

(b) minimize $z = x_1x_2 + 1/x_1x_2$
starting point $(1, 1)$

(c) minimize $z = x_1x_2 + 1/x_1^2x_2$
starting point $(1, 1)$

4.2. Apply the gradient projection method to the following constrained problems:

(a) minimize $z = x_1^2 + x_2^2 - 25x_1 - 10x_2 + 5$
subject to
$$\begin{aligned} x_1 + 2x_2 &\leq 6 \\ x_1 \quad &\leq 3 \\ x_2 &\leq 2 \\ x_1, x_2 &\geq 0 \end{aligned}$$
starting point $(3, 0)$

(b) maximize $z = 8x_1^2 + 2x_2^2$
subject to
$$\begin{aligned} x_1 + x_2 &\leq 9 \\ x_1 \quad &\leq 6 \\ x_1, x_2 &\geq 0 \end{aligned}$$
starting point $(2, 1/2)$

(c) minimize $z = (x_1 - 2)^2 + (x_2 - 1)^2$
subject to
$$\begin{aligned} -x_1 + x_2 &\geq 0 \\ x_1 + x_2 &\leq 2 \end{aligned}$$
starting point $(1, 1)$

(d) minimize $z = 4(x_1 - 5)^2 + 6(x_2 - 4)^2$
subject to
$$\begin{aligned} x_1 + x_2 &\leq 6 \\ x_1 - x_2 &\leq 1 \\ 2x_1 + x_2 &\geq 5 \\ x_1 \quad &\geq 1 \\ x_2 &\geq 0 \end{aligned}$$
starting point $(2, 1)$

(e) maximize $z = 5(x_1 - 2)^2 + 3(x_2 - 2)^2$
subject to
$$\begin{aligned} x_1 + x_2 &\leq 6 \\ x_1 - x_2 &\geq -2 \\ x_1 + x_2 &\geq 2 \end{aligned}$$

$$x_1 - 3x_2 \le 2$$
$$x_1, x_2 \ge 0$$

Use two starting points in this problem, (1/2, 2) and (2, 0).

4.3. The State Liquor Control Board (SLCB) has determined that it would like to sell only Scotch whiskey and gin and that 5.5 million ounces of Scotch and 7.5 million ounces of gin were sold last year. Therefore, the SLCB has expected purchase levels of about 6 million and 8 million ounces of Scotch and gin respectively this year. If X_1 million ounces of Scotch and X_2 million ounces of gin are purchased by the SLCB this year, $(X_1 - 6)^2$ and $(X_2 - 8)^2$ should be minimized because losses in state revenues occur if too little liquor is available, but there are inventory costs if not all the liquor is sold within a year. A university study in Pittsedelphia has shown that Scotch is more harmful to the human brain than gin, so the SLCB decides to put more importance on the gin control than the Scotch, hoping Scotch consumption falls below 6 million ounces. The SLCB selects the objective function

$$z = W_1(X_1 - 6)^2 + W_2(X_2 - 8)^2$$

and requires $W_2 > W_1$. The SLCB has decided it should

$$\text{minimize} \quad z = 4(X_1 - 6)^2 + 5(X_2 - 8)^2$$

The SLCB's budget constraint is \$32 million (M). Therefore, $X_1 p_1 + X_2 p_2 \le \$32$ M, when p_1 and p_2 are the SLCB's costs of Scotch and gin respectively. Because of unusual purchase arrangements in the state, the price of Scotch rises as the quantities purchased rise, but the price for gin is inversely related to the Scotch purchased. The price equations dictated to the SLCB are $p_1 = 8X_1 + X_2$ and $p_2 = 2X_2 - X_1$ so that its budget constraint is

$$X_1(8X_1 + X_2) + X_2(2X_2 - X_1) \le \$32\text{ M}$$

or $4X_1^2 + X_2^2 \le 16$, where $X_1 \ge 0$ and $X_2 \ge 0$ are required.

Using separable programming with $r_1 = 2$ and $r_2 = 4$, determine X_1 and X_2 for the SLCB to

$$\begin{aligned} \text{minimize} \quad & z = 4(X_1 - 6)^2 + 5(X_2 - 8)^2 \\ \text{subject to} \quad & 4X_1^2 + X_2^2 \le 16 \\ & X_1, X_2 \ge 0 \end{aligned}$$

4.4. Solve the problem in Example 4.2, ignoring the basis restriction.

4.5. In an effort to gain control of its budget, the government of Transylvania decides that it will set a target of having investment I at \$2 billion and government expenditure G at \$1 billion and that it will be equally undesirable for I or G to be below its target as it would be for I or G to be above its target. Therefore, a combination of squared deviations from the targets will be minimized. If the deviation of I from its target or G from its target is of equal significance, Transylvania's disutility from a deviation from a target is $DU = (I - 2)^2 + (G - 1)^2$ because each term will have an equivalent coefficient. As a result of a survey two important relationships between I and G have been discovered. First, to free the administration to adopt programs for promoting private investment, $I + 1 \geq 2G$ will be maintained. Second, by requiring that I and G be very confined

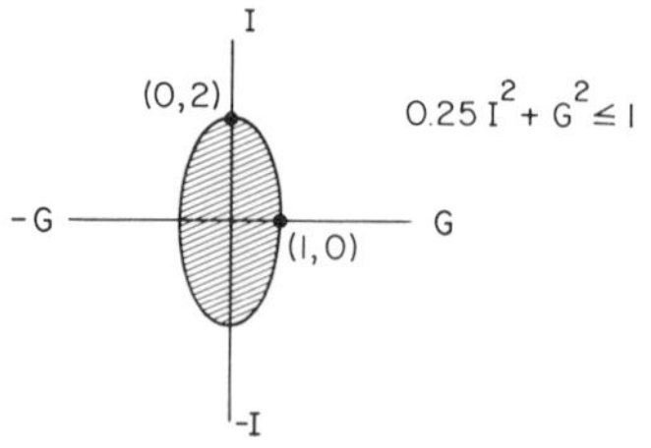

Fig. P4.5.

(within the shaded area in Fig. P4.5), I and G will not reach an inflationary level. The goal of the Transylvanian government is to

$$\begin{aligned}\text{minimize} \quad & DU = (I - 2)^2 + (G - 1)^2 \\ \text{subject to} \quad & I + 1 \geq 2G \\ & 0.25\, I^2 + G^2 \leq 1\end{aligned}$$

Determine the optimal investment and government expenditure levels to minimize Transylvania's disutility using separable programming, letting $r_1 = r_2 = 4$.

4.6. An investor currently holds \$2000 in bonds and \$2000 in stocks and cannot rearrange these investments. He has asked a broker to determine his maximum return z from total investments in stocks and bonds. If X_1 and X_2 are his total investments in bonds and stocks respectively, his new investments are Y_1 and Y_2; $Y_1 = X_1 - 2$ and $Y_2 = X_2 - 2$.

The investor knows that he incurs opportunity losses from not investing as well as brokerage costs when investing and tells his broker that $(X_1 - 2)^2$ and $(X_2 - 2)^2$ should be maximized, but the investor's tax status is such that returns on stocks have only 60 percent as much effect on him as returns on bonds. Therefore, the broker decides that he should

$$\begin{aligned}
\text{maximize} \quad & z = 5(X_1 - 2)^2 + 3(X_2 - 2)^2 \\
\text{subject to} \quad & X_1 + X_2 \leq 6 \\
& X_1 + 2 \geq X_2 \\
& X_1 + X_2 \geq 2 \\
& X_1 \leq 3X_2 + 2 \\
& X_1, X_2 \geq 0
\end{aligned}$$

Solve the broker's problem by separable programming using $r_1 = r_2 = 4$. Note that you may find some inconsistency in the logic of the constraints as expressed by the investor.

4.7. Consider the following problem:

$$\begin{aligned}
\text{maximize} \quad & z = x_1 x_2 e^{x_1 x_2} \\
\text{subject to} \quad & x_1 x_2 \geq 1 \\
& 3x_1 + 5x_2 \leq 15 \\
& x_1, x_2 \geq 0
\end{aligned}$$

By using transformation of variables, recast the problem so that it is amenable to separable programming resolution and, for $r_1 = r_2 = 4$, write down the initial tableau.

4.8. Apply the Taylor series algorithm discussed in Section 4.5 to the following problems:

(a)
$$\begin{aligned}
\text{maximize} \quad & z = 5x_1 + 6x_2 \\
\text{subject to} \quad & x_1 + 3x_2 \leq 6 \\
& 3x_1 + 4x_2 \leq 12 \\
& x_1, x_2 \geq 0 \\
\text{starting point} \quad & (0, 2)
\end{aligned}$$

(b)
$$\begin{aligned}
\text{minimize} \quad & z = x_1 + x_1 x_2 + x_2 \\
\text{subject to} \quad & x_1^2 + x_2 \geq 4 \\
& 0.5x_1 + x_2 \geq 2 \\
& x_1, x_2 \geq 0 \\
\text{starting point} \quad & (4, 4)
\end{aligned}$$

(c)
$$\begin{aligned}
\text{maximize} \quad & z = 5(x_1 - 2)^2 + 3(x_2 - 2)^2 \\
\text{subject to} \quad & x_1 + x_2 \leq 6 \\
& x_1 - x_2 \geq -2 \\
& x_1 + x_2 \geq 2
\end{aligned}$$

$$x_1 - 3x_2 \le 2$$
$$x_1, x_2 \ge 0$$
starting point (2, 0)

4.9. Solve by gradient projection:

$$\begin{aligned} \text{minimize} \quad & z = 2(x_1 - 4)^2 + (x_2 - 4)^2 \\ \text{subject to} \quad & 2x_1 + 3x_2 \le 12 \\ & x_1 + x_2 \ge 2 \\ & 3x_1 - 4x_2 \le 6 \\ & x_1 \ge 0 \end{aligned}$$

starting point (0, 2)

4.10. For the problem

$$\begin{aligned} \text{maximize} \quad & z = 6(x_1 - 2)^2 + 4(x_2 - 2)^2 \\ \text{subject to} \quad & 0 \le x_1 \le 4 \\ & 0 \le x_2 \le 4 \end{aligned}$$

solve by using gradient projection starting at the point (2, 2). Replace x_1 by 4 $\sin^2 y_1$ and x_2 by 4 $\sin^2 y_2$. Note by this transformation the resulting problem is unconstrained; solve by differential calculus. Compare the solutions generated by each method.

4.11. Consider the problem

$$\begin{aligned} \text{maximize} \quad & z = 4x_1 + 7x_2 \\ \text{subject to} \quad & x_1 + 2x_2 \le 8 \\ & 7x_1 + 4x_2 \le 28 \\ & x, x_2 \ge 0 \end{aligned}$$

By introducing an additional variable x_3, the problem may be recast as

$$\begin{aligned} \text{maximize} \quad & z = x_3 \\ \text{subject to} \quad & x_1 + 2x_2 \le 8 \\ & 7x_1 + 4x_2 \le 28 \\ & x_3 - 4x_1 - 7x_2 = 0 \\ & x_1, x_2 \ge 0 \end{aligned}$$

Show that both problems have the same solution by solving each, using the simplex algorithm.

5 QUADRATIC PROGRAMMING

5.1 INTRODUCTION

A quadratic programming problem is one in which a quadratic objective function is to be optimized. The constraints are assumed to be linear inequalities, and the variables in the problem must be nonnegative.

If all the functions in the model are linear, the methods described in Chapter 2 are suitable for solving the problem. If the constraints are not linear or if the objective function is neither linear nor quadratic, the problem is a nonlinear programming problem to which the material in Chapters 3 and 4 is relevant.

The emphasis here will not be on the algorithms that have been conceived for quadratic programming but on the Lagrangian multiplier method. This derivation produces the Kuhn-Tucker conditions for quadratic programming, using calculus. This approach is intended to provide a better understanding of the optimality conditions for quadratic programming and to give some insight into the interpretation of the Lagrangian multipliers. This point is of particular significance because so many applications of quadratic programming appear in the economics, management science, and operations research literature. Some of the implied quadratic programming problems and models are often overlooked. Examples of such cases occur when imperfectly competitive output markets are assumed and only nonnegative levels of inputs, outputs, and prices are permitted. These assumptions cannot be introduced into programming models without linear inequalities or more complex inequality constraints.

5

In this chapter all functions are assumed to be differentiable, and all variables take only real values (imaginary numbers are excluded). This is required since derivatives will be applied to find the optimality conditions.

Consider the constraints of the linear programming problem presented in Example 2.9;

$$4x_1 + x_2 \leq 35 \qquad x_1 + 4x_2 \leq 20 \qquad x_1, x_2 \geq 0$$

The feasible area determined by these constraints is the polygon $0ABC$ in Figure 5.1.

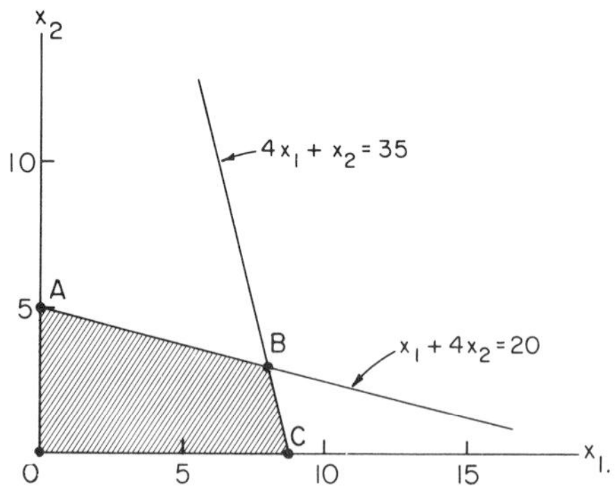

Fig. 5.1. Feasible area determined by constraints.

An optimal solution to a linear programming problem occurs at one of the corner points of the feasible polygon. In Example 2.9, $z = 10x_1 + 15x_2$ has been maximized, subject to the constraints that form the polygon in Figure 5.1. The maximum value of z, 125, occurs at $x_1 = 8$ and $x_2 = 3$, which is at point B.

The optimal solution to a quadratic programming problem does not necessarily occur at a point on the perimeter of the feasible polygon. Suppose the example presented above is changed by adding two quadratic terms, $-1.25x_1^2$ and $-3.75x_2^2$, to the objective function. Then the problem is to find x_1 and x_2 to

$$\begin{aligned} \text{maximize} \quad & w = 10x_1 + 15x_2 - 1.25x_1^2 - 3.75x_2^2 \\ \text{subject to} \quad & 4x_1 + x_2 \leq 35 \\ & x_1 + 4x_2 \leq 20 \\ & x_1, x_2 \geq 0 \end{aligned}$$

The maximum value of w when the quadratic function is unconstrained is $w = 35$, which occurs when $x_1 = 4$ and $x_2 = 2$; the first-order (necessary) conditions are

$$\frac{\partial w}{\partial x_1} = 10 - 2.5x_1 = 0 \qquad x_1 = 4$$

$$\frac{\partial w}{\partial x_2} = 15 - 7.5x_2 = 0 \qquad x_2 = 2$$

and the required second-order (sufficient) conditions are satisfied at $x_1 = 4$ and $x_2 = 2$.

Furthermore, $x_1 = 4$ and $x_2 = 2$ satisfies each of the inequalities

$$4x_1 + x_2 < 35 \qquad x_1 + 4x_2 < 20 \qquad x_1, x_2 > 0$$

Therefore, $x_1 = 4$ and $x_2 = 2$ is the optimal solution to the quadratic programming problem defined in the previous paragraph. This solution is not on the boundary of the feasible polygon since the inequality is strictly satisfied for each constraint. This is not often the case, as will be shown in Section 5.2, and more sophisticated methods are needed to solve the problem.

5.2 METHOD OF LAGRANGIAN MULTIPLIERS

Example 5.1

The method of Lagrangian multipliers for quadratic programming will be introduced by considering a simple example. Suppose that a firm produces two commodities in quantities x and z. The prices are assumed to be P_x and P_z, which define the respective average revenue functions for x and z.

$$P_x = 10 - x + 150/x \qquad P_z = 64 - 8z + 250/z$$

respectively. It costs \$200 to produce a unit of x and \$100 to produce a unit of z. The firm's budget is \$4700. Therefore, $200x + 100z \leq 4700$, or $2x + z \leq 47$, is the firm's budget constraint.

Suppose that the firm attempts to maximize its total revenue TR.

$$TR = xP_x + zP_z = 10x - x^2 + 64z - 8z^2 + 400$$

The optimization problem is

$$\begin{aligned} \text{maximize} \quad & TR = 10x - x^2 + 64z - 8z^2 + 400 \\ \text{subject to} \quad & 2x + z \leq 47 \\ & x, z \geq 0 \end{aligned}$$

This problem can be solved using the Lagrangian multiplier technique. The procedure is the same as the approach to linear programming described in Chapter 2 and as that to nonlinear models in Chapter 3.

The budget constraint is converted to an equation by adding the slack variable y; $2x + z + y = 47$. Each variable is set equal to the square of a real value; $x = v^2$, $z = w^2$, and $y = u^2$. Assuming that v, w, and u have real values implies that v^2, u^2, and w^2 are nonnegative. Therefore, x, z, and y are nonnegative values.

The Lagrangian function for this problem is

$$\begin{aligned} L = {} & 10x - x^2 + 64z - 8z^2 + 400 + \lambda(2x + z + y - 47) \\ & + \rho(x - v^2) + \phi(z - w^2) + \pi(y - u^2) \end{aligned}$$

This function includes four Lagrangian multipliers λ, ρ, ϕ, and π. One Lagrangian multiplier precedes each constraint in the problem, since there is a budget constraint and three nonnegative variables.

The first-order conditions for this problem are derived by determining ten partial derivatives and setting each equal to zero. These derivatives follow:

$$\frac{\partial L}{\partial x} = 10 - 2x + 2\lambda + \rho = 0$$

$$\frac{\partial L}{\partial z} = 64 - 16z + \lambda + \phi = 0$$

$$\frac{\partial L}{\partial \lambda} = 2x + z + y - 47 = 0$$

$$\frac{\partial L}{\partial y} = \lambda + \pi = 0 \qquad \lambda = -\pi$$

$$\left.\begin{aligned} \frac{\partial L}{\partial \rho} &= x - v^2 = 0 \qquad & x &= v^2 \\ \frac{\partial L}{\partial v} &= -2\rho v = 0 \qquad & \rho v^2 &= 0 \end{aligned}\right\} x\rho = 0$$

$$\left.\begin{aligned}\frac{\partial L}{\partial \phi} &= z - w^2 = 0 \qquad & z &= w^2 \\ \frac{\partial L}{\partial w} &= -2\phi w = 0 \qquad & \phi w^2 &= 0\end{aligned}\right\} z\phi = 0$$

$$\left.\begin{aligned}\frac{\partial L}{\partial \pi} &= y - u^2 = 0 \qquad & y &= u^2 \\ \frac{\partial L}{\partial u} &= -2\pi u = 0 \qquad & \pi u^2 &= 0\end{aligned}\right\} y\pi = 0$$

These results can be reduced to a set of three linear equations in six unknowns with three additional conditions. The variable π can be replaced by minus λ, and the equations that remain are

$$-2x + 2\lambda + \rho = -10$$

$$-16z + \lambda + \phi = -64$$

$$2x + z + y = 47$$

$$x\rho = 0 \qquad z\phi = 0 \qquad y\lambda = 0$$

Each possibility that the side conditions provide is a set of three linear equations in three unknowns. There are eight of these sets, $2^3 = 8$, since there are three side conditions and each has two variables.

The eight possible cases are indicated in Table 5.1. It should be noted that these are an exhaustive set of possibilities. Some of these eight provide infeasible solutions or solutions that are not unique.

Table 5.1. Possible Solutions

Case	x	z	y	ρ	ϕ	λ	TR
1	0	0	0				
2	0	0				0	
3	0		0		0		
4	0				0	0	
5		0	0	0			
6		0		0		0	
7			0	0	0		
8				0	0	0	

Each case from Table 5.1 is solved below and the results are summarized in Table 5.2. Case 8 provides the optimal solution.

The cases in which there is no unique, feasible solution are indicated by an asterisk in Table 5.2.

Case 1: $x = z = y = 0$ no solution
$$2\lambda + \rho = -10$$
$$\lambda + \phi = -64$$
$$0 = 47$$

Case 2: $x = z = \lambda = 0$ $TR = 400$
$$\rho = -10$$
$$\phi = -64$$
$$y = 47$$

Case 3: $x = y = \phi = 0$ $TR = -14264$
$$2\lambda + \rho = -10 \qquad \rho = -1386$$
$$-16z + \lambda = -64 \qquad \lambda = 688$$
$$z = 47 \qquad z = 47$$

Case 4: $x = \phi = \lambda = 0$ $TR = 528$
$$\rho = -10 \qquad z = 4$$
$$-16z = -64 \qquad y = 43$$
$$z + y = 47 \qquad \rho = -10$$

Case 5: $z = y = \rho = 0$ $TR = 82.75$
$$-2x + 2\lambda = -10 \qquad \lambda = 19.5$$
$$\lambda + \phi = -64 \qquad \phi = 83.5$$
$$2x = 47 \qquad x = 23.5$$

Case 6: $z = \rho = \lambda = 0$ $TR = 425$
$$-2x = -10 \qquad x = 5$$
$$\phi = -64 \qquad \phi = -64$$
$$2x + y = 47 \qquad y = 37$$

Case 7: $y = \rho = \phi = 0$ $TR = 289$
$$-2x + 2\lambda = -10 \qquad x = 21$$
$$-16z + \lambda = -64 \qquad z = 5$$
$$2x + z = 47 \qquad \lambda = 16$$

Case 8: $\rho = \phi = \lambda = 0$ $TR = 553$
$$-2x = -10 \qquad x = 5$$
$$-16z = -64 \qquad z = 4$$
$$2x + z + y = 47 \qquad y = 33$$

Only the first-order conditions have been considered for this example. The general quadratic programming problem and the required conditions are presented in Section 5.4.

Table 5.2. Solutions

Case	x	z	y	ρ	ϕ	λ	TR
1	0	0	0	*	*	*	*
2	0	0	47	−10	−64	0	400
3	0	47	0	−1386	0	688	−14,264
4	0	4	43	−10	0	0	528
5	23.5	0	0	0	83.5	195	82.75
6	5	0	37	0	−64	0	425
7	21	5	0	0	0	16	289
[8]	5	4	33	0	0	0	553

5.3 QUADRATIC FORMS

A quadratic form of the variables $x_1, x_2, \ldots, x_m$ is a polynomial in which each term is quadratic. This means x_j^2 or $x_j x_k$ must appear in each term. The equation

$$W = \sum_{j=1}^{m} \sum_{k=1}^{m} r_{jk} x_j x_k$$

is a general quadratic form that could also be written as

$$W = \sum_{j=1}^{m} r_{jj} x_j^2 + \sum_{\substack{j=1 \\ j \neq k}}^{m} \sum_{k=1}^{m} r_{jk} x_j x_k$$

Employing matrix notation, a matrix of the quadratic form is expressed as $W = \mathbf{x}'\mathbf{R}\mathbf{x}$, where $\mathbf{x}$ is an $m \times 1$ vector, $\mathbf{x}'$ is the transpose of the $\mathbf{x}$ vector, and $\mathbf{R}$ is an $m \times m$ matrix. If $\mathbf{R}$ is not a symmetric matrix, one exists that can be substituted for $\mathbf{R}$ to provide the same value for W.

Example 5.2

For example, let

$$W = [x_1, x_2, x_3] \begin{bmatrix} 3 & 2 & 2 \\ 4 & 5 & 2 \\ 5 & 3 & 7 \end{bmatrix} \begin{bmatrix} x_1 \\ x_2 \\ x_3 \end{bmatrix}$$

Then

$$W = 3x_1^2 + 5x_2^2 + 7x_3^2 + 6x_1x_2 + 7x_1x_3 + 5x_2x_3$$

However, the same result is obtained by letting

$$W = [x_1, x_2, x_3] \begin{bmatrix} 3 & 3 & 3.5 \\ 3 & 5 & 2.5 \\ 3.5 & 2.5 & 7 \end{bmatrix} \begin{bmatrix} x_1 \\ x_2 \\ x_3 \end{bmatrix}$$

Comparing the elements of the two coefficient matrices shows that the two matrices have the same main diagonals. Furthermore, the average of the two elements in row i, column j and row j, column i of the nonsymmetric matrix appears in both these locations in the symmetric matrix. For example, in the first matrix the elements in row 1, column 3 and row 3, column 1 are 2 and 5 respectively. In the second case, where the symmetric coefficient matrix is employed, both values are 3.5.

In general, if **R** is a nonsymmetric square matrix, a square symmetric matrix **R*** can be found to provide the same quadratic form, using two rules: (1) the elements of the main diagonal of **R** become the elements of the main diagonal of **R*** and (2) the elements from row i, column j and row j, column i from **R** are averaged, and this average becomes the element in both row i, column j and row j, column i of **R***.

Definite and Semidefinite Forms

If a quadratic programming problem has an objective function of $w = \mathbf{c}'\mathbf{x} + \mathbf{x}'\mathbf{R}\mathbf{x}$, sometimes the quadratic form $\mathbf{x}'\mathbf{R}\mathbf{x}$ will enable us to consider very simple second-order optimizing conditions depending on determinants formed from within the matrix **R**. Let R_i be the determinant of an $i \times i$ matrix. The determinant R_1 is formed by ignoring all rows and columns except the first element of the matrix **R**. The determinant R_2 is formed using only the first two rows and first two columns of the matrix **R**.

$$R_1 = | r_{11} |$$

$$R_2 = \begin{vmatrix} r_{11} & r_{12} \\ r_{21} & r_{22} \end{vmatrix}$$

$$R_3 = \begin{vmatrix} r_{11} & r_{12} & r_{13} \\ r_{21} & r_{22} & r_{23} \\ r_{31} & r_{32} & r_{33} \end{vmatrix}$$

$$\vdots$$

$$R_m = | \mathbf{R} |$$

The quadratic form $\mathbf{x}'\mathbf{R}\mathbf{x}$ is negative definite if $(-1)^{i+1} R_i < 0$, $i = 1, \ldots, m$; $R_1 < 0$, $R_2 > 0$, $R_3 < 0, \ldots$. If $(-1)^{i+1}R_i \leq 0$ and at least one of $(-1)^{i+1}R_i \neq 0$, the quadratic form $\mathbf{x}'\mathbf{R}\mathbf{x}$ is negative semidefinite. If $R_i > 0$, $i = 1, \ldots, m$, the quadratic form is positive definite. If $R_i \geq 0$, $i = 1, \ldots, m$, and at least one of $R_i \neq 0$, the quadratic form is positive semidefinite.

Variance and Covariance

One quadratic form of particular interest in quantitative economic models is the variance-covariance form. This will be needed to study portfolio models in the final section of this chapter.

Suppose that $\mathbf{C}$ is a matrix having m rows and n columns with elements c_{ij}, $i = 1, \ldots, m$, and $j = 1, \ldots, n$. If t_j is the sum of the elements in column j, then t_j/m is the average or mean of the elements in column j; let $u_j = t_j/m$.

Example 5.3

If

$$\mathbf{C} = \begin{bmatrix} 3 & 2 & 8 \\ 2 & 5 & 3 \\ 5 & 1 & 4 \\ 6 & 0 & 5 \end{bmatrix}$$

then $t_1 = 16$, $t_2 = 8$, $t_3 = 20$, $u_1 = 4$, $u_2 = 2$, and $u_3 = 5$.

The sample variance of the elements in column j of the matrix $\mathbf{C}$ is denoted as s_j^2, where

$$s_j^2 = \sum_{i=1}^{n} (c_{ij} - u_j)^2 \Big/ (n - 1)$$

For the first two columns of the matrix $\mathbf{C}$,

$$s_1^2 = \frac{(3 - 4)^2 + (2 - 4)^2 + (5 - 4)^2 + (6 - 4)^2}{4 - 1} = \frac{10}{3}$$

and

$$s_2^2 = \frac{(2 - 2)^2 + (5 - 2)^2 + (1 - 2)^2 + (0 - 2)^2}{4 - 1} = \frac{14}{3}$$

The sample covariance between the ith and jth columns of $\mathbf{C}$ is denoted as s_{ij} and defined by

$$s_{ij} = \sum_{k=1}^{n} (c_{ki} - u_i)(c_{kj} - u_j) \Big/ (n - 1)$$

For the first two columns of $\mathbf{C}$

$$s_{12} = \frac{(3-4)(2-2) + (2-4)(5-2) + (5-4)(1-2) + (6-4)(0-2)}{4-1} = -\frac{11}{3}$$

For the three columns of the matrix $\mathbf{C}$ there are three variances s_1^2, s_2^2, and s_3^3 and three covariances s_{12}, s_{13}, and s_{23}. These can all be represented in a symmetric matrix as

$$\begin{bmatrix} s_1^2 & s_{12} & s_{13} \\ s_{12} & s_2^2 & s_{23} \\ s_{13} & s_{23} & s_3^2 \end{bmatrix} = \begin{bmatrix} 10/3 & -11/3 & 0 \\ -11/3 & 14/3 & -5/3 \\ 0 & -5/3 & 14/3 \end{bmatrix}$$

The results are the matrix product of the quadratic form $(\hat{\mathbf{C}}'\hat{\mathbf{C}})/(n - 1)$, where the matrix $\hat{\mathbf{C}}$ is formed by subtracting the column mean from each element of the respective column.

$$\hat{\mathbf{C}} = \begin{bmatrix} (3-4) & (2-2) & (8-5) \\ (2-4) & (5-2) & (3-5) \\ (5-4) & (1-2) & (4-5) \\ (6-4) & (0-2) & (5-5) \end{bmatrix} = \begin{bmatrix} -1 & 0 & -3 \\ -2 & 3 & -2 \\ 1 & -1 & -1 \\ 2 & -2 & 0 \end{bmatrix}$$

Example 5.4

Suppose that $\mathbf{p}$ and $\mathbf{q}$ represent vectors of n prices and commodities respectively, so that

$$E = \sum_{i=1}^{n} p_i q_i = \mathbf{p}'\mathbf{q}$$

is the quadratic form of the expenditures on the n items. If the actual prices are unknown, the average price vector $\mathbf{u}$ may be used as an estimator for $\mathbf{p}$, and the expected or average expenditure E would be

$$\bar{E} = \sum_{i=1}^{n} u_1 q_i = \mathbf{u}'\mathbf{q}$$

Here u_1 represents the average of observations on the price of commodity 1. The variance of the expenditures $V(E)$ can also be specified if the variances and covariances among the prices are known. If σ_i^2 is the variance of p_i and σ_{ij} is the covariance between p_i and p_j,

$$V(E) = \sum_{i=1}^{n} \sigma_i^2 q_i^2 + \sum_{i=1}^{n} \sum_{\substack{j=1 \\ i \neq j}}^{n} \sigma_{ij} q_i q_j = \mathbf{q}'\mathbf{S}\mathbf{q}$$

where

$$\mathbf{S} = \begin{bmatrix} s_1^2 & s_{12} & \cdots & s_{1n} \\ s_{21} & s_2^2 & \cdots & s_{2n} \\ \vdots & \vdots & \cdots & \vdots \\ s_{n1} & s_{n2} & \cdots & s_n^2 \end{bmatrix}$$

is a symmetric variance-covariance matrix for the stock prices. The matrix form of the variance of expenditures is a quadratic form in the vector of variables in $\mathbf{q}$.

In portfolio models the most common objective function is a combination of the average expenditures $\bar{E}$ and the variance among the expenditures $V(E)$. Some of these models are presented in Section 5.8.

5.4 GENERAL QUADRATIC PROBLEM

In matrix notation the quadratic programming problem is

$$\text{optimize} \quad w(\mathbf{x}) = \mathbf{c}'\mathbf{x} + \mathbf{x}'\mathbf{R}\mathbf{x} \tag{5.1}$$

$$\text{subject to} \quad \mathbf{A}\mathbf{x} \le \mathbf{b} \tag{5.2}$$

$$\mathbf{x} \ge \mathbf{0} \tag{5.3}$$

where

$$\mathbf{c}' = [c_1, c_2, \ldots, c_m]$$
$$\mathbf{x}' = [x_1, x_2, \ldots, x_m]$$
$$\mathbf{b}' = [b_1, b_2, \ldots, b_n]$$

$$\mathbf{A} = \begin{bmatrix} a_{11} & a_{12} & \cdots & a_{1m} \\ a_{21} & a_{22} & \cdots & a_{2m} \\ \vdots & & & \vdots \\ a_{n1} & a_{n2} & \cdots & a_{nm} \end{bmatrix} \qquad \mathbf{R} = \begin{bmatrix} r_{11} & r_{12} & \cdots & r_{1m} \\ r_{21} & r_{22} & \cdots & r_{2m} \\ \vdots & & & \vdots \\ r_{m1} & r_{m2} & \cdots & r_{mm} \end{bmatrix}$$

As for the linear case let y_i, $i = 1, \ldots, n$, be defined as a slack variable so that $\mathbf{y}' = [y_1, y_2, \ldots, y_n]$, $\mathbf{y} \geq \mathbf{0}$, and

$$\mathbf{Ax} + \mathbf{Iy} = \mathbf{b} \tag{5.4}$$

Let

$$x_j = v_j^2 \qquad j = 1, \ldots, m \tag{5.5}$$

$$y_i = u_i^2 \qquad i = 1, \ldots, n \tag{5.6}$$

so that each x and each y has a positive value, since only real valued variables are considered. When the slack variables are included, their coefficients in the objective function are zero, so that the objective function is

$$w(\mathbf{x}, \mathbf{y}) = \mathbf{c}'\mathbf{x} + \mathbf{x}'\mathbf{R}\mathbf{x} + \mathbf{0}'\mathbf{y} \tag{5.7}$$

where $\mathbf{0}'$ is a row vector of zeroes for the slack variables.

To optimize (5.7) subject to (5.4)–(5.6), $2n + m$ Lagrangian multipliers must be introduced. Let

$$\boldsymbol{\lambda}' = [\lambda_1, \lambda_2, \ldots, \lambda_n] \qquad \boldsymbol{\phi}' = [\phi_1, \phi_2, \ldots, \phi_n] \qquad \boldsymbol{\rho}' = [\rho_1, \rho_2, \ldots, \rho_m]$$

be Lagrangian multipliers so that the Lagrangian function for this problem can be formed. The Lagrangian function is

$$L = \mathbf{c}'\mathbf{x} + \mathbf{x}'\mathbf{R}\mathbf{x} + \mathbf{0}'\mathbf{y} + \boldsymbol{\lambda}'(\mathbf{Ax} + \mathbf{Iy} - \mathbf{b})$$
$$+ \sum_{j=1}^{m} \rho_j(x_j - v_j^2) + \sum_{i=1}^{n} \phi_i(y_i - u_i^2)$$

or

$$L = \sum_{j=1}^{m} c_j x_j + \sum_{j=1}^{m} \sum_{k=1}^{m} r_{jk} x_j x_k$$

$$+ \sum_{i=1}^{n} \sum_{j=1}^{m} \lambda_i (a_{ij} x_j + y_i - b_i) + \sum_{j=1}^{m} \rho_j (x_j - v_j^2)$$

$$+ \sum_{i=1}^{n} \phi_i (y_i - u_i^2)$$

The first-order conditions for maximizing or minimizing $w(\mathbf{x})$ in (5.1) are the same. They are determined by setting the $4n + 3m$ possible first derivatives equal to zero. Thus

$$\frac{\partial L}{\partial x_j} = c_j + 2\sum_{k=1}^{m} r_{jk} x_k + \sum_{i=1}^{n} \lambda_i a_{ij} + \rho_j = 0 \qquad j = 1, \ldots, m \tag{5.8}$$

$$\frac{\partial L}{\partial y_i} = \lambda_i + \phi_i = 0 \qquad i = 1, \ldots, n \tag{5.9}$$

$$\frac{\partial L}{\partial v_j} = -2\rho_j v_j = 0 \qquad \rho_j v_j^2 = 0 \qquad j = 1, \ldots, m \tag{5.10}$$

$$\frac{\partial L}{\partial \mu_i} = -2\phi_i \mu_i = 0 \qquad \phi_i \mu_i^2 = 0 \qquad i = 1, \ldots, n \tag{5.11}$$

$$\frac{\partial L}{\partial \lambda_i} = \sum_{j=1}^{m} a_{ij} x_j + y_i - b_i = 0 \qquad i = 1, \ldots, n \tag{5.12}$$

$$\frac{\partial L}{\partial \rho_j} = x_j - v_j^2 = 0 \qquad j = 1, \ldots, m \tag{5.13}$$

$$\frac{\partial L}{\partial \phi_i} = y_i - \mu_i^2 = 0 \qquad i = 1, \ldots, n \tag{5.14}$$

Together, (5.10) and (5.13) imply

$$\rho_j x_j = 0 \qquad j = 1, \ldots, m \tag{5.15}$$

and (5.9), (5.11), and (5.14) provide

$$\lambda_i y_i = 0 \qquad i = 1, \ldots, n \tag{5.16}$$

The results of (5.8) and (5.12) are $n + m$ linear equations in $2n + 2m$ unknowns (n λ's, n y's, m x's, and m ρ's). However, $n + m$ of these variables have a zero value from (5.15) and (5.16). Thus (5.8), (5.12),

(5.15), and (5.16) provide 2^{n+m} sets of $n + m$ linear equations in $n + m$ variables. The results of (5.8), (5.12), (5.15), and (5.16) are the Kuhn-Tucker conditions for quadratic programming.

These conditions can be summarized as

$$\begin{bmatrix} 2\mathbf{R} & \mathbf{A}' & \mathbf{I} & \mathbf{0} \\ \mathbf{A} & \mathbf{0} & \mathbf{0} & \mathbf{I} \end{bmatrix} \begin{bmatrix} \mathbf{x} \\ \boldsymbol{\lambda} \\ \boldsymbol{\rho} \\ \mathbf{y} \end{bmatrix} = \begin{bmatrix} -\mathbf{c} \\ \mathbf{b} \end{bmatrix} \tag{5.17}$$

where (5.15) and (5.16) are required as side conditions.

To determine the optimal solution, each alternative solution is substituted into the objective function. The one that optimizes $w(\mathbf{x})$ is then identified from among the feasible alternatives. For a specific case, it is often best to form a table, as has been done for the complete description method of linear programming.

The first-order conditions in (5.15)–(5.17) are some of the necessary conditions to solve the problem delineated by (5.1)–(5.3). The first-order and second-order conditions are sufficient to solve the problem.

Properties of the Objective Function and Second-Order Conditions

The objective function $w(\mathbf{x}) = \mathbf{c}'\mathbf{x} + \mathbf{x}'\mathbf{R}\mathbf{x}$ may either be convex or concave in $\mathbf{x}$. Since a linear function is both convex and concave, the convexity or concavity of $w(\mathbf{x})$ depends solely on $\mathbf{x}'\mathbf{R}\mathbf{x}$; the results in Appendix B may be utilized to assess the convexity of this term. If $\mathbf{x}'\mathbf{R}\mathbf{x}$ is convex and $w(\mathbf{x})$ is to be minimized or if $\mathbf{x}'\mathbf{R}\mathbf{x}$ is concave and $w(\mathbf{x})$ is to be maximized, (5.1)–(5.3) constitute a convex programming problem; then the Kuhn-Tucker conditions are also sufficient for a global optimizing point $\mathbf{x}^*$. Therefore, the second-order conditions for global optimality in the quadratic programming model are convexity of $\mathbf{x}'\mathbf{R}\mathbf{x}$ when minimizing $w(\mathbf{x})$ and concavity of $\mathbf{x}'\mathbf{R}\mathbf{x}$ when maximizing $w(\mathbf{x})$. These conditions are equivalent to the quadratic form $\mathbf{x}'\mathbf{R}\mathbf{x}$ being positive definite or positive semidefinite when $w(\mathbf{x})$ is to be minimized and its being negative definite or negative semidefinite when $w(\mathbf{x})$ is to be maximized (see Section 5.3). If these conditions are not satisfied, an optimizing point $\mathbf{x}^*$ satisfying (5.8), (5.12), (5.15), and (5.16) may not provide a global optimum.

Example 5.5

Consider the quadratic programming problem

$$\text{maximize } w = 20x_1 - 6x_1^2 + 40x_2 - 8x_2^2 + 30x_3 - 15x_3^2 \tag{5.18}$$

$$\text{subject to} \quad 3x_1 + 2x_2 + 6x_3 \le 60 \qquad (5.19)$$
$$x_1, x_2, x_3 \ge 0 \qquad (5.20)$$

In the matrix notation used for (5.1)–(5.3), the problem is to

$$\begin{aligned} \text{maximize} \quad & w = \mathbf{c}'\mathbf{x} + \mathbf{x}'\mathbf{R}\mathbf{x} \\ \text{subject to} \quad & \mathbf{A}\mathbf{x} \le \mathbf{b} \\ & \mathbf{x} \ge \mathbf{0} \end{aligned}$$

where

$$\mathbf{c}' = [20 \quad 40 \quad 30]$$
$$\mathbf{A} = [3 \quad 2 \quad 6]$$
$$\mathbf{b} = [60]$$
$$\mathbf{x} = \begin{bmatrix} x_1 \\ x_2 \\ x_3 \end{bmatrix} \qquad \mathbf{R} = \begin{bmatrix} -6 & 0 & 0 \\ 0 & -8 & 0 \\ 0 & 0 & -15 \end{bmatrix}$$

The Lagrangian function for this problem is

$$\begin{aligned} L = {} & 20x_1 - 6x_1^2 + 40x_2 - 8x_2^2 + 30x_3 - 15x_3^2 \\ & + \lambda(3x_1 + 2x_2 + 6x_3 + y - 60) + \sum_{j=1}^{3} \phi_j(x_j - v_j^2) \\ & + \pi(y - u^2) \end{aligned}$$

This problem includes five Lagrangian multipliers λ, ϕ_1, ϕ_2, ϕ_3, and π. The four variables that must be nonnegative are x_1, x_2, x_3, and y; $x_j = v_j^2, j = 1, 2, 3$, and $y = u^2$.

This problem has sixteen possible cases and solutions. These are determined by Kuhn-Tucker conditions, which follow for this problem.

$$\begin{bmatrix} -12 & 0 & 0 & 3 & 1 & 0 & 0 & 0 \\ 0 & -16 & 0 & 2 & 0 & 1 & 0 & 0 \\ 0 & 0 & -30 & 6 & 0 & 0 & 1 & 0 \\ 3 & 2 & 6 & 0 & 0 & 0 & 0 & 1 \end{bmatrix} \begin{bmatrix} x_1 \\ x_2 \\ x_3 \\ \lambda \\ \phi_1 \\ \phi_2 \\ \phi_3 \\ y \end{bmatrix} = \begin{bmatrix} -20 \\ -40 \\ -30 \\ 60 \end{bmatrix}$$

$$x_j\phi_j = 0 \qquad j = 1, 2, 3 \qquad \lambda y = 0$$

The reader should compare these conditions with the general results provided by (5.15)–(5.17). The alternative solutions to this problem are given in Table 5.3. The optimal decision is that of Case 2.

Table 5.3. Alternative Solutions

Case	x_1	x_2	x_3	λ	ϕ_1	ϕ_2	ϕ_3	y	w
1	6.667	5.0	5.0	20.0	0	0	0	0	−358.333
[2]	1.667	2.5	1.0	0	0	0	0	44.0	81.667
3	14.167	8.75	0	50.0	0	0	−330.0	0	−1183.333
4	1.667	2.5	0	0	0	0	−30.0	50.0	66.667
5	7.949	0	6.026	25.128	0	−90.256	0	0	−583.974
6	1.667	0	1.0	0	0	−40.0	0	49.0	31.667
7	20.0	0	0	73.333	0	−186.667	−469.999	0	−1999.997
8	1.667	0	0	0	0	−40.0	−30.0	55.0	16.667
9	0	6.724	7.759	33.793	−121.379	0	0	0	−762.931
10	0	2.5	1.0	0	−20.0	0	0	49.0	65.0
11	0	30.0	0	220.0	−680.0	0	−1350.0	0	−6000.0
12	0	2.5	0	0	−20.0	0	−30.0	55.0	50.0
13	0	0	10.0	45.0	−155.0	−130.0	0	0	−11.999
14	0	0	1.0	0	−20.0	−40.0	0	54.0	15.0
15	0	0	0	60.0	−20.0	−40.0	−30.0	0	0
16	0	0	0	0	−20.0	−40.0	−30.0	60.0	0

This problem could have been solved more efficiently by implementing the Lagrangian algorithm delineated in Chapter 4. However, it is instructive to see the solutions for all 16 cases as illustrated in Table 5.3. Notice that only a small portion of these cases would be solved by the Lagrangian algorithm before locating the global solution (Case 2).

For linear models, the Lagrangian multipliers have been shown to be the dual variables. Together, the optimal allocations for the primal and dual linear programming problems identify a saddle point. In quadratic programming this same result is true; however, this feature does not generate further considerations such as a simple game. The interpretation of the Lagrangian multipliers for the quadratic problem is very similar to that for the linear case. Considering (5.9), (5.15), and (5.16), the key Lagrangian multipliers are the elements of λ. Either λ_i or y_i is zero. Since λ_i is the premultiplier of the linear constraint having the slack variable y_i, the interpretation for the linear programming model can be extended to the primal quadratic problem. The dual models are described in Section 5.6.

5.5 WOLFE'S SIMPLEX METHOD FOR QUADRATIC PROGRAMMING

A vast number of algorithms have been presented in the relevant literature of quadratic programming. Many have been developed to deal with problems having special features. The short form of the Wolfe algorithm [1959] may be applied to any quadratic programming problem in which the necessary second-order conditions are satisfied.

The Kuhn-Tucker conditions for quadratic programming are given by (5.15)–(5.17) or

$$\begin{aligned} 2\mathbf{Rx} \qquad + \mathbf{A}'\boldsymbol{\lambda} + \mathbf{I}\boldsymbol{\rho} &= -\mathbf{c} \\ \mathbf{Ax} + \mathbf{Iy} \qquad\qquad &= \mathbf{b} \end{aligned} \tag{5.21}$$

For any constraint among

$$\sum_{j=1}^{m} a_{ij}x_j + y_i = b_i \qquad i = 1, \ldots, n$$

where b_i is negative, the constraint is to be multiplied by minus one. The form $\mathbf{Ax} + \mathbf{Iy} = \mathbf{b}$ will be employed throughout this section, assuming that each b_i is positive but that the coefficient of y_i could be negative.

The Framework of the Algorithm

The short form of the Wolfe algorithm is a two-phase linear programming process. An artificial variable with a coefficient of plus one is added to each constraint of $\mathbf{Ax} + \mathbf{Iy} = \mathbf{b}$. Let $\hat{\boldsymbol{\pi}}$ denote an $n \times 1$ vector of artificial variables so that $\mathbf{I}\hat{\boldsymbol{\pi}}$ is a diagonal matrix. Then each constraint of the form $\mathbf{Ax} + \mathbf{Iy} + \mathbf{I}\hat{\boldsymbol{\pi}} = \mathbf{b}$ has a slack variable y_i and an artificial variable, $\hat{\pi}_i$. At least one of y_i or $\hat{\pi}_i$ has a coefficient of plus one.

If the ith linear constraint is

$$\sum_{j=1}^{m} a_{ij}x_j \geq b_i \qquad b_i > 0$$

the artificial variable will be needed for the initial basic feasible solution. Inequalities of this sort are the usual occurrence when a quadratic minimization problem is considered.

Two artificial variables π_j^a and π_j^b will be included in each condition among those in $2\mathbf{Rx} + \mathbf{A}'\boldsymbol{\lambda} + \mathbf{I}\boldsymbol{\rho} = -\mathbf{c}$. Let $\boldsymbol{\pi}^a$ and $\boldsymbol{\pi}^b$ be two-column vectors of m elements each so that $\mathbf{I}\boldsymbol{\pi}^a$ and $\mathbf{I}\boldsymbol{\pi}^b$ provide diagonal matrices. To provide an obvious initial basic feasible solution to $2\mathbf{Rx} + \mathbf{A}'\boldsymbol{\lambda} + \mathbf{I}\boldsymbol{\rho} = -\mathbf{c}$, the two diagonal matrices of artificial variables are included with opposite signs. Then $2\mathbf{Rx} + \mathbf{A}'\boldsymbol{\lambda} + \mathbf{I}\boldsymbol{\rho} + \mathbf{I}\boldsymbol{\pi}^a - \mathbf{I}\boldsymbol{\pi}^b = -\mathbf{c}$.

The constraints given by (5.21) have been transformed into

$$\begin{aligned} 2\mathbf{Rx} + \mathbf{A}'\boldsymbol{\lambda} + \mathbf{I}\boldsymbol{\rho} + \mathbf{I}\boldsymbol{\pi}^a - \mathbf{I}\boldsymbol{\pi}^b &= -\mathbf{c} \\ \mathbf{Ax} \qquad + \mathbf{Iy} + \mathbf{I}\hat{\boldsymbol{\pi}} &= \mathbf{b} \end{aligned} \tag{5.22}$$

The artificial variables in $\boldsymbol{\pi}^a$ and $\boldsymbol{\pi}^b$ are introduced so that the signs of the right-hand values (elements of $-\mathbf{c}$) need not be assumed. If $-c_j < 0$, then $-\pi_j^b = -c_j < 0$ and $\pi_j^b = c_j > 0$.

Setting $\pi_j^a = 0$ and $\pi_j^b = c_j$ provides one variable that may be part of an initial basic feasible solution to the first set of conditions given by (5.22). If $-c_j > 0$, then $\pi_j^a = -c_j > 0$ and $\pi_j^b = 0$ may be part of an initial feasible solution to the first set of conditions of (5.22).

PHASE I

In the first phase of the algorithm the linear programming problem to be solved is

$$\begin{aligned} \text{minimize} \quad & \hat{\psi} = \textstyle\sum_{i=1}^{n} \hat{\pi}_i \\ \text{subject to} \quad & 2\mathbf{Rx} + \mathbf{A}'\boldsymbol{\lambda} + \mathbf{I}\boldsymbol{\rho} + \mathbf{I}\boldsymbol{\pi}^a - \mathbf{I}\boldsymbol{\pi}^b = -\mathbf{c} \\ & \mathbf{Ax} \qquad + \mathbf{Iy} + \mathbf{I}\hat{\boldsymbol{\pi}} = \mathbf{b} \\ & x_j\rho_j = 0 \qquad j = 1, \ldots, m \\ & y_i\lambda_i = 0 \qquad i = 1, \ldots, n \\ & \mathbf{x}, \mathbf{y}, \boldsymbol{\pi}^a, \boldsymbol{\pi}^b, \hat{\boldsymbol{\pi}} \geq \mathbf{0} \end{aligned}$$

An initial basic feasible solution to this problem is given by $\hat{\pi}_i = b_i$, $i = 1, \ldots, n$, and $\pi_j^a = -c_j$, $-c_j > 0$, or $-\pi_j^b = -c_j$, $-c_j < 0$.

The solution to the Phase I linear programming problem appears when each $\hat{\pi}_j$ has been replaced by another variable from among $\mathbf{x}$, $\boldsymbol{\lambda}$, $\boldsymbol{\rho}$, and $\mathbf{y}$. The optimal value of $\hat{\psi}$ is zero. However, $\hat{\psi} = 0$ only if $\hat{\pi}_i = 0$, $i = 1, \ldots, n$. The Phase I linear programming problem can be solved using the simplex method.

The Phase I problem has $4m + 3n$ variables.

m x's	n λ's
m ρ's	n y's
m π^a's	n π's
m π^b's	

The optimal solution to Phase I will provide $\hat{\pi}_i = 0$, $i = 1, \ldots, n$, and either π_j^a or $\pi_j^b = 0, j = 1, \ldots, m$. This result is useful to begin Phase II.

PHASE II

The final simplex tableau from Phase I is employed to begin Phase II. First, $m + n$ of the variables that appeared in the Phase I linear programming problem can be eliminated. Since each $\hat{\pi}_i$ must be zero at the end of Phase I, each of these artificial variables may be eliminated at the end of Phase I. Also, the one of π_j^a or π_j^b that is zero, $j = 1, \ldots, m$, is eliminated before beginning the Phase II problem. Let $\boldsymbol{\pi}^*$ be an m-dimensional vector,

$$\boldsymbol{\pi}^* = \begin{bmatrix} \pi_1^* \\ \vdots \\ \pi_m^* \end{bmatrix}$$

The jth element of $\boldsymbol{\pi}^*$, π_j^*, is defined so that π_j^* is the nonzero one of π_j^a or π_j^b in the optimal solution to the Phase I problem.

The result of solving the Phase I linear programming problem is to attain a basic feasible solution to the conditions:

$$\begin{aligned} 2\mathbf{Rx} + \mathbf{A}'\boldsymbol{\lambda} + \mathbf{I}\boldsymbol{\rho} + \mathbf{E}\boldsymbol{\pi}^* &= -\mathbf{c} \\ \mathbf{Ax} \qquad\qquad + \mathbf{Iy} \qquad\qquad &= \mathbf{b} \\ x_j \rho_j = 0 \qquad j = 1, \ldots, m & \end{aligned}$$

The matrix $\mathbf{E}$ is taken from the final tableau of Phase I. The first column of $\mathbf{E}$ is the coefficient vector of the nonzero variable between π_1^a and π_1^b. In this manner all m of the columns of $\mathbf{E}$ are selected from the optimal solution to Phase I. Recall that either π_j^a or π_j^b is dropped from Phase II.

Phase II is completed upon solving another linear programming problem. The Phase II problem is

$$\begin{aligned} \text{minimize} \quad & \psi^* = \textstyle\sum_{i=1}^n \pi_i^* \\ \text{subject to} \quad & 2\mathbf{Rx} + \mathbf{A}'\boldsymbol{\lambda} + \mathbf{I}\boldsymbol{\rho} + \mathbf{E}\boldsymbol{\pi}^* = -\mathbf{c} \\ & \mathbf{Ax} \qquad\qquad + \mathbf{Iy} \qquad\qquad = \mathbf{b} \\ & x_j \rho_j = 0 \qquad j = 1, \ldots, m \\ & \lambda_i y_i = 0 \qquad i = 1, \ldots, n \\ & \mathbf{x}, \mathbf{y}, \boldsymbol{\pi}^* \geq 0 \end{aligned}$$

The optimal solution to this problem occurs when each π_i^* has a value of zero and $\psi^* = 0$. When $\sum_{i=1}^{n} \pi_i^* = 0$; $\rho_j x_j = 0$, $j = 1, \ldots, m$; and $\lambda_i y_i = 0$, $i = 1, \ldots, n$, the optimal levels of the x_j's, ρ_j's, λ_i's, and y_i's are determined.

Wolfe [1959] has proved a number of theorems that form the framework of the two-phase process. The interested reader should consult this source.

Special Cases

There are a number of important special cases of quadratic programming. Many of these can be solved using a modified version of the method presented above. Most of the special cases do not require the use of all the artificial variables that have been described along with Wolfe's algorithm.

Case 1: **R** is negative definite

$$\begin{aligned}
\text{maximize} \quad & w = \mathbf{c}'\mathbf{x} + \mathbf{x}'\mathbf{R}\mathbf{x} \\
\text{subject to} \quad & \mathbf{A}\mathbf{x} \le \mathbf{b} \\
& \mathbf{x} \ge \mathbf{0} \\
& c_j \ge 0 \qquad j = 1, \ldots, m \\
& b_i \ge 0 \qquad i = 1, \ldots, n
\end{aligned}$$

Artificial variables employed are

$$\begin{aligned}
\hat{\boldsymbol{\pi}} &\equiv \mathbf{0} \text{ since } \hat{\boldsymbol{\pi}} = \mathbf{y} = \mathbf{b} \ge \mathbf{0} \\
\boldsymbol{\pi}^a &\equiv \mathbf{0} \\
-\boldsymbol{\pi}^b &= -\mathbf{c} \qquad \boldsymbol{\pi}^b = \mathbf{c} \ge \mathbf{0}
\end{aligned}$$

Phase I: $\mathbf{y} = \mathbf{b} \qquad \boldsymbol{\pi}^b = \mathbf{c}$

Phase II:

$$\begin{aligned}
\text{minimize} \quad & \textstyle\sum_{j=1}^{m} \pi_j^b \\
\text{subject to} \quad & 2\mathbf{R}\mathbf{x} + \mathbf{A}'\boldsymbol{\lambda} + \mathbf{I}\boldsymbol{\rho} - \mathbf{I}\boldsymbol{\pi}^b = -\mathbf{c} \\
& \mathbf{A}\mathbf{x} + \mathbf{I}\mathbf{y} = \mathbf{b} \\
& x_j \rho_j = 0 \qquad j = 1, \ldots, m \\
& y_i \lambda_i = 0 \qquad i = 1, \ldots, n \\
& \mathbf{x}, \mathbf{y}, \boldsymbol{\pi}^b \ge \mathbf{0}
\end{aligned}$$

Case 2: **R** is positive definite

$$\text{minimize} \quad w = \mathbf{c}'\mathbf{x} + \mathbf{x}'\mathbf{R}\mathbf{x}$$

$$\begin{aligned}
\text{subject to} \quad & \mathbf{Ax} \geq \mathbf{b} \\
& \mathbf{x} \leq \mathbf{0} \\
& c_j \leq 0 \qquad j = 1, \ldots, m \\
& b_i \geq 0 \qquad i = 1, \ldots, n
\end{aligned}$$

Artificial variables employed are

$$\begin{aligned}
& \hat{\boldsymbol{\pi}} = \mathbf{b} \text{ since } \mathbf{Ax} - \mathbf{Iy} = \mathbf{b} \\
& \boldsymbol{\pi}^b \equiv \mathbf{0} \text{ and } \boldsymbol{\pi}^a \equiv \mathbf{0} \text{ since } \boldsymbol{\rho} = -\mathbf{c} \geq \mathbf{0}
\end{aligned}$$

Phase I:

$$\begin{aligned}
\text{minimize} \quad & \textstyle\sum_{i=1}^{n} \hat{\pi}_i \\
\text{subject to} \quad & 2\mathbf{Rx} + \mathbf{A}'\boldsymbol{\lambda} + \mathbf{I}\boldsymbol{\rho} = -\mathbf{c} \\
& \mathbf{Ax} - \mathbf{Iy} + \hat{\boldsymbol{\pi}} = \mathbf{b} \\
& x_j \rho_j = 0 \qquad j = 1, \ldots, m \\
& y_i \lambda_i = 0 \qquad i = 1, \ldots, n
\end{aligned}$$

Phase II: not required since no $\hat{\pi}_i$'s remain at the end of Phase I.

Example 5.6

At the end of this chapter several portfolio models will be described to illustrate one of the best known applications of quadratic programming. Here the Wolfe algorithm will be applied to a simple problem where a fixed asset of cash $b = 1$ is to be allocated by proportions among three projects x_1, x_2, and x_3.

Consider the quadratic programming problem

$$\begin{aligned}
\text{maximize} \quad & w = \mathbf{c}'\mathbf{x} + \mathbf{x}'\mathbf{Rx} \\
\text{subject to} \quad & \mathbf{Ax} \leq \mathbf{b} \\
& \mathbf{x} \geq \mathbf{0}
\end{aligned}$$

where

$$\mathbf{c}' = [6, 9, 12]_{1\times 3} \qquad \mathbf{A} = [1, 1, 1]_{1\times 3} \qquad \mathbf{b} = [1]_{1\times 1}$$

$$\mathbf{R} = \begin{bmatrix} -8 & -10 & -9.5 \\ -10 & -16 & -11 \\ -9.5 & -11 & -30 \end{bmatrix}_{3\times 3} \qquad \mathbf{x} = \begin{bmatrix} x_1 \\ x_2 \\ x_3 \end{bmatrix}_{3\times 1}$$

The value of x_j is the percentage of the total cash resource that is to be allocated to project j. The constraint $\mathbf{Ax} \leq \mathbf{b}$ is given by $x_1 + x_2 + x_3 \leq 1$. If $x_1 + x_2 + x_3 < 1$, part of the cash available is not allocated to any of the three projects.

The Kuhn-Tucker conditions for this example are given by (5.15), (5.16), and (5.21). These conditions are

$$x_1 + x_2 + x_3 + y = 1 \tag{5.23}$$

since $\mathbf{Ax} + \mathbf{Iy} = \mathbf{b}$, and

$$\begin{aligned} -16x_1 - 20x_2 - 19x_3 + \lambda + \rho_1 &= -6 \\ -20x_1 - 32x_2 - 22x_3 + \lambda + \rho_2 &= -9 \\ -19x_1 - 22x_2 - 60x_3 + \lambda + \rho_3 &= -12 \end{aligned}$$

since $2\mathbf{Rx} + \mathbf{A}'\boldsymbol{\lambda} + \mathbf{I}_3\boldsymbol{\rho} = -\mathbf{c}$.

Because the elements of $-\mathbf{c}$ are negative, the last three constraints are multiplied by minus one.

$$\begin{aligned} 16x_1 + 20x_2 + 19x_3 - \lambda - \rho_1 &= 6 \\ 20x_1 + 32x_2 + 22x_3 - \lambda - \rho_2 &= 9 \\ 19x_1 + 22x_2 + 60x_3 - \lambda - \rho_3 &= 12 \end{aligned} \tag{5.24}$$

For Phase I of the Wolfe algorithm seven artificial variables must be employed. Since the quadratic programming model has only one constraint, one artificial variable $\hat{\pi}$ is introduced into (5.23). Thus $x_1 + x_2 + x_3 + y + \hat{\pi} = 1$ is one constraint for the first phase of the Wolfe algorithm.

Six artificial variables are to be introduced into (5.24). The revised constraints are

$$\begin{aligned} 16x_1 + 20x_2 + 19x_3 - \lambda - \rho_1 + \pi_1^a - \pi_1^b &= 6 \\ 20x_1 + 32x_2 + 22x_3 - \lambda - \rho_2 + \pi_2^a - \pi_2^b &= 9 \\ 19x_1 + 22x_2 + 60x_3 - \lambda - \rho_3 + \pi_3^a - \pi_3^b &= 12 \end{aligned}$$

For Phase I of the Wolfe algorithm the problem is

minimize $\psi = \hat{\pi}$
subject to

$$\begin{aligned} 16x_1 + 20x_2 + 19x_3 - \lambda - \rho_1 + \pi_1^a - \pi_1^b &= 6 \\ 20x_1 + 32x_2 + 22x_3 - \lambda - \rho_2 + \pi_2^a - \pi_2^b &= 9 \\ 19x_1 + 22x_2 + 60x_3 - \lambda - \rho_3 + \pi_3^a - \pi_3^b &= 12 \\ x_1 + x_2 + x_3 + y + \hat{\pi} &= 1 \\ \hat{\boldsymbol{\pi}}, \mathbf{y}, \mathbf{x}, \boldsymbol{\pi}^a, \boldsymbol{\pi}^b &\geq \mathbf{0} \end{aligned}$$

One initial basic feasible solution to this problem is given by $\pi_1^a = 6, \pi_2^a = 9, \pi_3^a = 12, \hat{\pi} = 1$, and all other variables have zero values. An optimal solution to Phase I is $\pi_1^a = 6$, $\pi_2^a = 9$, $\pi_3^a = 12, y = 1$. All other variables have zero values.

For Phase II, four of the artificial variables that appeared in Phase I of the problem are removed. Recall that only the artificial variables that are nonzero for the optimal solution in Phase I must be carried along to Phase II (π_1^a, π_2^a, π_3^a).

The goal in Phase II is to minimize the sum of the remaining artificial variables. The constraints for the Phase II problem are shown in the final tableau from the optimal solution to Phase I.

The problem for Phase II is

$$\begin{aligned} \text{minimize} \quad & \psi = \pi_1^a + \pi_2^a + \pi_3^a \\ \text{subject to} \quad & 16x_1 + 20x_2 + 19x_3 - \lambda + \pi_1^a = 6 \\ & 20x_1 + 32x_2 + 22x_3 - \lambda + \pi_2^a = 9 \\ & 19x_1 + 22x_2 + 60x_3 - \lambda + \pi_3^a = 12 \\ & x_1 + x_2 + x_3 + y = 1 \end{aligned}$$

The initial basic feasible solution is $\pi_1^a = 6$, $\pi_2^a = 9$, $\pi_3^a = 12$, $y = 1$. All other variables are zero. This is the optimal solution to the Phase I problem. For Phase II the optimal solution is

$$x_2 = 0.1922 \qquad y = 0.6783$$
$$x_3 = 0.1295 \qquad \rho_1 = 0.3050$$

The final decision is to invest 19.22% and 12.95% of the cash in projects 2 and 3 respectively. Of the total funds available, 67.83% are held in cash since the slack variable has this value.

5.6 QUADRATIC PROGRAMMING AND DUALITY

The consideration of duality and its importance in quadratic programming is slight when compared to the role of duality in linear programming models. Just as in linear programming, the dual to a maximization problem is a minimization problem and vice versa. However, some of the primal variables appear in the dual to a quadratic programming model. Recall that the solution to a primal linear programming problem provides information about some variables (such as imputed costs in models of the firm) that appear in the dual and might not be observable. Since each primal variable appears in the dual in quadratic cases, sometimes nothing significant is learned from formulating and solving a dual problem.

The dual to the problem of (5.1)–(5.3) is

$$\begin{aligned} \text{optimize} \quad & z(\mathbf{x}, \boldsymbol{\lambda}) = -\mathbf{x}'\mathbf{R}\mathbf{x} - \boldsymbol{\lambda}'\mathbf{b} \\ \text{subject to} \quad & 2\mathbf{R}\mathbf{x} + \mathbf{A}'\boldsymbol{\lambda} \geq -\mathbf{c} \\ & \mathbf{x} \geq \mathbf{0} \end{aligned}$$

where $\mathbf{b}$ is the vector of constants defined for the primal problem and $\boldsymbol{\lambda}$ is the vector of Lagrangian multipliers that precede the linear constraints $\mathbf{A}\mathbf{x} \leq \mathbf{b}$ in the Lagrangian function. Hadley [1964, p. 238] shows the relationship between the first-order conditions for the primal (5.17) and the dual problems. From (5.17)

$$\begin{aligned} \mathbf{A}\mathbf{x} + \mathbf{y} &= \mathbf{b} \\ \boldsymbol{\lambda}'\mathbf{A}\mathbf{x} + \boldsymbol{\lambda}'\mathbf{y} &= \boldsymbol{\lambda}'\mathbf{b} \\ \boldsymbol{\lambda}'\mathbf{A}\mathbf{x} &= \boldsymbol{\lambda}'\mathbf{b} - \boldsymbol{\lambda}'\mathbf{y} \end{aligned} \tag{5.25}$$

and

$$\begin{aligned} 2\mathbf{R}\mathbf{x} + \mathbf{A}'\boldsymbol{\lambda} + \boldsymbol{\rho} &= -\mathbf{c} \\ \mathbf{c} + 2\mathbf{R}\mathbf{x} + \mathbf{A}'\boldsymbol{\lambda} + \boldsymbol{\rho} &= \mathbf{0} \\ \mathbf{x}'\mathbf{c} + 2\mathbf{x}'\mathbf{R}\mathbf{x} + \mathbf{x}'\mathbf{A}'\boldsymbol{\lambda} + \mathbf{x}'\boldsymbol{\rho} &= \mathbf{0} \\ \mathbf{c}'\mathbf{x} + 2\mathbf{x}'\mathbf{R}\mathbf{x} + \boldsymbol{\lambda}'\mathbf{A}\mathbf{x} + \boldsymbol{\rho}'\mathbf{x} &= \mathbf{0} \end{aligned} \tag{5.26}$$

Substituting (5.25) into (5.26) leaves

$$\mathbf{c}'\mathbf{x} + 2\mathbf{x}'\mathbf{R}\mathbf{x} + \boldsymbol{\lambda}'\mathbf{b} - \boldsymbol{\lambda}'\mathbf{y} + \boldsymbol{\rho}'\mathbf{x} = \mathbf{0}$$

From (5.9), (5.15), and (5.16) $\boldsymbol{\lambda}'\mathbf{y} = \mathbf{0}$, and $\boldsymbol{\rho}'\mathbf{x} = \mathbf{0}$. Therefore, $\mathbf{c}'\mathbf{x} + \mathbf{x}'\mathbf{R}\mathbf{x} = -\mathbf{x}'\mathbf{R}\mathbf{x} - \boldsymbol{\lambda}'\mathbf{b}$, and the values of the objective functions for the primal and dual are proved to be equal; $w(\mathbf{x}) = z(\mathbf{x}, \boldsymbol{\lambda})$.

Dorn [1960, 1961] and Hadley [1964] have suggested numerous points of interest about duality conditions for nonlinear problems. Some of their results will be employed to construct the conditions that have been mentioned in this section.

Example 5.7

Recall (5.18)–(5.20). In matrix notation the problem is

$$\text{maximize} \quad w = \mathbf{c}'\mathbf{x} + \mathbf{x}'\mathbf{R}\mathbf{x}$$

$$\text{subject to} \quad \mathbf{Ax} \leq \mathbf{b}$$
$$\mathbf{x} \geq \mathbf{0}$$

where

$$\mathbf{c}' = [20, 40, 30]$$
$$\mathbf{A} = [3, 2, 6] \qquad \mathbf{R} = \begin{bmatrix} -6 & 0 & 0 \\ 0 & -8 & 0 \\ 0 & 0 & -15 \end{bmatrix}$$
$$\mathbf{x}' = [x_1, x_2, x_3]$$
$$\mathbf{b} = [60]$$

The dual to this problem is

$$\begin{aligned} \text{minimize} \quad & \psi = 6x_1^2 + 8x_2^2 + 15x_3^2 - 60\lambda \\ \text{subject to} \quad & -12x_1 + 3\lambda \geq -20 \\ & -16x_2 + 2\lambda \geq -40 \\ & -30x_3 + 6\lambda \geq -30 \\ & x_1, x_2, x_3 \geq 0 \end{aligned}$$

The optimal solution to this problem is the same as that given in Case 2 of Table 5.3: $x_1 = 1.667$, $x_2 = 2.5$, $x_3 = 1.0$, $\lambda = 0$, $w = 81.667$.

One distinct feature of the dual to a quadratic programming problem should be noted. The matrix **R** that provides the second-order conditions for the primal problem appears in both the objective function and the constraints of the dual problem.

5.7 APPLICATIONS OF QUADRATIC PROGRAMMING

Elementary Models of the Firm

Example 5.8

The applications of quadratic programming are considerably more extensive than usually have been stated in the relevant literature. Most models such as those to which elementary calculus is applied should have inequality constraints added to them. Three examples of this are where a simple revenue, cost, or profit function is to be optimized and only nonnegative output levels are acceptable.

Suppose that the firm's total revenue, total cost, and profit functions are

$$TR = -10q^2 - 100q \tag{5.27}$$

$$TC = 5q^2 + 200q + 1125 \tag{5.28}$$

$$\pi = TR - TC = -15q^2 - 300q - 1125 \tag{5.29}$$

respectively. If differential calculus is applied to optimize these functions, the results are: $q = -5$ maximizes TR, $q = -20$ minimizes TC, and $q = -10$ maximizes π.

The second-order conditions are satisfied for these results. However, if $q \geq 0$ is added to each of (5.27)–(5.29), each problem becomes a quadratic programming problem. The Lagrangian functions are

$$L^R = -10q^2 - 100q + \lambda(q - v^2)$$
$$L^C = 5q^2 + 200q + 1125 + \lambda(q - v^2)$$
$$L^\pi = -15q^2 - 300q - 1125 + \lambda(q - v^2)$$

The first-order conditions for maximizing the total revenue are

$$\frac{\partial L^R}{\partial q} = -20q - 100 + \lambda = 0$$
$$\frac{\partial L^R}{\partial \lambda} = q - v^2 = 0$$
$$\frac{\partial L^R}{\partial v} = -2\lambda v = 0$$

or $-20q + \lambda = 100$ and $q\lambda = 0$.

The alternative solutions are $q = 0$ and $\lambda = 100$ or $\lambda = 0$ and $q = -5$, but the second is not feasible since $q \geq 0$ is violated. The optimal solution is $q = 0$ and $\lambda = 100$, which suggests that the firm is better off not to produce any output. If the firm is considered in a dynamic framework, this may well be an optimal solution in any particular short-run circumstance. Similar results are determined for minimizing the total cost function given by (5.28) and maximizing the total profit function of (5.29). In each instance the optimal solution is to set the output level at zero. These examples are provided mainly to emphasize the possibility that quadratic programming algorithms, rather than simple differential calculus, should be applied to many models of the firm.

Consumer Behavior

Another quadratic programming model occurs for a consumer who has a quadratic utility function and is restricted by a linear budget constraint. The consumer does not purchase negative quantities. Furthermore, if the consumer is permitted to save part of his income, his budget constraint is an inequality such as $p_1q_1 + p_2q_2 \leq Y^0$.

Consider the consumer whose behavior is to

$$\begin{aligned} \text{maximize} \quad & U(q_1, q_2) = a_1q_1^2 + a_2q_2^2 + b_1q_1 + b_2q_2 \\ \text{subject to} \quad & p_1q_1 + p_2q_2 \leq Y^0 \\ & q_1, q_2 \geq 0 \end{aligned}$$

The Lagrangian function is

$$\begin{aligned} L = {} & a_1q_1^2 + a_2q_2^2 + b_1q_1 + b_2q_2 + \lambda_0(p_1q_1 + p_2q_2 + S - Y^0) \\ & + \lambda_1(q_1 - v_1^2) + \lambda_2(q_2 - v_2^2) + \lambda_3(S - v_3^2) \end{aligned}$$

The slack variable S in the income constraint is the consumer's savings. The first-order conditions to maximize the consumer's utility are

$$\frac{\partial L}{\partial q_1} = 2a_1q_1 + b_1 + \lambda_0 p_1 + \lambda_1 = 0$$

$$\frac{\partial L}{\partial q_2} = 2a_2q_2 + b_2 + \lambda_0 p_2 + \lambda_2 = 0$$

$$\frac{\partial L}{\partial S} = \lambda_0 + \lambda_3 = 0$$

$$\frac{\partial L}{\partial \lambda_0} = p_1q_1 + p_2q_2 + S - Y^0 = 0$$

$$\frac{\partial L}{\partial \lambda_1} = q_1 - v_1^2 = 0$$

$$\frac{\partial L}{\partial \lambda_2} = q_2 - v_2^2 = 0$$

$$\frac{\partial L}{\partial \lambda_3} = S - v_3^2 = 0$$

$$\frac{\partial L}{\partial v_1} = -2\lambda_1 v_1 = 0 \qquad \lambda_1 v_1^2 = 0$$

$$\frac{\partial L}{\partial v_2} = -2\lambda_2 v_2 = 0 \qquad \lambda_2 v_2^2 = 0$$

$$\frac{\partial L}{\partial v_3} = -2\lambda_3 v_3 = 0 \qquad \lambda_3 v_3^2 = 0$$

These results can be reduced to

$$2a_1q_1 + \lambda_0 p_1 + \lambda_1 = -b_1 \tag{5.30}$$

$$2a_2q_2 + \lambda_0 p_2 + \lambda_2 = -b_2 \tag{5.31}$$

$$p_1q_1 + p_2q_2 + S = Y^0 \tag{5.32}$$

$$\lambda_0 + \lambda_3 = 0 \tag{5.33}$$

$$\lambda_1 q_1 = 0 \tag{5.34}$$

$$\lambda_2 q_2 = 0 \tag{5.35}$$

$$\lambda_3 S = 0 \tag{5.36}$$

Equations (5.30)–(5.33) provide four linear equations in seven unknowns ($q_1, q_2, S, \lambda_0, \lambda_1, \lambda_2$, and λ_3); however, (5.34)–(5.36) provide for three of the seven variables to have zero values and eight cases.

Example 5.9

Suppose, for example, that one consumer's utility function is $U = -0.05q_1^2 - 0.10q_2^2 + 100q_1 + 150q_2$ and the fixed prices of commodities 1 and 2 are \$3 and \$4 respectively. If the consumer has an income of \$8000, his utility function is constrained by $3q_1 + 4q_2 \leq 8000$, and $q_1, q_2 \geq 0$.

The necessary linear conditions for this consumer to maximize his utility subject to these constraints are those given in (5.30)–(5.33), where $a_1 = -0.50$, $a_2 = -0.10$, $b_1 = 100$, $b_2 = 150$, $p_1 = 3$, $p_2 = 4$, and $Y^0 = 8000$. Therefore,

$$\begin{aligned} -0.10q_1 + 3\lambda_0 + \lambda_1 &= -100 \\ -0.20q_2 + 4\lambda_0 + \lambda_2 &= -150 \\ 3q_1 + 4q_2 + S &= 8000 \\ \lambda_0 + \lambda_3 &= 0 \\ \lambda_1 q_1 = 0 \qquad \lambda_2 q_2 = 0 \qquad \lambda_3 S &= 0 \end{aligned} \tag{5.37}$$

Replacing λ_0 by $-\lambda_3$ from (5.37) leaves

$$-0.10q_1 - 3\lambda_3 + \lambda_1 = -100$$
$$-0.20q_2 - 4\lambda_3 + \lambda_2 = -150$$
$$3q_1 + 4q_2 + S = 8000$$

and $\lambda_1 q_1 = 0$, $\lambda_2 q_2 = 0$, and $\lambda_3 S = 0$.

The eight alternative solutions to these constraints are provided in Table 5.4. The optimum is case 4 where $q_1 = 0$, $q_2 = 750$, and $S = 5000$.

Table 5.4. Alternative Solutions to Constraints

Case	x_1	x_2	x_3	q_1	q_2	S	U
1	0	0	0	1000	750	2000	−343,750.00
2	0	0	−11.76	1352.94	985.29	0	−729,217.13
3	0	−150	0	1000	0	5000	−400,000.00
[4]	−100	0	0	0	750	5000	56,250.00
5	0	−372.22	−55.56	2666.67	0	0	−9,533,333.33
6	−287.5	0	−62.5	0	2000	0	−100,000
7	−100	−150	0	0	0	8000	0
8†	*	*	*	0	0	0	*

†Not a unique feasible case.

Matrix Models of the Firm

The models presented in the beginning of this section suggest the multitude of quadratic programming problems that should be integrated with the neoclassical theory of the firm. When the quantity demanded **q** is a linear function of the price **p**, total revenue TR is a quadratic function. If **p** and **q** are vectors and the demand functions are given by

$$\mathbf{p} = \mathbf{B}\mathbf{q} \tag{5.38}$$

then

$$TR = \mathbf{q}'\mathbf{p} = \mathbf{q}'\mathbf{B}\mathbf{q} \tag{5.39}$$

is the firm's total revenue function. In the short run, output levels may often be linearly related to the vector of inputs **x** so that

$$\mathbf{q} = \mathbf{A}\mathbf{x} \tag{5.40}$$

The matrix **A** is a matrix of marginal products.

Suppose that **b** is a vector of the quantities of the resources available to the firm. If the elements of the matrix $\mathbf{C}$ represent the input-output coefficients,

$$\mathbf{Cq} \leq \mathbf{b} \tag{5.41}$$

When there is a minimum output level, a constraint such as

$$\mathbf{q} \geq \mathbf{k} \tag{5.42}$$

may be relevant. Then

$$\mathbf{Ax} \geq \mathbf{k} \tag{5.43}$$

from (5.40) and (5.42).

Just as the outputs might be sold in markets in which the prices and quantities are related, this may be true for the firm's input purchases. If $\mathbf{r}$ is a vector of input prices,

$$\mathbf{r} = \mathbf{Dx} \tag{5.44}$$

may represent the firm's input demand functions, and the firm's total costs TC are determined as

$$TC = \mathbf{x}'\mathbf{r} = \mathbf{x}'\mathbf{Dx} \tag{5.45}$$

These functions (5.38)–(5.45) may be used to construct any number of models for the firm. In (5.38), (5.40), and (5.44) intercepts can be included by letting the first element of $\mathbf{q}$, $\mathbf{x}$, or $\mathbf{x}$ respectively be one.

If the firm maximizes its profits, $\pi = TR - TC$, subject to the production restraints, the problem may be formulated as

$$\begin{aligned} \text{maximize} \quad & \pi = \mathbf{q}'\mathbf{Bq} - \mathbf{x}'\mathbf{Dx} \\ \text{subject to} \quad & \mathbf{q} = \mathbf{Ax} \\ & \mathbf{q}, \mathbf{x} \geq \mathbf{0} \end{aligned}$$

If the resources to be applied are fixed at $\mathbf{b}$ and the firm maximizes sales revenue rather than profits, the problem is

$$\begin{aligned} \text{maximize} \quad & TR = \mathbf{q}'\mathbf{Bq} \\ \text{subject to} \quad & \mathbf{Cq} \leq \mathbf{b} \\ & \mathbf{q} \geq \mathbf{0} \end{aligned}$$

Another interesting case appears for the firm that sells its outputs at fixed prices and depends upon minimizing its total costs. The optimization problem for such a firm is

$$\begin{aligned} \text{minimize} \quad & TC = \mathbf{x}'\mathbf{D}\mathbf{x} \\ \text{subject to} \quad & \mathbf{A}\mathbf{x} \geq \mathbf{k} \\ & \mathbf{x} \geq \mathbf{0} \end{aligned}$$

These three optimization problems for the firm are provided to emphasize to the reader that some of the neoclassical theory of the firm falls within the scope of quadratic programming. It has already been suggested that input and output levels for the firm may well be set at zero levels for any short-run period. For the multiproduct firm that employs a variety of inputs, it may be impossible to produce all the products in any single period.

Restricted Least Squares

Example 5.10

An important problem from econometrics also requires the application of quadratic programming. This is the problem of selecting least squares estimators, the elements of **B**, to

$$\text{minimize} \quad \mathbf{e}'\mathbf{e} = \mathbf{Y}'\mathbf{Y} - 2\mathbf{B}'\mathbf{X}'\mathbf{Y} + \mathbf{B}'\mathbf{X}'\mathbf{X}\mathbf{B} \tag{5.46}$$

when there are linear restrictions on the elements of **B**. Goldberger [1964] and Zellner [1963] have each considered cases of this model. Often when economic relationships are to be estimated, the signs and sum of the parameters are given. An example appears when we try to estimate the parameters of a Cobb-Douglas production function. The problem is to determine A, B_1, B_2, and B_3, where output $\mathbf{y}$ is a function of the inputs $\mathbf{x}_1, \mathbf{x}_2$, and $\mathbf{x}_3$ so that

$$\mathbf{y} = A\mathbf{x}_1^{B_1}\mathbf{x}_2^{B_2}\mathbf{x}_3^{B_3}\mathbf{E} \tag{5.47}$$

and **E** is a multiplicative statistical error term. Let

$$\begin{aligned} \mathbf{Y} &= \log \mathbf{y} & \mathbf{X}_3 &= \log \mathbf{x}_3 \\ \mathbf{X}_1 &= \log \mathbf{x}_1 & \mathbf{e} &= \log \mathbf{E} \\ \mathbf{X}_2 &= \log \mathbf{x}_2 & B_0 &= A \end{aligned}$$

Then the production function is estimated by minimizing $\mathbf{e}'\mathbf{e}$ in (5.46).

From (5.47)

$$\frac{\partial \mathbf{y}}{\partial \mathbf{x}_1} = B_1 A \mathbf{X}_1^{B_1 - 1} \mathbf{X}_2^{B_2} \mathbf{X}_3^{B_3} \mathbf{E}$$

and

$$\frac{\partial \mathbf{y}}{\partial \mathbf{x}_1} \frac{\mathbf{x}_1}{\mathbf{y}} = \frac{B_1 A \mathbf{x}_1^{B_1} \mathbf{x}_2^{B_2} \mathbf{x}_3^{B_3} \mathbf{E}}{\mathbf{y}} = B_1$$

Likewise,

$$B_2 = \frac{\partial \mathbf{y}}{\partial \mathbf{x}_2} \frac{\mathbf{x}_2}{\mathbf{y}} \qquad B_3 = \frac{\partial \mathbf{y}}{\partial \mathbf{x}_3} \frac{\mathbf{x}_3}{\mathbf{y}}$$

Each B_i is an elasticity of input substitution. Therefore,

$$B_1, B_2, B_3 \geq 0 \tag{5.48}$$

is a reasonable requirement.

The value of $B_1 + B_2 + B_3$ determines whether there are increasing, constant, or decreasing returns to the scale of the production process from changing the inputs by a fixed proportion. If each $\mathbf{x}_i$ in (5.47) is multiplied by k and $\mathbf{y}^*$ replaces $\mathbf{y}$,

$$\mathbf{y}^* = A(k\mathbf{x}_1)^{B_1}(k\mathbf{x}_2)^{B_2}(k\mathbf{x}_3)^{B_3}\mathbf{E}$$

or

$$\mathbf{y}^* = A\mathbf{x}_1^{B_1}\mathbf{x}_2^{B_2}\mathbf{x}_3^{B_3}\mathbf{E}k^{B_1+B_2+B_3}$$
$$\mathbf{y}^* = \mathbf{y}k^{B_1+B_2+B_3}$$

If

$$B_1 + B_2 + B_3 > 1 \qquad k > 1 \tag{5.49}$$

$\mathbf{y}^* > k\mathbf{y}$, and there are increasing returns to scale. Increasing each input by the percentage k increases the output by more than k percent.

If there are constant returns to scale, $\mathbf{y}^* = k\mathbf{y}$ and

$$B_1 + B_2 + B_3 = 1 \tag{5.50}$$

or, if there are decreasing returns to scale, $\mathbf{y}^* < k\mathbf{y}$ and

$$B_1 + B_2 + B_3 < 1 \tag{5.51}$$

Thus it is likely that (5.48) and one of (5.49), (5.50), or (5.51) form the relevant constraints subject to which the error sum of squares, $\mathbf{e}'\mathbf{e}$ in (5.46), is minimized. It may also be necessary to identify an upper limit for one or more of the B's. If one of the inputs is applied in relatively inelastic terms, the exponent of the ith input must be less than one. Failure to recognize constraints such as these often leads statisticians to determine meaningless parameter estimates when considering the statistical properties, while ignoring the economic content of regression models.

5.8 BASIC PORTFOLIO MODELS

Markowitz [1959] has presented an interesting application of quadratic programming to financial analysis. The model is one in which a fixed amount of cash is to be allocated for investment into N stocks. Sharpe [1970] has extended the work by Markowitz to consider capital market theory. Similar models have also been employed to analyze allocations among projects at the sector or regional level for macroeconomic investment planning.

Suppose that N stocks are to be considered for possible inclusion in a client's portfolio and that x_i is the percentage of the available money that will be invested in stock i. If $x_i = 0$, nothing is invested in stock i. If $x_i = 1$, 100% of the money is invested in stock i. In general, the constraints on the allocation process are $\sum_{i=1}^{N} x_i \leq 1$ and $x_i \geq 0$, $i = 1, \ldots, N$.

If p_i is the price of the ith stock, the weighted average or value of the portfolio can be represented by $P = \sum_{i=1}^{N} p_i x_i$. For the case of the portfolio of three stocks where $p_1 = 5$, $p_2 = 7$, $p_3 = 4$, $x_1 = 0.4$, $x_2 = 0.1$, and $x_3 = 0.5$, $P = (5)(0.4) + (7)(0.1) + (4)(0.5) = 4.7$. However, in most cases the prices of the stocks vary. The usual approach for analyzing stock portfolios under this condition is to concentrate on the mean or average value of P and the variance of P.

The mean and variance of P, $\bar{P}$ and $V(P)$ respectively, can be defined as

$$\bar{P} = \sum_{i=1}^{N} u_i x_i \tag{5.52}$$

$$V(P) = \sum_{i=1}^{N} \sum_{j=1}^{N} \sigma_{ij} x_i x_j \tag{5.53}$$

Here u_i is the average value for the price of stock i, σ_{ij} is the covariance between the prices of stocks i and j, and $\sigma_{ii} = \sigma_i^2$ represents the variance in the price of stock i.

In the best known portfolio models, which have been solved using quadratic programming, the objective function is a combination of the mean $\bar{P}$ and the variance $V(P)$ of the portfolio value. The usual approach is to define

$$W = \sum_{i=1}^{N} u_i x_i - \rho \sum_{i=1}^{N} \sum_{j=1}^{N} \sigma_{ij} x_i x_j \tag{5.54}$$

as the objective function to be maximized for the client whose portfolio is being managed. From (5.52)–(5.54), $W = \bar{P} - \rho V(P)$.

Each of the parameters in the objective function (5.54) has been defined and explained except for ρ. The value of ρ is a measure of the willingness of the client to accept risk in his portfolio. The client can act as a gambler and have the expected value of the portfolio maximized if he sets $\rho = 0$. The problem that remains when $\rho = 0$ is the linear programming problem,

$$\begin{aligned} \text{maximize} \quad & W = \textstyle\sum_{i=1}^{N} u_i x_i \\ \text{subject to} \quad & \textstyle\sum_{i=1}^{N} x_i \leq 1 \\ & x_i \geq 0 \qquad i = 1, \ldots, N \end{aligned}$$

The investor who is unwilling to accept any risk in his portfolio may employ the objective function given by (5.54) if he sets $\rho = \infty$. Then maximizing

$$W = \bar{P} - \infty V(P) \tag{5.55}$$

is virtually maximizing $-\infty V(P)$ or minimizing $\infty V(P)$, since $\bar{P}$ be comes insignificant in (5.55).

The general portfolio problem for the investor who seeks a large expected value at the expense of some risk is

$$\begin{aligned} \text{maximize} \quad & W = \textstyle\sum_{i=1}^{N} u_i x_i - \rho \sum_{i=1}^{N} \sum_{j=1}^{N} \sigma_{ij} x_i x_j \\ \text{subject to} \quad & \textstyle\sum_{i=1}^{N} x_i \leq 1 \\ & x_i \geq 0 \qquad i = 1, \ldots, N \end{aligned}$$

Given values for the u_i's and σ_{ij}'s, an optimal allocation of the cash among the N stocks can be determined as a function of ρ.

Example 5.11

If the average prices are given by

$$\begin{bmatrix} u_1 \\ u_2 \\ u_3 \end{bmatrix} = \begin{bmatrix} 6 \\ 8 \\ 9 \end{bmatrix}$$

and the variance-covariance matrix among the stock prices is

$$\mathbf{V} = \begin{bmatrix} \sigma_{11} & \sigma_{12} & \sigma_{13} \\ \sigma_{12} & \sigma_{22} & \sigma_{23} \\ \sigma_{13} & \sigma_{23} & \sigma_{33} \end{bmatrix} = \begin{bmatrix} 8 & 5 & 7 \\ 5 & 16 & 6 \\ 7 & 6 & 30 \end{bmatrix}$$

the investor should

$$\begin{aligned} &\text{maximize} \\ &W = 6x_1 + 8x_2 + 9x_3 \\ &\quad -\rho(8x_1^2 + 16x_2^2 + 30x_3^2 + 10x_1x_2 + 14x_1x_3 + 12x_2x_3) \\ &\text{subject to} \quad x_1 + x_2 + x_3 \leq 1 \\ &\qquad\qquad\quad x_1, x_2, x_3 \geq 0 \end{aligned}$$

The optimal allocations for this client are presented in Table 5.5 for four values of ρ.

Table 5.5. Optimal Allocations

ρ	x_1	x_2	x_3	W
0.5	0.4356	0.3130	0.1358	3.1696
2	0.1359	0.0825	0	0.7379
5	0.0436	0.0313	0.0136	0.3170
10	0.0218	0.0156	0.0068	0.1585

The results presented in Table 5.5 illustrate the significance of the parameter ρ in the investment portfolio model. As the value of ρ is increased, the investor accepts less risk in his portfolio, and the value of the objective function W decreases. For values of ρ larger than 10, the value of W falls below 0.1585. If the investor decides to ignore the risk in the model, he sets $\rho = 0$ and maximizes the linear objective function $W = 6x_1 + 8x_2 + 9x_3$; the optimal solution is (x_1^*, $x_2^* = 0$, $x_3^* = 1$, $W^* = 9$).

In general, there are an infinite number of possible values for ρ. If the optimal allocation among x_1, x_2, and x_3 is determined for every value of ρ, an efficient set of portfolios is identified as a function of the investor's concern for risk.

In this chapter and Chapter 2 the programming problems have been reduced to various sets of solutions of n linear equations in n unknowns. When there has been risk or uncertainty involved in the parameters of the model, this issue has been circumvented. Dealing with risk and uncertainty in models in economics and business will be the main concern in Chapter 10.

PROBLEMS

5.1. Find the optimal value of the objective function and x in each of the following cases. Are the second-order conditions satisfied?

(a) maximize $z = 45x - 15x^2$
subject to $0 \leq x \leq 6$

(b) minimize $\psi = 6x^2 - 30x$
subject to $x \geq 4$

5.2. Solve each of the following problems using the Lagrangian multiplier method. Check to see if the second-order conditions are satisfied.

(a) maximize $z = 25x_1 + 36x_2 - 3x_1^2 + 5x_1x_2 - 9x_2^2$
subject to
$$2x_1 + 5x_2 \leq 30$$
$$5x_1 + 2x_2 \leq 40$$
$$x_1, x_2 \geq 0$$

(b) minimize $\psi = 15y_1^2 + 4y_1y_2 + 6y_2^2 - 2y_1 - 3y_2$
subject to
$$3y_1 + 6y_2 \geq 20$$
$$y_1 + y_2 \geq 4$$
$$y_1, y_2 \geq 0$$

5.3. A firm has two plants in which costs per hour are c_1 and c_2 respectively. The outputs of the two plants are x_1 and x_2 respectively.

$$c_1 = 80 + 2x_1 + 0.001x_1^2$$
$$c_2 = 90 + 1.5x_2 + 0.002x_2^2$$
$$x_1, x_2 \geq 0$$

The corporate manager has decided that 2000 units of output must be produced. Therefore, $x_1 + x_2 = 2000$. Use the Lagrangian multiplier method to determine the optimal levels of x_1 and x_2 to minimize total costs and satisfy the restrictions. Check to see if the second-order conditions are satisfied.

5.4. A monopolist produces two products at levels x_1 and x_2. Find the prices p_1 and p_2 that should be charged for the firm to maximize profits. Check to see that the second-order conditions are satisfied.

$$\begin{aligned} &\text{cost function:} && c = x_1 + 2x_2 + 1300 \\ &\text{demand curves:} && x_1 = 600\,p_2 - 400\,p_1 \\ &&& x_2 = 1800 - 100\,p_1 - 300\,p_2 \\ &&& x_1, x_2, p_1, p_2 \geq 0 \end{aligned}$$

5.5. Given

$$\mathbf{e} = \begin{bmatrix} e_1 \\ e_2 \\ e_3 \\ e_4 \end{bmatrix} \quad \mathbf{y} = \begin{bmatrix} 2 \\ 4 \\ 7 \\ 9 \end{bmatrix} \quad \mathbf{w} = \begin{bmatrix} 3 \\ 6 \\ 4 \\ 2 \end{bmatrix} \quad \mathbf{z} = \begin{bmatrix} 5 \\ 3 \\ 2 \\ 1 \end{bmatrix}$$

(a) Find a and b that are estimators of $\mathbf{y} = a + b\mathbf{x} + \mathbf{e}$ such that $a + b \leq 10$. The value of $\mathbf{e}'\mathbf{e}$ must be a minimum.

(b) Find b and c that are estimators of $\mathbf{y} = b\mathbf{w} + c\mathbf{z} + \mathbf{e}$ such that $b \geq 0$. The value of $\mathbf{e}'\mathbf{e}$ must be a minimum.

5.6. Set up the dual quadratic programming problems for 5.1(a), 5.1(b), 5.2(a), and 5.2(b). Do not solve the problems.

5.7. Consider the portfolio model in which two stocks are to be considered. Any money not allocated to one of the stocks is held in a savings account at 4% annual interest. The investor's data follows.

Stock	Annual rate of return	Variance of return
1	12%	20
2	9%	10

The stocks are independent: $\sigma_{12} = 0$.

(a) Construct the programming model for the investor who maximizes his return and ignores the variance in his portfolio.

(b) Construct and solve the programming problem for the investor who minimizes the variance in his portfolio. The investor demands an annual return of at least 10%.

5.8. A firm produces two products at levels x and y. The firm's revenue and cost functions are $R(x,y)$ and $C(x,y)$ respectively.

$$R(x,y) = -3x^2 - 600x - 8y^2 + 400y + 6xy$$
$$C(x,y) = 2x^2 + 100x + y^2 - 80y - 2xy$$

The firm's profit p is $R(x,y) - C(x,y)$.
(a) Find the values of x and y that maximize p.
(b) Find the values of x and y that maximize $R(x,y)$.
(c) Find the values of x and y that minimize $C(x,y)$.
Check to see that the second-order conditions are satisfied for each case.

5.9. Consider the quadratic programming to

$$\begin{aligned} \text{maximize} \quad & w = \mathbf{c}'\mathbf{x} + \mathbf{x}'\mathbf{A}\mathbf{x} \\ \text{subject to} \quad & \mathbf{R}\mathbf{x} \leq \mathbf{s} \\ & \mathbf{x} \geq \mathbf{0} \end{aligned}$$

where

$$\mathbf{c}' = [4 \;\; 7 \;\; 13] \qquad \mathbf{x}' = [x_1 \;\; x_2 \;\; x_3] \qquad \mathbf{R} = [1 \;\; 1 \;\; 1] \qquad \mathbf{s} = [1]$$

$$\mathbf{A} = \begin{bmatrix} -75 & -90 & -135 \\ -90 & -525 & -375 \\ -135 & -375 & -2400 \end{bmatrix}$$

Solve this problem using both the Wolfe algorithm and the Lagrangian multiplier technique.

6 INTEGER PROGRAMMING

6.1 INTRODUCTION

In the previous chapters it has been assumed that in all mathematical programming problems the variables are continuous. However, in many problems, particularly in business and economics, noninteger solutions to a mathematical programming problem are not desirable because of the interpretation given to the solution. As an example, suppose the requirement is to optimally allocate six trucks for transporting goods to or from three different metropolitan areas. The optimal solution might require that two trucks be allocated for one area, three and one-half to a second area, and one-half to a third area, which would be infeasible.

The requirement for an integer-valued solution may arise in either a linear or nonlinear mathematical programming problem. In this chapter, only the integer linear programming problems are considered in detail. A method to solve the integer nonlinear problem is suggested; dynamic programming is applied to provide a solution method for this case in Chapter 7. Before presenting algorithms for solving the integer linear problem, situations in which such a model may arise will be considered.

6.2 INTEGER PROGRAMMING MODELS

Model 6.1. The Transportation Model

Suppose there are N origins (for example, warehouses) that will supply M destinations (retailers) with a specified product. Let a_i represent the amount of the product available at the ith origin, b_j the amount required at the jth destination, and c_{ij} the cost of shipping one unit of the product from the ith origin to the jth destination. Let x_{ij} be the number of units of the product shipped from the ith origin to the jth destination. It is desired to select the optimum shipping pattern (x_{ij}, $i = 1, 2, \ldots, N$; $j = 1, 2, \ldots, M$) by minimizing the total shipping cost, assuming that shipping costs are proportional to the amount shipped. To resolve this problem, the following integer linear programming model is formulated.

$$\begin{aligned} \text{minimize} \quad z &= \textstyle\sum_{i=1}^{N} \sum_{j=1}^{M} c_{ij} x_{ij} \\ \text{subject to} \quad \textstyle\sum_{j=1}^{M} x_{ij} &= a_i \qquad i = 1, 2, \ldots, N; a_i > 0 \quad (6.1) \\ \textstyle\sum_{i=1}^{N} x_{ij} &= b_j \qquad j = 1, 2, \ldots, M; b_j > 0 \quad (6.2) \\ x_{ij} &= 0, 1, 2, \ldots \text{ for all } i \text{ and } j \quad (6.3) \end{aligned}$$

The first N constraints (6.1) require that the total number of units shipped from any one of the origins to each of the destinations must equal the number of units available at the origin. The M constraints (6.2) require that the total number of units shipped from each of the origins to any given destination must equal the amount requested by the destination. In most applications it is impossible to ship a fractional unit to a destination, and this restriction is specified in the model by (6.3), which makes it an integer programming problem.

It is possible to generalize the model by replacing (6.1) with

$$\sum_{j=1}^{M} x_{ij} \leq a_i \qquad i = 1, 2, \ldots, N; a_i > 0 \qquad (6.4)$$

or by replacing (6.2) with

$$\sum_{i=1}^{N} x_{ij} \geq b_j \qquad j = 1, 2, \ldots, M; b_j > 0$$

The N constraints in (6.4) perhaps more realistically depict the origin-destination problem than (6.1). This is certainly the case when the origins are warehouses and the destinations are retail outlets.

If the a_i and b_j of (6.1) and (6.2) respectively are integer-valued, the simplex algorithm will produce a solution satisfying

(6.3); i.e., the optimal values of the decision variables will all be integers. However, if one or more of the a_i or b_j are fractional, the simplex algorithm may not produce a solution satisfying (6.3), and one of the integer programming algorithms developed in this chapter should be used to solve the problem.

Model 6.2. The Assignment Model

Suppose it is desired to optimally assign N men to N jobs such that one and only one man can be assigned to each job and every man can do any of the N jobs. Define

c_{ij} = the performance measure for the ith man doing the jth job (may be anticipated costs, profits, or performance scores predicted by the personnel department, for example)

x_{ij} = 1 if the ith man is assigned to the jth job
= 0 if the ith man is not assigned to the jth job

Further, the assignments of the men to the jobs is to be made so that the sum of the performance measures is maximized; i.e.,

$$\text{maximize} \quad z = \textstyle\sum_{i=1}^{N} \sum_{j=1}^{N} c_{ij}x_{ij}$$

subject to the condition that each man does one and only one job:

$$\textstyle\sum_{j=1}^{N} x_{ij} = 1 \qquad i = 1, 2, \ldots, N$$

and one job is assigned to each man

$$\textstyle\sum_{i=1}^{N} x_{ij} = 1 \qquad j = 1, 2, \ldots, N$$

Notice that this is a special case of the transportation model presented in Model 6.1 ($a_i = b_i = 1$ for all i and j, and $M = N$).

Model 6.3. The Fixed Charge Model

In many business operations it is common to have a fixed charge or setup cost when ordering one or more units of a required product. Suppose, for example, that a company is concerned with ordering M products from a wholesaler. Let x_i be the number of units of the ith product ordered, c_i the cost of each unit of the ith product, and l_i the fixed charge incurred if at least one unit of the ith product is ordered. Further assume that the company can specify bounds on the number of units of the ith product required:

$$a_i \leq x_i \leq b_i \qquad a_i, b_i \geq 0 \qquad a_i \leq b_i$$

If it is desired to minimize total cost, the appropriate model is

$$\begin{aligned} \text{minimize} \quad & z = f_1(x_1) + f_2(x_2) + \cdots + f_M(x_M) \\ \text{subject to} \quad & a_i \leq x_i \leq b_i \qquad i = 1, 2, \ldots, M \\ & x_i = 0, 1, 2, \ldots \qquad i = 1, 2, \ldots, M \end{aligned} \tag{6.5}$$

where

$$\begin{aligned} f_i(x_i) &= l_i + c_i x_i \text{ if } x_i > 0 \\ &= 0 \text{ if } x_i = 0 \end{aligned}$$

In this model it is assumed there is no discount as x_i becomes large; c_i is constant for all i. If $l_i \equiv 0$ for each i, the model would become a linear programming model amenable to resolution by the simplex algorithm. However, any positive fixed charges induce nonlinearity in the objective function, since there will be jumps in z (e.g., $f_1(x_1) = 0$ when $x_1 = 0$ and jumps to $l_1 + c_1 x_1$ when $x_1 > 0$).

The above model can be converted to a linear model by the following method. First notice that the objective function z may be rewritten in the following form:

$$z = \sum_{i=1}^{M} (c_i x_i + l_i y_i)$$

where

$$\begin{aligned} y_i &= 1 \text{ if } x_i > 0 \\ &= 0 \text{ if } x_i = 0 \end{aligned}$$

If constraints can be introduced that insure that y_i, $i = 1, 2, \ldots, M$, will take on its proper binary value, the model can be formulated in terms of $2M$ variables (x_i, y_i; $i = 1, 2, \ldots, M$). The binary property of y may be insured by introducing the following $2M$ constraints:

$$y_i \leq 1, y_i \geq 0 \qquad i = 1, 2, \ldots, M$$

$$\text{subject to} \quad \mathbf{Ax} \leq \mathbf{b} \tag{6.11}$$
$$x_i = 0, 1, 2, \ldots \qquad i = 1, 2, \ldots, n \tag{6.12}$$

where $\mathbf{c}' = [c_1, c_2, \ldots, c_n]$, $\mathbf{x}' = [x_1, x_2, \ldots, x_n]$, $\mathbf{b}' = [b_1, b_2, \ldots, b_m]$, and

$$\mathbf{A} = \begin{bmatrix} a_{11} & a_{12} & \cdots & a_{1n} \\ a_{21} & a_{22} & \cdots & a_{2n} \\ \vdots & & & \\ a_{m1} & a_{m2} & \cdots & a_{mn} \end{bmatrix} \tag{6.13}$$

with the elements of **c**, **b**, and **A** being known constants.

The above model is a linear programming problem if the integer requirements (6.12) are dropped and replaced by $x_i \geq 0$, $i = 1, 2, \ldots, n$. Since the model is very similar to the linear programming model, it might be suspected that the optimal integer problem can be achieved by rounding to integers the solution of the analogous continuous variable problem solved by the simplex algorithm. This is not the case, as will be demonstrated by considering the following example.

Example 6.1

$$\text{maximize} \quad z = 21x_1 + 16x_2$$
$$\text{subject to} \quad 4x_1 + 6x_2 \leq 24 \tag{6.14}$$
$$7x_1 + 4x_2 \leq 28 \tag{6.15}$$
$$0 \leq x_1 \leq 3.75$$
$$x_1, x_2 = 0, 1, 2, \ldots \tag{6.16}$$

The problem is presented graphically, in Figure 6.1 where it is clear that there are only 14 feasible integer solutions. These solutions are given with their associated functional values in Table 6.1.

From Table 6.1 it is seen that the optimal integer solution occurs at the point ($x_1^* = 3$, $x_2^* = 1$) with an objective function value $z^* = 79$. Dropping the integer restrictions (6.16) and applying the simplex algorithm to this problem, an optimal solution with ($x_1 = 2.77$, $x_2 = 2.15$), and $z^* = 92.61$ is achieved. The continuous variable solution point (2.77, 2.15) occurs at the intersection of the first and second constraints (6.14) and (6.15) and is not a feasible solution to the integer problem. After rounding the simplex solution to integer values, the solution be-

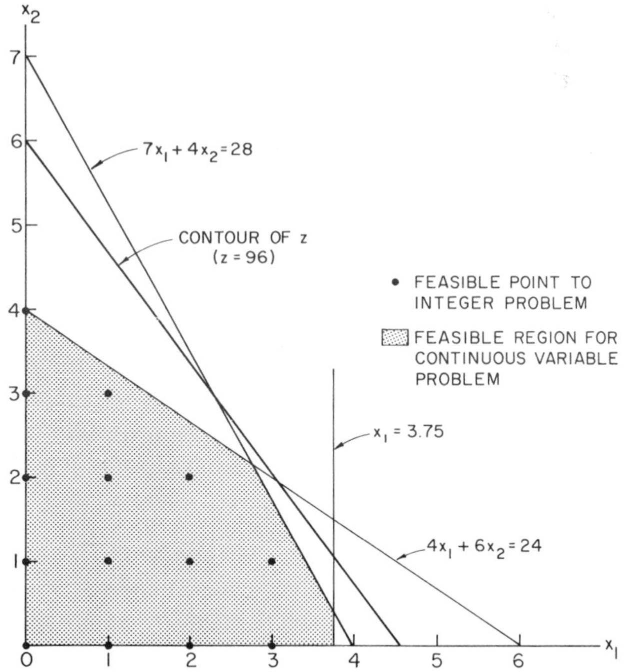

Fig. 6.1. Graphic presentation of the Example 6.1 problem.

Table 6.1. Feasible Integer Solutions and Associated Functional Values

(x_1, x_2)	z	(x_1, x_2)	z
(0, 0)	0	(1, 2)	53
(0, 1)	16	(1, 3)	69
(0, 2)	32	(2, 0)	42
(0, 3)	48	(2, 1)	58
(0, 4)	64	(2, 2)	74
(1, 0)	21	(3, 0)	63
(1, 1)	37	(3, 1)	79

comes ($x_1 = 3$, $x_2 = 2$), which violates the second constraint (6.15) since $7(3) + 4(2) > 28$.

Thus using the simplex solution with rounding fails to provide a feasible integer solution to the problem. Indeed, it is possible in some problems to determine a feasible but suboptimal integer solution by rounding the simplex solution (see Example 6.4).

However, the above discussion does suggest a way to use the rounded simplex solution. Exploring around the solution of ($x_1 = 3$, $x_2 = 2$) by moving up and down each axis from this point by one unit (3, 1), (3, 3), (2, 2), (4, 2), the optimal integer solution is encountered in this problem. Searching in the vicinity of the rounded simplex solution by this method will not always provide the optimal integer solution but will usually supply a good solution. More will be said about integer searching methods in the vicinity of the continuous variable solution point later in this chapter.

Before proceeding, note an obvious fact. If the simplex solution to the integer programming problem provides an integer solution, it is optimal. This fact is important and is occasionally overlooked in solving integer linear programming problems. Unless it is clear that the simplex solution will not be integral, as a first attempt in solving the integer problem apply the simplex algorithm. Even if the simplex solution is not integral, some valuable information will generally be gained about the problem.

Since the simplex algorithm will usually fail to provide a feasible integer solution to the general integer linear programming problem, special algorithms are required for solving it; some that are commonly utilized will now be considered.

If the problem is bivariate, graphing will most often provide the most efficient solution method. From Figure 6.1 it is quickly determined that the optimal integer solution is (3, 1). Graphic methods can be employed for three-dimensional problems but usually not without difficulty.

As Table 6.1 suggests, a total enumeration of all feasible integer points will guarantee locating the optimal integer solution. In the Example 6.1 problem, only fourteen points had to be enumerated. However, in large problems, total enumeration may require considerable computational effort in locating each and every feasible integer point and evaluating the objective function at each. Before presenting more efficient solution methods than total enumeration, an algorithm that will locate all feasible integer points will be delineated.

If the problem contains nonnegative restrictions on the variables, begin by selecting the feasible point $\mathbf{x}_1' = [0, 0, \ldots, 0]$ and evaluating the objective function at this point. Now increment by one unit the last variable value forming $\mathbf{x}_2' = [0, 0, \ldots, 1]$ and check this new point for feasibility. If it is feasible, evaluate the objective function at $\mathbf{x}_2'$. Continue to increase the value of the nth variable in this way until the $\mathbf{x}_{i+1}'$ point violates a constraint. Next, select the $\mathbf{x}_{i+1}' = [0, 0, \ldots, 1, 0]$ and proceed to increment the last variable as before. Continue this process of lexicographically incrementing each variable by one unit until the first variable can no longer be increased without producing an infeasible point. In Table 6.1 the pattern of the above enumeration scheme is indicated for the example problem.

Although this method appears cumbersome, the computer can very quickly form each new point, check for feasibility against the constraints, and evaluate a linear function. The check for feasibility will normally take the most time. However, if a point is infeasible, not all constraints need to be checked. Suppose there are m constraints and the rth ($r < m$) is violated by the point $\mathbf{x}'_s$, then the $m - r$ remaining constraints need not be checked and the $\mathbf{x}'_{s+1}$ point may be formed immediately. Fortunately, there are more efficient methods for solving the integer linear programming problem. Two of these will now be considered in detail.

6.3 CUTTING PLANE METHODS

Cutting plane methods, as the name suggests, attempt to cut or reduce the constraint region in the integer programming problem without deleting any feasible integer solutions. The cuts are achieved by introducing new constraints to the original constraint set of the problem. In a typical cutting plane algorithm, each iteration involves the addition of a constraint that will reduce the feasible region at the expense of enlarging the problem by one constraint. The algorithm terminates when, in an iteration, an all-integer solution is encountered from solving the continuous variable problem by the simplex algorithm.

Example 6.2

To illustrate the Gomory [1963] algorithm, it will be applied to Example 6.1. Recall that the simplex solution to this problem is $x_1 = 2.77$ and $x_2 = 2.15$. A method is needed for generating a new constraint that will reduce the original feasible region by eliminating noninteger feasible solutions. The Gomory method utilizes the current constraints to construct the cutting planes at each iteration. See Table 6.2 for the initial tableau.

Table 6.2. Initial Tableau

Cost coefficient:	21	16	0	0	0		
Variable:	x_1	x_2	s_1	s_2	s_3	Index	Basis
	4	6	1	0	0	24	s_1
	7	4	0	1	0	28	s_2
	[4]	0	0	0	1	15	s_3
$c_j - z_j$:	21	16	0	0	0		

Notice that the third constraint $x_1 \leq 3.75$ has been recast as $4x_1 \leq 15$. The reason for multiplying through to clear out the decimal portion of the constraint will become apparent shortly. The terminal solution via the simplex problem is given in Table 6.3.

Table 6.3. Final Tableau

Cost coefficient:	21	16	0	0	0		
Variable:	x_1	x_2	s_1	s_2	s_3	Index	Basis
	0	0	0.62	−0.92	1	3.92	s_3
	0	1	0.27	−0.15	0	2.15	x_2
	1	0	−0.15	0.23	0	2.77	x_1
$c_j - z_j$:	0	0	−1.17	−2.43	0		

The first constraint in the final tableau (Table 6.3) will now be used to generate a cutting plane. Write this constraint as

$$(0)x_1 + (0)x_2 + (0 + 0.62)s_1 + (-1 + 0.08)s_2 + (1 + 0)s_3 = 3 + 0.92 \qquad (6.17)$$

No feasible solution to the original problem may violate (6.17). Now rewrite this constraint as

$$(0)x_1 + (0)x_2 + (0.62)s_1 + (0.08)s_2 + (0)s_3 = 0.92 + [3 - (0)s_1 + (1)s_2 - (1)s_3] \qquad (6.18)$$

Suppose that x_1, x_2, s_1, s_2, and s_3 all take on integer values. Then the left-hand side of (6.18) is positive, which implies that the right-hand side must also be positive. The term in brackets must be integral (0, 1, . . .), and the left-hand side is therefore bounded below by 0.92; i.e., $(0)x_1 + (0)x_2 + (0.62)s_1 + (0.08)s_2 + (0)s_3 \geq 0.92$.

This expression must hold for all feasible integer solutions. Written more compactly, the new constraint is

$$0.62s_1 + 0.08s_2 \geq 0.92 \qquad (6.19)$$

Since $4x_1 + 6x_2 + s_1 = 24$ and $7x_1 + 4x_2 + s_2 = 28$, (6.19) may be written in terms of x_1 and x_2 as $(0.62)(24 - 4x_1 - 6x_2) + (0.08)(28 - 7x_1 - 4x_2) \geq 0.92$, or

$$3x_1 + 4x_2 \leq 16 \tag{6.20}$$

For convenience of presentation, the numbers in the simplex tableaus have been rounded to two decimal places. For example, in Table 6.3 the 0.62, −0.92, and 3.92 in the first row are 0.6154, −0.9231, and 3.9231 respectively when carried to four decimal places. Thus in forming (6.20), for example, the proper multipliers are 0.6154 and 0.0769.

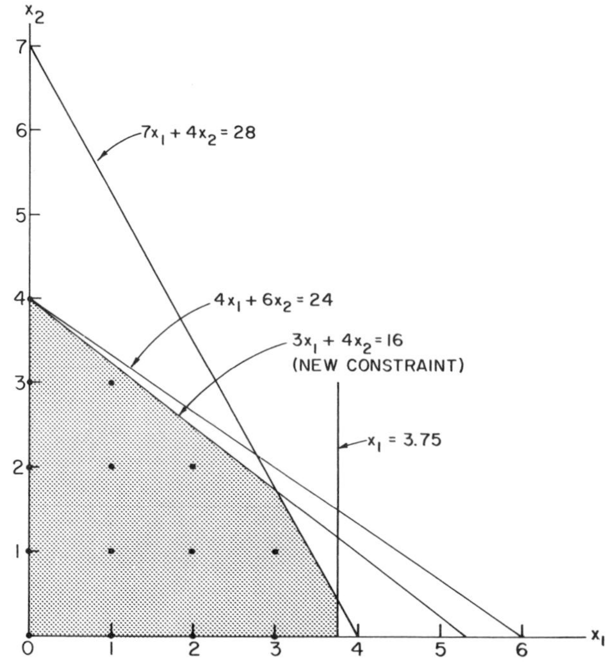

Fig. 6.2. Effect of adding a constraint.

From Figure 6.2 the effect of adding the constraint in (6.20) to the problem is clear. No integer solutions have been eliminated, but the feasible region for the continuous variable problem has been reduced in size. Rather than adjoining (6.20) to the original problem, (6.19) is used by first adding a slack and an artificial variable giving $0.62s_1 + 0.08s_2 - s_4 + s_5 = 0.92$. The initial tableau for the next iteration is given in Table 6.4. After applying the simplex algorithm to this tableau, see the final tableau (Table 6.5).

Table 6.4. Initial Tableau

Cost coefficient:	21	16	0	0	0	0	$-M$		
Variable:	x_1	x_2	s_1	s_2	s_3	s_4	s_5	Index	Basis
	0	0	0.62	−0.92	1	0	0	3.92	s_3
	0	1	0.27	−0.15	0	0	0	2.15	x_2
	1	0	−0.15	0.23	0	0	0	2.77	x_1
	0	0	[0.62]	0.08	0	−1	1	0.92	s_5
$c_j - z_j$:	0	0	$0.62M - 1.17$	$0.08M - 2.43$	0	$-M$	0		

Table 6.5. Final Tableau

Cost coefficient:	21	16	0	0	0	0	$-M$		
Variable:	x_1	x_2	s_1	s_2	s_3	s_4	s_5	Index	Basis
	0	0	0	−1	1	1	−1	3.00	s_3
	0	1	0	−0.18	0	0.44	−0.44	1.75	x_2
	1	0	0	0.25	0	−0.25	0.25	3.00	x_1
	0	0	1	0.12	0	−1.62	1.62	1.50	s_1
$c_j - z_j$:	0	0	0	−2.37	0	−1.79	$-M + 1.79$		

Since x_1 is now integral, one of the other constraints is selected to form the new cutting plane. If the second constraint is chosen,

$$0x_1 + 1x_2 + 0s_1 - 0.18s_2 + 0s_3 + 0.44s_4 - 0.44s_5 = 1.75 \tag{6.21}$$

In forming the new constraint, the method used above can be applied. However, the new cutting plane equation can be written immediately, since each coefficient in the constraint is replaced by the smallest positive number congruent to that coefficient. (A is congruent to B if $A - B$ is zero or an integer.) Consider the constraint in (6.21). The inequality achieved by replacing each coefficient by the smallest positive number (including zero) that is congruent to it is

$$(0)x_1 + (0)x_2 + (0.82)s_2 + (0.44)s_4 + (0.56)s_5 \geq 0.75 \tag{6.22}$$

Thus the new constraint is $0.82s_2 + 0.44s_4 + 0.56s_5 \geq 0.75$; and after adding a slack and artificial variable $0.82s_2 + 0.44s_4 + 0.56s_5 - s_6 + s_7 = 0.75$.

The constraint in (6.22) may be written in terms of x_1 and x_2 as

$$7x_1 + 5x_2 \leq 29 \tag{6.23}$$

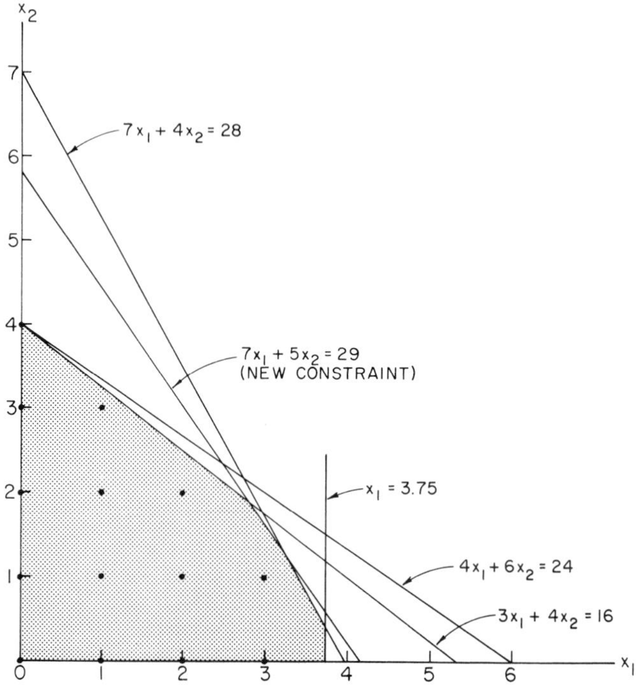

Fig. 6.3. Effect of adjoining the cutting plane.

The effect of adjoining the cutting plane in (6.23) to the problem is illustrated in Figure 6.3, the initial tableu in the second iteration is given in Table 6.6, and the final tableau after applying the simplex algorithm is given in Table 6.7.

Table 6.6 Initial Tableau

Cost coefficient:	21	16	0	0	0	0	$-M$	0	$-M$		
Variable:	x_1	x_2	s_1	s_2	s_3	s_4	s_5	s_6	s_7	Index	Basis
	0	0	0	-1	1	1	-1	0	0	3.00	s_3
	0	1	0	-0.18	0	0.44	-0.44	0	0	1.75	x_2
	1	0	0	0.25	0	-0.25	0.25	0	0	3.00	x_1
	0	0	1	0.12	0	-1.62	1.62	0	0	1.50	s_1
	0	0	0	[0.82]	0	0.44	0.56	-1	1	0.75	s_7
$c_j - z_j$:	0	0	0	$0.82M - 2.37$	0	$0.44M - 1.79$	$-0.44M + 1.79$	$-M$	0		

Table 6.7. Final Tableau

Cost coefficient:	21	16	0	0	0	0	$-M$	0	$-M$		
Variable:	x_1	x_2	s_1	s_2	s_3	s_4	s_5	s_6	s_7	Index	Basis
	0	0	0	0	1	1.54	-1.54	-1.23	1.23	3.92	s_3
	0	1	0	0	0	0.54	-0.54	-0.23	0.23	1.92	x_2
	1	0	0	0	0	-0.38	0.38	0.31	-0.31	2.77	x_1
	0	0	1	0	0	-1.69	1.69	0.15	-0.15	1.38	s_1
	0	0	0	1	0	0.54	-0.54	-1.23	1.23	0.92	s_2
$c_j - z_j$:	0	0	0	0	0	-0.66	$-M + 0.66$	-2.83	$-M + 2.83$		

Since x_1 and x_2 are noninteger, the process must continue. Select constraint number one from Table 6.7:

$$(0)x_1 + (0)x_2 + (0)s_1 + (0)s_2 + (1)s_3 + (1.54)s_4 - (1.54)s_5 - (1.23)s_6 + (1.23)s_7 = 3.92$$

which becomes the cutting plane constraint

$$(0)x_1 + (0)x_2 + (0)s_1 + (0)s_2 + (0)s_3 + (0.54)s_4 + (0.46)s_5 + (0.77)s_6 + (0.23)s_7 \geq 0.92$$

or more simply, $(0.54)s_4 + (0.46)s_5 + (0.77)s_6 + (0.23)s_7 \geq 0.92$.

The process of adding cutting planes is continued in the same manner until termination upon experiencing an all-integer solution to the iterative problem via the simplex algorithm. For this problem, the Gomory algorithm required 15 iterations before settling upon the optimal solution ($x_1^* = 3, x_2^* = 1$), with $z^* = 79$.

Prior to discussing the properties of this algorithm, the method for selecting the cutting plane at a given iteration is formalized. In the model formulation of the integer programming problem given in (6.10)–(6.13), let $\mathbf{A}^*$ represent the matrix of constant coefficients achieved by adding slack and artificial variables required to form an initial basis; $\mathbf{A}^*_{m \times s} = [\mathbf{A}, \mathbf{S}]$, where $\mathbf{A}$ is the $m \times n$ matrix in (6.13) and $\mathbf{S}$ is an $m \times (s - n)$ matrix containing coefficients for the slack and artificial variables. Now write $\mathbf{A}^*$ as $\mathbf{A}^* = [\mathbf{B}, \mathbf{E}]$, where $\mathbf{B}$ is the $m \times m$ basis matrix formed from $\mathbf{A}^*$ by placing the columns associated with the m basic variables in the first m columns of $\mathbf{A}^*$ and $\mathbf{E}$ is the $m \times (s - m)$ matrix of coefficients associated with the $s - m$ nonbasic variables. Let $\mathbf{x}^{*\prime} = [\mathbf{x}_B', \mathbf{x}_E']$ where $\mathbf{x}_B$ is the $m \times 1$ column vector of basic variables and $\mathbf{x}_E$ is the $(s - m) \times 1$ vector containing the nonbasic variables. Then since $\mathbf{A^*x^*} = \mathbf{b}$,

$$[\mathbf{B}, \mathbf{E}] \begin{bmatrix} \mathbf{x}_B \\ \mathbf{x}_E \end{bmatrix} = \mathbf{b}$$

and $\mathbf{B}\mathbf{x}_B + \mathbf{E}\mathbf{x}_E = \mathbf{b}$. But since $\mathbf{B}$ is the $m \times m$ basis matrix, $\mathbf{B}^{-1}$ exists and

$$\mathbf{x}_B = \mathbf{B}^{-1}[\mathbf{b} - \mathbf{E}\mathbf{x}_E] = \mathbf{B}^{-1}\mathbf{b} - \mathbf{B}^{-1}\mathbf{E}\mathbf{x}_E$$

Now let $\mathbf{Z} = \mathbf{B}^{-1}\mathbf{b}$ and $\mathbf{W} = \mathbf{B}^{-1}\mathbf{E}$. Thus $\mathbf{x}_B$ may be written in terms of the $(m \times 1)$ vector $\mathbf{Z}$ and the $m \times (s - m)$ matrix $\mathbf{W}$ as $\mathbf{x}_B = \mathbf{Z} - \mathbf{W}\mathbf{x}_E$.

Let $\mathbf{a}_i$ be the ith $(m \times 1)$ column vector of $\mathbf{E}$;

$$\mathbf{E} = [\mathbf{a}_1, \mathbf{a}_2, \ldots, \mathbf{a}_i, \ldots \mathbf{a}_{s-m}]_{m \times (s-m)} \tag{6.24}$$

Define by $\mathbf{W}_i$ the ith $m \times 1$ column vector of $\mathbf{W}$. From (6.24) it follows that $\mathbf{W}_i = \mathbf{B}^{-1}\mathbf{a}_i$, and $\mathbf{x}_B$ may be written in terms of $\mathbf{W}_i$, $i = 1, 2, \ldots, (s - m)$, as

$$\mathbf{x}_B = \mathbf{Z} - \sum_{i \in E} \mathbf{W}_i x_i \tag{6.25}$$

where the summation in (6.25) covers the nonbasic variables x_i, $i \in E$. Notice that for the current basis $x_i = 0$, $i \in E$, so that $\mathbf{x}_B = \mathbf{Z}$. Thus if the jth basic variable x_{Bj}, $j \in B$, is noninteger, so is the jth component of $\mathbf{Z}$, denoted by Z_j. Now write Z_j as the sum of an integer f_j and a fractional part h_j:

$$Z_j = f_j + h_j \tag{6.26}$$

where $f_j \geq 0$ and $0 < h_j < 1$. Additionally, write the jth element of $\mathbf{W}_i$, denoted by W_{ji}, as the sum of an integer f_{ji} and a fractional part h_{ji}:

$$W_{ji} = f_{ji} + h_{ji} \tag{6.27}$$

where $i \in E$ and $0 \leq h_{ji} < 1$ for each i and j. Using (6.26) and (6.27), the jth element of (6.25) is

$$\begin{aligned} x_{Bj} &= (f_j + h_j) - \sum_{i \in E} (f_{ji} + h_{ji})x_i \\ &= (f_j - \sum_{i \in E} f_{ji}x_i) + (h_j - \sum_{i \in E} h_{ji}x_i) \end{aligned} \tag{6.28}$$

If the new solution $\mathbf{x}_B$ is to be an integer, at least one x_i, $i \in E$, must become positive and enter the new basis. The first term on the right-hand side of (6.28) must be an integer, since f_j and f_{ji} for $i \in E$ are by definition integral and each x_i, $i \in E$, is an integer (or zero). Thus the jth element of the new integer solution may be written as

$$x_{Bj} = I_j + \left(h_j - \sum_{i \in E} h_{ji} x_i\right) \tag{6.29}$$

where I_j is an integer (positive or negative). If x_{Bj} is an integer, the second term on the right-hand side of (6.29) must be an integer. But $0 < h_j < 1$ and $x_i \geq 0$. Therefore, this term will be integer only if

$$h_j - \sum_{i \in E} h_{ji} x_i \leq 0 \tag{6.30}$$

If, for example, the left-hand side of (6.30) was greater than zero, this would imply that

$$h_j > \sum_{i \in E} h_{ji} x_i \tag{6.31}$$

But since $0 < h_j < 1$ and $\sum_{i \in E} h_{ji} x_i > 0$, if (6.31) were satisfied, the term $h_j - \sum_{i \in E} h_{ji} x_i$ could not be an integer. Thus for this term to be an integer, (6.30) must hold. However, it should be noted that the condition in (6.30) certainly does not *guarantee* that $\mathbf{x}_B$ will be integral; rather, it is only a necessary condition. It is also clear that any integer solution to the problem will satisfy (6.30), but the current solution $\mathbf{x}_B$ will not, since $x_i = 0, i \in E$. Thus by augmenting the constraint (6.30) to the original problem, we are forced to move from the current solution point $\mathbf{x}_B$. The continuous variable constraint space is consequently reduced without eliminating any feasible integer points.

Backtrack momentarily and consider the reason for changing the third constraint from $x_1 \leq 3.75$ to $4x_1 \leq 15$ in the Example 6.2 problem. In the development of the cutting plane above, notice that it is required that all variables in $\mathbf{x}^*$ be integral including the slack and artificial variables. Writing $x_1 + s_3^* = 3.75$, it is clear that s_3^* cannot be an integer if x_1 is integral. Certainly, the two constraints $x_1 \leq 3.75$ and $4x_1 \leq 15$ are inequality equivalents. Therefore, care must be taken to insure that all variables in $\mathbf{x}^*$ can be integral before applying the Gomory algorithm to an integer programming problem.

After reflecting upon the cuts made in the first two iterations (see Figures 6.2 and 6.3), it might be asked why more efficient cuts could not be made. For example, if the initial cuts had been $x_1 + x_2 \leq 4$ and

$x_1 \leq 3$, no integer solution would have been deleted, and the simplex algorithm would have identified the solution in two iterations. Other methods have been suggested for forming cuts, e.g., Glover [1968], but none of these have been shown to enjoy significant computational efficiency over the Gomory method.

The above comment conveniently leads to a discussion of considerations made in employing the Gomory algorithm and its properties. Suppose there is more than one noninteger valued variable in the current solution. Which constraint should be selected to form the plane cut in (6.30)? An operating rule might be to take the constraint associated with the variable that has the largest fractional part in the current solution. No theoretical development has been proffered for selecting the variable and associated constraint from which the cutting plane is formed. However, the decision can significantly affect the computational efficiency on occasion. Regardless of the selection mechanism for forming the Gomory cutting plane (apart from always using a basis variable which is noninteger in the current solution), Gomory has shown that the algorithm will converge in a finite number of steps to the optimal integer solution.

A disadvantage of the Gomory algorithm is its failure to produce integer solutions at intermediate iterations. Recall that the process stops when the first integer solution is encountered. In some problems many cuts may be required before termination, and the computational cost may become excessive. Other integer programming algorithms such as the branch and bound method presented in Section 6.4 will provide good all-integer solutions if they are terminated before encountering the optimal integer solution. It is also extremely difficult to gauge how many cuts will be required for solving a particular problem by the Gomory algorithm. For example, in the bivariate problem sixteen cuts were required for termination, while in other bivariate problems perhaps only one cut may be required (see Problem 6.2). Normally, if the Gomory algorithm will solve a problem within a reasonable number of iterations, it will do so by making only a very few cuts. If a large number of cuts are made without resolving the problem, it is recommended that the problem be recast by changing the order of the constraints, multiplying one or more constraints by a positive constant, or changing the number of constraints (e.g., adding redundant constraints).

An additional disadvantage of the Gomory algorithm appears to be that the number of constraints increases by one in each iteration, creating a sizable linear programming problem if many iterations are required. However, with a little extra effort redundant constraints at each iteration may be eliminated from the current tableau so that normally no more than three or four additional constraints to the number in the original problem need to be carried along from iteration to iteration.

6.4 BRANCH AND BOUND ALGORITHM

The branch and bound algorithm (see Agin [1966]) intelligently searches the feasible region for integer solutions by implicitly enumerating most such solutions and enumerating only a few of them explicitly. To describe the algorithm, it will be applied to the problem in Example 6.2; the simplex solution is given in Figure 6.4.

Example 6.3

Consider the following:

$$\begin{aligned} \text{maximize} \quad & z = 21x_1 + 16x_2 \\ \text{subject to} \quad & 4x_1 + 6x_2 \leq 24 \\ & 7x_1 + 4x_2 \leq 28 \\ & 0 \leq x_1 \leq 3.75 \\ & x_1, x_2 = 0, 1, 2 \ldots \end{aligned}$$

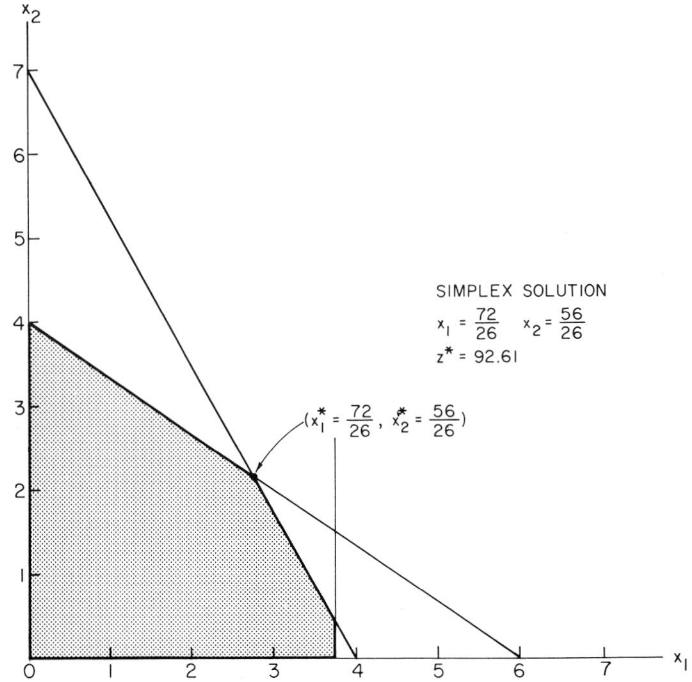

Fig. 6.4. Simplex solution to the Example 6.2 problem.

Since the simplex solution for variable x_1 is 2.77 and the optimal integer solution for x_1 cannot be between 2 and 3, the original problem can be separated into Subproblem I with $x_1 \leq 2$ and Subproblem II with $x_1 \geq 3$. Subproblems I and II are illustrated in Figures 6.5 and 6.6, where their simplex solutions are given. Each subproblem is further dichotomized by now

$$\frac{\text{I}}{x_1 \leq 2}$$

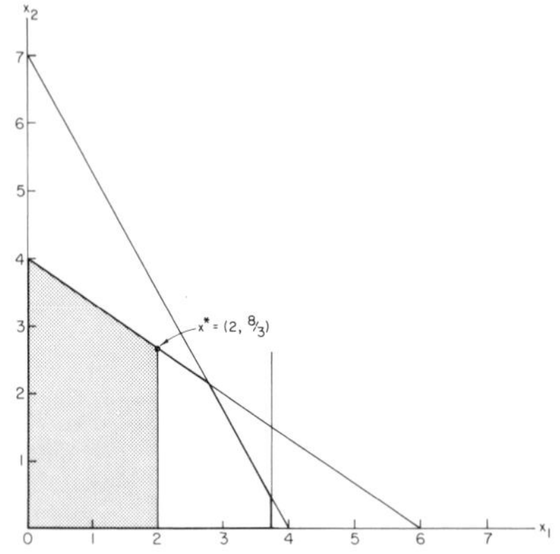

Simplex solution:

$(x_1^* = 2, x_2^* = 8/3)$
$z^* = 84\ 2/3$

Fig. 6.5. Subproblem I.

$$\frac{\text{II}}{x_1 \geq 3}$$

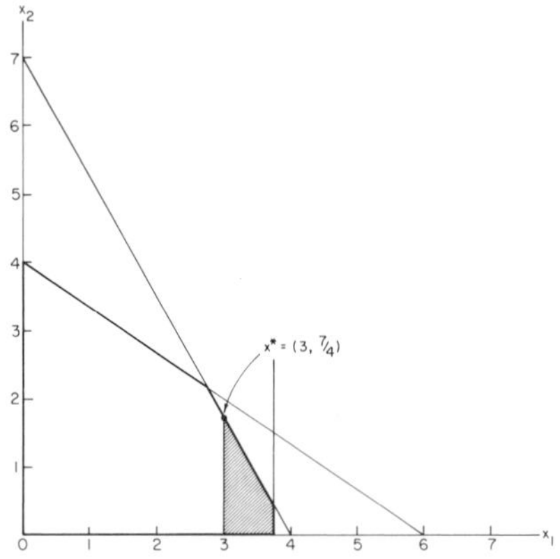

Simplex solution:

$(x_1^* = 3, x_2^* = 7/4)$
$z^* = 91$

Fig. 6.6. Subproblem II.

branching on x_2. By using $x_2 \leq 2$ and $x_2 \geq 3$ in I and $x_2 \leq 1$ and $x_2 \geq 2$ in II, the branches and subsequent node problems are developed and illustrated, with their simplex solutions, in Figures 6.7 and 6.8. The path leading to I(1) terminates at this node since there cannot be a better integer solution with node I(1) as an ancestor. Hence the integer solution points (0, 0), (1, 0), (0, 1), (1, 1), (2, 1), and (1, 2) have been enumerated implicitly and are bounded by node I(1), since none of these points can give a better solution than the point (2, 2). The path through I(2) must be continued, since it may be possible to find an integer solution along this path that is better than the one encountered at I(1). Similarly, the path through II(1) must be continued, but we are through with II(2) since it has carried us out of the feasible region.

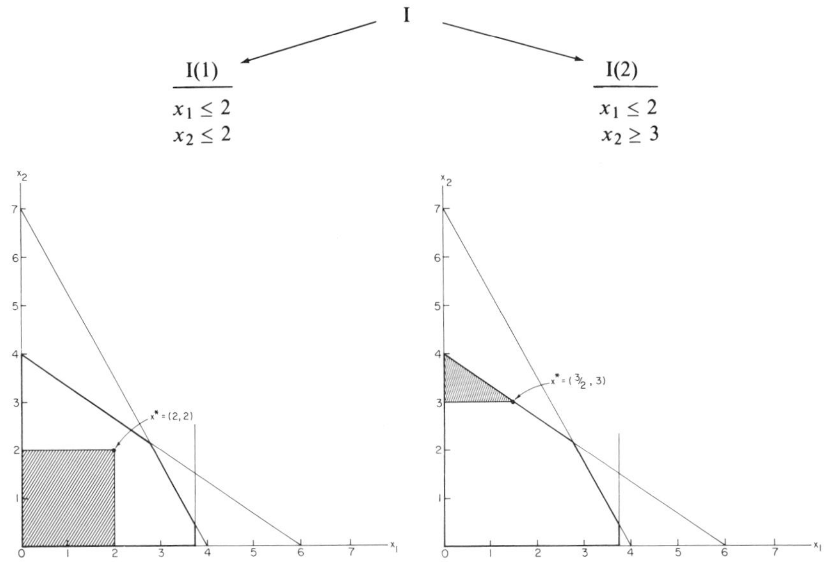

Solution:
$(x_1^* = 2, x_2^* = 2)$
$z^* = 74$

Solution:
$(x_1^* = 3/2, x_2^* = 3)$
$z^* = 79.5$

Fig. 6.7. Branches and nodes of Subproblem I.

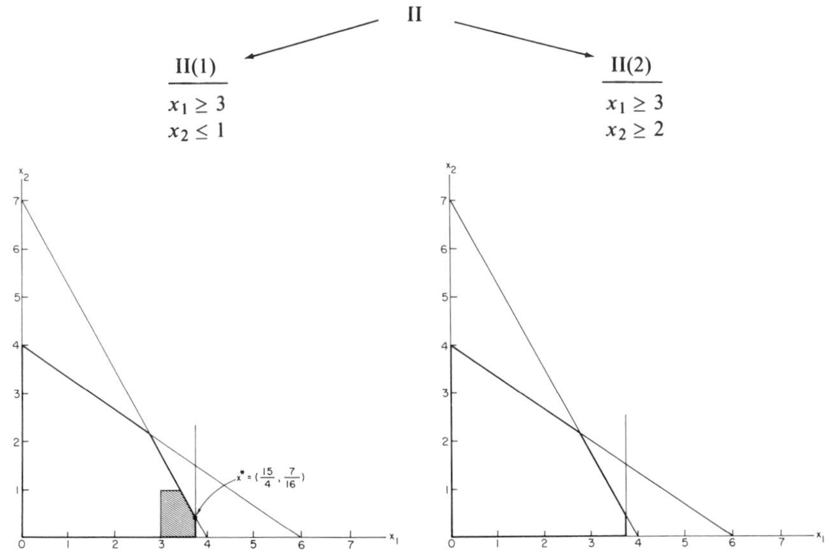

Solution:
$(x_1^* = 15/4, x_2^* = 7/16)$
$z^* = 85.75$

Solution:
Infeasible

Fig. 6.8. Branches and nodes of Subproblem II.

To continue the paths, branch again on x_1. Since at node I(2), x_1 must be less than or equal to 2, x_2 must be greater than or equal to 3, and the simplex solution for x_1 at this node is 1.5. This region can be further broken down by requiring that $x_1 \leq 1$ or $x_1 \geq 2$. Thus nodes I(2, 1) and I(2, 2) are formed as branches from I(2). These two new nodes with their simplex solutions are illustrated in Figure 6.9.

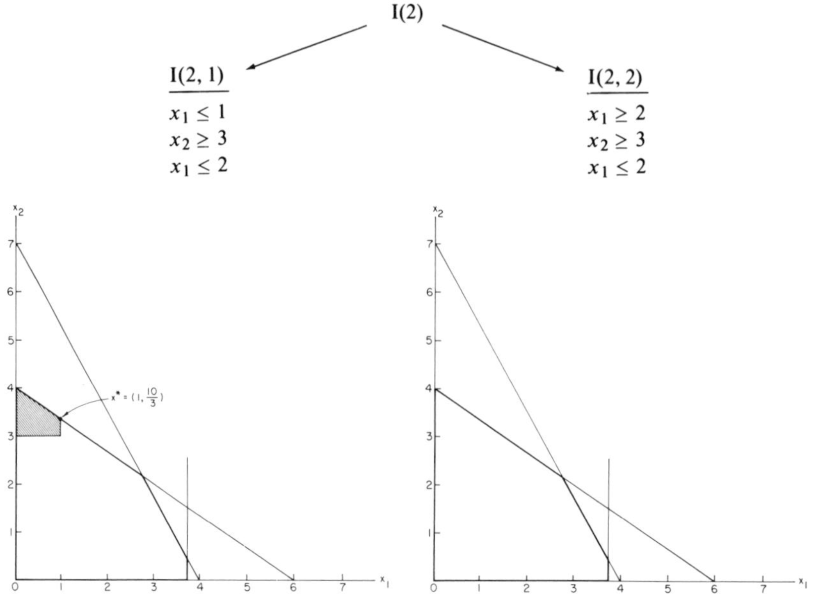

Solution:

($x_1^* = 1, x_2^* = 10/3$)
$z^* = 74.33$

Solution:

Infeasible

Fig. 6.9. Branches and nodes for Subproblem I(2).

A terminal solution point occurs at node I (2, 2) since this path has gone outside the feasible region. A path through I(2, 1) is possible, but first return to II(1) and complete the second iteration. At this node the optimal integer solution is $x_1^* = 3.75$ and $x_2^* = 7/16$, and it is required that $x_1 \geq 3$ and $x_2 \leq 1$. Thus branch on x_1 using $x_1 \leq 3$ and $x_1 \geq 4$. The two new nodes II(1, 1) and II(1, 2) are illustrated with their simplex solutions in Figure 6.10.

II(1)

II(1, 1)	II(1, 2)
$x_1 \leq 3$	$x_1 \geq 4$
$x_2 \leq 1$	$x_2 \leq 1$
$x_1 \geq 3$	$x_1 \geq 3$

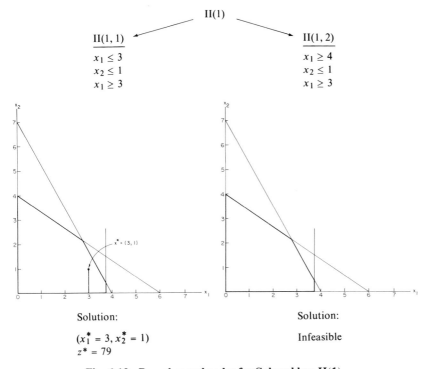

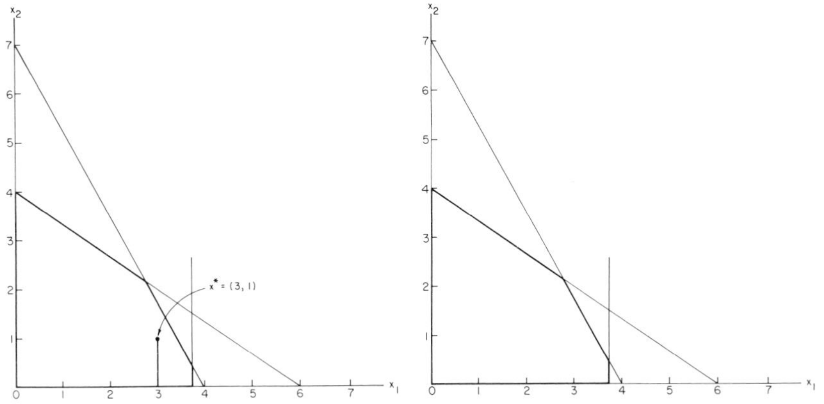

Solution:	Solution:
$(x_1^* = 3, x_2^* = 1)$ $z^* = 79$	Infeasible

Fig. 6.10. Branches and nodes for Subproblem II(1).

Node II(1, 1) clearly bounds all other nodes. Since the objective function at II(1, 1) exceeds that at I(2, 1), there can be no better integer solution descending from I(2, 1). Likewise, the solution at II(1, 1) is better than the one at I(1). Therefore, the problem is solved, and the optimal integer solution is $x_1^* = 3, x_2^* = 1, z^* = 79$.

The branch and bound algorithm may now be delineated by the following steps:

1. Solve the original problem by the simplex algorithm. If the solution point is integral, this is the optimal integer programming solution. Otherwise, go on to step (2).
2. Select one of the n variables, say x_i, whose value from the simplex tableau is nonintegral. Denote this simplex-determined value of x_i by c. Now construct two new problems as descendents of the current problem by adding the constraint $x_i \geq (c + 1)$ to one descendent and the constraint $x_i \leq (c)$ to the other descendent where (c) de-

notes the integer part of c. Solve each of these problems via the simplex algorithm. If a solution point at a node is all integer and is better than any other current bound, it is the optimal integer solution point. Otherwise go to step (3).

3. Since each path is bounded by the current node (no descendent can produce a better solution than the one at the current node), continue as in step (2), treating each node as the current problem until an all-integer solution is experienced that is superior to the current bounds on all paths.

In step (3) discontinue a path as soon as it leads to an infeasible solution or is bounded by an all-integer solution arising from another path. Also, notice that step (3) does not provide a rule for determining which of the current nodes should be chosen first for branching. An intuitive rule might be to select the node whose simplex solution is the best, but this rule may not increase the efficiency of the algorithm.

In the example problem, notice that besides the optimal solution, only one other integer solution, II(1): ($x_1^* = 2, x_2^* = 2$), has been encountered explicitly by the branch and bound algorithm. All other integer solutions are enumerated implicitly with the solution at node II(1) bounding most of them. As the size (number of constraints and variables) of the problem increases, the ratio of the number of solutions enumerated explicitly to the total number of feasible solutions decreases.

6.5 MIXED INTEGER–CONTINUOUS VARIABLE PROBLEM

Occasionally, there is a linear programming problem where some but not all of the variables are constrained to integer values; an example is the fixed charge model discussed in Section 6.1. Formally, the mixed integer–continuous variable linear programming model is

$$\begin{aligned} &\text{minimize} \quad z = \mathbf{c}'\mathbf{x} \\ &\text{subject to} \quad \mathbf{A}\mathbf{x} \le \mathbf{b} \\ &\qquad x_i \ge 0 \qquad i = 1, 2, \ldots, n \\ &\qquad x_i \text{ integer for } i \in I \end{aligned} \tag{6.32}$$

where I is the subset of indices from the set $i = 1, 2, \ldots, n$ associated with the integer decision variables.

Gomory [1960a] has adapted his all-integer cutting plane algorithm to solve the mixed integer–continuous variable problem; the modification is a modest one. The steps in the Gomory mixed integer programming algorithm are basically the same as those for the all-integer algorithm presented in Section 6.3:

1. Solve the problem using the simplex algorithm. If the solution contains integer values for the variables whose indices belong to I, the optimal mixed integer solution has been encountered. If this is not the case, proceed to step (2).
2. Select a row from the current optimum tableau of the linear programming problem to be the generating row in making a Gomory cut.
3. Use the generating row to derive a cut that is added to the bottom of the current tableau.
4. Apply the simplex algorithm to the augmented tableau. If the solution contains integer values for the variables whose indices belong to I, stop. Otherwise return to step (2).

The only modification in the all-integer Gomory algorithm is in step (3); the cut is derived from the generating row in a slightly different manner than before. Recall that the generating row is selected by identifying a basic variable in the current optimal linear programming tableau; let this variable be x_{Bj} as in (6.28). For the mixed integer programming problem $j \in I$ select a row, associated with a basic variable, having a current noninteger solution whose value is constrained to be an integer by (6.32). In the all-integer algorithm, the constraint in (6.30) is augmented to the current optimal simplex tableau. Instead of using (6.30) in the mixed integer programming problem, add the following constraint:

$$h_j - \sum_{i \in E} h_{ji}^* x_i \leq 0 \tag{6.33}$$

where h_{ji}^* = w_{ji} if $w_{ji} \geq 0$ and x_i is restricted to be nonintegral
= $[h_j/(h_j - 1)]w_{ji}$ if $w_{ji} < 0$ and x_i is restricted to be nonintegral
= h_{ji} if $h_{ji} \leq h_j$ and x_i is restricted to be integral
= $[h_j/(1 - h_j)]\,(1 - h_{ji})$ if $h_{ji} > h_j$ and x_i is restricted to be integral

Gomory [1960a] has proved that by making the cut in (6.33), the cutting plane algorithm will converge to the optimal mixed integer solution in a finite number of iterations (cuts). The computational experience with this algorithm is similar to the all-integer algorithm; it is difficult to predict how many cuts may be required before the optimal solution is encountered. If many cuts have been employed without success, the problem can occasionally be solved by using some or all of the suggestions at the end of Section 6.3 (such as multiplying one or more constraints by a constant). For a complete derivation of Gomory's integer programming algorithms, the reader is referred to Hu [1969 Chs. 13–15].

The branch and bound algorithm may also be extended to solve the mixed integer–continuous variable linear programming model. Consider the example in the previous sections with the modification that x_1 must be an integer:

$$\text{maximize} \quad z = 21x_1 + 16x_2 \tag{6.34}$$
$$\text{subject to} \quad 4x_1 + 6x_2 \leq 24 \tag{6.35}$$
$$7x_1 + 4x_2 \leq 28 \tag{6.36}$$
$$0 \leq x_1 \leq 3.75 \tag{6.37}$$
$$x_1 = 0, 1, \ldots \qquad x_2 \geq 0 \tag{6.38}$$

In Subproblem II (Fig. 6.6) the simplex solution is $x_1^* = 3$ and $x_2^* = 7/4$ with $z^* = 91$. This solution clearly bounds the path begun by Subproblem I (Fig. 6.5) and hence is the optimal solution to (6.34)–(6.38). Suppose that (6.38) is replaced by the constraint

$$x_1 \geq 0 \qquad x_2 = 0, 1, \ldots \tag{6.39}$$

Can the optimal solution to the new problem, (6.34)–(6.37) and (6.39), be determined from the network developed in Section 6.4? The best solution with an integral value of x_2 in the network occurs at node I(2) in Figure 6.7, where $x_1^* = 3/2$, $x_2^* = 3$, and $z^* = 79.5$, but this solution is far from optimal for this problem. The optimal solution occurs at the point $x_1^* = 20/7$ and $x_2^* = 2$ with $z^* = 92$. From Figures 6.5 and 6.6, the region containing the optimal x_1^* value 20/7 has been eliminated in constructing the two subproblems. Notice that the optimal corner point (the intersection of $4x_1 + 6x_2 = 24$ and $7x_1 + 4x_2 = 28$ where $x_1 = 72/26$ and $x_2 = 56/26$) for the linear programming problem has been eliminated by the first branch.

There is a clear message here: the network developed from the branch and bound algorithm for a problem where all the variables are constrained to integer values cannot generally be used to determine an optimal solution to the same problem when one or more of the integer value constraints has been dropped. For the mixed integer problem the branch and bound algorithm should be applied so that the branches are made only on the basis of those variables whose values are constrained to be integers. For additional information on applying the branch and bound algorithm to a mixed integer problem, the reader is referred to Beale and Small [1965].

After gaining computational experience in solving integer linear programming problems by such algorithms as those of Gomory and the branch and bound, it becomes apparent that the solution time required increases dramatically as the size (number of constraints and variables) of the problem increases. Indeed, frequently it is not computationally feasible to solve an integer problem by using these algorithms. In this case, there is little recourse but to apply the simplex algorithm to the problem (linear programming problems with thousands of variables and constraints have been solved) and round the solution to integer values. Frequently, the optimal integer solution can be determined from the rounded simplex solution by evaluating the objective function at integer points in the neighborhood of the rounded simplex solution. For example, try the following method. Suppose there are n variables, and let c_i be the optimal integer or noninteger simplex solution for the ith variable x_i. Let $x_i^1 = (c_i)$ and $x_i^2 = (c_i) + 1$ where (c_i) is the integral part of c_i. Evaluate the objective function at the 2^n points $[x_1^j, x_2^j, \ldots, x_n^j]$ where $j = 1$ or 2 for $i = 1, 2, \ldots, n$ and select as the optimal integer solution the optimum feasible solution from these 2^n evaluations. The method can be easily extended to move more than one unit from the optimal simplex solution along each dimension. How good a rounded simplex solution is, or a solution determined by the above method, is not generally known.

6.6 BINARY PROGRAMMING

Binary or zero-one programming has been of considerable interest because of its applicability to problems that are not inherently binary and because of the efficiency of newly devised algorithms to solve the binary problem. The binary linear programming problem is

$$\begin{aligned} &\text{minimize} \quad z = \mathbf{c}'\mathbf{x} \\ &\text{subject to} \quad \mathbf{Ax} \le \mathbf{b} \\ &\qquad\qquad x_i = 0, 1 \qquad i = 1, 2, \ldots, n \end{aligned}$$

Balas [1965] has developed an implicit enumeration algorithm specifically designed to solve this problem. Geoffrion [1969] has also developed an efficient implicit enumeration algorithm that has been programmed; the computer code developed for the algorithm is called RIP30-C[1]. Some mathematical programming problems such as the assignment problem (Model 6.2) are inherently binary. However, the renewed interest in binary programming is principally due to the ability to convert integer linear and nonlinear programming problems to the binary programming model.

The conversion of an integer linear programming problem to a binary problem is a straightforward task. Consider the problem:

$$\begin{aligned} \text{maximize} \quad & z = 21x_1 + 16x_2 \\ \text{subject to} \quad & 4x_1 + 6x_2 \le 24 \qquad (6.40) \\ & 7x_1 + 4x_2 \le 28 \\ & x_1 \le 3.75 \qquad (6.41) \\ & x_1, x_2 = 0, 1, 2, \ldots \end{aligned}$$

Due to (6.41), the largest integer value x_1 can assume is 3. Now expand x_1 by the following function:

$$x_1 = 2^0 y_{10} + 2^1 y_{11} = y_{10} + 2y_{11} \qquad (6.42)$$

where y_{10} and y_{11} are binary variables. It is apparent that the integer values of x_1 can be recovered by allowing the pair y_{10} and y_{11} to assume all its possible values: (0, 0), (1, 0), (0, 1), and (1, 1). Since x_2 is bounded to integer values less than or equal to 4 by (6.40),

$$x_2 = 2^0 y_{20} + 2^1 y_{21} + 2^2 y_{22} = y_{20} + 2y_{21} + 4y_{22} \qquad (6.43)$$

Using (6.42) and (6.43), the problem may now be rewritten as:

$$\begin{aligned} \text{maximize} \quad & z = 21y_{10} + 42y_{11} + 16y_{20} + 32y_{21} + 64y_{22} \\ \text{subject to} \quad & 4y_{10} + 8y_{11} + 6y_{20} + 12y_{21} + 24y_{22} \le 24 \quad (6.44) \\ & 7y_{10} + 14y_{11} + 4y_{20} + 8y_{21} + 16y_{22} \le 28 \quad (6.45) \\ & y_{10} + 2y_{11} \le 3 \quad (6.46) \\ & \text{all } y_{ij} = 0, 1 \end{aligned}$$

By solution of this problem, the coefficients of the powers of two of the binary expansions of x_1 and x_2 in (6.42) and (6.43) are determined. Since there are five binary variables in this problem, $2^5 = 32$ possible solutions exist, 18 of which are infeasible due to (6.44)–(6.46). The optimal solution is $y_{10}^* = y_{11}^* = y_{20}^* = 1, y_{21}^* = y_{22}^* = 0$ with $f^* =$

---

1. A user's guide, Memorandum RN-5657-PR (May 1968), is available from the RAND Corporation, Santa Monica, Calif.

79. From (6.42) and (6.43), the solution to the original problem variables is $x_1^* = (1) + 2(1) = 3$; $x_2^* = (1) + 2(0) + 4(0) = 1$. This is the same solution to this problem determined earlier in this chapter by the Gomory and branch and bound algorithms. The Balas algorithm or the RIP30-C computer code could be applied to the problem and would produce the answer very efficiently without having to enumerate all the 14 feasible solutions.

One obvious difficulty with using the binary expansion method to solve integer linear programming problems is the large number of binary variables that may be generated in the conversion process. In general, let U_i be the upper bound on the ith variable x_i in the original problem. Then choose k such that $2^{k+1} - 1 \geq U_i$, and let $x_i = \sum_{j=1}^{k} 2^j y_{ij}$. If U_i is large, the number of binary variables $k + 1$ required to represent x_i may become large. If this is true for all x_i, $i = 1, 2, \ldots, n$, the resulting binary problem may be quite sizable and costly to solve.

It is possible to convert an integer nonlinear programming problem to a binary problem in a similar fashion. Consider the problem

$$\begin{aligned} \text{maximize} \quad & z = x_1^2 + x_1 x_2 + 2x_1 \\ \text{subject to} \quad & 2x_1 + 3x_2 \leq 6 \\ & x_1, x_2 = 0, 1, 2, \ldots \end{aligned}$$

As before, expand x_1 and x_2 as a function of binary variables: $x_1 = y_{10} + 2y_{11}$; $x_2 = y_{20} + 2y_{21}$. Now the problem can be rewritten as:

$$\begin{aligned} \text{maximize} \quad z &= (y_{10} + 2y_{11})^2 + (y_{10} + 2y_{11})(y_{20} + 2y_{21}) \\ &\quad + 2(y_{10} + 2y_{11}) \\ &= y_{10}^2 + 2y_{10}y_{11} + 4y_{11}^2 + y_{10}y_{20} + 2y_{10}y_{21} \\ &\quad + 2y_{11}y_{20} + 4y_{11}y_{21} + 2y_{10} + 4y_{11} \\ \text{subject to} \quad & 2y_{10} + 4y_{11} + 3y_{20} + 6y_{21} \leq 6 \\ & \text{all } y_{ij} = 0, 1 \end{aligned}$$

The terms y_{ij}^2 may be set to y_{ij} since each is a binary variable. This leaves the cross product terms $y_{ij}\,y_{rs}$ to deal with before the problem is linearized into a binary linear programming problem. Let $y_p = y_{ij}\,y_{rs}$ and write

$$y_{ij} + y_{rs} - y_p \leq 1 \tag{6.47}$$

$$-y_{ij} - y_{rs} + 2y_p \leq 0 \tag{6.48}$$

$$y_p = 0, 1 \tag{6.49}$$

By adding (6.47)–(6.49), y_p will assume the proper value of the cross product term $y_{ij}y_{rs}$:

1. If $y_{ij} = y_{rs} = 0$, then (6.48) will cause $y_p = 0$.
2. If either $y_{ij} = 0$ or $y_{rs} = 0$, but not both, (6.48) will cause $y_p = 0$.
3. If $y_{ij} = y_{rs} = 1$, then (6.47) and (6.48) will cause $y_p = 1$.

Now in the above problem let $y_1 = y_{10}y_{11}$, $y_2 = y_{10}y_{20}$, $y_3 = y_{10}y_{21}$, $y_4 = y_{11}y_{20}$, and $y_5 = y_{11}y_{21}$. The binary programming problem is therefore

$$\text{maximize} \quad z = y_{10} + 2y_1 + 4y_{11} + y_2 + 2y_3 + 2y_4 + 4y_5 + 2y_{10} + 4y_{11}$$

$$\text{subject to} \quad 2y_{10} + 4y_{11} + 3y_{20} + 6y_{21} \le 6$$

$$\begin{aligned} y_{10} + y_{11} - y_1 &\le 1 & y_{10} + y_{20} - y_2 &\le 1 \\ -y_{10} - y_{11} + 2y_1 &\le 0 & -y_{10} - y_{20} + 2y_2 &\le 0 \end{aligned}$$

$$\begin{aligned} y_{10} + y_{21} - y_3 &\le 1 \\ -y_{10} - y_{21} + 2y_3 &\le 0 \end{aligned}$$

$$\begin{aligned} y_{11} + y_{20} - y_4 &\le 1 & y_{11} + y_{21} - y_5 &\le 1 \\ -y_{11} - y_{20} + 2y_4 &\le 0 & -y_{11} - y_{21} + 2y_5 &\le 0 \end{aligned}$$

$$\text{all } y_p, y_{ij} = 0, 1$$

In construction of a binary equivalent problem, a binary variable raised to any positive integer exponent n (say z^n) may be replaced by z. Similarly, a cross product term of the form $z_1^{n1} z_2^{n2} z_3^{n3} \ldots z_t^{nt}$ may be replaced by $z_1 z_2 z_3 \ldots z_t$ for any positive set of integer exponents $n1$, $n2, \ldots, nt$. A cross product term involving t binary variables, $z_1 z_2 \ldots z_t$ may be replaced by one term (say z_p) and three constraints of the form

$$\begin{aligned} z_1 + z_2 + \cdots + z_t - z_p &\le t - 1 \\ -z_1 - z_2 - \cdots - z_t + t z_p &\le 0 \\ z_p &= 0, 1 \end{aligned}$$

This method of dealing with cross product apparently was first published by Watters [1967].

It is apparent that the conversion of an integer nonlinear programming problem to a binary problem can result in a very sizable binary problem, since not only are new variables generated but new constraints must also be added when the original problem has cross product terms.

The interested reader is directed to Glover and Woolsey [1974] for a discussion of the process of converting a nonlinear problem to the binary programming model form. It should be evident that the above procedure for dealing with nonlinear terms will be successful only if the terms are of a polynomial form. A term such as e^{x_1} cannot be dealt with in this fashion. An easy-to-follow explanation of the Balas algorithm and a chapter of applications of binary programming are given by Plane and McMillan [1971].

6.7 EXAMPLES

Example 6.4. An Investment Problem

The Nittany Life Insurance Company is interested in investing \$500,000 during the next planning period and has determined that three projects are suitable for investment. Each unit in the first project requires an investment of \$2500, and the investment counselor estimates that Nittany will enjoy a profit of \$240 per unit over the next period. There are 125 units in the first project available for investment. Each unit in the second project requires an investment of \$1500 and will yield an estimated profit of \$150 per unit. There are 260 units in the second project available for investment. Each unit in the third project requires an investment of \$500 and returns a profit of \$40 per unit; there are 5 units available for investment.

The investment counselor can determine his optimal investment strategy for the next period by solving the following integer linear programming problem:

$$\begin{aligned}
\text{maximize} \quad & z = 240x_1 + 150x_2 + 40x_3 \\
\text{subject to} \quad & 2500x_1 + 1500x_2 + 500x_3 \le 500{,}000 \\
& x_1 \le 125 \\
& x_2 \le 260 \\
& x_3 \le 5 \\
& x_1, x_2, x_3 = 0, 1, 2, \ldots
\end{aligned}$$

If the linear programming simplex algorithm is applied to this problem, the optimal solution of ($x_1^* = 114.28$, $x_2^* = 142.86$, $x_3^* = 0$) is encountered after four iterations. Rounding the simplex solution, leaves $x_1^* = 114$, $x_2^* = 143$, and $x_3^* = 0$, which gives a profit of \$48,810. This is not the optimal integer solution, however. If the Gomory cutting plane algorithm is applied to this problem, the optimal integer solution is encountered after making two cuts ($x_1^* = 116$, $x_2^* = 140$, $x_3^* = 0$); the profit at this point is \$48,840. The investment counselor probably would not be heartbroken if he used the rounded simplex solution and lost the additional \$30 profit that he would have enjoyed by employing the Gomory solution. Indeed, for large problems the savings in the cost of using the simplex algorithm as opposed to branch and bound or the Gomory cutting plane algorithm may offset any lost profit by failing to secure the globally optimal solution. If the values of the decision variables are large (> 20), the rounded simplex solution usually will be reasonably close to the optimum integer solution, as occurred in this example.

Example 6.5. A Transportation Problem

The Acme Carpet Company has three distribution centers to serve four large retail outlets that sell mostly to apartments, motels, and hotels. In Table 6.8, the number of 12′ × 15′ carpets required by the outlets, the number of these carpets each distribution center can supply, and the anticipated profit per carpet adjusted for price differences at the outlets and shipping costs are specified for the next planning period.

Table 6.8. Acme Carpet Company Data

Retail outlet	Distribution center 1	2	3	Number required
1	\$75	80	69	75
2	59	68	52	50
3	80	88	76	55
4	62	70	58	120
Number available	100	70	130	

The optimal shipping pattern may be determined by solving the following integer linear programming problem:

$$\text{maximize} \quad z = 75x_{11} + 80x_{12} + 69x_{13} + 59x_{21} + 68x_{22} + 52x_{23} + 80x_{31} + 88x_{32} + 76x_{33} + 62x_{41} + 70x_{42} + 58x_{43}$$

$$\text{subject to}\quad \begin{aligned} x_{11} + x_{12} + x_{13} &= 75 \\ x_{21} + x_{22} + x_{23} &= 50 \\ x_{31} + x_{32} + x_{33} &= 55 \\ x_{41} + x_{42} + x_{43} &= 120 \\ x_{11} + x_{21} + x_{31} + x_{41} &= 100 \\ x_{12} + x_{22} + x_{32} + x_{42} &= 70 \\ x_{13} + x_{23} + x_{33} + x_{43} &= 130 \\ x_{11}, x_{12}, \ldots, x_{43} &= 0, 1, 2, \ldots \end{aligned}$$

The problem is a linear programming problem except for the integer requirements on the decision variables x_{ij}, $i = 1, 2, 3, 4; j = 1, 2, 3$. If the simplex algorithm is applied to this problem, the solution for the decision variables in Table 6.9 is determined.

Table 6.9. Decision Variables

Retail outlet	Distribution center 1	2	3
1	$x_{11}^* = 75$	$x_{12}^* = 0$	$x_{13}^* = 0$
2	$x_{21}^* = 0$	$x_{22}^* = 50$	$x_{23}^* = 0$
3	$x_{31}^* = 0$	$x_{32}^* = 0$	$x_{33}^* = 55$
4	$x_{41}^* = 25$	$x_{42}^* = 20$	$x_{43}^* = 75$

Since the simplex solution is integer valued for all the decision variables, the optimal integer solution to the transportation problem has been encountered. The profit incurred for the shipping plan in Table 6.9 is \$20,505. It is not suprising to find that the simplex solution is an integer solution in this case since the problem is a standard transportation model. All standard transportation models will have an optimal integer solution when solved by the simplex algorithm if the supplies and demands are integer valued.

Occasionally the simplex algorithm will produce an all-integer solution, in which case the use of more time-consuming algorithms such as the Gomory cutting plane method may be obviated. Most computer codes for integer programming automatically check to see if the simplex solution to the original problem is integer valued; but some do not, so that the above point should be kept in mind when solving integer problems.

PROBLEMS

6.1. The county high school, which has \$180,000 to allocate for teaching French and mathematics for 1975, has learned that it can hire desirable full-time French teachers for \$15,000 per year but it will cost \$30,000 per year to hire a good mathematics teacher full time. The teachers' union has exacted an agreement from the county that part-time teachers will not be hired. If the number of French and mathematics teachers hired are represented by x_1 and x_2 respectively, $\$15{,}000x_1 + \$30{,}000x_2 \leq \$120{,}000$, or $x_1 + 2x_2 \leq 8$, is the school's budget constraint for French and mathematics teachers.

The high school has 140 classrooms it can allocate for teaching French and mathematics. However, because the school board has ruled that French classes will have 35 students and mathematics will have 20 students and one teacher is assigned to each classroom, $35x_1$ classrooms are needed to teach French and $20x_2$ classrooms are needed to teach mathematics. Therefore, $35x_1 + 20x_2 \leq 140$, or $7x_1 + 4x_2 \leq 28$, represents the classroom constraint.

The new school superintendent has decided that the rates of return on teaching mathematics and French are 4 units of satisfaction and 7 units of satisfaction respectively because few students have majored in college mathematics after graduation, and several French linguists have graduated from this school.

Therefore, the optimization problem that faces the superintendent is

$$\begin{aligned} \text{maximize} \quad & z = 4x_1 + 7x_2 \\ \text{subject to} \quad & x_1 + 2x_2 \leq 8 \\ & 7x_1 + 4x_2 \leq 28 \\ & x_1, x_2 = 0, 1, 2, \ldots \end{aligned}$$

(a) Solve this problem graphically.

(b) Solve this problem by using the branch and bound algorithm. Show the network of branches and subsequent node problems, as done in the text.

6.2. Solve the following problem (a) by branch and bound and (b) by using the Gomory cutting plane algorithm (make no more than three cuts, the problem should require only one).

$$\text{maximize} \quad z = x_1 + x_2$$

$$\begin{aligned} \text{subject to} \quad & 2x_1 + x_2 \le 8 \\ & x_1 + 2x_2 \le 8 \\ & x_1, x_2 = 0, 1, 2 \ldots \end{aligned}$$

6.3. Solve the following problem (a) by branch and bound and (b) by using the Gomory cutting plane algorithm (make no more than three cuts; the problem should require only one).

$$\begin{aligned} \text{maximize} \quad & z = 5x_1 + 6x_2 \\ \text{subject to} \quad & x_1 + 2x_2 \le 10 \\ & 2x_1 + x_2 \le 12 \\ & x_1, x_2 = 0, 1, 2, \ldots \end{aligned}$$

6.4. Solve the following problem by using the Gomory cutting plane algorithm (make no more than three cuts; the problem should require two).

$$\begin{aligned} \text{minimize} \quad & z = x_1 + x_2 + x_3 \\ \text{subject to} \quad & 3x_1 - 4x_2 - 6x_3 \le 2 \\ & x_1 + x_2 - 2x_3 \le 5 \\ & 2x_1 + x_2 + 2x_3 \ge 11 \\ & x_1, x_2, x_3 = 0, 1, 2, \ldots \end{aligned}$$

6.5. A manufacturer of electronic components can ship units of a particular component out in either large or medium containers. The net profit on each large container shipped out is \$200, on each medium container the net profit is \$100. The large container can be packed with 20 units of the component and the medium container can hold 10 units. In the next planning period the manufacturer must ship out at least 75 units to meet the expected demand. Each large container requires 12 man-hours to pack and each medium container requires 7 man-hours. No more than 55 man-hours are available for packing the containers. Each large container weighs 25 pounds when packed and each medium container weighs 10 pounds. The total shipment can weigh no more than 90 pounds. Let x_1 be the number of large containers to pack and let x_2 be the number of medium containers. The packing problem the manufacturer must solve is

$$\begin{aligned} \text{maximize} \quad & z = 200x_1 + 100x_2 \\ \text{subject to} \quad & 12x_1 + 7x_2 \le 55 \\ & 25x_1 + 10x_2 \le 90 \\ & 20x_1 + 10x_2 \ge 75 \\ & x_1, x_2 = 0, 1, 2, \ldots \end{aligned}$$

(a) Find the solution to the problem by graphing.

(b) Solve the problem by using the Gomory algorithm. Make no more than three cuts. If by three cuts the solution has not been determined, show graphically how the cuts are homing in on the optimal solution determined in (a).

6.6. Consider the nonlinear integer problem

$$
\begin{aligned}
\text{maximize} \quad & z = 3x_1^2 + x_2^2 \\
\text{subject to} \quad & 3x_1 + 4x_2 \leq 12 \\
& 2x_1 + x_2 \leq 6 \\
& x_1, x_2 = 0, 1, 2, \ldots
\end{aligned}
$$

Solve this problem by (a) constructing the equivalent binary linear programming problem and (b) enumerating all the solutions to the binary problem.

6.7. Solve the following mixed integer–continuous variable problem by (a) branch and bound and (b) by using the Gomory cutting plane algorithm.

$$
\begin{aligned}
\text{maximize} \quad & z = 2x_1 + 5x_2 \\
\text{subject to} \quad & 2x_1 + x_2 \geq 2 \\
& 5x_1 + 3x_2 \leq 15 \\
& x_2 \leq 4 \\
& x_1 = 0, 1, 2, \ldots \qquad x_2 \geq 0
\end{aligned}
$$

6.8. Solve the following fixed charge problem by branch and bound:

$$
\begin{aligned}
\text{minimize} \quad & z = f_1(x_1) + f_2(x_2) \\
\text{subject to} \quad & 5x_1 + 4x_2 \geq 122 \\
& 0 \leq x_1 \leq 30 \\
& 0 \leq x_2 \leq 40 \\
& x_1, x_2 = 0, 1, 2, \ldots
\end{aligned}
$$

where

$$
\begin{aligned}
f_1(x_1) &= 100 + 75x_1 \text{ if } x_1 > 0 \\
&= 0 \text{ if } x_1 = 0 \\
f_2(x_2) &= 50 + 20x_2 \text{ if } x_2 > 0 \\
&= 0 \text{ if } x_2 = 0
\end{aligned}
$$

6.9. Solve the following transportation problem by branch and bound:

$$\text{minimize} \quad z = \textstyle\sum_{i=1}^{2} \sum_{j=1}^{2} c_{ij} x_{ij}$$

subject to $\sum_{j=1}^{2} x_{ij} = a_i \quad i = 1, 2$
$\sum_{i=1}^{2} x_{ij} = b_j \quad j = 1, 2$
$x_{ij} = 0, 1, 2, \ldots \quad i = 1, 2; j = 1, 2$

where $c_{11} = \$50$, $c_{12} = \$40$, $c_{21} = \$45$, $c_{22} = \$55$, $a_1 = 24$, $a_2 = 36$, $b_1 = 20$, and $b_2 = 40$.

6.10. The Tactical Toy Corporation manufactures two toy automobile models. It costs them \$3 to produce one unit of each model. Let x_1 and x_2 be (in thousands) the numbers of units of models 1 and 2 respectively produced this year on a budget of \$108,000. Tactical's budget constraint is $\$3(1000)x_1 + \$3(1000)x_2 \leq \$108{,}000$ or $x_1 + x_2 \leq 36$. It has also been established that no more than 10,000 units of model 1 should be produced so that $x_1 \leq 10$. Further, due to other factors such as marketing and production considerations, it is determined that the numbers of units of model 1 and model 2 to be produced must satisfy the constraint: $-4x_1 - 16x_1^{-1} + x_2 \leq 2$. Tactical's average profits on the two models are r_1 and r_2 respectively, and the firm's total profit is $z = r_1x_1 + r_2x_2$, where $r_1 = 2x_1 - 3x_2$ and $r_2 = 3x_1 + 10x_2$.

If Tactical's goal is to maximize its profit, it faces the optimization problem:

$$\begin{aligned} \text{maximize} \quad & z = 2x_1^2 + 10x_2^2 \\ \text{subject to} \quad & x_1 + x_2 \leq 36 \\ & -4x_1 - 16x_1^{-1} + x_2 \leq 2 \\ & x_1 \leq 10 \\ & x_1, x_2 = 0, 1, 2, \ldots \end{aligned}$$

(a) Solve the problem graphically.
(b) Use the branch and bound algorithm to solve the problem. Discuss possible modifications of the algorithm that might be made to solve this and other nonlinear integer programming problems for the *global* optimum.

7 DYNAMIC PROGRAMMING

7.1 INTRODUCTION

In the previous five chapters, the emphasis has been on model characterization (linear, nonlinear, quadratic and integer linear programming models, specifically) and on algorithms for solving problems involving these models. In this chapter, an optimization method is discussed that separates a problem into a sequence of interrelated subproblems, each of which may be solved with reasonable effort from the knowledge gained by solving its predecessors in the sequence. The final solution to a problem is achieved by solving the first in the sequence of subproblems, then solving each of the remaining subproblems in the sequence and backtracking from the solution in the final subproblem to the solution of the original problem.

Dynamic programming is not a method nor an algorithm per se; rather, it is a solution strategy that involves the implicit enumeration of all possible solutions to a problem. The means by which a solution is achieved sets the method apart from others: dynamic programming methods rely on similar appearing recursive formulas and are applied to problems that must have specific characteristics.

7.2 CHARACTERISTICS OF A DYNAMIC PROGRAMMING PROBLEM

A problem must have three primary characteristics to be amenable to resolution by dynamic programming:

1. The problem can be split into subproblems grouped into *stages,* where in each stage it must be possible to solve a particular sub-

problem based only on the solution to the subproblem in the previous stage. Frequently, the splitting of a problem into stages will occur naturally (e.g., the sequence of grouped subproblems may be identified with successive time intervals); but the splitting may be somewhat abstract, such as using the indices of the decision variables in an optimization problem as stages.

2. Associated with the stages is a *state variable* (there may be more than one variable) whose values represent the possible conditions or states that the process may be in at a particular stage. The state variables are selected in such a way that they convey useful information about the subproblems and are used to link each subproblem with its predecessors in preceding stages and its descendents in succeeding stages.
3. Each subproblem has a parameter called a *decision variable* that can be used to obtain an optimal solution for the subproblem. The grouping of subproblems into stages is ordinarily accomplished so that the same decision variable applies to every subproblem in a given stage.

The condition that each subproblem may be solved based only on the solutions to its immediate predecessors is called the *principle of optimality* in dynamic programming. This principle must be satisfied by a problem before it can be solved by dynamic programming; but as will be shown, the principle is sufficiently ambiguous so that most problems can be cast in such a way that the principle is satisfied by appropriately defining the stages, state, and decision variables of a problem. In this sense applying the dynamic programming solution strategy frequently is an art that involves considerable manipulation of the problem until it can be arranged in a way that satisfies the above three characteristics.

To see how the solution strategy of dynamic programming is applied, consider the following classical dynamic programming problem.

Example 7.1. The Stagecoach Problem

A Fulla Brush salesman, Herman Klutz, in the days of cowboys and Indians, is instructed by his superior to leave his secure office in Boston and traverse the country, visiting localities in predetermined regions for the purpose of creating local sales forces. The regions and localities are illustrated in Figure 7.1. Each locality is denoted by a two-digit number. The second digit is the number of the region in which the locality occurs. The first digit is the office number within the region. To help the reader

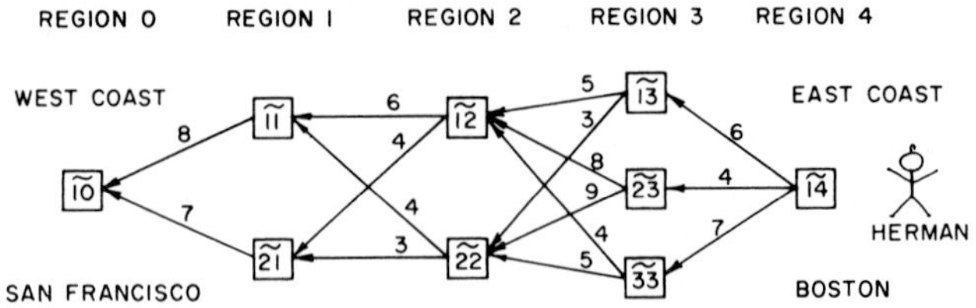

Fig. 7.1. Insurance rates for alternative pathways.

distinguish a locality from the other coefficients in this problem, a tilde, (~) will always appear above a locality number.

Herman's superior requests that he visit one and only one locality in each region as he moves from the East Coast to the West Coast. Herman instinctively senses the opportunity to exercise some independent thinking, for his superior's instructions allow him to select one of many possible paths (12 in all) in traversing the country. The responsibility of making a unilateral decision on which route to take is amplified by the fact that certain routes between regions are frequently targets for stagecoach robberies by outlaws and attacks by Indians. Herman, therefore, decides to select the safest path from the home office $\widetilde{14}$ to the San Francisco office $\widetilde{10}$. After days of brooding over the difficulty of determining the safeness of a pathway, he is struck with the answer.

A new insurance company, Crudential, is insuring stagecoach travelers against theft, death, and the like. He quickly sees his local Crudential agent and procures the latest insurance rates, which are shown for the pathways in Figure 7.1. For example, the cost of insurance along the pathway from locality $\widetilde{23}$ to $\widetilde{22}$ is \$9. Since there are only 12 possible paths from Boston to San Francisco, Herman can easily evaluate the total insurance cost of each path. He will, of course, select the path with the least insurance cost. However, since Herman considers himself to be quite a good mathematician, he wonders if there is not a way to solve this problem without enumerating all possible routes. He first notes that the most economical route cannot be determined by choosing the best pathway for connecting successive regions (e.g., $\widetilde{14} \rightarrow \widetilde{23} \rightarrow \widetilde{12} \rightarrow \widetilde{21} \rightarrow \widetilde{10}$). By losing a little on the first pathway, gains can be experienced thereafter; $\widetilde{14} \rightarrow \widetilde{23} \rightarrow \widetilde{12} \rightarrow \widetilde{21}$ costs \$16, while $\widetilde{14} \rightarrow \widetilde{13} \rightarrow \widetilde{22} \rightarrow \widetilde{21}$ cost \$12, a savings of \$4, although the first pathway $\widetilde{14} \rightarrow \widetilde{13}$ costs \$2 more in the latter path than $\widetilde{14} \rightarrow \widetilde{23}$ does in the former.

After considerable study of the problem, Herman settles upon the following solution method:

1. Identify the regions as *stages* of the problem. Let stage 1 be associated with $\widetilde{10}$; stage 2 with $\widetilde{11}$ and $\widetilde{21}$; stage 3 with $\widetilde{12}$ and $\widetilde{22}$; stage 4 with $\widetilde{13}$, $\widetilde{23}$, and $\widetilde{33}$; and stage 5 with $\widetilde{14}$.
2. Choose as the *decision variables,* x_n, $n = 1, 2, 3, 4$, the immediate destination when there are n stages to go. The route selected is therefore $x_5 = \widetilde{14} \rightarrow x_4 \rightarrow x_3 \rightarrow x_2 \rightarrow x_1 = \widetilde{10}$.
3. Let the *state variable* s denote the locality in a particular stage. Thus when Herman has one more stage to go, s can take on the values $\widetilde{11}$ or $\widetilde{21}$, for example.

Let $f_n(s, x_n)$ be the total cost of the best policy for the last n stages, given that Herman is in state s and selects x_n as his immediate destination. Let x_n^* be the value of x_n that minimizes $f_n(s, x_n)$ for fixed values of n and s, and let $f_n^*(s)$ be the minimum value of $f_n(s, x_n)$. Herman's goal, therefore, is to find f_4^* and the optimal values of the decision variables x_1^*, x_2^*, x_3^*, and x_4^*. The strategy Herman will employ is to find f_4^* by first successively finding $f_1^*(s), f_2^*(s), f_3^*(s)$, and then $f_4^*(\widetilde{14})$.

Herman begins solving the problem working backward by considering that he only has one stage to go; i.e., he is in locality $\widetilde{11}$ or $\widetilde{21}$ from which he will move to $\widetilde{10}$. In this one-stage problem his route is determined by his final destination $\widetilde{10}$. By inspection the solution to the single-stage problem is as given in Table 7.1. Given that he is presently in locality $\widetilde{11}$, his best (and only) choice of the decision variable to minimize $f_1(s = \widetilde{11}, x_1)$ is $x_1^* = \widetilde{10}$. Similarly, $x_1^* = \widetilde{10}$ will minimize $f_1(s = \widetilde{21}, x_1)$, given that he is in locality $\widetilde{21}$.

Table 7.1. Single-Stage Solution

State variable s	Optimal solution $f_1^*(s)$	Decision variable x_1^*
$\widetilde{11}$	8	$\widetilde{10}$
$\widetilde{21}$	7	$\widetilde{10}$

Suppose that Herman has two more stages to go; i.e., he is presently in locality $\widetilde{12}$ or $\widetilde{22}$ from which he must move to $\widetilde{11}$ or $\widetilde{21}$. Assume first that he is in locality $\widetilde{12}$. If he decides to move to locality $\widetilde{11}$, the insurance cost will be \$6; the additional insurance cost of moving from $\widetilde{11}$ to $\widetilde{10}$ in the last stage is \$8, so that the total insurance cost of this route is \$6 + \$8 = \$14.

Since by moving from $\widetilde{12}$ to $\widetilde{21}$ the total cost will be only $\$11$ ($\$4 + \$7$), he will choose $x_2^* = \widetilde{21}$, given that he is presently in locality $\widetilde{12}$, for this choice of the decision variable will minimize the two-stage cost $f(s = \widetilde{12},\ x_2)$. Notice that $f_2(s, x_2)$ can be written as the sum of the cost of moving from $\widetilde{12}$ or $\widetilde{22}$ to a value of x_2 ($\widetilde{11}$ or $\widetilde{21}$) and the minimum additional cost $f_1^*(x_2)$ required to complete the journey once a value x_2 has been selected. Now assume he is in locality $\widetilde{22}$ and decides to move to $\widetilde{11}$ ($x_2 = \widetilde{11}$). The cost of this move $c_{\widetilde{11}}$ is $\$4$ plus a minimum additional cost $f_1^*(x_2 = \widetilde{11})$ of $\$8$ to complete the journey for a total cost of $\$12$. If he decides to go to $\widetilde{21}$ from $\widetilde{22}$, the cost is $c_{\widetilde{21}} = \$3$ plus the additional cost of $f_1^*(x_2 = \widetilde{21}) = \$7$ to reach $\widetilde{10}$. Thus the total cost is $\$10$ for this path, and he would therefore select $x_2^* = \widetilde{21}$ if he were presently at locality $\widetilde{22}$. The results of the two-stage problem are summarized in Table 7.2. If Herman is presently at locality $\widetilde{12}$, his best choice of x_2 is $x_2^* = \widetilde{21}$, in which case his total insurance cost with two stages to go will be minimized with $f_2^*(s = \widetilde{12}) = \$11$. If Herman is presently at locality $\widetilde{22}$, his optimal choice of x_2 is $x_2^* = \widetilde{21}$.

Table 7.2. Two-Stage Solution

State variable s	Total cost $f_2(s, x_2) = c_{x_2} + f_1^*(x_2)$: $x_2 = \widetilde{11}$	$x_2 = \widetilde{21}$	Minimum cost $f_2^*(s)$	Optimal decision variable x_2^*
$\widetilde{12}$	$6 + 8 = 14$	$4 + 7 = 11$	$\$11$	$\widetilde{21}$
$\widetilde{22}$	$4 + 8 = 12$	$3 + 7 = 10$	10	$\widetilde{21}$

Now proceed to the situation where three stages remain. If Herman is at locality $\widetilde{13}$ and goes to $\widetilde{12}$ ($x_3 = \widetilde{12}$), his total cost is $c_{\widetilde{12}} = \$5$ plus the minimum additional cost $f_2^*(\widetilde{12}) = \$11$, of moving from $\widetilde{12}$ to $\widetilde{10}$, the latter number being secured from Table 7.2. The total cost for selecting $x_3 = \widetilde{12}$ is therefore $\$5 + \$11 + \$16$. If Herman moves from $\widetilde{13}$ to $\widetilde{22}$ ($x_3 = \widetilde{22}$), then $c_{22} = \$3$ and $f_2^*(\widetilde{22}) = \$10$ from Table 7.2, so that the total cost is $\$3 + \$10$ to move from $\widetilde{13}$ to $\widetilde{10}$ along this route. The complete three-stage results are given in Table 7.3. If Herman is at $\widetilde{13}$, his best move is to $\widetilde{22}$; if he is at $\widetilde{23}$ or $\widetilde{33}$, he can move to either $\widetilde{12}$ or $\widetilde{22}$ to secure a minimum total insurance cost over the last three stages.

Table 7.3. Three-Stage Solution

State variable s	Total cost $f_3(s, x_3) = c_{x_3} + f_2^*(s)$		Minimum cost $f_3^*(s)$	Optimal decision variable x_3^*
	$x_3 = \widetilde{12}$	$x_3 = \widetilde{22}$		
$\widetilde{13}$	$5 + 11 = 16$	$3 + 10 = 13$	$\widetilde{13}$	$\widetilde{22}$
$\widetilde{23}$	$8 + 11 = 19$	$9 + 10 = 19$	$\widetilde{19}$	$\widetilde{12}$ or $\widetilde{22}$
$\widetilde{33}$	$4 + 11 = 15$	$5 + 10 = 15$	$\widetilde{15}$	$\widetilde{12}$ or $\widetilde{22}$

Finally, consider the four-stage problem. As above, the optimal strategy (minimum insurance cost), given an immediate destination, is determined by summing the cost of the first move and the minimum cost thereafter. The four-stage results are given in Table 7.4. If Herman should move from $\widetilde{14}$ to $\widetilde{13}$, his total minimum insurance cost in going from $\widetilde{14}$ to $\widetilde{10}$ is given by $f_4^*(\widetilde{14}) = \$19$. But what is the optimal route? It is determined by tracing back through the tables to determine the optimal solutions for the decision variables. Recall that the route in terms of the decision variables is $14 \rightarrow x_4 \rightarrow x_3 \rightarrow x_2 \rightarrow x_1 = \widetilde{10}$. Tracing through Tables 7.4, 7.3, 7.2, and 7.1, $x_4^* = \widetilde{13}$, $x_3^* = \widetilde{22}$, $x_2^* = \widetilde{21}$, and $x_1^* = \widetilde{10}$. Therefore, the optimal route or path is $\widetilde{14} \rightarrow \widetilde{13} \rightarrow \widetilde{22} \rightarrow \widetilde{21} \rightarrow \widetilde{10}$, which gives a minimum insurance cost of \$19.

Table 7.4. Four-Stage Solution

State variable s	Total cost $f_4(s, x_4) = c_{x_4} + f_3^*(s)$			Minimum cost $f_4^*(s)$	Optimal decision variable x_4^*
	$x_4 = \widetilde{13}$	$x_4 = \widetilde{23}$	$x_4 = \widetilde{33}$		
$\widetilde{14}$	$6 + 13 = 19$	$4 + 19 = 23$	$7 + 15 = 22$	19	13

Reconsider the salient features of Herman's solution method (dynamic programming) to the stagecoach problem. The problem clearly has stages (traveling from region 4 to 3, 3 to 2, 2 to 1, and 1 to 0—the legs of the journey), a decision variable (x_n = immediate destination with n more stages to go), and a state variable (a locality in a region). Furthermore, the decision at each stage (optimal selection of x_n) links the current state with a state in the next stage. The principle of optimality is satisfied; given the current state, the optimal solution for the remaining stages is independent of the decisions made in previous stages. Once Herman reaches a state in a particular stage, the best route from the current point to his final destination $\widetilde{10}$ is independent of how he reached the current point. A recursive relationship identifies the optimal decision for each state with n stages remaining, given the optimal decision with $n - 1$ stages remaining. In the stagecoach problem the mathematical expression for this recursive relationship is

$$f_n^*(s) = \min_{x_n}[c_{x_n} + f_{n-1}^*(x_n)]$$

(This relationship is not related to the methods applied to recursive programming models presented in Chapter 8.) Thus with three stages remaining, for example, minimize $c_{x_3} + f_2^*(x_3)$ with respect to x_3 for each state in the nth stage. This involves optimally solving three subproblems; i.e., minimizing $f_3(s, x_3)$ with respect to x_3 for states $\widetilde{13}$, $\widetilde{23}$, and $\widetilde{33}$ (Table 7.3).

The solution method for the stagecoach problem moves backward stage by stage, using the recursive relationship until the optimal decision is determined when beginning at the initial stage. The solution to the one-stage problem is usually trivial, as Table 7.1 illustrates. The recursive formula will provide solutions at each stage, moving from the one-stage problem to the n-stage problem. The optimal solution for the decision variables, $x_n, n = 1, 2, \ldots,$ can be determined by tracing through the tables giving the optimal decisions for all states in each stage once the n-stage table has been specified (and x_n^* consequently determined.)

A fair question to Herman from his superior might be why he used such a sophisticated solution method for a problem that could have been readily solved by exhaustive explicit enumeration of all possible routes. Herman explains that this solution strategy can conveniently be applied to much larger problems, affording more efficient solutions (less computations) than exhaustive explicit enumeration. The potential saving in solution costs immediately appeals to his superior who is envisioning the future proliferation of Fulla Brush sales forces throughout the country; Herman is immediately promoted and assured that he will be a fixture in the Boston home office if he returns safely from his trip to the West Coast.

The principle of optimality employs the recursive relationship to reduce the magnitude of a single massive problem having T stages, with the ith stage having N_i states (subproblems). The original problem has N_i subproblems to solve for the ith stage; and if the stages are not linked somehow, there are $\Pi_{i=1}^{T} N_i$ total problems to solve. However, the recursive relationship in conjunction with the principle of optimality guarantees that no more than $\sum_{i=1}^{T} N_i$ optimization subproblems will be required to solve all T stages of the problem. For example, in the stagecoach problem if the problem is now viewed as being linked by the recursive relationship in (7.1), there would be (2)(2)(3)(1) = 12 subproblems to solve, while the dynamic programming solution strategy solves 2 + 2 + 3 + 1 = 8 subproblems. There is a difference in the nature of the subproblems in these two cases; in the former a solution to a subproblem involves forming the total of four insurance costs, while in the latter case a solution to a subproblem involves the optimal selection of a value for the decision variable, and the solution to a small optimization problem. Therefore, the numbers of subproblems, 12 and 8, cannot be compared by magnitude so easily, since the computations required to solve a subproblem are clearly different in the two cases. However, as the number of stages and states within stages increases, dynamic programming will increasingly involve less computational effort than exhaustive explicit enumeration for most problems.

In employing dynamic programming to solve a problem, there are optimization problems as the subproblems in each stage. In the stagecoach example the subproblem optimizations are conducted very easily. For each value of the appropriate decision variable in a given stage, the total insurance cost is computed and the minimum cost chosen as a solution to the subproblem. In general, these subproblems may be complicated mathematical programming problems, which may be solved by employing one or more of the algorithms presented in Chapters 2–6. For example, the subproblem may be a linear programming problem, in which case the simplex algorithm can be employed to solve it. If the problem is nonlinear, the possibility of local optima occurring as solutions to the problem must be considered. Since dynamic programming is ideally suited for problems where the subproblem optimizations are discrete (the decision variables $x_n, n = 1, 2, \ldots,$ are discrete), the problem of nonlinearity can become critical, for it is generally extremely difficult to assess nonlinear functions of discrete variables for local optimality. No theories for discrete optimization are analogous to those for continuous variable optimization problems (e.g., the Kuhn-Tucker theorem and the constraint qualification). In Section 7.3, a continuous variable problem will be discretized so that it can be resolved by dynamic programming. If the continuous variable problem is nonlinear, there is the problem of determining whether the solution by dynamic programming is globally optimal.

The form of the recursive relationship may differ from problem to problem. Indeed, its specification is ordinarily dictated by the requirement that the problem must satisfy the three primary requirements of a dynamic programming problem discussed above—most notably, iple of optimality. In this sense, for a particular problem there be a unique recursive relationship; the same problem might be y dynamic programming, using two entirely different sets of ns for the stages, state variables, decision variables, and recursive relationship. However, the recursive relationship will always have the following basic form:

$$f_n^*(s) = \max \text{ or } \min\,[f_n(s, x_n)]$$

where $f_n(s, x_n)$ is written in terms of s, x_n, and $f_{n-1}^*(\cdot)$. Since dynamic programming is a solution strategy that may take on a diversity of forms subject only to the three problem characteristics and the principle of optimality, perhaps the best way to develop competence in using the strategy is to first see it in action. In this section it is applied to the knapsack problem in Model 6.4; in Section 7.4 two more example dynamic programming problems will be presented. For additional example problems the reader is directed to books on dynamic programming by Bellman and Dreyfus [1962], Hadley [1964], and Nemhauser [1966].

Example 7.2. The Knapsack Problem

A bush pilot of a single-engine aircraft on a return trip to civilization from a remote and desolate region of Lower Slovakia has the opportunity to bring back products made by the natives. Specifically, he can select from three products but is limited in the number of each by the cargo weight restriction of his aircraft. He has determined that his plane has the capability of returning with 130 pounds of native products. Each unit of the first product weighs 30 pounds, and he anticipates a net profit of \$50 from selling one unit of this upon his return. Each unit of the second product weighs 40 pounds and will return a profit of \$60. Finally, each unit of the third product weighs 60 pounds and will return a profit of \$80. Let x_i be the number of units of the ith product packed in his knapsack (the airplane). The pilot can determine the optimal number of the product units to pack by solving the following integer linear programming problem:

$$\text{maximize} \quad z = 50x_1 + 60x_2 + 80x_3 \tag{7.1}$$

$$\text{subject to} \quad 30x_1 + 40x_2 + 60x_3 \le 130 \tag{7.2}$$

$$x_1, x_2, x_3 = 0, 1, 2, \ldots \tag{7.3}$$

Since the values of x_1, x_2, and x_3 must be integer valued, the problem is an integer programming problem. Additionally, constraint (7.2) may be replaced with the equivalent inequality constraint

$$3x_1 + 4x_2 + 6x_3 \leq 13 \tag{7.4}$$

Exhaustive explicit enumeration can be employed to solve this problem (as well as the algorithms in Chapter 6). But can this problem be solved by dynamic programming? Yes it can, by judiciously defining a stage, the state variable, and the decision variable. The variables x_1, x_2, and x_3 will be used as decision variables, and they also will be used to identify stages. The sequence of stages will begin with solving a problem depending only on x_1, in the next stage x_2 will be augmented, and the final stage will adjoin x_3. The state variable will be defined as the slack in the constraint (7.4) for a given value of the decision variable and stage.

To see how these definitions relate to the problem, suppose for the moment that the optimal values of x_2 and x_3 are known and denoted by x_2^* and x_3^* respectively. Then (7.1)–(7.4) could be reduced to the following one-variable problem:

$$\begin{aligned} \text{maximize} \quad & z = 50x_1 + [60x_2^* + 80x_3^*] && (7.5)\\ \text{subject to} \quad & 3x_1 \leq [13 - 4x_2^* - 6x_3^*] = s && (7.6)\\ & x_1 = 0, 1, 2, \ldots && (7.7) \end{aligned}$$

Clearly, $s = [13 - 4x_2^* - 6x_3^*] \geq 0$ if there is a feasible solution to the problem. Also, notice that the term $60x_2^* + 90x_3^*$ in (7.5) is a constant if x_2^* and x_3^* are known; hence it may be dropped from the objective function since it will not contribute to the maximization. Thus the problem in (7.5)–(7.7) may be recast as

$$\begin{aligned} \text{maximize} \quad & f_1(s, x_1) = 50x_1 && (7.8)\\ \text{subject to} \quad & 3x_1 \leq s && (7.9)\\ & x_1 = 0, 1, 2, \ldots && (7.10) \end{aligned}$$

Thus if x_2^* and x_3^* are known and only one stage remains (optimizing over x_1 alone), the optimal solution to the original problem in (7.1)–(7.4) may be determined by solving this univariate linearly constrained optimization problem. The state variable is the slack s in (7.9), which depends on the values of x_2^* and x_3^*; the decision variable is x_1, whose possible values are $0, 1, 2, 3, 4$; and x_1 cannot exceed 4 since s takes on a maximum value of 13 when $x_2 = x_3 = 0$. Thus as with the stagecoach problem, begin by assuming there is only one stage to go; the optimization problem depends only on x_1 and is given by (7.8)–(7.10). In Table 7.5 the solutions to the subproblems for the possible values of the state variable are given. As before, the maximum of $f_1(s, x_1)$ with respect to x_1 is denoted as $f_1^*(s)$. For example, if $s = 10$, the optimal value of x_1 is $x_1^* = 3$, which gives a maximum solution of $f_1^*(3) = 150$.

Table 7.5. First Stage

State variable s	Optimal solution $f_1^*(s)$	Optimal decision variable x_1^*
0	0	0
1	0	0
2	0	0
3	50	1
4	50	1
5	50	1
6	100	2
7	100	2
8	100	2
9	150	3
10	150	3
11	150	3
12	200	4
13	200	4

The next step is to consider the problem where two stages remain, x_1 and x_2. For the moment assume that the value of x_3 is known. The original problem then becomes

$$\text{maximize} \quad z = 50x_1 + 60x_2 + [80x_3^*] \tag{7.11}$$
$$\text{subject to} \quad 3x_1 + 4x_2 \leq [13 - 6x_3^*] = s \tag{7.12}$$
$$x_1, x_2 = 0, 1, 2, \ldots \tag{7.13}$$

By use of the information in Table 7.5, the optimal solution to this problem can be determined by considering only the possible values of x_2, which are 0, 1, 2, and 3. To illustrate this, suppose $x_2 = 0$. Then the problem (7.11)–(7.13) becomes

$$\text{maximize} \quad f(x_1) = 50x_1$$
$$\text{subject to} \quad 3x_1 \leq s$$
$$x_1 = 0, 1, 2, \ldots$$

The solution to this problem can be determined from Table 7.5 once s is specified. If, for example, $s = 13$, the optimal solution is $f_1^*(13) = 200$ and $x_1^* = 4$ from Table 7.5. To build the second-stage table the recursive relationship relating stage n with stage $n - 1$ must be specified. This will provide the rules for determining x_2^* and $f_2^*(s)$ for an arbitrary value of s. First, the range of values for x_2 must be specified:

$$x_2 = 0, 1, \ldots (s/4) \tag{7.14}$$

where $(s/4)$ is the integer part of the ratio $s/4$ (i.e., the largest integer $\leq s/4$). For each value of x_2, the reduced problem of (7.11)–(7.13) is

$$\begin{aligned} \text{maximize} \quad & z = 50x_1 + [60x_2] & (7.15) \\ \text{subject to} \quad & 3x_1 \leq s - [4x_2] & (7.16) \\ & x_1 = 0, 1, \ldots \end{aligned}$$

For a specified value of x_2 the bracketed quantities in (7.15) and (7.16) will be constants. Notice that the term $80x_3^*$ in (7.11) has been dropped in the current objective function (7.15) since, for known x_3^*, it will be a constant. Now the optimal solution for the reduced problem may be obtained from Table 7.5 when the constant term (the first-stage state variable s) is equal to $s - 4x_2$. Hence the associated objective function value is given by $f_1^*(s - 4x_2)$ and may also be read from Table 7.5. The value of $f_2(s, x_2)$ for the two-variable problem is therefore

$$f_2(s, x_2) = f_1^*(s - 4x_2) + 60x_2 \tag{7.17}$$

For a given value of s, by maximizing (7.17) with respect to x_2, the optimal value $f_2^*(s)$ can be determined for the second stage. Thus the stage two problem can be restated as

$$\begin{aligned} \text{maximize} \quad & f_2(s, x_2) = f_1^*(s - 4x_2) + 60x_2 & (7.18) \\ \text{subject to} \quad & 0 \leq x_2 \leq (s/4) & (7.19) \\ & x_2 = 0, 1, 2, \ldots & (7.20) \end{aligned}$$

The constraint (7.19) follows from (7.14). Notice that the constraint $3x_1 \leq s$, (7.9) in the first-stage problem, could have been written as $0 \leq x_1 \leq (s/3)$. The second-stage results are given in Table 7.6.

Table 7.6. Second Stage

State variable s	Optimal solution $f_2^*(s)$	Optimal decision variable x_2^*
0	0	0
1	0	0
2	0	0
3	50	0
4	60	1
5	60	1
6	100	0
7	110	1
8	120	2
9	150	0
10	160	1
11	170	2
12	200	0
13	210	1

Two of the optimal subproblem solutions are verified in Table 7.6. If $s = 3$, the problem in (7.18)—(7.20) becomes

$$\begin{aligned} &\text{maximize} \quad f_2(3, x_2) = f_1^*(3 - 4x_2) + 60x_2 \\ &\text{subject to} \quad 0 \le x_2 \le (3/4) = 0 \qquad (7.21) \\ &\qquad x_2 = 0, 1, 2, \ldots \end{aligned}$$

Thus from (7.21) x_2 must be 0 so that $f_2(3, x_2) = f_1^*(3 - 4(0)) + 60(0) = f_1^*(3)$, and from Table 7.5 $f_1^*(3) = 50$. Thus $f_2^*(3) = 50$ and $x_2^* = 0$.

If $s = 8$, the problem in (7.18)–(7.20) becomes:

$$\begin{aligned} &\text{maximize} \quad f_2(8, x_2) = f_1^*(8 - 4x_2) + 60x_2 \\ &\text{subject to} \quad 0 \le x_2 \le (8/4) = 2 \qquad (7.22) \\ &\qquad x_2 = 0, 1, 2, \ldots \qquad (7.23) \end{aligned}$$

From (7.22) and (7.23) there are now three possible values of the decision variable x_2 (0, 1, and 2):

$$f_2(8, 0) = f_1^*(8 - 0) + 60(0) = f_1^*(8) = 100$$
$$f_2(8, 1) = f_1^*(8 - 4) + 60(1) = f_1^*(4) + 60 = 50 + 60 = 110$$
$$f_2(8, 2) = f_1^*(8 - 8) + 60(2) = f_1^*(0) + 120 = 0 + 120 = 120$$

Thus the best selection for the decision variable is $x_2^* = 2$, for which $f_2^*(8) = 120$. The reader should verify the remaining entries in Table 7.6.

Now that the form of the recursive relationship is clear, the optimization problem for the subproblem in the third stage can be written as

$$\text{maximize} \quad f_3(s, x_3) = f_2^*(s - 6x_3) + 80x_3 \qquad (7.24)$$
$$\text{subject to} \quad 0 \le x_3 \le (s/6) \qquad (7.25)$$
$$x = 0, 1, 2, \ldots \qquad (7.26)$$

This problem may be used to determine the solutions for the third-stage subproblem given in Table 7.7.

Table 7.7. Third Stage

State variable s	Optimal solution $f_3^*(s)$	Optimal decision variable x_3^*
0	0	0
1	0	0
2	0	0
3	50	0
4	60	0
5	60	0
6	100	0
7	110	0
8	120	0
9	150	0
10	160	0
11	170	0
12	200	0
13	210	0

For example, when $s = 12$, the problem (7.24)—(7.26) becomes:

$$\text{maximize} \quad f_3(12, x_3) = f_2^*(12 - 6x_3) + 80x_3$$
$$\text{subject to} \quad 0 \le x_3 \le (12/6) = 2 \qquad (7.27)$$
$$x = 0, 1, 2, \ldots \qquad (7.28)$$

From (7.27) and (7.28) the possible values of x_3 are 0, 1, and 2:

$$f_3(12, 0) = f_2^*(12 - 0) + 80(0) = f_2^*(12) + 0 = 200 + 0 = 200$$
$$f_3(12, 1) = f_2^*(12 - 6) + 80(1) = f_2^*(6) + 80$$
$$= 100 + 80 = 180$$
$$f_3(12, 2) = f_2^*(12 - 12) + 80(2) = f_2^*(0) + 160$$
$$= 0 + 160 = 160$$

Thus when $s = 12$, the best choice of x_3 is $x_3^* = 0$, for which $f_3^*(12) = 200$.

From Table 7.7 the maximum value of the objective function in the original problem (7.1) is $f^*(s) = 210$. The optimal solution of the decision variables can be determined by tracing back through Tables 7.7, 7.6, and 7.5. From Table 7.7, $x_3^* = 0$ [associated with $f^*(s) = 210$]. Since $x_3^* = 0$, s in Table 7.6 will be 13 (no slack is absorbed by x_3^*); and $x_2^* = 1$ (corresponding to $s = 13$). Since $x_2^* = 1$, 4 units of slack have been absorbed; and s in Table 7.5 will therefore be $13 - 4 = 9$. For $s = 9$ in Table 7.5, $x_1^* = 3$. The optimal variable values are $x_1^* = 3$, $x_2^* = 1$, and $x_3^* = 0$, for which $f^*(s) = 210$.

The dynamic programming solution to the general knapsack problem can now be delineated. The general problem is

$$\begin{aligned} &\text{maximize} \quad z = \textstyle\sum_{j=1}^{n} c_j x_j \\ &\text{subject to} \quad \textstyle\sum_{j=1}^{n} a_j x_j \le b \\ &\qquad x_j = 0, 1, 2, \ldots \qquad j = 1, 2, \ldots, n \end{aligned}$$

The kth-stage subproblem corresponding to this general form of the knapsack problem is:

$$\begin{aligned} &\text{maximize} \quad z_k = \textstyle\sum_{j=1}^{k} c_j x_j \\ &\text{subject to} \quad \textstyle\sum_{j=1}^{k} a_j x_j \le s = b - \sum_{j=k+1}^{n} a_j x_j^* \\ &\qquad x_j = 0, 1, 2, \ldots \qquad j = 1, 2, \ldots, k \end{aligned}$$

From the above development the kth stage subproblem may be rewritten, using the recursive relationship as

$$\begin{aligned} &\max_{x_k} \, [f_{k-1}(s - a_k x_k) + c_k x_k] \\ &\text{subject to} \quad 0 \le x_k \le (s/a_k) \\ &\qquad x_k = 0, 1, 2, \ldots \end{aligned} \tag{7.29}$$

When $k = 1$, define (7.29) as

$$f_0(s) = 0 \text{ for all } s \tag{7.30}$$

The redefinition of f_{k-1} when $k = 1$ is necessary, since there is no previous stage that contributes to the current objective function value. The definition in (7.30) is frequently called a dynamic programming *initialization condition.*

Five characteristics of the knapsack example problem should be mentioned. The ensuing discussion is intended to provide some insight into the process of applying dynamic programming to a mathematical programming problem.

1. Notice that the objective function in the knapsack problem (7.1) is a linear function in the three variables x_1, x_2, and x_3. The objective function certainly can be nonlinear; each subproblem in this case will be a nonlinear optimization problem, however.
2. The objective function is integer valued. If x_1, x_2, and x_3 had been continuous variables, the state variable s could assume an infinite number of values and the nth-stage tables could not be formed as for the knapsack problem. In the Section 7.3, a method for dealing with continuous variables is discussed for solving a continuous variable mathematical programming problem by dynamic programming.
3. Most important, the objective function in the knapsack problem is separable; the n-variable function $f(x_1, x_2, \ldots, x_n)$ can be written as the following linear combination of univariate functions:

$$f(x_1, x_2, \ldots, x_n) = f_1(x_1) + f_2(x_2) + \cdots + f_n(x_n)$$

   Recall from Section 4.3 that any n-variable linear function is separable. What if the objective function is nonlinear and not separable? To develop a simple recursive relationship relating the nth stage subproblems with those in the $n - 1$st stage, the objective function must be separable since staging is on the variables $x_1, x_2, \ldots, x_n$. As pointed out in Section 4.3, with some effort most functions can be separated; but this will almost always result in new constraints that must be augmented to the problem's original constraint set. This suggests the most serious restriction in applying dynamic programming to mathematical programming problems.
4. The knapsack problem has only one constraint. Suppose that a programming problem has two constraints; this will ordinarily necessitate adding another state variable, say t. Consequently, the recursive relationship must depend on two variables s and t; as a result the nth-stage table will be a matrix of $f_n^*(s,t)$ values, as shown in Table 7.8. If f_n were a function of three state parameters, a three-dimensional array would be required at the nth stage. The restriction on the number of state parameters is called the *dimensionality restriction* in dynamic programming; it is virtually impossible computationally to solve most problems with more than two state variables. It should be pointed out here that the constraints are not simple upper and lower bound constraints on the variables $x_1, x_2, \ldots, x_n$. Indeed, the inclusion of upper and lower bounds for the variables will tend to

Table 7.8. Matrix of $f_n^*(s,t)$ Values

s \ t	0	1	2	...
0	$f_n(0,0)$	$f_n(0,1)$	$f_n(0,2)$	...
1	$f_n(1,0)$	$f_n(1,1)$	$f_n(1,2)$	...
2	$f_n(2,0)$	$f_n(2,1)$	$f_n(2,2)$	...
⋮	⋮	⋮	⋮	

reduce the size of the tables; hence the bound constraints are considered an advantage in dynamic programming. If a problem has more than two constraints, by employing Lagrangian multipliers, it may be possible to recast the problem so that it has two or less constraints. The addition of variables will increase the number of stages required to solve the problem.

5. The last comment is on the *computational efficiency* of dynamic programming. In the knapsack problem there were 39 subproblems to solve, 13 in each stage (although it is not necessary to solve all 13 subproblems in the third stage for this problem). There is a maximum of (5)(4)(3) = 60 subproblems that must be solved if all possible solutions are enumerated explicitly, but some combinations of variable values such as $x_1 = 4$, $x_2 = 3$, $x_3 = 2$ will not be feasible, so that the actual number of explicit enumerable solutions is some number less than 60. As the number of variables increases, dynamic programming is certain to enjoy an increasing computational advantage over the exhaustive explicit enumeration of all possible solutions. The amount of computation required to solve a problem by dynamic programming is proportional to the number of variables, whereas the amount of computations in enumerating all possible solutions increases exponentially with an increase in the number of variables. The size of the problem solved by dynamic programming is somewhat restricted by the amount of computer storage required by the tables of $f_n^*(s)$ values. This restriction is becoming less crucial due to advances in digital computing hardware, however.

7.3 CONTINUOUS VARIABLE CASE

Consider the problem

$$\begin{aligned} &\text{maximize} \quad f(x_1, x_2, \ldots, x_n) = \textstyle\sum_{i=1}^{n} f_i(x_i) \qquad (7.31)\\ &\text{subject to} \quad \textstyle\sum_{i=1}^{n} a_i x_i \leq b \\ &\qquad\qquad x_i \geq 0 \qquad i = 1, 2, \ldots, n \end{aligned}$$

Notice that the integer restrictions on the n variables have been dropped. Proceeding as in Section 7.2, the recursive relationship for the kth stage is

$$f_k^*(s) = \max_{x_k} [f_{k-1}^*(s - a_k x_k) + f_k(x_k)]$$

The difference between the present continuous variable problem and that of Section 7.2 occurs in the process of minimizing $f_{k-1}^*(s - a_k x_k) + f_k(x_k)$ with respect to x_k. Suppose that the value of s is fixed and the corresponding row in the kth-stage table must be constructed. The optimization subproblem that remains is

$$\begin{aligned} \text{maximize} \quad & f(s, x_k) = f_{k-1}^*(s - a_k x_k) + f_k(x_k) \\ \text{subject to} \quad & 0 \leq x_k \leq b/a_k \end{aligned} \tag{7.32}$$

Notice that x_k is bounded above by the ratio b/a_k, not the integer part of this ratio as before. Assuming that the function $f_{k-1}^*(\cdot)$ is known, $f(s, x_k)$ is defined over all real values of x_k between 0 and b/a_k. This function might appear as the one illustrated in Figure 7.2. The value $f(s, x_k)$ of x_k that maximizes $f(s, x_k)$ over this interval must be determined.

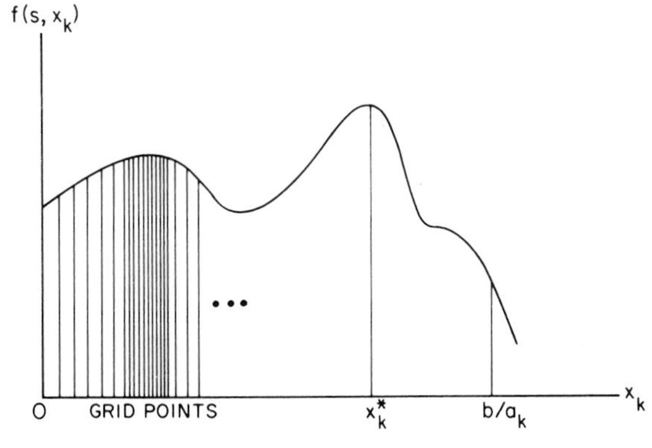

Fig. 7.2. Maximization of a nonconvex and continuous kth stage function $f(s, x_k)$.

Since x_k takes on a continuum of values, unless it is known that the function $f(s, x_k)$ does not have local maxima, there is little choice but to search for the global maximum over the interval $(0, b/a_k)$. In general, the subproblem may be a nonconvex programming problem such as the one whose function is illustrated in Figure 7.2. Hadley [1964] recommends first conducting a rough grid search over the interval, followed by a finer grid search over promising subintervals of the interval $(0, b/a_k)$. It may also be possible to employ a nonlinear programming algorithm or at least a more efficient search procedure (e.g., the Hooke-Jeeves algorithm); see Chapter 4 for a discussion of nonlinear algorithms. The overbearing restriction is the time involved in solving the subproblem. If each and every subproblem involves maximizing a nonconvex objective function, a solution algorithm that is very time-consuming cannot be employed or the time required to solve all N stages of the problem may be infeasibly large. Therefore, determining x_k^* for a specified value of s can present a difficult task. Generally, attaining a good optimizing value of x_k is only a dream if the subproblem function $f_k(s, x_k)$ is viciously nonlinear. This is only part of the problem, though.

The subproblem optimization must be repeated for different values of the state variable s to complete the kth stage table. Since $x_1, x_2, \ldots, x_n$ are continuous variables, s can also take on a continuum of values over the interval $(0, b)$. An obvious solution is to form a grid of s values over this interval, as has been done for x_k over its interval $(0, b/a_k)$. However, the value of s $(s - a_k x_k)$ that is needed from the $(k - 1)$st table may not be one of the grid values in that table. To find $f_{k-1}^*(s - a_k x_k)$, it may be necessary to interpolate in the $(k - 1)$st table. This is another problem that is going to add to the computational time and contribute to the problem round-off error. To interpolate in these tables, the function $f_k(s, x_k)$ must be continuous in s. Unless this is the case, there is an additional serious complication. Hadley [1964] proves that if each $f_i(x_i)$ in (7.31) is a continuous function of x_i, then $f_k(s, x_k)$ will be continuous in s.

If the functions $f_i(x_i)$ in (7.31) are either all convex or all concave in x_i, the computations in forming the tables can be reduced considerably. If each $f_i(x_i)$ is a continuous convex function of x_i, then $f_k(s, x_k)$ will be a convex function in s (see Bellman [1957]). Since the maximum of a convex function over a closed convex set will always occur at one of the extreme points, the maximum of $f_k(s, x_k)$ in (7.32) for fixed s will occur at either $x_k = 0$ or $x_k = b/a_k$. Thus no searching over the interval $(0, b/a_k)$ is necessary in this case; it is only necessary to compute $f_k(s, 0)$ and $f_k(s, b/a_k)$ and select as $f_k^*(s)$ the maximum of these two numbers. If each $f_i(x_i)$ is a continuous concave function of x_i, then $f_k(s, x_k)$ will be a concave function in s. Then at least for a fixed value of s, any local maximum of $f_k(s, x_k)$ will also be a global maximum. The procedure for searching out the optimizing point x_k for fixed s could at least be modified to take advantage of this fact.

Fortunately, in most applications of dynamic programming the decision variables $x_1, x_2, \ldots, x_n$ are naturally discrete; the resolution of a continuous variable problem by dynamic programming is, at best, an arduous task and, at worst, infeasible computationally. For more information on the continuous variable applications of dynamic programming, the interested reader is directed to Bellman [1957], Bellman and Dreyfus [1962], Hadley [1964], and White [1969].

7.4 EXAMPLES

Example 7.3. An Application to the Firm

The XYZ Corporation is considering a government offer to produce a special piece of equipment for a three-year period. The government agrees to purchase as many units x as the firm produces, up to a fixed number. Because of the size of the plant the corporation is not capable of producing as many units as the upper limit the government will accept in one period. The firm must deliver all units produced in time period t in that interval. The formula by which the government will determine how much to pay the company in period t is

$$f_t(x_t) = -2.5x_t^3 + 35.4x_t^2 - 4x_t + 10\delta_t$$

$$x_t \text{ an integer} \qquad x_t \geq 0$$

$$\delta_t = 0 \text{ if } x_t = 0 \qquad \delta_t = 1 \text{ if } x_t > 0$$

The value of $f_t(x_t)$ is in thousands of dollars.

Notice that this is a form of the fixed-charge problem discussed in Section 6.2. The company feels that it can allocate \$44,000 of its budget over a three-year period to produce the item in question. Because of other commitments by the firm, the costs of producing the item will vary during the three one-year periods. In period two it will be necessary to pay heavy costs in overtime wages, and renting several extra machines will be required. In period three the extra machines will not be needed, but some overtime will be required of the work force. The cost of producing one unit of the item is \$2000 in period one, \$8000 in period two, and \$4000 in period three. For the three periods, the firm's budget constraint is $2000x_1 + 8000x_2 + 4000x_3 \leq 44{,}000$ or $x_1 + 4x_2 + 2x_3 \leq 22$.

Over the three-year period, the firm can determine the optimal number of units to produce in each period (x_1^*, x_2^*, and

x_3^*) and its maximum payment from the government z by solving the following mathematical programming problem:

$$\begin{aligned} \text{maximize} \quad & z = \textstyle\sum_{t=1}^{3} f_t(x_t) \\ \text{subject to} \quad & x_1 + 4x_2 + 2x_3 \leq 22 \\ & x_t \text{ an integer} \\ & x_t \geq 0 \qquad t = 1, 2, 3 \\ \delta_t & = 0 \text{ if } x_t = 0 \qquad t = 1, 2, 3 \\ \delta_t & = 1 \text{ if } x_t > 0 \qquad t = 1, 2, 3 \end{aligned}$$

where

$$f_t(x_t) = -2.5x_t^3 + 35.4x_t^2 - 4x_t + 10\delta_t$$

Table 7.9. Results of Subproblem Optimizations

State variable	Stage one		Stage two		Stage three	
s	$f_1^*(s)$	x_1^*	$f_2^*(s)$	x_2^*	$f_3^*(s)$	x_3^*
0	0	0	0	0	0	0
1	38.9	1	38.9	0	38.9	0
2	123.6	2	123.6	0	123.6	0
3	249.1	3	249.1	0	249.6	0
4	400.4	4	400.4	0	400.4	0
5	562.5	5	562.5	0	562.5	0
6	720.4	6	720.4	0	720.4	0
7	859.1	7	859.1	0	859.1	0
8	963.6	[8]	963.6	[0]	963.6	0
9	1018.9	9	1018.9	0	1018.9	0
10	1018.9	9	1018.9	0	1018.9	0
11	1018.9	9	1018.9	0	1057.8	1
12	1018.9	9	1018.9	0	1087.2	2
13	1018.9	9	1057.8	1	1142.5	2
14	1018.9	9	1057.8	1	1212.7	3
15	1018.9	9	1057.8	1	1268.0	3
16	1018.9	9	1087.2	2	1364.0	4
17	1018.9	9	1142.5	2	1421.6	5
18	1018.9	9	1142.5	2	1526.1	5
19	1018.9	9	1142.5	2	1581.4	5
20	1018.9	9	1212.7	3	1684.0	6
21	1018.9	9	1268.0	3	1739.3	6
22	1018.9	9	1268.0	3	[1822.7]	7

To solve this problem, the results of Section 7.2 will be employed. In the first stage the optimization subproblem has the form

$$\begin{aligned}
\text{maximize} \quad & f_1(s,x_1) = -2.5x_1^3 + 35.4x_1^2 - 4x_1 + 10\delta_1 \\
\text{subject to} \quad & 0 \le x_1 \le (s/1) = s \\
& x_1 = 0, 1, 2, \ldots \\
& \delta_1 = 1 \text{ if } x_1 > 0 \\
& \quad\;\, = 0 \text{ if } x_1 = 0
\end{aligned}$$

The state variable s can take on 23 values $0, 1, 2, \ldots, 22$ so that the quadratic optimization problem must be solved 23 times for x_1^* and $f_1^*(s)$. The results of the first-stage subproblem optimizations are given in Table 7.9. Notice that when $x_1^* \ge 1$, $\delta_1 = 1$ so that 10 is contributed to $f_1(s, x_1)$ resulting in $f_1^*(1) = 38.9$, for example.

In the second stage the appropriate form of the optimization subproblem is:

$$\begin{aligned}
\text{maximize} \quad & f(s,x_2) = f_1^*(s - 4x_2) + f_2(x_2) \\
\text{subject to} \quad & 0 \le x_2 \le (s/4) \\
& x_2 = 0, 1, 2, \ldots \\
& \delta_2 = 1 \text{ if } x_2 > 0 \\
& \quad\;\, = 0 \text{ if } x_2 = 0
\end{aligned}$$

The results of the second-stage subproblem optimizations are given in Table 7.9.

In the third stage the form of the optimization subproblem is:

$$\begin{aligned}
\text{maximize} \quad & f(s,x_3) = f_2^*(s - 2x_3) + f_3(x_3) \\
\text{subject to} \quad & 0 \le x_3 \le (s/2) \\
& x_3 = 0, 1, 2, \ldots \\
& \delta_3 = 1 \text{ if } x_3 > 0 \\
& \quad\;\, = 0 \text{ if } x_3 = 0
\end{aligned}$$

The results of the third-stage subproblem optimizations are also given in Table 7.9.

From Table 7.9, the maximum payment is $f_3^*(s) =$ \$1822.70. The value of the decision variable x_3 corresponding to this functional value is $x_3^* = 7$. Since each unit of x_3 will expend two of the 22 total resource units, only $s = 8$ units will remain after x_3 has apportioned 14 of them. For stage two, from across the row $s = 8$ of Table 7.9, $x_2^* = 0$. There are $s = 8$ resource units left for x_1, and its optimal value is determined to be $x_1^* = 8$ from the same row in the stage-one column. Hence the complete solution is $x_1^* = 8$, $x_2^* = 0$, and $x_3^* = 7$, with a maximum objective function value of $f(x^*) = \$1822.70$.

If the firm suddenly found that it could only budget \$34,000 for the project over the three years, the optimal solution would be (from Table 7.9):

$$\max\left[\sum_{t=1}^{3} f_t(x_t)\right] = \$1421.6$$

$$x_3^* = 5 \qquad x_2^* = 0 \qquad x_1^* = 7$$

Example 7.4. A Nonlinear Integer Programming Problem

Consider the following bivariate problem:

$$\begin{aligned} \text{maximize} \quad & z = (x_1 + 3)^2 + (x_2 + 4)^2 \\ \text{subject to} \quad & 3x_1 + 4x_2 \leq 12 \\ & x_1 \leq 3 \\ & x_1, x_2 = 0, 1, 2, \ldots \end{aligned}$$

This is a nonconvex integer programming problem. Although this example is quite easy to solve (e.g., graphic methods could be used), dynamic programming will be applied to indicate its ability to solve one of the most difficult types of mathematical programming problems, those with integer-valued variables and a nonconvex objective function.

The first-stage optimization subproblem is:

$$\begin{aligned} \text{maximize} \quad & f_1(s, x_1) = (x_1 + 3)^2 \\ \text{subject to} \quad & 0 \leq x_1 \leq (s/3) \\ & x_1 \leq 3 \\ & x_1 = 0, 1, 2, \ldots \end{aligned} \tag{7.33}$$

Notice that the constraint $x_1 \leq 3$ dominates the upper bound in (7.33) when $s \geq 10$. The solutions to the stage one subproblems are given in Table 7.10.

The second-stage optimization subproblem is:

$$\begin{aligned} \text{maximize} \quad & f_2(s, x_2) = f_1^*(s - 4x_2) + (x_2 + 4)^2 \\ \text{subject to} \quad & 0 \leq x_2 \leq (s/4) \\ & x_2 = 0, 1, 2, \ldots \end{aligned}$$

Table 7.10. Solutions to Subproblems

State variable	Stage one		Stage two	
s	$f_1^*(s)$	x_1^*	$f_2^*(s)$	x_2^*
0	9	0	$9 + 16 = 25$	0
1	9	0	$9 + 16 = 25$	0
2	9	0	$9 + 16 = 25$	0
3	16	1	$16 + 16 = 32$	0
4	16	1	$9 + 25 = 34$	1
5	16	1	$9 + 25 = 34$	1
6	25	2	$25 + 16 = 41$	0
7	25	2	$25 + 16 = 41$ $16 + 25 = 41$	0 or 1
8	25	2	$9 + 36 = 45$	2
9	36	3	$36 + 16 = 52$	0
10	36	3	$36 + 16 = 52$	0
11	36	3	$36 + 16 = 52$ $16 + 36 = 52$	0 or 2
12	36	3	$9 + 49 = 58$	3

The solutions to the stage two subproblems are given in Table 7.10 also.

From Table 7.10, $f^*(s, x_2) = 58$ and $x_2^* = 3$. When $x_2 = 3$, there are no remaining resource units, so that x_1^* is determined by entering row one ($s = 0^*$ in Table 7.10) for stage one, giving $x_1^* = 0$. The complete solution is $x_1^* = 0$, $x_2^* = 3$, and $f^*(s, x_2) = 58$. Notice that the next best objective function value is 52, which occurs when $x_2^* = 0$ (and $x_1^* = 3$) or when $x_2^* = 2$ (and $x_1^* = 1$). An additional benefit from applying dynamic programming to integer problems is that the solution process will supply several good answers as well as the optimal one.

PROBLEMS

7.1. Find by dynamic programming the least expensive route from $\widetilde{15}$ to $\widetilde{10}$. The costs of each leg of the journey are shown on the lines connecting the nodes (see Fig. P7.1).

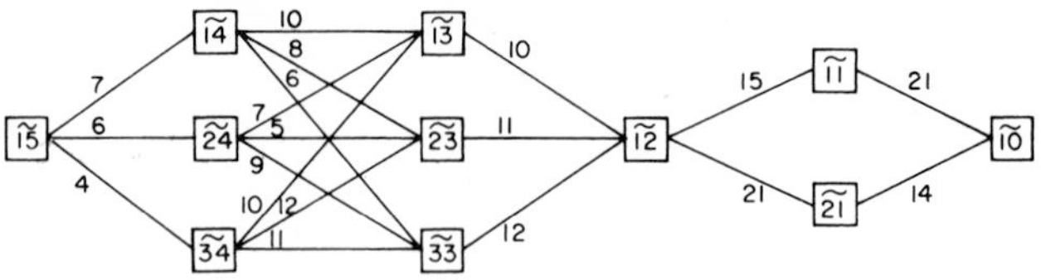

Fig. P7.1.

7.2. Solve the following nonlinear integer programming problem by dynamic programming:

$$\begin{aligned} \text{maximize} \quad & z = \textstyle\sum_{t=1}^{3} 3 \log (x_t + 1) \\ \text{subject to} \quad & \textstyle\sum_{t=1}^{3} x_t \leq 5 \\ & x_t = 0, 1, 2, \ldots \qquad t = 1, 2, 3 \end{aligned}$$

7.3. A representative of radio station WXYZ has approached the sales manager at Townwide Motors with the following proposal. Townwide buys x_t units of advertising each three months (quarter), and in each advertisement the public is told that WXYZ will pay the car dealer \$200 per quarter if any car purchaser says he made the purchase as a result of the radio ad; Townwide will give that purchaser 50% of the \$200. The car dealer, who attempts to maximize his additional sales revenue as a result of advertising for a year z, knows that additional car sales revenue in the tth quarter S_t is a function of advertising units x_t and that $S_t = -2.5x_t^3 + 25.4x_t^2 - 4x_t + (200 - 100)\delta_t$, where $\delta_t = 0$ if no customer makes a purchase as a result of the WXYZ advertisement ($x_t = 0$) but $\delta_t = 1$ if $x_t \geq 1$.

For the year the car dealer's goal is to determine the number of minutes of advertising to buy from WXYZ (x_t) for each quarter ($t = 1, 2, 3, 4$). He wants to

$$\begin{aligned} \text{maximize} \quad & z = \textstyle\sum_{t=1}^{4} (-2.5x_t^3 + 25.4x_t^2 - 4x_t + 100\delta_t) \\ \text{subject to} \quad & x_1 + 4x_2 + 2x_3 + 5x_4 \leq A \\ & x_1 \geq 1 \\ & x_t \geq 0 \qquad t = 2, 3, 4 \\ & x_t = 0, 1, 2, \ldots \qquad t = 1, 2, 3, 4 \\ & \delta_t = 0 \text{ if } x_t = 0 \\ & \quad\; = 1 \text{ if } x_t \geq 1 \end{aligned}$$

where the first constraint provides the advertising budget and at least one unit of advertising time is purchased in the first quarter.

The optimization table for this problem is

A	$f_1^*(s)$	s^*	$f_2^*(s)$	s^*	$f_3^*(s)$	s^*	$f_4^*(s)$	s^*
1	118.9	1	118.9	0	118.9	0	118.9	0
2	173.6	2	173.6	0	173.6	0	173.6	0
3	249.1	3	249.1	0	249.1	0	249.1	0
4	330.4	4	330.4	0	330.4	0	330.4	0
5	402.5	5	402.5	0	402.5	0	402.5	0
6	450.4	6	450.4	0	450.4	0	450.4	0
7	459.1	7	459.1	0	521.4	1	521.4	0
8	459.1	7	459.1	0	569.3	1	569.3	0
9	459.1	7	521.4	1	578.0	1	578.0	0
10	459.1	7	569.3	1	624.0	2	624.0	0
11	459.1	7	578.0	1	651.6	3	651.6	0
12	459.1	7	578.0	1	699.5	3	699.5	0
13	459.1	7	578.0	1	732.9	4	732.9	0

Find the optimal value of z and x_t, $t = 1, 2, 3, 4$, for annual advertising budgets of $A = 6, 8, 10$, and 13.

7.4. Solve the following knapsack problem by dynamic programming:

$$\begin{aligned} \text{maximize} \quad & z = \textstyle\sum_{j=1}^{4} c_j x_j \\ \text{subject to} \quad & \textstyle\sum_{j=1}^{4} a_j x_j \le b \\ & x_j = 0, 1, 2, \ldots \qquad j = 1, 2, 3, 4 \end{aligned}$$

where $c_1 = 50$, $c_2 = 75$, $c_3 = 60$, $c_4 = 100$, $a_1 = 100$, $a_2 = 150$, $a_3 = 120$, $a_4 = 200$, and $b = 730$.

Solve the following three nonlinear integer programming problems by dynamic programming:

7.5.
$$\begin{aligned} \text{Maximize} \quad & z = 10x_1(x_1 - 2x_2) + 20x_2(2x_2 + x_1) \\ \text{subject to} \quad & 50x_1 + 100x_2 \le 500 \\ & x_1, x_2 = 0, 1, 2, \ldots \end{aligned}$$

7.6.
$$\begin{aligned} \text{Maximize} \quad & z = 4x_1 + 8x_2 \\ \text{subject to} \quad & 4x_1^2 + 6x_2^2 \le 96 \\ & x_1, x_2 = 0, 1, 2, \ldots \end{aligned}$$

7.7.
$$\begin{aligned} \text{Maximize} \quad & z = 4x_1 + 2x_2 + 6x_3 + x_4^2 \\ \text{subject to} \quad & 3x_1 + x_2 + x_3 + 5x_4 \le 12 \\ & x_1 \ge 2 \\ & x_2 \ge 2 \\ & x_1, x_2, x_3, x_4 = 0, 1, 2, \ldots \end{aligned}$$

7.8. Solve the following continuous variable mathematical programming problem by dynamic programming (grid the x_1 and x_2 axes from 0 in steps of 0.5 to convert the problem to a discrete dynamic programming problem):

$$\begin{aligned} \text{minimize} \quad & z = (x_1 - 4)^2 + (x_2 - 4)^2 \\ \text{subject to} \quad & 3x_1 + 2x_2 \leq 6 \\ & x_1, x_2 \geq 0 \end{aligned}$$

7.9. Solve the following integer problem with two constraints by dynamic programming:

$$\begin{aligned} \text{maximize} \quad & z = 4x_1 + 8x_2 \\ \text{subject to} \quad & 2x_1 + x_2 \leq 4 \\ & x_1 + 3x_2 \leq 6 \\ & x_1, x_2 = 0, 1, 2, \ldots \end{aligned}$$

7.10. A plant manager is required to meet the schedule given in the table below for the delivery of a specific product in the next four periods.

Period	Units to be delivered
1	3
2	5
3	2
4	4
Total	14

No units of the product are presently in inventory and the manager prefers to have none after the four periods. He has the option of producing the 14 units in any periods he desires as long as the delivery schedule is met (e.g., he could produce all 14 units in period 1). Two costs are associated with the production and delivery of the units: setup and inventory. Suppose the setup cost required to produce one or more units in each period is \$100. Each unit left in inventory from a past period costs \$50. Let y_n be the number of units stored in inventory at the end of the nth period, d_n the number of units to be delivered at the end of the nth period, and x_n the number of units produced in the nth period. The manager can determine his optimal production schedule by solving the following problem:

$$\begin{aligned} \text{minimize} \quad & z = \textstyle\sum_{j=1}^{4} c_j \\ \text{subject to} \quad & y_n = y_{n-1} + x_n - d_n \\ & y_0 = 0 \\ & y_4 = 0 \\ & y_n, d_n, x_n = 0, 1, 2, \ldots \qquad n = 0, 1, 2, 3, 4 \end{aligned}$$

where $c_j = \$100\,\delta_j + 50y_j, j = 1, 2, 3, 4$, and

$$\begin{aligned} \delta_j &= 1 \text{ if } x_j > 0 \\ &= 0 \text{ if } x_j = 0 \end{aligned}$$

Solve this problem by dynamic programming.

8 RECURSIVE PROGRAMMING

8.1 INTRODUCTION

Recursive programming models are dynamic models in which time is a discrete variable and the parameters in the model may change over time. A recursive programming model may be a linear or nonlinear optimization problem to which the algorithms presented in Chapters 2–6 should be applicable. This differs from dynamic programming, which represents a computational technique and optimization model, and from calculus of variations, which usually requires more specialized techniques from the dynamics of calculus.

Many business and economic decision models have some recursive aspects because decisions for a particular period are somewhat determined by those made previously, and the data for any particular decision must be updated over time. Recursive programming models are virtually unique because the parameters in the model for any particular period can be adjusted as a result of the optimal allocations of previous periods.

Recursive programming problems require the selection of decisions for time period t based upon data and decisions in previous periods. The optimal allocation vector in period $t, \mathbf{x}_t^0$, is often determined as $\mathbf{x}_t^0 = f_t(\mathbf{B}_t, \mathbf{x}_{t-1}^0)$, where $\mathbf{x}_{t-1}^0$ is the optimal vector of allocations in period $t - 1, \mathbf{B}_t$ is the parameter or data matrix for period t, and f_t represents a scalar function. Then

$$\mathbf{x}_t^0 = f_t(\mathbf{B}_t, \mathbf{x}_{t-1}^0) = f_t[\mathbf{B}_t, f_{t-1}(\mathbf{B}_{t-1}, \mathbf{x}_{t-2}^0)]$$

This process can be extended back as many periods as we may choose.

The differences between this recursive allocation process and the dynamic processes delineated in Chapter 7 are twofold: (1) the parameter matrix $\mathbf{B}_t$ is permitted to change between time periods and (2) the "principle of optimality" discussed with respect to dynamic programming is not required for the recursive allocations discussed here.

The simplest recursive process for an activity in period t, defined by x_t, is described by a first-order difference equation, $x_t = ax_{t-1}$, with a single constant coefficient a. Then

$$x_t = ax_{t-1} = a(ax_{t-2}) = aa(ax_{t-3}) = \cdots = a^N x_{t-N}$$

is a solution to the difference equation, where the value of x_t is completely determined by the value of x in N periods prior to period t and the value of the constant coefficient raised to the Nth power.

When the constant is permitted to vary over time,

$$x_t = a_t x_{t-1} = a_t a_{t-1} x_{t-2} = a_t a_{t-1} a_{t-2} x_{t-3} = \cdots = \prod_{j=0}^{N-1} a_{t-j} x_{t-N}$$

In this case x_t is partially determined by the initial allocation in period $t - N$ and, implicitly, each allocation between period $t - N$ and period t. If a_t is also influenced by values of x's prior to period t, the result is a system of recursive relationships,

$$x_t = a_t x_{t-1} \qquad a_t = f_t(x_{t-j})$$

which can be represented as $x_t = h_t(x_{t-1}, x_{t-j})$ but cannot always be solved easily.

In many instances the optimal solution to a recursive programming model will consist of a series of recursive relationships. It is usually easier to deal with these equations and inequalities by solving sequential programming models rather than attempting to determine the general solution.

The "principle of optimality" for dynamic programming requires that each optimal allocation after the first one be an optimal policy with regard to decisions for stages resulting from the first optimal decision. This principle does not allow for the revision of parameters and the introduction of new data gained after the first optimal decision is selected. In reality, revised data sets should be expected for each time period on the basis of what has been learned in the past. This special feature of recursive programming models is illustrated throughout this chapter.

8.2 MODELS OF RECURSIVE PROGRAMMING

Wood has presented a model in which he assumed "that in no time period shall the level of the activity exceed the level in the previous time period by a ratio greater than $1 + \alpha$" [1951, p. 217]. If x is the activity, $x_t \leq x_{t-1}(1 + \alpha)$, $\alpha > 0$, for such a model.

This type of constraint was not included in a mathematical programming model until Day did so; most of the early models and applications of recursive programming have been presented by him. Day and his associates have applied recursive programming to the business and agricultural firm. In a recursive programming model at least one of the activities is constrained by its level during the previous period.

In his first study Day [1963] analyzed an agricultural production model. He adapted Henderson's [1959] model of the Mississippi Delta area into a linear programming model. The problem is

$$\text{maximize} \quad w_t = \mathbf{c}_t'\mathbf{x}_t \tag{8.1}$$
$$\text{subject to} \quad \mathbf{x}_t \leq \mathbf{B}_t\mathbf{x}_{t-1} \tag{8.2}$$
$$\mathbf{x}_t \geq \mathbf{b}_t\mathbf{x}_{t-1} \tag{8.3}$$
$$\textstyle\sum_{j=1}^{m} x_{jt} \leq \bar{x} \tag{8.4}$$

where $\bar{x}$ = the total amount of land available
$\mathbf{x}_t$ = an $m \times 1$ vector with a typical element of x_{jt}
$\mathbf{c}_t$ = an $m \times 1$ vector with a typical element of c_{jt}
$\mathbf{b}_t$ = an $m \times 1$ vector with a typical element of b_{jt}
b_{jt} = 1 − the maximum percentage by which the production of crop j can be decreased from period $t - 1$ to t
$\mathbf{B}_t$ = an $m \times 1$ vector, with a typical element of B_{jt}
B_{jt} = 1 + the maximum percentage by which the production of crop j can be increased from period $t - 1$ to period t

Since $B_{jt} = 1 +$ (a percentage) and $b_{jt} = 1 -$ (a percentage), $B_{jt} - b_{jt} > 0$.

Let $\mathbf{y}_t$ and $\mathbf{z}_t$ be $m \times 1$ vectors of slack variables such that

$$\mathbf{x}_t + \mathbf{y}_t = \mathbf{B}_t\mathbf{x}_{t-1} \tag{8.5}$$

$$\mathbf{x}_t - \mathbf{z}_t = \mathbf{b}_t\mathbf{x}_{t-1} \tag{8.6}$$

Equations (8.5) and (8.6) replace (8.2) and (8.3) in the programming model. Let u_t be a single slack variable such that

$$\sum_{j=1}^{m} x_{jt} + u_t = \overline{x} \tag{8.7}$$

The $2m + 1$ restrictions of the linear programming model involve $3m + 1$ variables ($2m + 1$ slacks and m originals).

If this linear programming problem is not degenerate, m of the $3m + 1$ variables must be zero, and $2m + 1$ variables are not zero; the $2m + 1$ nonzero variables provide a basic feasible solution. If c_{jt} is positive, $c_{jt}x_{jt}$ is largest if x_{jt} is as great as the restrictions permit. If c_{jt} is negative, $c_{jt}x_{jt}$ contributes most to maximize w_t when x_{jt} is as small as possible. Thus one should expect the following results:

$$\begin{aligned} \text{if } c_{jt} > 0, x_{jt} &= B_{jt}x_{j\,t-1} \\ y_{jt} = 0, z_{jt} &= (B_{jt} - b_{jt})x_{jt} \end{aligned} \tag{8.8}$$

but

$$\begin{aligned} \text{if } c_{jt} < 0, x_{jt} &= b_{jt}x_{j\,t-1} \\ z_{jt} = 0, y_{jt} &= (B_{jt} - b_{jt})x_{jt} \end{aligned} \tag{8.9}$$

This would be the solution if it were not for the restriction on the available resource $\overline{x}$ in (8.4). If the allocations given by (8.8) and (8.9) do not violate (8.4), then (8.8) and (8.9) provide the optimal solution. These are difference equations of the first order and are summarized in Table 8.1.

Table 8.1. Summary of Optimal Difference Equations

For all $c_j > 0$	$x_{jt}^0 = (B_{jt})^t x_{j0}$	$y_{jt}^0 = 0$	$z_{jt}^0 = (B_{jt} - b_j)(B_{jt})^t x_{j0}$
For all $c_j < 0$	$x_{jt}^0 = (b_{jt})^t x_{j0}$	$z_{jt}^0 = 0$	$y_{jt}^0 = (B_{jt} - b_j)(b_{jt})^t x_{j0}$
$\sum_{j=1}^{m} x_{jt}^0 < \overline{x}$ $w_t^0 = \sum_{j=1}^{m} c_j x_{jt}^0$			

If the optimal allocations defined in Table 8.1 for x_{jt}^0 violate the restriction on the resource available, one of the x_j's will not be at one of its boundaries. Then the equality $\sum_{j=1}^{m} x_{jt} = \overline{x}$ is specified, and $u_t = 0$. Then $m - 1$ of the $2m$ y_j's and z_j's will be zero, and $2m + 1$ variables are not zero (m x's, $m + 1$ slacks). Thus, one element of $\mathbf{x}_t$ is between its upper and lower bound. The questions that remain are to identify which x_{jt} is not at a boundary point and what the value of that activity will be.

Suppose that the minimum positive c_{jt} is $c_{\bar{j}t}$ and that most negative c_{jt} is $c_{\underline{j}t}$. (Overbars and underbars are also used to denote the corresponding x_{jt}'s, b_{jt}'s, and B_{jt}'s.) Then if $\sum_{j=1}^{m} x_{jt} = \bar{x}$, one of two cases occurs.

CASE 1

$$x_{\underline{j}t}^0 = B_{\underline{j}t} x_{\underline{j}\,t-1} \qquad b_{\bar{j}t} x_{\bar{j}\,t-1} < x_{\bar{j}t}^0 < B_{\bar{j}t} x_{\bar{j}\,t-1}$$

$$x_{\bar{j}t}^0 = \bar{x} - \sum_{\substack{j=1 \\ j \neq \bar{j}}}^{m} x_{jt}^0$$

and define

$$p = c_{\underline{j}t} B_{\underline{j}t} x_{\underline{j}\,t-1} + c_{\bar{j}t} \left(\bar{x} - \sum_{\substack{j=1 \\ j \neq \bar{j}}}^{m} x_{jt}^0 \right)$$

CASE 2

$$x_{\bar{j}t}^0 = b_{\bar{j}t} x_{\bar{j}\,t-1} \qquad b_{\underline{j}t} x_{\underline{j}\,t-1} < x_{\underline{j}t}^0 < B_{\underline{j}t} x_{\underline{j}\,t-1}$$

$$x_{\underline{j}t}^0 = \bar{x} - \sum_{\substack{j=1 \\ j \neq \underline{j}}}^{m} x_{jt}^0$$

and define

$$q = c_{\underline{j}t} \left(\bar{x} - \sum_{\substack{j=1 \\ j \neq \underline{j}}}^{m} x_{jt}^0 \right) + c_{\bar{j}t} b_{\bar{j}t} x_{\bar{j}\,t-1}$$

If $p > q$, Case 1 contributes more toward maximizing w; but if $p < q$, Case 2 contributes more. Thus for the optimal solution when $\sum_{j=1}^{m} x_{jt}^0 = \bar{x}$, the results are summarized in Table 8.2. If $p = q$, either Case 1 or Case 2 may be used, since each gives the same optimal w_t.

Models with Nonlinear Objective Functions

When the objective function is nonlinear, $w_t = f(\mathbf{x}_t) = f(x_{1t}, \ldots, x_{mt})$, and the restrictions are those given by (8.2)–(8.4), some

Table 8.2. Summary of Optimal Solution

For all $c_j > 0$ except $c_j = c_{\bar{J}}$	$x_{jt}^0 = (B_{jt})^t x_{j0}$	$y_{jt}^0 = 0$	$z_{jt}^0 = (B_{jt} - b_{jt})(B_{jt})^t x_{j0}$
For all $c_j < 0$ except $c_j = c_{\underline{j}}$	$x_{jt}^0 = (b_{jt})^t x_{j0}$	$y_{jt}^0 = 0$	$z_{jt}^0 = (B_{jt} - b_{jt})(b_{jt})^t x_{j0}$
	$\sum_{j=1}^{m} x_{jt}^0 = \bar{x}$		
$c_j = c_{\bar{J}}$	$x_{\underline{j}t}^0 = B_{\underline{j}t} x_{\underline{j}\,t-1}$	$x_{jt}^0 = \bar{x} - \sum_{\substack{j=1 \\ j \neq \bar{J}}}^{m} x_{jt}^0$	$p > q$
$c_j = c_{\underline{j}}$	$x_{\bar{j}t}^0 = b_{\bar{j}t} x_{\bar{j}\,t-1}$	$x_{\underline{j}t}^0 = \bar{x} - \sum_{\substack{j=1 \\ j \neq \underline{j}}}^{m} x_{jt}^0$	$p < q$

insight is gained by considering

$$\frac{\partial w_{jt}}{\partial x_{jt}} \qquad j = 1, \ldots, m$$

Let

$$\frac{\partial w_{jt}}{\partial x_{jt}} = f_j \qquad j = 1, \ldots, m$$

$$x_{jt}^* = B_{jt} x_{j\,t-1} \qquad j = 1, \ldots, m \tag{8.10}$$

$$\bar{x}_{jt} = b_{jt} x_{j\,t-1} \qquad j = 1, \ldots, m \tag{8.11}$$

and $F_j = f_j(x_{jt}^*) - f_j(\bar{x}_{jt}), j = 1, \ldots, m$.

If $F_j > 0$, the optimal value of x_{jt} is as close to x_{jt}^* as possible. If $F_j < 0$, the optimal value of x_{jt} is as close to $\bar{x}_{jt}$ as possible. If the constraints on the elements of $\mathbf{x}_t$ are only those of (8.2) and (8.3), the allocations given by (8.10) and (8.11) are optimal depending upon F_j. If there are constraints other than those of (8.2) and (8.3) but the results of (8.10) and (8.11) do not violate the other constraints, (8.10) and (8.11) still provide optimal decisions. If the results of (8.10) and (8.11) violate a constraint of the model, these solutions may provide a useful initial point from which to search for an optimal feasible solution.

Suppose that $w_t = f(\mathbf{x}_t)$ is a quadratic function such as $w_t = \mathbf{k}'\mathbf{x}_t + \mathbf{x}_t'\mathbf{K}\mathbf{x}_t$, where $\mathbf{k}$ is an $m \times 1$ vector and $\mathbf{K}$ is an $m \times m$ matrix. Then w_t can be expressed as

$$w_t = \sum_{j=1}^{m} w_{jt} = \sum_{j=1}^{m} k_j x_{jt} + \sum_{j=1}^{m} \sum_{h=1}^{m} K_{jh} x_{jt} x_{ht}$$

so that

$$w_{jt} = k_j x_{jt} + \sum_{h=1}^{m} K_{jh} x_{jt} x_{ht} \qquad j = 1, \ldots, m$$

Then

$$\frac{\partial w_{jt}}{\partial x_{jt}} = f_j = k_j + 2 \sum_{h=1}^{m} K_{jh} x_{ht} \qquad j = 1, \ldots, m$$

is a linear function having a constant slope with respect to each x_{ht}. If K_{jh} is positive, x_{jt}, $j = 1, \ldots, m$, should be made to approach its upper (lower) bound to maximize (minimize) w_{jt} and w_t. If K_{jh} is negative, x_{jt}, $j = 1, \ldots, m$, should be made to approach its lower (upper) boundary point to maximize (minimize) w_t.

A Linear Model for the Firm

Other applications of recursive programming have been completed by Day and Tinney. One of these [1969a] will be summarized here because it illustrates several of the conclusions they offer elsewhere. Consider a problem for the firm in time period t where the requirement is

$$\text{maximize} \quad \pi_t = \mathbf{c}_t'\mathbf{x}_t \tag{8.12}$$
$$\text{subject to} \quad \mathbf{A}\mathbf{x}_t \leq \mathbf{b}_t \tag{8.13}$$
$$\mathbf{x}_t \geq \mathbf{0} \tag{8.14}$$

The dual to this problem is

$$\text{minimize} \quad \psi_t = \mathbf{b}_t'\mathbf{r}_t$$
$$\text{subject to} \quad \mathbf{A}'\mathbf{r}_t \geq \mathbf{c}_t$$
$$\mathbf{r}_t \geq \mathbf{0}$$

where $\mathbf{x}_t$ = an $m \times 1$ vector of output levels for the m products
$\mathbf{c}_t$ = an $m \times 1$ vector of net revenues (profits) for the respective outputs
$\mathbf{b}_t$ = an $n \times 1$ vector of resource availabilities
$\mathbf{r}_t$ = an $n \times 1$ vector of accounting values or shadow prices for the inputs
$\mathbf{A}$ = an $n \times m$ matrix of input-output coefficients
π_t = the profit level achieved for a particular level of x
ψ_t = the value of the inputs for an optimal allocation

Static models similar to this one and their interpretations have been considered in detail in Chapter 2. That analysis will not be repeated here. However, Day suggests a periodic analysis that is a contribution. Suppose that the model is studied before period t. Then $\mathbf{c}_t$ must be a vector of expected net returns (profits) where $\mathbf{c}_t = \mathbf{p}_t - \mathbf{k}_t$. In this formulation, $\mathbf{p}_t$ is the expected price vector, which is not necessarily known, and $\mathbf{k}_t$ is the cost vector, which is assumed to be given. Since $\mathbf{p}_t$ is the expected price vector, $\mathbf{p}_t = \mathbf{p}_{t-1}$ is often assumed. Then

$$\mathbf{c}_t = \mathbf{p}_{t-1} - \mathbf{k}_t \tag{8.15}$$

Let a vector of linear demand curves be defined as

$$\mathbf{p}_t = \mathbf{a} + \mathbf{B}\mathbf{x}_t^0 \qquad \mathbf{p}_t \geq \mathbf{0} \tag{8.16}$$

If $\mathbf{a} + \mathbf{B}\mathbf{x}_t^0$ is negative, $\mathbf{p}_t = \mathbf{0}$. In (8.16), $\mathbf{x}_t^0$ is the optimal solution to the problem given by (8.12)–(8.14), $\mathbf{a}$ is an $m \times 1$ vector, and $\mathbf{B}$ is an $m \times m$ matrix. Combining (8.12), (8.15), and (8.16),

$$\pi_t = \mathbf{c}_t'\mathbf{x}_t = (\mathbf{p}_{t-1}' - \mathbf{k}_t')\mathbf{x}_t = (\mathbf{a} + \mathbf{B}\mathbf{x}_{t-1}^0 - \mathbf{k}_t)'\mathbf{x}_t$$

or

$$\pi_t = \mathbf{a}'\mathbf{x}_t + (\mathbf{x}_{t-1}^0)'\mathbf{B}'\mathbf{x}_t - \mathbf{k}_t'\mathbf{x}_t \tag{8.17}$$

Suppose the first restriction of (8.13) is a budget constraint, $\sum_{j=1}^m a_{1j}x_{jt} \leq b_{1t}$. Then b_{1t} = (income of period $t - 1$) − (overhead expenses of period t)

$$b_{1t} = \sum_{j=1}^{m} p_{jt-1}x_{jt-1}^0 - h_t$$

$$b_{1t} = \mathbf{p}_{t-1}'\mathbf{x}_{t-1}^0 - h_t \tag{8.18}$$

or

$$b_{1t} = (\mathbf{a} + \mathbf{B}\mathbf{x}_{t-1}^0)'\mathbf{x}_{t-1}^0 - h_t$$

after (8.16) is substituted into (8.18)

The new primal linear programming problem is

$$\begin{aligned} \text{maximize} \quad & \pi_t = \mathbf{a}'\mathbf{x}_t + (\mathbf{x}_{t-1}^0)'\mathbf{B}'\mathbf{x}_t - \mathbf{k}_t'\mathbf{x}_t \\ \text{subject to} \quad & \textstyle\sum_{j=1}^m a_{1j}x_{jt} \le [\mathbf{a}'\mathbf{x}_{t-1}^0 + (\mathbf{x}_{t-1}^0)'\mathbf{B}'\mathbf{x}_{t-1}^0 - h_t] \qquad (8.19) \\ & \textstyle\sum_{j=1}^m a_{ij}x_{jt} \le b_{it} \qquad i = 2, \ldots, n \qquad (8.20) \\ & x_{jt} \ge 0 \qquad j = 1, \ldots, m \end{aligned}$$

The value of $\mathbf{x}_{t-1}^0$ is data for period t. This is a recursive programming problem since the allocations and profits in period t are dependent upon the allocations in period $t - 1$.

The dual to this problem is

$$\begin{aligned} \text{minimize} \quad & \psi_t = [\mathbf{a}'\mathbf{x}_{t-1}^0 + (\mathbf{x}_{t-1}^0)'\mathbf{B}'\mathbf{x}_{t-1}^0 - h_t]r_{1t} + \textstyle\sum_{i=2}^n b_{it}r_{it} \\ \text{subject to} \quad & \mathbf{A}'\mathbf{r}_t \ge \mathbf{a} + \mathbf{B}\mathbf{x}_{t-1}^0 - \mathbf{k}_t \qquad (8.21) \\ & \mathbf{r}_t \ge \mathbf{0} \end{aligned}$$

For the case where $n = m = 2$ for the problem presented in the previous paragraph, Day and Tinney [1969a] extend the analysis to consider the questions: Do equilibria exist? Are they stable? Do cycles exist? Are they stable? Can growth occur? The analysis proceeds by studying the possibilities with regard to the constraints of the programming model as shown in Tables 8.1 and 8.2.

Example 8.1

To illustrate these linear models of the firm, consider the model defined by equations (8.12)–(8.14), where

$$\mathbf{c}_t' = [3,\ 4,\ 3.5,\ 2] \qquad \mathbf{A} = \begin{bmatrix} 2 & 5 & 7 & 2.5 \\ 3 & 4 & 6 & 2 \\ 1.5 & 2 & 0.7 & 0.5 \end{bmatrix} \qquad \mathbf{b}_t = \begin{bmatrix} 60 \\ 45 \\ 25 \end{bmatrix}$$

Then the problem for period t is

$$\begin{aligned} \text{maximize} \quad & \pi_t = 3x_{1t} + 4x_{2t} + 3.5x_{3t} + 2x_{4t} \\ \text{subject to} \quad & 2x_{1t} + 5x_{2t} + 7x_{3t} + 2.5x_{4t} \le 60 \\ & 3x_{1t} + 4x_{2t} + 6x_{3t} + 2x_{4t} \le 45 \\ & 1.5x_{1t} + 2x_{2t} + 0.7x_{3t} + 0.5x_{4t} \le 25 \\ & x_{jt} \ge 0 \qquad j = 1, 2, 3, 4 \end{aligned}$$

The optimal solution to this problem is ($\pi_t^0 = 112.50$, $x_{1t}^0 = 0$, $x_{2t}^0 = 0$, $x_{3t}^0 = 0$, $x_{4t}^0 = 22.5$).

Assuming the demand functions

$$\begin{aligned} p_{1t} &= 120 - 0.5x_{1t}^0 \qquad & p_{3t} &= 80 - 0.2x_{3t}^0 \\ p_{2t} &= 50 - 0.1x_{2t}^0 \qquad & p_{4t} &= 70 - 0.3x_{4t}^0 \end{aligned}$$

then $p_{1t}^0 = 120$, $p_{2t}^0 = 50$, $p_{3t}^0 = 80$, and $p_{4t}^0 = 63.25$.

The problem for period $t + 1$ is defined by (8.17), (8.19), and (8.20). Suppose that the first constraint is a budget constraint and that $k_1 = 4$, $k_2 = 2$, $k_3 = 3$, $k_4 = 1$, and $h = 1200$. Then, following the previous model, the problem for period $t + 1$ is:

$$\begin{aligned} \text{maximize} \quad \pi_{t+1} = {} & 120x_{1t+1} + 50x_{2t+1} + 80x_{3t+1} \\ & + 70x_{4t+1} - x_{1t}^0(0.5)x_{1t+1} \\ & - x_{2t}^0(0.1)x_{2t+1} - x_{3t}^0(0.2)x_{3t+1} \\ & - x_{4t}^0(0.3)x_{4t+1} - 4x_{1t+1} - 2x_{2t+1} \\ & - 3x_{3t+1} - x_{4t+1} \end{aligned}$$

or

maximize

$$\pi_{t+1} = 116x_{1t+1} + 48x_{2t+1} + 77x_{3t+1} + 62.25x_{4t+1}$$

$$\begin{aligned} \text{subject to} \quad 2x_{1t+1} + 5x_{2t+1} + 7x_{3t+1} + 2.5x_{4t+1} &\le k \\ 3x_{1t+1} + 4x_{2t+1} + 6x_{3t+1} + 2x_{4t+1} &\le 45 \\ 1.5x_{1t+1} + 2x_{2t+1} + 0.7x_{3t+1} + 0.5x_{4t+1} &\le 25 \\ k = p_{1t}^0 x_{1t}^0 + p_2^0 x_{2t}^0 + p_3^0 x_{3t}^0 & \\ + p_4^0 x_{4t}^0 - 1200 &= 223.125 \\ x_{jt+1} \ge 0 \qquad j = 1, 2, 3, 4 & \end{aligned}$$

The optimal p_t^0's and x_t^0's are obtained from the previous paragraph in which the problem for period t is solved.

For period $t + 1$ the optimal results are:

$$\begin{aligned} x_{1t+1}^0 &= 15 \qquad & x_{4t+1}^0 &= 0 \qquad & p_{3t+1}^0 &= 80 \\ x_{2t+1}^0 &= 0 \qquad & p_{1t+1}^0 &= 112.5 \qquad & p_{4t+1}^0 &= 70 \\ x_{3t+1}^0 &= 0 \qquad & p_{2t+1}^0 &= 50 \qquad & \pi_{t+1}^0 &= 1740.00 \end{aligned}$$

This example illustrates an important result from recursive programming that distinguishes this approach from some other optimization techniques over time. In recursive programming the optimal allocations for period t are data for the optimization model in period $t + 1$, and the parameters are not necessarily the same for each period.

8.3 APPLICATIONS OF RECURSIVE PROGRAMMING

Two new applications of recursive programming will be presented here. In each instance the emphasis is on the main feature of recursive programming: an optimal decision in period t can only be determined after an optimal decision for period $t - 1$ is identified. Optimization problems in which the constraints are a model of an economy or firm often are recursive programming problems. In a macroeconomic model where investment and capital stock are considered, the feasible stock in any period is the previous stock plus new stock (investment) minus the stock that is worn out during the period (depreciation). For the firm, models often include analogous functions for the inventory levels of material inputs and outputs that can be stored.

Example 8.2. A Multiproduct Firm in a Dynamic Setting

Consider the firm that produces two products at levels q_{jt}, $j = 1, 2$, in period t and sells these outputs at prices p_{jt}, $j = 1, 2$. The price of each product is set by the firm according to how much is produced in period t. The firm's revenue from product j in period t is $p_{jt}q^s_{jt}$, where q^s_{jt} is the quantity of output j that is sold in period t. The beginning inventory level of product j in period $t + 1$, q^*_{jt+1}, is the beginning inventory q^*_{jt}, plus the quantity produced q_{jt}, minus the quantity sold q^s_{jt}. Thus

$$TR_t = p_{1t}q^s_{1t} + p_{2t}q^s_{2t} \tag{8.22}$$

$$p_{jt} = s_j + S_{j1}q_1 + S_{j2}q_2 \qquad j = 1, 2 \tag{8.23}$$

$$q^*_{jt+1} = q^*_{jt} + q_{jt} - q^s_{jt} \qquad j = 1, 2 \tag{8.24}$$

Suppose a single material is applied to produce each product, and the other inputs, labor and machinery, are employed in fixed proportions to the materials used. Then the production relationships can be specified by

$$q_j = a_j x^U \qquad j = 1,2 \tag{8.25}$$

$$x_j^L = A_j^L x^U \qquad j = 1,2 \tag{8.26}$$

$$x_j^M = A_j^M x^U \qquad j = 1,2 \tag{8.27}$$

where x^U = the quantity of materials used
x_j^L = the labor applied to product j
x_j^M = the number of machine hours used to produce q_j and a_j, A_j^L, and A_j^M are constants for $j = 1,2$

The inventory relationship on the input side is given by

$$x_{t+1}^* = x_t^* + x_t^p - x^U \tag{8.28}$$

where $x_{t+1}^*(x_t^*)$ is the beginning materials inventory in period $t + 1$ (t), and x_t^p is the quantity of materials purchased in period t.

The total cost TC is assumed to have three components. The first is the production cost TPC, the second is the inventory cost TIC, and the third is the fixed cost TFC.

$$TC_t = TPC_t + TIC_t + TFC_t \tag{8.29}$$

The production costs are the costs of purchasing inputs in period t. If simple demand curves are assumed for the factor inputs,

$$r_{jt}^L = b_j^L + B_j^L x_{jt}^L \qquad j = 1,2 \tag{8.30}$$

$$r_{jt}^M = b_j^M + B_j^M x_{jt}^M \qquad j = 1,2 \tag{8.31}$$

$$r_t^P = b^P + B^P x_t^P \tag{8.32}$$

The average costs of labor and materials applied to product j are r_j^L and r_j^M respectively. Allowing two cases in (8.30) and (8.31) permits the possibility that a laborer is paid a different wage for working on each product. This may occur because of the danger or the precision required to produce one product and not the other. The cost allocated to a machine can also depend on the item to which the machine is applied. The average cost of the material is r^P. It is assumed that the b's are positive constants and that the B's are negative constants.

The total production cost is given by

$$TPC_t = r_{1t}^L x_{1t}^L + r_{2t}^L x_{2t}^L + r_{1t}^M x_{1t}^M + r_{2t}^M x_{2t}^M + r_t^P x_t^P \quad (8.33)$$

An inventory cost is assessed if an ending inventory is different from the beginning inventory ($x_{t+1}^* \neq x_t^*$ and $q_{jt+1}^* \neq q_{jt}^*$, $j = 1, 2$). The cost is a linear function of the inventory change. Let

$$\bar{c}_1 = c_1(q_{1t+1}^* - q_{1t}^*) \quad (8.34)$$

$$\bar{c}_2 = c_2(q_{2t+1}^* - q_{2t}^*) \quad (8.35)$$

$$\bar{c}_3 = c_3(x_{t+1}^* - x_t^*) \quad (8.36)$$

be the per unit respective inventory cost, so that

$$TIC_t = \bar{c}_1(q_{1t+1}^* - q_{1t}^*) + \bar{c}_2(q_{2t+1}^* - q_{2t}^*) + \bar{c}_3(x_{t+I}^* - x_t^*)$$

or

$$TIC_t = c_1(q_{1t+1}^* - q_{1t}^*)^2 + c_2(q_{2t+1}^* - q_{2t}^*)^2 + c_3(x_{t+1}^* - x_t^*)^2 \quad (8.37)$$

after substituting (8.34)–(8.36).

Among the firm's fixed costs are the warehouse needed to store the inventory (if there is no change) and machinery depreciation charges. Let

$$TFC_t = K_t \quad (8.38)$$

The firm's profit, which is to be maximized, is

$$\pi_t = TR_t - TPC_t - TIC_t - TFC_t \quad (8.39)$$

The model delineated by (8.22)–(8.39) can be optimized as a recursive programming problem. Given initial values for $q_{jt}^*, j = 1, 2$, and x_t^*, optimal values for $j = 1, 2$ or the other variables are determined. This enables us to find the optimal values for π, TR, TPC, and TIC.

This application and model of recursive programming is very close to real-world decision making and planning. In planning for the future, the entrepreneur must realize that each allocation depends heavily on the previous ones. In any period the choices and decisions available to a firm are very much restricted by the inventory levels and available resources.

Example 8.3. A Recursive Macrodynamic Problem

An interesting macroeconomic recursive programming problem occurs when an objective function is assumed for a simple macrodynamic model with upper and lower bounds on income, consumption, and investment. Suppose that it is desirable to maximize z_t, the national utility,

$$z_t = w_1 C_t + w_2 I_t + w_3 G_t \tag{8.40}$$

where C_t = aggregate consumption in period t
I_t = aggregate investment in period t
G_t = aggregate government expenditure in period t
w_1, w_2, w_3 = nonnegative constants

It is assumed that the economy is described by the following equations.

$$C_{t-1} \leq C_t \leq C_{t-1}(1 + r_C) \tag{8.41}$$

$$I_{t-1} \leq I_t \leq I_{t-1}(1 + r_I) \tag{8.42}$$

$$Y_{t-1} \leq Y_t \leq Y_{t-1}(1 + r_Y) \tag{8.43}$$

$$C_{t-1} = a_1 + a_2(Y_{t-2} - T_{t-2}) \tag{8.44}$$

$$T_{t-2} = b_1 + b_2 Y_{t-3} \tag{8.45}$$

$$I_{t-1} = g_1 + g_2 Y_{t-2} \tag{8.46}$$

$$Y_t = C_t + I_t + G_t \tag{8.47}$$

The constants a_1, a_2, b_1, b_2, g_1, and g_2 will be provided. The selected values of r_I, r_C, and r_Y are maximum rates of growth for that aggregate. Such a rate is to reflect the maximum increases that would not be inflationary.

$$r_C = \max \frac{1}{C_t}\frac{dC_t}{dt} \qquad r_I = \max \frac{1}{I_t}\frac{dI_t}{dt} \qquad r_Y = \max \frac{1}{Y_t}\frac{dY_t}{dt}$$

It is also expected that the levels of C, I, and Y in period t will be at least as large as the respective values in the previous period.

The programming problem to be solved will be to

$$\text{maximize} \quad z = w_3 Y_t + (w_1 - w_3)C_t + (w_2 - w_3)I_t \tag{8.48}$$

subject to the model given by (8.41)–(8.47). Equation (8.48) is the result of substituting (8.47) into (8.40) to eliminate G_t. The optimal level of G_t can be determined by substituting the optimal levels of Y_t, C_t, and I_t into (8.47).

For this exposition suppose that appropriate estimates and data are those for the U.S. economy from a study by Koot and Walker [1970]; $t = 0$ is the first quarter of 1969. Then $a_1 = 12.30$, $a_2 = 0.69$, $b_1 = -9.93$, $b_2 = 0.12$, $g_1 = 3.92$, and $g_2 = 0.15$.

Suppose $r_C = 0.01625$ (6.5% per year)
$r_I = 0.02$ (8% per year)
$r_Y = 0.01725$ (7% per year)
$Y_0 = 988.4$ (1970, 4th quarter)
$Y_{-1} = 983.5$ (1970), 3rd quarter)
$Y_{-2} = 968.5$ (1970, 2nd quarter)
$Y_{-3} = 956.0$ (1970, 1st quarter)

Optimal values for Y_t, C_t, I_t, and G_t can be determined without specifying the w's, but it is necessary to assume $w_1 > w_2$ or $w_1 < w_2$. If it is assumed that $w_1 > w_2$, there is greater utility in increasing C_t by one unit than by increasing I_t by one unit. If $w_1 < w_2$, the implication is that there is greater national utility from one unit of I than one unit of G.

For the assumed parameters the optimal allocations are given in Table 8.3. Regardless of whether $w_1 > w_2$ or $w_1 < w_2$, the optimal levels of Y_t^0 are the same. The difference appears in the method of achieving this level of gross national product.

This application of recursive programming to macrodynamics can be compared to the first study by Day where he sets upper and lower limits on each activity [see (8.1)–(8.7)]. In this macroeconomics model there are bounds on Y, C, and I as well as linear functions that provide a hypothesized behavioral model.

Table 8.3. Optimal Allocations

		$w_1 > w_2$			$w_1 < w_2$		
t*	Y^0	C^0	I^0	G^0	C^0	I^0	G^0
1	1005.7	616.5	151.4	237.8	606.6	154.4	244.7
2	1023.3	618.7	152.2	252.4	608.8	155.2	259.3
3	1041.2	630.7	154.8	255.7	620.3	157.9	263.0
4	1059.4	641.4	157.4	260.6	631.1	160.5	267.8
5	1078.0	652.4	160.1	265.5	642.0	163.3	272.7
6	1096.8	663.7	162.8	270.3	653.1	166.1	277.6
7	1116.0	675.4	165.6	275.0	664.6	168.9	282.5
8	1135.5	687.0	168.4	280.1	676.0	171.8	287.7
9	1155.4	698.9	171.3	285.2	687.7	174.7	293.0
10	1175.6	711.3	174.2	290.1	699.7	177.7	298.2
11	1196.2	723.4	177.2	295.6	711.8	180.7	303.7
12	1217.1	736.0	180.3	300.8	724.2	183.9	309.0

Note: The data in this table are at annual rates.
*t = 1 for 1971, 1st quarter.

Several applications of recursive programming have appeared in the literature. Among these are a recursive bank asset management model by Walker [1972] and a book of applications edited by Day [1975]. The book includes applications to agricultural development, the coal industry, the petroleum refining industry, the development of an urban area, bank management, and numerous multisector models.

8.4 RECURSIVE AND DYNAMIC PROGRAMMING CONTRASTED

The difference between recursive and dynamic programming is subtle and important. Some of the problems that involve time differences and have been studied using dynamic programming might deserve reexamination in the recursive programming framework.

In a comment and reply, Leontief has considered the difference between recursive and dynamic economic growth models [1958, 1959]. He points out that "economic growth is described as a continuing, unending process, the path of which is determined by a never-ending sequence of choices. Particularly important, from this point of view, is the fact that the explicit time-horizon of each one of these successive choices is much shorter, in principle infinitely shorter, than is the span of time covered by the dynamic process as a whole. Thus, while each step, being determined by a conscious act of choice, satisfies certain maximizing conditions, this sequence as a whole does not. Its path can be compared to the course of a dog running across a field toward his master, while the master walks along a road. The dog's path will usually describe a gentle arc, while the fastest way of joining his master would be to run along a straight, properly aimed intercepting line" [1959, p. 1041].

Leontief's analogy of a dog running to meet his master points out an especially important feature of recursive programming that dynamic programming cannot offer. Leontief's references to a sequence of choices and successive choices is a recursive framework. His sequence as a whole would be a dynamic programming framework. When a decision is made at each of several stages, the course can be reversed or varied as desired at each stage. This is true of models to which recursive programming is applied. The solution to a dynamic programming problem in a single period predetermines all future decisions.

The solution to a dynamic programming problem in period t, $\mathbf{x}_t^{0D}$, requires that all decisions be optimal with regard to the first optimal decision, $\mathbf{x}_t^{0D} = f(\mathbf{x}_1^0)$. The solution to a dynamic programming problem is determined by a single optimizing decision. The optimal solution to a recursive programming problem in period t, $\mathbf{x}_t^{0R}$, is optimal with respect to the optimal solution in period $t - 1$, $\mathbf{x}_t^{0R} = h(\mathbf{x}_{t-1}^0)$. The solution to a recursive programming problem is determined by a sequence of optimal allocations with parameters that may change between time periods as do the right-hand sides of equations (8.19), (8.21), and (8.41)–(8.43).

Day [1967, p. 2] suggests that the purpose of recursive programming is to analyze performance under well-defined conditions, while the purpose in dynamic programming is to discover normative rules for action. Day implies that the difference in purpose is the major distinction between recursive and dynamic programming. He has emphasized the possibility of shifting from one phase to another with different optimizing rules in recursive programming. The solution to a dynamic programming problem provides a single optimizing rule.

For a linear model, the recursive programming problem over N periods would be

$$\text{optimize} \quad w_t = \mathbf{c}_t'\mathbf{x}_t \tag{8.49}$$

$$\text{subject to} \quad \mathbf{A}_t\mathbf{x}_t + \mathbf{K}_{t-1}\mathbf{x}_{t-1}^0 \leq \mathbf{b}_t \tag{8.50}$$

$$\mathbf{x}_t \geq \mathbf{0} \tag{8.51}$$

where $\mathbf{x}_{t-1}^0$ = an optimal solution to the problem for the previous period

$\mathbf{c}_t'$ and $\mathbf{b}_t$ = appropriate sized vectors
$\mathbf{A}_t$ and $\mathbf{K}_{t-1}$ = appropriate sized matrices; N problems are specified by equations (8.49)–(8.51) for $t = 1, \ldots, N$

The dynamic programming model would be

$$\begin{aligned}
\text{optimize} \quad & z = \mathbf{c}_N'\mathbf{x}_N + \mathbf{c}_{N-1}'\mathbf{x}_{N-1} + \cdots + \mathbf{c}_1'\mathbf{x}_1 \\
\text{subject to} \quad & \mathbf{A}_1^1\mathbf{x}_1 \leq \mathbf{b}_1 \\
& \mathbf{A}_2^1\mathbf{x}_2 + \mathbf{A}_1^2\mathbf{x}_1 \leq \mathbf{b}_2 \\
& \mathbf{A}_3^1\mathbf{x}_3 + \mathbf{A}_2^2\mathbf{x}_2 + \mathbf{A}_1^3\mathbf{x}_1 \leq \mathbf{b}_3 \\
& \qquad \vdots \qquad\qquad\qquad\qquad \vdots \\
& \mathbf{A}_N^1\mathbf{x}_N + \mathbf{A}_{N-1}^2\mathbf{x}_{N-1} + \cdots + \mathbf{A}_N^N\mathbf{x}_1 \leq \mathbf{b}_N \\
& \mathbf{x}_t \geq 0 \qquad t = 1, \ldots, N
\end{aligned}$$

where $\mathbf{x}_t$ = the vector of activities in period t, $t = 1, \ldots, N$
$\mathbf{c}_t'$ and $\mathbf{b}_t$ = coefficient vectors for period t, $t = 1, \ldots, N$
$\mathbf{A}_T^\tau$ = the coefficient matrix for the vector $\mathbf{x}$ in period τ

Thus $\mathbf{A}_3^4$ is the third coefficient matrix in the fourth period.

Now the reader should be able to recognize a programming problem in which time is a factor and identify whether there is one or a series of problems to be solved. If there is a series of problems, the methods of this chapter may be applied. If a single optimizing decision is to be identified, the methods described in Chapter 7 should be helpful.

PROBLEMS

8.1 A firm produces two products at levels x_{1t} and x_{2t}. A single resource, of which 20 units are available, is employed. Find the optimal values for x_{jt} and $x_{j\,t+1}$, $j = 1, 2$, and w_t and w_{t+1} to

$$\begin{aligned}
\text{maximize} \quad & w_t = 3x_{1t} + 4x_{2t} \\
\text{subject to} \quad & x_{1t} \leq 1.5\,x_{1\,t-1} \\
& x_{2t} \leq 1.6\,x_{2\,t-1} \\
& x_{1t} \geq 0.75\,x_{1\,t-1} \\
& x_{2t} \geq 0.80\,x_{2\,t-1} \\
& x_{1t} + x_{2t} \leq 20
\end{aligned}$$

Assume that $x_{1\,t-1} = 4$ and $x_{2\,t-1} = 5$.

8.2 Consider the quadratic recursive programming problem to

$$
\begin{aligned}
\text{maximize} \quad & w_t = 20x_{1t} + 30x_{2t} - 8x_{1t}^2 - 9x_{2t}^2 \\
\text{subject to} \quad & x_{1t} \le 1.5\, x_{1\,t-1} \\
& x_{2t} \le 1.6 x_{2\,t-1} \\
& x_{1t} \ge 0.75\, x_{1\,t-1} \\
& x_{2t} \ge 0.80\, x_{2\,t-1} \\
& x_{1t} + x_{2t} \le 20
\end{aligned}
$$

Assume that $x_{1\,t-1} = 2$ and $x_{2\,t-1} = 3$. Find the optimal values of x_{jt} and $x_{j\,t+1}$, $j = 1, 2$, and w_t and w_{t+1}.

8.3 Suppose that the same constraints are imposed as those in Problem 8.2. Assume that the requirement is to

$$
\text{minimize} \quad w_t = 20x_{1t} + 30x_{2t} - 4x_{1t}^2 - 3x_{2t}^2
$$

Assume that $x_{1\,t-1} = 2$ and $x_{2\,t-1} = 3$. Find the optimal values of x_{jt} and $x_{j\,t+1}$, $j = 1, 2$, and w_t and w_{t+1}.

8.4 For the linear programming model

$$
\begin{aligned}
\text{maximize} \quad & w_t = 3x_t - 2x_{t-1} \\
\text{subject to} \quad & x_t - 4x_{t-1} \le 10 \\
& x_t \ge 0 \\
& x_0 = 1
\end{aligned}
$$

find the optimal values for x_t and w_t, $t = 1, 2, 3$.

9 CALCULUS OF VARIATIONS

9.1 INTRODUCTION

The calculus of variations (also called variational calculus) is an optimization technique that can only be applied to models having continuous variables and time as a continuous variable. This is different from the applications of dynamic and recursive programming in the previous chapters.

The attention of economists to variational calculus must be considered recent when we realize that mathematicians first developed and applied this method of analysis in the late seventeenth and eighteenth centuries. The early applications of these mathematicians are similar to the types of problems recent economists have studied; therefore, two of the problems studied in the late 1600s are described here first.

The first variational calculus problem to be suggested was the classical brachistochrone problem. Johann Bernoulli challenged fellow mathematicians with this in 1696. The brachistochrone problem is to define a path that joins two fixed points (A and B), not lying on the same vertical line, such that a particle following the path would move from point A to point B in the shortest possible time. In Figure 9.1 the straight dotted line from A to B describes the shortest distance to be traveled; however, the curved solid line may require less time to traverse because of gravity and the momentum that the particle gains soon after moving away from A. Therefore, a straight line between two points does not necessarily define the path that can be traveled most quickly between the two points.

An analogous economic problem is encountered for the nation

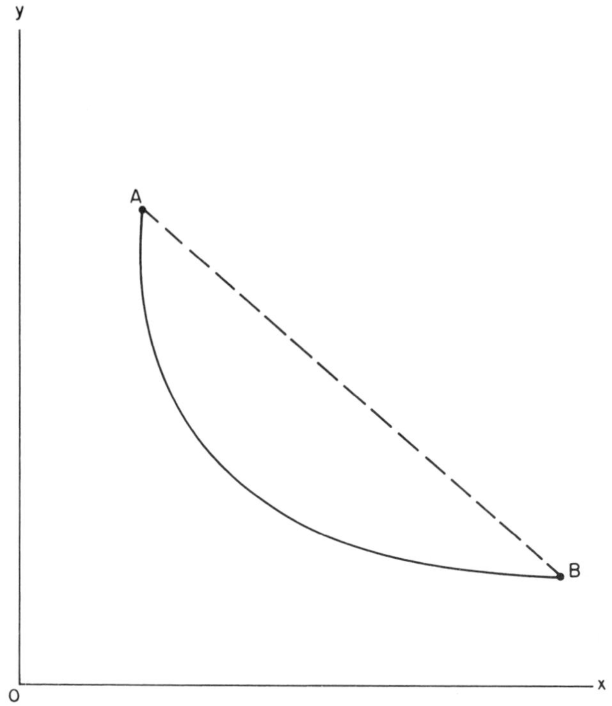

Fig. 9.1. The brachistochrone problem.

whose current total consumption expenditure is $\overline{C}$ if the goal is to raise consumption to a higher level C^* as soon as possible. If consumption is related to national income Y by $C = f(Y)$, $dC/dY > 0$, where $C = \overline{C}$ when $Y = \overline{Y}$ and $C = C^*$ when $Y = Y^*$, the problem could be solved by increasing national income from $\overline{Y}$ to Y^* in as short a time as possible. Later in this chapter a similar problem is solved, but there are additional constraints in the model.

The earliest variational calculus problem in which the objective function is constrained is the problem of geodesics. The goal is to find the shortest arc on a surface $g(x, y, z) = 0$ that joins two points on the surface, (x_1, y_1, z_1) and (x_2, y_2, z_2). All the arcs that connect points on a surface are called geodesics. Although the geodesics problem was also solved by Johann Bernoulli in 1697, a more significant contribution was made later by Euler and Lagrange because their method provided a general procedure for solving constrained variational calculus problems.

Such a constrained problem arises for the nation that is trying to increase consumption from $\overline{C}$ to C^* if, in addition to the consumption function, the level of income is somehow constrained. For example, suppose an investment constraint is added to the model. Then the problem of increasing consumption becomes more complicated, since the level of income is constrained by the investment function. A more general macroeconomics problem and a problem from the theory of the firm are studied in detail after the methodology of variational calculus is described.

9.2 DEFINITIONS AND PROPERTIES OF THE PROBLEM

Functions of Time and Their Derivatives

Before describing the specific techniques of variational calculus, several definitions must be given. These are useful in describing and analyzing the methodology. If x is a variable in a variational calculus problem, x with k dots above it denotes the kth derivative of x with respect to time t: $\dot{x} = dx/dt$, $\ddot{x} = d^2x/dt^2$, $\dddot{x} = d^3x/dt^3$, and so on.

These continuous time derivatives have discrete counterparts. Thus it is possible to approximate continuous time derivatives by discrete time differences with future or past observations at any given time. For the forward case with future observations:

$$\dot{x} = \frac{dx}{dt} \doteq \frac{\Delta x_t}{\Delta t} = x_{t+1} - x_t$$

$$\ddot{x} = \frac{d^2x}{dt^2} \doteq \frac{\Delta^2 x_t}{\Delta t^2} = \frac{\Delta}{\Delta t}\frac{\Delta x_t}{\Delta t} = \frac{\Delta}{\Delta t}(x_{t+1} - x_t)$$

$$= \frac{\Delta x_{t+1}}{\Delta t} - \frac{\Delta x_t}{\Delta t} = x_{t+2} - 2x_{t+1} + x_t$$

$$\dddot{x} = \frac{d^3x}{dt^3} \doteq \frac{\Delta^3 x_t}{\Delta t^3} = \frac{\Delta}{\Delta t}\frac{\Delta^2 x_t}{\Delta t^2} = \frac{\Delta}{\Delta t}(x_{t+2} - 2x_{t+1} + x_t)$$

$$= \frac{\Delta x_{t+2}}{\Delta t} - 2\frac{\Delta x_{t+1}}{\Delta t} + \frac{\Delta x_t}{\Delta t} = x_{t+3} - 3x_{t+2} + 3x_{t+1} - x_t$$

and so on. In general for the forward case:

$$\frac{d^i x}{dt^i} \doteq \frac{\Delta^i x_t}{\Delta t^i} = \sum_{k=0}^{i} \frac{i!\,(-1)^k}{(i-k)!\,k!} x_{t+i-k}$$

The past case is similar:

$$\dot{x} = \frac{dx}{dt} \doteq \frac{\Delta x_t}{\Delta t} = x_t - x_{t-1}$$

$$\ddot{x} = \frac{d^2x}{dt^2} \doteq \frac{\Delta^2 x_t}{\Delta t^2} = \frac{\Delta}{\Delta t}\frac{\Delta x_t}{\Delta t} = \frac{\Delta}{\Delta t}(x_t - x_{t-1})$$

$$= x_t - 2x_{t-1} + x_{t-2}$$

etc. The general formulation for the backward case is

$$\frac{d^ix}{dt^i} \doteq \frac{\Delta^i x_t}{\Delta t^i} = \sum_{k=0}^{i} \frac{i!\,(-1)^{k+1}}{(i-k)!\,k!} x_{t+k-i}$$

The importance of these two representations of d^ix/dt^i should become clear shortly.

Functionals

In variational calculus problems it is desired to find the optimum of a functional. A functional is a relation between a set of functions and a variable. This is different from a relation between a variable and a function of another function, although the two situations are sometimes confused. Suppose, for example, that x, y, and t are variables. Let f and g be single-valued functions such that

$$y = f(x) \tag{9.1}$$

$$x = g(t) \tag{9.2}$$

This formulation implies that for any value of t, there is one corresponding value of x that determines one value for y. In place of (9.1) and (9.2), y is often written as the composite function

$$y = f[g(t)] \tag{9.3}$$

A functional covers considerably more mathematical territory. Again consider the variables x, y, and t and the single-valued function f such that $y = f(x)$. However, now suppose that x is related to t in one of n possible ways so that $x = g_1(t)$; $x = g_2(t)$; ...; or $x = g_n(t)$. Then $x = g(t)$, where $g(t) \in [g_1(t), g_2(t), \ldots, g_n(t)]$. Now the relationship between y and t cannot be clearly identified as is the case in (9.3). Only after a particular relationship between x and t is established can a value for y be determined from a value of t; i.e., one of the g's must be selected.

To specify that y and t are related by a functional, write

$$y = f(x) \qquad x = g(t)$$
$$g(t) \in [g_1(t), g_2(t), \ldots, g_n(t)]$$

or

$$y = h(t) \qquad h(t) \in [h_1(t), h_2(t), \ldots, h_n(t)] \tag{9.4}$$

Comparing (9.4) and (9.3) should clarify the difference between a functional and a composite function. Problems in variational calculus are ones in which the purpose is to select the appropriate member of $g_1(t), \ldots, g_n(t)$ to achieve a prescribed goal. This is the issue in the brachistochrone problem, the problem of geodesics, and the economic growth problem described earlier.

The Optimization Problem

Usually, functionals in variational calculus problems are implicit in the statement or formulation of the problem. Suppose, for example, that a problem requires that W be optimized, where

$$W = \int_{\tau}^{T} F\left(x, \frac{dx}{dt}, \frac{d^2x}{dt^2}, \ldots, \frac{d^kx}{dt^k}, t\right) dt \tag{9.5}$$

and F is a functional. Then x is obviously a function of time (otherwise the derivatives are all zero), and the need is to find the function $x = g(t)$ that optimizes W. A closed time interval on t, $\tau \leq t \leq T$, is being considered. At least one of the boundary points is assumed to be fixed. The reason for this is explained in Section 9.3.

The value of W is a scalar since it is the level of the objective function. If there is only one variable in that function, x is that variable. If there are m variables, x represents a column vector of m elements,

$$\mathbf{x} \equiv [x_1, x_2, \ldots, x_m]'$$

and

$$\frac{\partial^i \mathbf{x}}{\partial \mathbf{t}^i} \equiv \left[\frac{\partial^i x_1}{\partial t^i}, \ldots, \frac{\partial^i x_m}{\partial t^i}\right]' \qquad i = 1, \ldots, k$$

As long as at least one variable appears in F at a nonzero level, the problem described by (9.5) is an optimization problem. However, if all the derivatives with respect to time are zero for all the variables, the variables cannot be functions of time. The problem of selecting the appropriate relation between x and t would disappear, since a set such as the one accompanying (9.4) does not exist. Another way of expressing the same idea is to state that the problem involves only a function $W = \int f(x,t)$ where x and t need not be related. Sometimes t may not appear explicitly in a functional. However, if

$$W = \int_{\tau}^{T} F\left(x, \frac{dx}{dt}, \ldots, \frac{d^k x}{dt^k}\right) dt \tag{9.6}$$

the variable t is implicitly related to x, and F is a functional. This is known, since x must be a function of t for the time derivatives to exist.

If the problem is one in which an objective function is to be optimized, subject to one or more constraints, the objective function need not involve derivatives with respect to time. Then, for a functional to be present, one or more of the constraints must involve one or more time derivatives of one or more variables.

9.3 TECHNIQUES FOR FINDING SOLUTIONS

Euler Equations

The solution to a problem of the form of (9.5) or (9.6) and not accompanied by any constraints is determined by an Euler equation or a similar form called an Euler-Poisson equation. The one to be used depends on which time derivatives of x appear in the problem.

If

$$W = \int_{\tau}^{T} F\left(x, \frac{dx}{dt}, t\right) dt \qquad \text{and} \qquad \frac{d^i x}{dt^i} = 0 \qquad \text{for all } i \geq 2$$

the solution is given by the Euler equation. This is

$$\frac{\partial F}{\partial x} - \frac{d}{dt}\frac{\partial F}{\partial \dot{x}} = 0, \qquad \text{where} \qquad \dot{x} \equiv \partial x / \partial t \tag{9.7}$$

Example 9.1

Suppose, for example, that

$$F\left(x, \frac{dx}{dt}, t\right) = 3x^2 + \left(\frac{dx}{dt}\right)^2$$

Then

$$\frac{\partial F}{\partial x} = 6x \qquad \frac{\partial F}{\partial \dot{x}} = 2\,\frac{dx}{dt} \qquad \frac{d}{dt}\,\frac{\partial F}{\partial \dot{x}} = 2\,\frac{d^2x}{dt^2}$$

Since

$$\frac{\partial F}{\partial x} - \frac{d}{dt}\,\frac{\partial F}{\partial \dot{x}} = 0$$

$$6x - 2\,\frac{d^2x}{dt^2} = 0 \qquad \text{or} \qquad \frac{d^2x}{dt^2} = 3x$$

One method for solving this differential equation is by trial and error. Suppose that $x = A\,e^{t\sqrt{3}}$. Then

$$3x = 3A\,e^{t\sqrt{3}} \qquad \frac{dx}{dt} = \sqrt{3}A\,e^{t\sqrt{3}} \qquad \frac{d^2x}{dt^2} = 3A\,e^{t\sqrt{3}}$$

Since $d^2x/dt^2 = 3x$ when $x = A\,e^{t\sqrt{3}}$, the solution to the differential equation is $x = A\,e^{t\sqrt{3}}$.

A more general problem is considered when

$$W = \int_{\tau}^{T} F\left(x, \frac{dx}{dt}, \frac{d^2x}{dt^2}, \ldots, \frac{d^kx}{dt^k}, t\right) dt \tag{9.8}$$

such that $d^ix/dt^i \neq 0$ for at least one value of $i, 2 \leq i \leq k$. Although it is likely that this condition holds for several values of i between 2 and k, as long as this condition holds for at least one value of i, the solution to (9.8) is given by

$$\frac{\partial F}{\partial x} + \sum_{i=1}^{k} (-1)^i \frac{d^i}{dt^i}\left(\frac{\partial F}{\partial^i x/\partial t^i}\right) = 0 \tag{9.9}$$

which is the Euler-Poisson equation.

This is a result that can be derived from the fundamental lemma of variational calculus (see Elsgolc [1961, p. 27]). This lemma shows that (9.9) is a necessary condition for the optimal solution to the problem described by (9.7) or (9.8). The difficulties in finding sufficient conditions are discussed in Section 9.5. The Euler equation (9.7) is a special case of the Euler-Poisson equation (9.9) when $k = 1$. Thus it is adequate to study the solutions to the Euler-Poisson equation. Nemhauser [1966, pp. 231–35] has followed a derivation by Bellman and several others to show that the Euler equation can be derived from a dynamic programming framework. However, several special assumptions are required.

Solving a Simple Linear Differential Equation

Recall that $\mathbf{x}$ and $\partial^i\mathbf{x}/\partial\mathbf{t}^i$ may represent vectors of m elements.

Therefore, the Euler-Poisson equation (9.9) is actually a set of m relationships—one for each x, $\partial^i x/\partial t^i$, $i = 1, \ldots, k$. Each of these equations is usually a differential equation to be solved; the order of this equation is at most $k + 1$. If the differential equation is linear, it is of the form

$$\sum_{i=1}^{k+1} a_i \frac{d^i x}{dt^i} + a_0 x + b_1 + b_2 e^{rt} = 0 \tag{9.10}$$

where $r, b_1, b_2, a_0, a_1, \ldots, a_{k+1}$ are constants depending on the parameters of the problem. Some of these parameters may be zero. Note that if $\mathbf{x}$ is a vector of m elements, then (9.10) represents m differential equations, and the problem may become extremely complex.

The solution to each differential equation such as (9.10) is of the form

$$x = \sum_{i=1}^{k+1} A_i e^{\lambda_i t} + B_1 + B_2 e^{\rho t} \tag{9.11}$$

where A_i's = constants of integration
B_1, B_2, and ρ = constants determined by the a's, b's, and r
λ's = characteristic roots

It can be shown that

$$B_1 = -b_1/a_0 \tag{9.12}$$

$$B_2 = -b_2\Bigg/\left(\sum_{i=1}^{k+1} a_i r^i + a_0\right) \tag{9.13}$$

$$\rho = r \tag{9.14}$$

First, however, the determination of the characteristic roots $\lambda_1, \ldots, \lambda_{k+1}$ shall be described.

The $k + 1$ λ's are obtained when all the roots of the following $(k + 1)$st degree equation are determined.

$$\sum_{i=1}^{k+1} a_i \lambda^i + a_0 = 0 \tag{9.15}$$

This is the characteristic equation associated with the differential equation (9.10). If (9.10) represents a set of m equations, (9.11) represents a set of m solutions, and (9.15) provides m sets of characteristic roots. The reader should be relieved to learn that computer programs are available to provide all the roots (real or complex, unique or multiplicative) of a polynomial of any degree below the ninty-ninth.

Example 9.2

To better understand this method for solving a linear differential equation, consider, for example, the third-order linear differential equation

$$\frac{d^3x}{dt^3} - 8\frac{d^2x}{dt^2} + 17\frac{dx}{dt} - 10x + 20 = 0 \tag{9.16}$$

Here $k + 1 = 3$, $a_3 = 1$, $a_2 = -8$, $a_1 = 17$, $a_0 = -10$, $b_1 = 20$, and $b_2 = 0$. The general form of the solution to this problem is

$$x = A_1 e^{\lambda_1 t} + A_2 e^{\lambda_2 t} + A_3 e^{\lambda_3 t} + B_1$$

where

$$B_1 = -b_1/a_0 = 2$$
$$a_3\lambda^3 + a_2\lambda^2 + a_1\lambda + a_0 = 0$$

Then

$$\lambda^3 - 8\lambda^2 + 17\lambda - 10 = 0$$

or

$$(\lambda - 5)(\lambda - 2)(\lambda - 1) = 0$$

Thus $x = A_1 e^{5t} + A_2 e^{t} + A_3 e^{2t} + 2$.

Now it must be proved that (9.11)–(9.15) provide a solution to (9.10). This can be shown by substituting (9.12)–(9.14) into (9.11) to obtain

$$x = \sum_{i=1}^{k+1} A_i e^{\lambda_i t} - \frac{b_1}{a_0} - b_2 e^{rt} \Big/ \left(\sum_{i=1}^{k+1} a_i r^i + a_0 \right) \tag{9.17}$$

and applying (9.10). From (9.17)

$$\frac{d^j x}{dt^j} = \sum_{i=1}^{k+1} A_i \lambda_i^j e^{\lambda_i t} - b_2 r^j e^{rt} \Big/ \left(\sum_{i=1}^{k+1} a_i r^i + a_0 \right) \tag{9.18}$$

Substituting (9.17) and (9.18) into the left-hand side of (9.10) should give zero on the left-hand side, forming the identity 0 = 0 if (9.17) is a solution to (9.10). From (9.10) and (9.18)

$$\begin{aligned}
\sum_{i=1}^{k+1} a_i \frac{d^i x}{dt^i} + a_0 x + b_1 + b_2 e^{rt} \\
&= \left(\sum_{j=1}^{k+1} a_j \right) \left(\sum_{i=1}^{k+1} A_i \lambda_i^j e^{\lambda_i t} \right) - \left(\sum_{i=1}^{k+1} a_i r^i b_2 e^{rt} \right) \Big/ \left(\sum_{i=1}^{k+1} a_i r^i + a_0 \right) \\
&\quad + a_0 \sum_{i=1}^{k+1} A_i e^{\lambda_i t} - b_1 - a_0 b_2 e^{rt} \Big/ \left(\sum_{i=1}^{k+1} a_i r^i + a_0 \right) + b_1 + b_2 e^{rt} \\
&= b_2 e^{rt} + \sum_{j=1}^{k+1} (a_j \lambda_i^j + a_0) \sum_{i=1}^{k+1} A_i e^{\lambda_i t} - b_1 + b_1 \\
&\quad - \left(\sum_{i=1}^{k+1} a_i r^i + a_0 \right) b_2 e^{rt} \Big/ \left(\sum_{i=1}^{k+1} a_i r^i + a_0 \right) \\
&= \sum_{j=1}^{k+1} (a_j \lambda_i^j + a_0) \sum_{i=1}^{k+1} A_i e^{\lambda_i t} = 0
\end{aligned}$$

Each step but the last is purely algebraic. The last step holds because the first term is zero from (9.15). Therefore, (9.17) is a solution to (9.10).

The Constants of Integration

The solution to the differential equation (9.10) includes at least one constant of integration because at least one $d^i x/dt^i$, $1 \leq i \leq k + 1$, must be different from zero. This is a condition for F to be a functional and, therefore, for the problem described by (9.5), to be one of variational calculus. To determine the constants of integration $A_1, A_2, \ldots, A_{k+1}$, it is necessary to draw on the assumption (stipulated in Section 9.2) that the horizon over which F is optimized has at least one fixed boundary point. If the problem is to reach a particular level of x at a future point in time, the end point $x(T)$ is predetermined. To increase x as soon as possible to as high a level as possible, it would be logical to assume an initial fixed level, $x(\tau)$. To increase x from one known level to a given level in the shortest possible time suggests fixing both $x(T)$ and $x(\tau)$.

To determine the $k + 1$ A_i's uniquely, it is sufficient to find $k + 1$ linearly independent equations in these $k + 1$ unknowns, $A_1, \ldots, A_{k+1}$. The $k + 1$ equations can be found using the solution to the problem given by (9.17). Then solve for the A_i's.

Suppose that one boundary point $t = \bar{t}$ is fixed and that $x = \bar{x}$ at that boundary. Then

$$\bar{x} = \sum_{i=1}^{k+1} A_i e^{\lambda_i \bar{t}} - \frac{b_1}{a_0} - b_2 e^{r\bar{t}} \Bigg/ \left(\sum_{i=1}^{k+1} a_i r^i + a_0 \right) \tag{9.19}$$

is an equation with only the constants of integration as the unknowns; k remaining equations can be found by evaluating

$$\frac{d^j x}{dt^j} = \sum_{i=1}^{k+1} A_i \lambda_i^j e^{\lambda_i \bar{t}} - b_2 r e^{r\bar{t}} \Bigg/ \left(\sum_{i=1}^{k+1} a_i r^i + a_0 \right) \tag{9.20}$$

for $j = 1, 2, \ldots, k$ at $t = \bar{t}$. Only a numerical representation for each $d^j x/dt^j$ is needed to make the k equations of (9.20) k equations in the $k + 1$ unknowns. Equations (9.19) and (9.20) together give $k + 1$ equations in $k + 1$ unknowns.

If the boundary $\bar{t}$ is an initial point, observations on x are assumed to be available for periods prior to $t = \bar{t}$. In this case the backward representation of $d^j x/dt^j$, as described in Section 9.2, can be used to obtain a value of $d^j x/dt^j$. If $t = \bar{t}$ is an end point or upper boundary, it is assumed that k points beyond the boundary can be stipulated. Then the forward representation of $d^j x/dt^j$, $j = 1, \ldots, k$, can be used, with t replaced by $\bar{t}$.

For the backward case

$$\frac{d^j x}{dt^j} \doteq \sum_{k=0}^{j} \frac{j!(-1)^{k+1}}{(j - k)!k!} x_{\bar{t}+k-j}$$

For the forward case

$$\frac{d^j x}{dt^j} \doteq \sum_{k=0}^{j} \frac{j!(-1)^k}{(j-k)!k!} x_{\bar{t}+j-k}$$

One example of this process is given here. Suppose that $k = 2$, and there is only one variable to be considered ($m = 1$). Then

$$x = A_1 e^{\lambda_1 t} + A_2 e^{\lambda_2 t} + A_3 e^{\lambda_3 t} - \frac{b_1}{a_0} - \frac{b_2 e^{rt}}{a_0 + a_1 r + a_2 r^2 + a_3 r^3}$$

$$\dot{x} = \lambda_1 A_1 e^{\lambda_1 t} + \lambda_2 A_2 e^{\lambda_2 t} + \lambda_3 A_3 e^{\lambda_3 t} - \frac{r b_2 e^{rt}}{a_0 + a_1 r + a_2 r^2 + a_3 r^3}$$

$$\ddot{x} = \lambda_1^2 A_1 e^{\lambda_1 t} + \lambda_2^2 A_2 e^{\lambda_2 t} + \lambda_3^2 A_3 e^{\lambda_3 t} - \frac{r^2 b_2 e^{rt}}{a_0 + a_1 r + a_2 r^2 + a_3 r^3}$$

Identifying x at $\bar{t}$ by $x(\bar{t})$, $\dot{x}$ at $\bar{t}$ by $\dot{x}(\bar{t})$, and $\ddot{x}$ at $\bar{t}$, by $\ddot{x}(\bar{t})$, as three predetermined values at the boundary point $\bar{t}$, the previous three equations represent three equations in three unknowns A_1, A_2, and A_3. The solutions are

$$\begin{bmatrix} A_1 \\ A_2 \\ A_3 \end{bmatrix} = \begin{bmatrix} e^{\lambda_1 \bar{t}} & e^{\lambda_2 \bar{t}} & e^{\lambda_3 \bar{t}} \\ \lambda_1 e^{\lambda_1 \bar{t}} & \lambda_2 e^{\lambda_2 \bar{t}} & \lambda_3 e^{\lambda_3 \bar{t}} \\ \lambda_1^2 e^{\lambda_1 \bar{t}} & \lambda_2^2 e^{\lambda_2 \bar{t}} & \lambda_3^2 e^{\lambda_3 \bar{t}} \end{bmatrix}^{-1} \begin{bmatrix} x(\bar{t}) + \dfrac{b_1}{a_0} + \dfrac{b_2 e^{rt}}{a_0 + a_1 r + a_2 r^2 + a_3 r^3} \\ \dot{x}(\bar{t}) + \dfrac{b_2 r e^{rt}}{a_0 + a_1 r + a_2 r^2 + a_3 r^3} \\ \ddot{x}(\bar{t}) + \dfrac{b_2 r^2 e^{rt}}{a_0 + a_1 r + a_2 r^2 + a_3 r^3} \end{bmatrix}$$

Example 9.3

Suppose for the problem given by (9.16) the following data are available: for 1970 $x(t) = 45$, for 1969 $x(t) = 33$; and for 1968 $x(t) = 25$. Then

$$x = x_t = 45$$

$$\dot{x} = x_t - x_{t-1} = 45 - 33 = 12$$

$$\ddot{x} = x_t - 2x_{t-1} + x_{t-2} = 45 - 2(33) + 25 = 4$$

and

$$x = A_1 e^{5t} + A_2 e^{t} + A_3 e^{2t} + 2 = 45$$

$$\dot{x} = (5)A_1 e^{5t} + (1)A_2 e^{t} + (2)A_3 e^{2t} = 12$$

$$\ddot{x} = (5)^2 A_1 e^{5t} + (1)^2 A_2 e^{t} + (2)^2 A_3 e^{2t} = 4$$

If the goal is to plan for years beyond 1970, let $t = 0$ for 1970. Since $e^{\lambda 0} = 1$,

$$\begin{aligned} A_1 + A_2 + A_3 &= 43 \\ 5A_1 + A_2 + 2A_3 &= 12 \\ 25A_1 + A_2 + 4A_3 &= 4 \end{aligned}$$

or

$$\begin{bmatrix} 1 & 1 & 1 \\ 5 & 1 & 2 \\ 25 & 1 & 4 \end{bmatrix} \begin{bmatrix} A_1 \\ A_2 \\ A_3 \end{bmatrix} = \begin{bmatrix} 43 \\ 12 \\ 4 \end{bmatrix}$$

Then

$$\begin{bmatrix} A_1 \\ A_2 \\ A_3 \end{bmatrix} = \begin{bmatrix} 1 & 1 & 1 \\ 5 & 1 & 2 \\ 25 & 1 & 4 \end{bmatrix}^{-1} \begin{bmatrix} 43 \\ 12 \\ 4 \end{bmatrix} = \begin{bmatrix} 27/6 \\ 777/6 \\ -294/6 \end{bmatrix}$$

Sometimes the solutions for the characteristic roots may be multiple or complex roots. In neither situation is the problem irresolvable. If $\lambda = \lambda^*$ appears as the first $p\,(p < k)$ multiple roots to the characteristic equation,

$$x = \sum_{i=1}^{p} A_i t^i e^{\lambda^* t} + \sum_{i=p+1}^{k} A_i e^{\lambda_i t} - \frac{b_1}{a_0} - b_2 e^{rt} \Bigg/ \left(\sum_{i=1}^{k+1} a_i r^i + a_0 \right)$$

If $\lambda_h = a - bi$ is a complex root ($i = \sqrt{-1}$, $i^2 = -1$) that appears in solving the characteristic equation, another complex root appears, and this is the complex conjugate of λ_h. (The complex conjugate of $a - bi$ is $a + bi$).

When λ_j and λ_h are complex conjugate characteristic roots, the constants of integration that precede the terms having $e^{\lambda_j t}$ and $e^{\lambda_h t}$ are also complex conjugates of one another. (This is proved in Appendix D.) Thus if $\lambda_h = a - bi$ and $\lambda_j = a + bi$, then $A_h = c + di$ and $A_j = c - di$. It can be shown that y is a real (not complex) number where $y = A_j e^{\lambda_j t} + A_h e^{\lambda_h t}$ since

$$\begin{aligned} e^{\lambda_j t} &= e^{at} e^{bit} = e^{at}(\cos bt + i \sin bt) \\ e^{\lambda_h t} &= e^{at} e^{-bit} = e^{at}(\cos bt - i \sin bt) \end{aligned}$$

Thus

$$\begin{aligned} y &= (c - di)e^{at}(\cos bt + i\sin bt) + (c + di)e^{at}(\cos bt - i\sin bt) \\ &= e^{at}(2c\cos bt - 2di^2\sin bt + ic\sin bt - ic\sin bt - di\cos bt \\ &\qquad + di\cos bt) \\ &= e^{at}(2c\cos bt + 2d\sin bt) \end{aligned}$$

The value of y must be a real number since e^{at}, c, d, $\sin bt$, and $\cos bt$ are all real values. Thus it can be assured that x, as defined by (9.17), is always a real number.

Constrained Models

The problem suggested by (9.5) is a very general one. Only the simplest type has been considered. Usually the functional to be optimized is restricted or constrained in some manner. The restrictions may range from a simple identity to a model defined by a system of equations.

To solve problems that require the optimization of a constrained functional, the Lagrangian multiplier technique is applied en route to determining the Euler or Euler-Poisson equation. Suppose the problem is to optimize W (9.8), subject to h constraints defined by $V_1, V_2, \ldots, V_h$. It is required that $V_j = 0$, as is the usual case when applying Lagrangian multipliers. This form can always be obtained by transposing all the terms to one side of each equation.

Consider

$$F^* = F + \mathbf{UV} \tag{9.21}$$

where F is the objective function defined by (9.6), $\mathbf{V}$ is an $h \times 1$ vector of constraints whose typical element is V_j, and $\mathbf{U}$ is a $1 \times h$ vector whose typical element is u_j. Therefore, F^*, F, and $\mathbf{UV}$ have scalar values involving constants and variables. Each u_j is a Lagrangian multiplier in the usual context. Now the problem is to determine a system of $m + h$ Euler-Lagrange equations. The Euler-Lagrange equation is determined on F^* just as the Euler equations are determined on F. The first m Euler-Lagrange equations are determined using (9.9), provided that F is replaced by F^* in (9.9). This very general result was discovered by Euler and Lagrange while they were solving the problem of geodesics. The other h equations are determined by finding

$$\frac{\partial F^*}{\partial \mathbf{U}} = \mathbf{V} \qquad \text{or} \qquad \frac{\partial F^*}{\partial u_j} = V_j \qquad j = 1, \ldots, h$$

and setting the resulting equations equal to zero. There are h equations of the form $V_j = 0, j = 1, \ldots, h$, and these are the original constraints of the problem.

Example 9.4

Consider for example, the problem

$$\begin{aligned} &\text{optimize} \quad W = \int_\tau^T (3x_1^2 - 2x_2)dt \\ &\text{subject to} \quad \dot{x}_2 = 3x_1 - 2\dot{x}_1 - 0.5x_2 \end{aligned}$$

Then

$$F^* = 3x_1^2 - 2x_2 + u(\dot{x}_2 - 3x_1 + 2\dot{x}_1 + 0.5x_2)$$

$$\frac{\partial F^*}{\partial x_1} = 6x_1 - 3u$$

$$\frac{d}{dt}\frac{\partial F^*}{\partial \dot{x}_1} = \frac{d}{dt}(2u) = 2\dot{u}$$

and $6x_1 - 3u = 2\dot{u}$. Also,

$$\frac{\partial F^*}{\partial x_2} = -2 + 0.5u \qquad \frac{d}{dt}\frac{\partial F^*}{\partial \dot{x}_2} = \frac{d}{dt}u = \dot{u}$$

and $\dot{u} = -2 + 0.5u$. Therefore, $u^0 = A_1e^{0.5t} + 4$.

Since

$$6x_1 = 2\dot{u} + 3u \text{ and } \dot{u}^0 = 0.5A_1e^{0.5t}$$

$$6x_1^0 = 2(0.5A_1e^{0.5t}) + 3(A_1e^{0.5t} + 4)$$

$$x_1^0 = (2/3)A_1e^{0.5t} + 2$$

Also,

$$\frac{\partial F^*}{\partial u} = \dot{x}_2 - 3x_1 + 2\dot{x}_1 + 0.5x_2$$

$$\frac{d}{dt}\frac{\partial F^*}{\partial \dot{u}} = 0$$

so that

$$\dot{x}_2 + 0.5x_2 = 3x_1 - 2\dot{x}_1$$

or

$$\dot{x}_2 + 0.5x_2 = (4/3)A_1e^{0.5t} + 6$$

Applying the results of (9.10)–(9.14) where $a_1 = 1$, $a_0 = 0.5$, $b_1 = -6$, $b_2 = (-4/3)A_1$, and $r = 0.5$, leaves

$$x_2^0 = A_2e^{-0.5t} + 12 + (4/3)A_1e^{0.5t}$$

Now time paths that satisfy the first-order conditions have been obtained. They are labeled x_1^0, x_2^0, and u^0.

If $x_1 = 10$ and $x_2 = 35$ at $t = 0$, then

$$x_1^0 = (2/3)A_1e^{0.5(0)} + 2$$
$$x_2^0 = A_2e^{-0.5(0)} + 12 + (4/3)A_1e^{0.5(0)}$$

Substituting $x_1 = 10$ and $x_2 = 35$ leaves

$$10 = (2/3)A_1 + 2$$
$$35 = A_2 + 12 + (4/3)A_1$$

Then $A_1 = 12$ and $A_2 = 7$, and the optimal time paths are

$$x_1^0 = 8e^{0.5t} + 2 \qquad x_2^0 = 16e^{0.5t} + 7e^{-0.5t} + 12$$

Henceforth the Euler equation, the Euler-Poisson equation, and the Lagrangian form of each, are referred to as Euler equations. To determine the optimal time path for each variable, the $m + h$ Euler equations must be solved. The solutions to these differential equations and the time derivatives of the solutions can be substituted into the objective function $\int_{\tau}^{t} F\,dt$ to find the optimum W. The difficulty of solving a variational calculus problem varies greatly depending on the nature of the problem, but in theory a method for selecting optimal time paths to optimize the constrained or unconstrained problems of (9.5), (9.6), or (9.21) has now been described. All that has been delineated so far is subject to some very important limitations. These are described in Section 9.5.

In Section 9.4, two of the many possible applications of variational calculus to economic models are presented. Through these illustrations the techniques should become clearer. The applications are presented to show the reader cases to which variational calculus is applicable.

9.4 APPLICATIONS

The following two cases illustrate problems from the theory of the firm and macroeconomics. Other applications are described in some of the references at the end of the book.

Example 9.5. A Macrodynamic Model

In the first example a social welfare function is to be optimized. The restrictions on the objective function are the relationships from a macroeconomic model. Optimal time paths are to be derived for gross national product (GNP) consumption, investment, and government expenditure. The model is a version of the Phillips multiplier-accelerator model. Using the notation from Allen [1960], the model is

$$Z = C + I + G \tag{9.22}$$

$$Y = [b/(D + b)]Z \tag{9.23}$$

$$C = (1 - s)Y \tag{9.24}$$

$$I = v\dot{Y} + A \tag{9.25}$$

where C = consumption
I = total investment
G = government expenditure
A = autonomous investment
Z = aggregate demand
Y = aggregate supply (GNP)
s = the savings ratio (S/Y)
b = the speed of response of supply to demand
v = the accelerator
D = the linear differential operator (d/dt)
$DY = \dot{Y}$

By substituting (9.24) and (9.25) into (9.22) and the resulting equation into (9.23), the model can be reduced to $(1 - bv)\dot{Y} + bsY - bG - bA = 0$.

The goal will be to minimize the sum of the squared deviations of two variables from their target levels during a time interval from τ to T. Suppose that the target or desired level of GNP is Y^* and the desired level of government expenditure is G^*. Furthermore, suppose that weights are attached to the two factors to suggest that any deviation of Y from Y^* is k times as costly or undesirable as the same deviation of G from G^*. The problem is then to

$$\text{minimize} \quad W = \int_\tau^T [a_1(Y - Y^*)^2 + a_2(G - G^*)^2] dt$$

subject to the multiplier-accelerator model [(9.22)–(9.25)] where $a_1 = ka_2$ and the values of v, b, and s are known constants. The welfare function is the well-known quadratic preference function that has been studied by Simon [1956], Theil [1964], and others in their work on quadratic programming.

The variational calculus problem is to determine optimal time paths for Y, C, I, and G to

$$\begin{aligned} &\text{minimize} \quad \int_\tau^T [a_1(Y - Y^*)^2 + a_2(G - G^*)^2] dt \\ &\text{subject to} \quad (1 - bv)\dot{Y} + bsY - bG - bA = 0 \end{aligned}$$

or to

$$\begin{aligned} \text{minimize} \quad \int_\tau^T \{ & a_1(Y - Y^*)^2 + a_2(G - G)^2 \\ & + u[(1 - bv)\dot{Y} + bsY - bG - bA]\} dt \end{aligned}$$

where u is a Lagrangian multiplier. In the framework of Section 9.3, the corresponding formulation is

$$\begin{aligned} F^* &= F + UV \\ F &= a_1(Y - Y^*)^2 + a_2(G - G^*)^2 \\ U &= u \\ V &= (1 - bv)\dot{Y} + bsY - bG - bA = 0 \end{aligned}$$

The Euler equation for Y is determined by computing $\partial F^*/\partial Y$ and $d(\partial F^*/\partial \dot{Y})dt$ separately and equating the results.

$$\begin{aligned} \frac{\partial F^*}{\partial Y} &= 2a_1(Y - Y^*) + bsu \\ \frac{d}{dt}\left(\frac{\partial F^*}{\partial \dot{Y}}\right) &= (1 - bv)\dot{u} \end{aligned}$$

Therefore

$$\dot{u} = [2a_1(Y - Y^*) + bsu]/(1 - bv) \qquad (9.26)$$

The Euler equations for G and u are similarly determined.

$$\frac{\partial F^*}{\partial G} = 2a_2(G - G^*) - bu \qquad \frac{d}{dt}\frac{\partial F^*}{\partial \dot{G}} = 0$$

$$2a_2(G - G^*) - bu = 0 \tag{9.27}$$

$$\frac{\partial F^*}{\partial u} = (1 - bv)\dot{Y} + bsY - bG - bA \qquad \frac{d}{dt}\left(\frac{\partial F^*}{\partial \dot{u}}\right) = 0$$

$$(1 - bv)\dot{Y} + bsY - bG - bA = 0 \tag{9.28}$$

Equations (9.26)–(9.28) are a system of three equations in three unknowns (Y, G, and u) and their time derivatives.

First solve (9.27) explicitly for G,

$$G = (b/2a_2)u + G^* \tag{9.29}$$

and then substitute for G in (9.28) to obtain

$$(1 - bv)\dot{Y} + bsY - (b^2u/2a_2) - bG^* - bA = 0 \tag{9.30}$$

Solving (9.26) for Y gives

$$Y = Y^* + (1 - bv)(\dot{u}/2a_1) - (bs/2a_1)u \tag{9.31}$$

Differentiating (9.31) with respect to t provides an equivalent for $\dot{Y}$:

$$\dot{Y} = [(1 - bv)\ddot{u}/2a_1] - (bs\dot{u}/2a_1) \tag{9.32}$$

Substituting (9.31) and (9.32) into (9.30) for Y and $\dot{Y}$ respectively gives

$$(1 - bv)^2\ddot{u} - b^2[s^2 + (a_1/a_2)]u = -2a_1b(sY^* - G^* - A)$$

This is a differential equation of the type described in (9.10). The solution is of the form suggested in (9.11):

$$u = A_1e^{\lambda_1 t} + A_2e^{\lambda_2 t} + B_1 \tag{9.33}$$

where

$$B_1 = \frac{2a_1(sY^* - G^* - A)}{b[s^2 + (a_1/a_2)]}$$

The lambdas are the characteristic roots determined by solving

$$(1 - bv)^2\lambda^2 - b^2[s^2 + (a_1/a_2)] = 0$$

for its two solutions, λ_1 and λ_2:

$$\lambda_1 = [b/(1 - bv)]\sqrt{s^2 + (a_1/a_2)}$$
$$\lambda_2 = [-b/(1 - bv)]\sqrt{s^2 + (a_1/a_2)}$$

The optimal paths for Y and G can now be determined. From (9.33) $\dot{\mu}$ is found to be

$$\dot{u} = A_1\lambda_1 e^{\lambda_1 t} + A_2\lambda_2 e^{\lambda_2 t} \qquad (9.34)$$

The optimal path for Y is determined by substituting (9.33) and (9.34) into (9.31). The optimal path for G is found by putting (9.33) into (9.29).

$$Y^0 = \left[\frac{\lambda_1(1 - bv) - bs}{2a_1}\right]A_1 e^{\lambda_1 t} + \left[\frac{\lambda_2(1 - bv) - bs}{2a_1}\right]A_2 e^{\lambda_2 t}$$
$$+ s\left[\frac{A + G^* - sY^*}{s^2 + (a_1/a_2)}\right] + Y^* \qquad (9.35)$$

$$G^0 = \frac{bA_1 e^{\lambda_1 t}}{2a_2} + \frac{bA_2 e^{\lambda_2 t}}{2a_2} + \frac{a_1}{a_2}\left[\frac{sY^* - G^* - A}{s^2 + (a_1/a_2)}\right] + G^*$$

The optimal paths for consumption and investment are found by substituting (9.35) into (9.24) and (9.25) respectively.

$$C^0 = (1 - s)Y^0 \qquad I^0 = v\dot{Y}^0 + A$$

where

$$\dot{Y}^0 = \lambda_1\left[\frac{\lambda_1(1 - bv) - bs}{2a_1}\right]A_1 e^{\lambda_1 t} + \lambda_2\left[\frac{\lambda_2(1 - bv) - bs}{2a_1}\right]A_2 e^{\lambda_2 t}$$

Assume that any deviation of GNP from Y^* in any period is twice as serious as the same deviation of government expenditure from G^*. This means that $a_1/a_2 = 2$. Furthermore, let $a_1 = 2$ and $a_2 = 1$. The speed of response of aggregate supply to aggregate demand is b. Note that another way of writing

(9.23) would be $\dot{Y} = b(Z - Y)$. Then $Z - Y$ is the excess demand when $Z > Y$, and $Z - Y$ is the excess supply when $Z < Y$. If $Z = Y$, $\dot{Y} = 0$ regardless of the value of b. For illustrative purposes it is assumed that there is a three-month lag in the response. Therefore, $b = 0.25$ years.

The values of s, v, and A have been estimated using ordinary least squares regression. Annual observations on Y (GNP), C (total consumption), I (aggregate investment) and $\dot{Y}$ (the change in GNP) have been employed for 1900 to 1970. (See Table 9.1.) Applying these estimates, the assumption about the speed of response, and the values for a_1 and a_2, values can be determined for λ_1 and λ_2.

Table 9.1. Estimated Values of s, v, and A

Equation	Parameter	Estimate	t value	R^2
$C = (1 - s)Y$	$1 - s$	0.63	201.40*	0.99
$I = v\dot{Y} + A$	A	10.02	3.22*	0.69
	v	1.63	12.36*	

*Significant at the 0.01 level.

In summary, the parameters that have been assumed or estimated are

$$\begin{array}{rlrlrl} b = & 0.25 & s = & 0.37 & a_1 = & 1 \\ A = & 10.02 & bs = & 0.09 & \lambda_1 = & 0.30 \\ v = & 1.63 & s^2 = & 0.02 & \lambda_2 = & -0.30 \\ 1 - s = & 0.63 & a_2 = & 2 & & \end{array}$$

The specific optimal paths are

$$Y^0 = 0.04A_1e^{0.30t} - 0.13A_2e^{-0.30t} + 0.71G^* + 0.74\,Y^* + 7.13 \tag{9.36}$$

$$C^0 = 0.63\,Y^0$$

$$I^0 = 0.02A_1e^{0.30t} - 0.06A_2e^{-0.30t} + 1.16\,\frac{dG^*}{dt} + 1.21\,\frac{dY^*}{dt} + 10.02$$

$$G^0 = 0.06A_1e^{0.30t} + 0.06A_2e^{-0.30t} + 0.36\,Y^* + 0.29G^* - 9.63 \tag{9.37}$$

Suppose that the national goals provide target values for both Y and G. These are the values for Y^* and G^* respectively. Assume that the policymaker selects a target rate of growth of 7% for the economy of the United States from the initial time period $t = \tau$, henceforth. Then $Y^* = Y_\tau e^{0.07t}$ where Y_τ is assumed to be known. According to Oshima's study [1957], the postwar average tax percent for the United States is approximately 26%. For the purpose of maintaining a nearly balanced budget, it is assumed that the policymaker sets the target for government expenditure to be $G^* = 0.26Y^*$.

Finally, the two constants of integration, A_1 and A_2, must be determined. Recall that to find $k + 1$ constants of integration, $k + 1$ conditions or boundary points must be selected. Suppose that the policymaker has been asked to specify the optimal policies in January 1972 for future years. To determine the boundary points, there are several approaches he might adopt. One would be to take 1970 as his initial period and select observations on two of the variables (Y, C, I, and G) for that year. Alternatively, he could use an initial observation for one variable and a fixed end point for that or another variable.

For this example the planning horizon is assumed to begin in 1971, i.e., $t = 1$. Initial values are presumed for Y and G; $Y(1970) = Y(t = \tau) = 974.0$ and $G(1970) = G(t = \tau) = 219.4$ in billions of current dollars are used as initial values. Substituting $Y^* = 974.0e^{0.07t}$, $G^* = 0.26(974.0e^{0.07t})$, $Y = 974.0$, $G = 219.4$, and $t = 0$ into (9.36) and (9.37), we find two equations in which the unknowns are the constants of integration. The solutions are $A_1 = -1154.96$ and $A_2 = -2095.87$.

Finally, the empirically specified optimal paths are

$$
\begin{aligned}
Y^0 &= -46.20e^{0.30t} + 272.46e^{-0.30t} + 900.56e^{0.07t} + 7.13 \\
C^0 &= -29.10e^{0.30t} + 171.65e^{-0.30t} + 576.35e^{0.07t} + 2.20 \\
I^0 &= -23.10e^{0.30t} + 125.75e^{-0.30t} + 89.06e^{0.07t} + 10.02 \\
G^0 &= -69.30e^{0.30t} - 125.75e^{-0.30t} + 424.08e^{0.07t} - 9.63
\end{aligned}
$$

For future planning, the policymaker sets $t = 1$ for 1971, $t = 2$ for 1972, etc., so that $t = h$ for the year $t = h + 1970$.

Example 9.6. Case for a Two-Product Firm

Another application of variational calculus is to the theory of the firm. Suppose that a hypothetical firm is producing two products a and b using two inputs, capital and labor. Capital is separated into new capital acquired during the year in question and capital acquired prior to that year. The distinction is necessary assuming that newly acquired capital is used less efficiently. Shortly, the new equipment should be highly productive.

The firm's short-run production functions are assumed to be linear. The demand function has the price of each output being determined by the quantity sold for that item. Optimal paths for both prices and quantities are to be derived. It is assumed that the quantity produced is the same as the quantity sold. No inventories are held.

The firm's total revenue is the sum of the price of a, p_a, times the quantity of a, q_a, plus the price of b, p_b, times the quantity of b, q_b. On the production and revenue side the variables and relationships are

q_a = output of product a
q_b = output of product b
p_a = price of output a
p_b = price of output b
L = labor hours used
K = capital stock on hand prior to the period
I = additional capital acquired during the period
TR = firm's total revenue

Revenue function: $TR = p_a q_a + p_b q_b$
Production functions: $q_a = a_1K + a_2L + a_3I + a_4$
$q_b = b_1K + b_2L + b_3I + b_4$
Demand functions: $p_a = A_1 + A_2 q_a$
$p_b = B_1 + B_2 q_b$

On the cost side, there are to be assessments for the wage bill, the depreciation of capital, and the installation of new capital. Therefore, the total costs can be written as $TC = \gamma_1 L + \gamma_2 K + \gamma_3 I$, where γ_1, γ_2, and γ_3 are the per unit costs of the respective inputs.

Three functions remain to be specified; two are simple definitions. The first is that investment in new capital in any period is the addition to capital stock. Therefore,

$$I_t = K_t - K_{t-1} = \dot{K} \tag{9.38}$$

The second is that profits, denoted by π, are the difference between total revenues and total costs: $\pi = TR - TC$.

Finally, there is an allowance for a difference between the firm's demand for labor L and the community's supply of labor N. Besides this, the community's supply of the type of labor the firm utilizes may be increasing, decreasing, or constant over time. The nature of the firm and the size and location of the community determine a community's supply of this resource and its rate of change. Let N be the labor available at any time t. Let r be the rate of change of the size of the labor force in the community, where labor force connotes the potential employees the firm might hire. If the firm requires highly skilled labor, N may represent a small part of the actual inhabitants. The firm's demand and the community's supply of labor (of the type needed by the firm) may be described as follows:

L = firm's demand for labor
N = community's supply of labor
$\bar{N}$ = community's initial supply of labor
r = rate of change of the supply of labor
ϕ = average percentage of the community's supply of labor that the firm hires

Supply of labor: $N = \bar{N} e^{rt}$
Demand for labor: $L = \phi N$
Equilibrium condition in the labor market: $L = \phi \bar{N} e^{rt}$ (9.39)

In summary, the model of the firm is

$$\pi = TR - TC$$
$$TR = p_a q_a + p_b q_b$$
$$q_a = a_1 K + a_2 L + a_3 I + a_4 \quad (9.40)$$

$$q_b = b_1 K + b_2 L + b_3 I + b_4 \quad (9.41)$$
$$p_a = A_1 + A_2 q_a$$
$$p_b = B_1 + B_2 q_b$$

$$TC = \gamma_1 L + \gamma_2 K + \gamma_3 I \quad (9.42)$$

$$I = \dot{K} \quad (9.43)$$

$$L = \phi \bar{N} e^{rt} \quad (9.44)$$

This model can be reduced and simplified by eliminating I, L, p_a, and p_b. This can be done by substituting (9.43) and (9.44) into (9.40)–(9.42). Thus

$$\begin{aligned} q_a &= a_1 K + a_2 \phi \overline{N} e^{rt} + a_3 \dot{K} + a_4 \\ q_b &= b_1 K + b_2 \phi \overline{N} e^{rt} + b_3 \dot{K} + b_4 \\ TC &= \gamma_1 \phi \overline{N} e^{rt} + \gamma_2 K + \gamma_3 \dot{K} \\ TR &= A_1 q_a + A_2 q_a^2 + B_1 q_b + B_2 q_b^2 \\ \pi &= TR - TC \end{aligned}$$

so that

$$\begin{aligned} \pi = {} & A_1(a_1 K + a_2 \phi \overline{N} e^{rt} + a_3 \dot{K} + a_4) \\ & + B_1(b_1 K + b_2 \phi \overline{N} e^{rt} + b_3 \dot{K} + b_4) \\ & + A_2(a_1 K + a_2 \phi \overline{N} e^{rt} + a_3 \dot{K} + a_4)^2 \\ & + B_2(b_1 K + b_2 \phi \overline{N} e^{rt} + b_3 \dot{K} + b_4)^2 \\ & - (\gamma_1 \phi \overline{N} e^{rt} + \gamma_2 K + \gamma_3 \dot{K}) \end{aligned} \qquad (9.45)$$

The goals of firms have been the subject of much concern. Shubik [1961], among others, has argued that single-purpose firms no longer exist. In a given period, profits may not indicate or measure the accomplishments of a firm. However, over several periods, the firm's activities should be reflected in the profits of all the periods. For this reason the goal is assumed to be to maximize profits π over the time horizon τ to T. This problem can be formulated mathematically to

$$\text{maximize} \quad W = \int_\tau^T \pi \, dt$$

The variational calculus problem is to maximize this function subject to the reduced model of the firm. Note that in (9.45), π is a function of only K, $\dot{K}$, and t. The Euler equation and optimal time path for K is

$$\begin{aligned} \frac{\partial \pi}{\partial K} = {} & A_1 a_1 + B_1 b_1 + 2a_1 A_2(a_1 K + a_2 \phi \overline{N} e^{rt} + a_3 \dot{K} + a_4) \\ & + 2b_1 B_2(b_1 K + b_2 \phi \overline{N} e^{rt} + b_3 \dot{K} + b_4) - \gamma_2 \\ \frac{d}{dt}\frac{\partial \pi}{\partial \dot{K}} = {} & 2A_2 a_3(a_1 \dot{K} + r a_2 \phi \overline{N} e^{rt} + a_3 \ddot{K}) \\ & + 2B_2 b_3(b_1 \dot{K} + b_2 r \phi \overline{N} e^{rt} + b_3 \ddot{K}) \end{aligned}$$

Setting

$$\frac{\partial \pi}{\partial K} = \frac{d}{dt}\frac{\partial \pi}{\partial \dot{K}}$$

gives

$$\begin{aligned}(2A_2a_3^2 + 2B_2b_3^2)\ddot{K} - (2A_2a_1^2 + 2B_2b_1^2)K \\ = A_1a_1 + B_1b_1 + 2a_1a_4A_2 + 2b_1b_4B_2 - \gamma_2 \\ + [2A_2a_2(a_1 - ra_3) + 2B_2b_2(b_1 - rb_3)]\phi\bar{N}e^{rt}\end{aligned}$$

The solution is

$$K^0 = C_1e^{\rho_1 t} + C_2e^{\rho_2 t} + C_3 + C_4e^{rt} \tag{9.46}$$

where ρ_1 and ρ_2 are the solutions to the characteristic equation,

$$(2A_2a_3^2 + 2B_2b_3^2)\rho^2 - (2A_2a_1^2 + 2B_2b_1^2) = 0$$

C_1 and C_2 are the constants of integration, and

$$C_3 = -\frac{(A_1a_1 + B_1b_1 - \gamma_2) + 2(A_2a_1a_4 + B_2b_1b_4)}{2(A_2a_1^2 + B_2b_1^2)}$$

$$C_4 = \frac{[A_2a_2(a_1 - ra_3) + B_2b_2(b_1 - rb_3)]\phi\bar{N}}{A_2(r^2a_3^2 - a_1^2) + B_2(r^2b_3^2 - b_1^2)}$$

The optimal paths can be determined by substituting (9.46) into the original model.

$$\begin{aligned}I^0 &= \frac{dK^0}{dt} = C_1\rho_1e^{\rho_1 t} + C_2\rho_2e^{\rho_2 t} + rC_4e^{rt} \\ q_a^0 &= (a_1 + a_3\rho_1)C_1e^{\rho_1 t} + (a_1 + a_3\rho_2)C_2e^{\rho_2 t} \\ &\quad + (a_1C_4 + a_3rC_4 + a_2\phi\bar{N})e^{rt} + a_1C_3 + a_4 \\ q_b^0 &= (b_1 + b_3\rho_1)C_1e^{\rho_1 t} + (b_1 + b_3\rho_2)C_2e^{\rho_2 t} \\ &\quad + (b_1C_4 + rb_3C_4 + b_2\phi\bar{N})e^{rt} + b_1C_3 + b_4 \\ p_a^0 &= A_1 + A_2q_a^0 \\ p_b^0 &= B_1 + B_2q_b^0 \\ L^0 &= (1/a_2)(q_a^0 - a_1K^0 - a_3I^0 - a_4) \\ &= (1/b_2)(q_b^0 - b_1K^0 - b_3I^0 - b_4) \\ \pi^0 &= p_a^0q_a^0 + p_b^0q_b^0 - \gamma_1L^0 - \gamma_2K^0 - \gamma_3I^0\end{aligned}$$

For illustrative purposes the following parameter values are assumed:

$a_1 = 0.9$	$a_2 = 0.5$	$a_3 = 0.1$	$a_4 = 15$
$b_1 = 0.8$	$b_2 = 0.5$	$b_3 = 0.2$	$b_4 = 20$
$\gamma_1 = 2$	$\gamma_2 = 12$	$\gamma_3 = 15$	
$A_1 = 50$	$A_2 = -2$	$B_1 = 60$	$B_2 = -3$
$\phi = 0.015$	$\bar{N} = 100{,}000$	$r = 0.01$	

Note that $a_i = 1/MPP_{a,i}$ and $b_i = 1/MPP_{b,i}$ for $i = 1, 2, 3$, where $MPP_{j,i}$ represents the marginal physical product to the jth output from the ith input.

From these data we find

$\rho_1 = 5.03$	$\rho_2 = -5.03$	$C_3 = -9.75$
$C_4 = 888.14$	$rC_4 = 8.88$	$\phi\bar{N} = 1500$

The specific forms of the optimal time paths are

$$K^0 = C_1 e^{5.03t} + C_2 e^{-5.03t} + 888.14e^{0.01t} - 9.75$$

$$I^0 = 5.03C_1 e^{5.03t} - 5.03C_2 e^{-5.03t} + 8.88e^{0.01t}$$

$$q_a^0 = 1.403C_1 e^{5.03t} + 0.397C_2 e^{-5.03t} + 1550.21e^{0.01t} + 6.225$$

$$q_b^0 = 1.806C_1 e^{5.03t} - 0.206C_2 e^{-5.03t} + 1462.29e^{0.01t} + 12.20$$

$$p_a^0 = -2.803C_1 e^{5.03t} - 0.794C_2 e^{-5.03t} - 3100.42e^{0.01t} + 37.55 \qquad (9.47)$$

$$p_b^0 = -5.418C_1 e^{5.03t} + 0.618C_2 e^{-5.03t} - 4386.87e^{0.01t} + 23.40 \qquad (9.48)$$

$$L^0 = q_a^0 - 1.8K^0 - 0.2I^0 - 30.00$$

Just as in the previous example, there are two constants of integration and therefore two initial conditions to be selected. From the prospective markets, substitute goods, and economic conditions, the firm's management has decided to set the initial price of product a at \$30 and the initial price of b at \$35. After substituting $t = 0$, $p_a^0 = 30$, and $p_b^0 = 35$ into (9.47) and (9.48), the constants of integration are found to be

$$C_1 = 2102.64 \qquad C_2 = 1723.04$$

These are two of many applications of variational calculus to economic and business models that might have been offered to the reader. Other applications can be found throughout the economics, management science, and operations research literature. Some are cited in the references at the end of the text. See especially Allen [1956, Ch. 20], Arrow et al. [1958, Chs. 4 and 5], Ramsey [1928], and Sengupta and Walker [1964]. Dorfman [1969] has discussed models from capital theory in the framework of optimal control theory, which he calls the modern version of calculus of variations.

9.5 LIMITATIONS

Several major limitations of variational calculus should not be ignored. These limitations make variational calculus an inapplicable optimization technique for some problems.

A variational calculus problem is one in which an integral of a functional is to be optimized. It has been presumed so far that the Euler equations can be determined and that the differential equations can be solved. If the differential equations cannot be solved, "direct methods" can sometimes be utilized. Discussions of these methods have been emphasized in literature of Russian mathematics. One excellent summary is provided by Elsgolc [1961, Ch. 5].

Unfortunately, the equations determined in (9.9) are only necessary conditions for an optimal solution. Sometimes the sufficient conditions are quite complex. Elsgolc [1961, Ch. 3], Bolza [1961, Ch. 3], and Morse [1934] provide these conditions for various cases. The role of the Euler equations in solving a variational calculus problem is similar to that of the first derivative in a calculus problem. The sufficient conditions for local extrema to be global extrema in a calculus problem are well known; one alternative to the sufficient conditions is that a convex (concave) function be minimized (maximized). It would be difficult to prove that a similar alternative to the sufficient conditions holds for must variational calculus problems. However, this appears to be logical extension from basic calculus.

There are two instances where simple, calculus-type, second-order conditions exist for variational calculus models. Consider the problem

$$\begin{array}{ll} \text{optimize} & W = \int_{\tau}^{T} F(X, \dot{X}, t)\,dt \\ \text{subject to} & G(X, \dot{X}, t) = 0 \end{array}$$

If W were to be maximized, $\partial F/\partial X \geq 0$, and F and G are concave functions of X, the second-order conditions will be satisfied for the variational calculus problem. If W were to be minimized, $\partial F/\partial X \leq 0$, and F and G are convex functions of X, the required second-order conditions are satisfied. Often in economics and business optimal control problems, the constraints are linear functions of X so that only the convexity or concavity of the objective function must be considered for the second-order conditions of the simplest type.

Usually sufficient conditions must be considered most seriously when the Euler equations cannot be solved. However, the sufficient conditions are sometimes stronger than those needed to solve the problem.

To determine whether or not the solution that is determined using the necessary conditions is in fact optimal, it is easier to look at the values of the objective function in the neighborhood of the solution. If the objective function cannot be made more extreme by varying any of the values of the variables for which solutions have been determined, the necessary conditions are also sufficient for a local optimum.

In two situations where it is possible to determine and solve the Euler equations, the solutions may have little meaning or applicability. One is when the Euler equation is a degenerate case; then neither the variable x nor any of its time derivatives $d^i x/dt^i$ appear. This occurs when x and all the $d^i t/dt^i$ that appear in the objective function and the constraints do so as linear terms.

On the other hand, the solutions to the Euler equations may be unstable. If (1) all the real parts of the characteristic roots are positive and larger than one, (2) the upper boundary has not been fixed, and (3) the coefficients of integration are all positive, the solutions to the Euler equations may explode over time. In this situation $\lim_{t \to \infty} x_t \to \infty$, and the optimal solution is said to be unstable in the Laplace sense. Detailed discussions of stability characteristics are given by Davis [1960], Lefscheta [1956], Struble [1962], and Samuelson [1963].

Finally, the problem may be one to which variational calculus should not be applied. For many of these situations the problem can be adapted to the form required to apply dynamic programming or the Pontryagin principle. Three instances when such an adaptation should be considered are: if there are many variables for which optimal paths are needed, if some of the variables in the problem cannot or should not be approximated by continuous variables, or if the problem involves inequalities. (Arrow et al. [1958, Chs. 4 and 5] have applied variational calculus to several problems involving inequalities. However, they refer to their method as one where "patch work" is required to handle them.) If these or other limitations of variational calculus make the technique inappropriate or unsuitable, another should be considered.

PROBLEMS

9.1. For each of the following cases, identify the order and degree of the differential equation. Find the characteristic roots for each instance.
(a) $2\dot{y} - 8y + 6 = 0$
(b) $3\ddot{y} - 11\dot{y} + 10y - 15 = 0$
(c) $\ddot{y} + 4\dot{y} + 3y - 12 = 0$
(d) $\ddot{y} - 1.25\dot{y} + 0.375y + 0.375 = 0$

9.2. Suppose that $y_0 = 0$ and $y_1 = 3$.
(a) Find the constants of integration for the four cases in Problem 1.
(b) Give the complete solution for each of the differential equations.
(c) As many time periods pass, which if any of the time paths for y are stable?

9.3. Consider the constrained variational calculus problem:

$$\begin{aligned} \text{maximize} \quad & \int_\tau^T \log C\, dt \\ \text{subject to} \quad & Y = C + I \\ & I = b\dot{Y} \end{aligned}$$

Find the optimal time paths for Y, C, and I.

9.4. For the problem:

$$\begin{aligned} \text{minimize} \quad & \int_\tau^T (C - C^*)^2 dt \\ \text{subject to} \quad & Z = C + I + A \\ & Y = [a/(D + a)]Z \\ & I = b\dot{Y} \end{aligned}$$

find the optimal time paths Y and I. The variables are defined as those for (9.22)–(9.25).

9.5. Find the optimal time path for y to

$$\text{optimize} \quad W = \int_\tau^T [16y^2 - (\ddot{y})^2]dt$$

9.6. An entrepreneur has decided he would like to maximize his total revenue TR subject to a budget constraint. The firm produces two products at levels x_1 and x_2 respectively. The prices of the finished goods are p_1 and p_2. The dollars to be allocated are C. It costs k_1 dollars to produce a unit of product 1 and k_2 dollars to produce a unit of product 2. The problem is to find x_1 and x_2 to

$$\begin{aligned} \text{maximize} \quad & W = \int_{\tau}^{T} TR\,dt \\ \text{subject to} \quad & TR = p_1 x_1 + p_2 x_2 \\ & p_1 = a_1 x_1 + b_1 \dot{x}_1 \\ & p_2 = a_2 x_2 + b_2 \dot{x}_2 \\ & k_1 x_1 + k_2 x_2 = C \end{aligned}$$

10 STOCHASTIC PROGRAMMING

10.1 INTRODUCTION

When the parameters in a mathematical programming model are presumed to be random variables rather than constants, a stochastic programming problem must be solved. These problems involve risk if the probability distributions of the random variables are known, or they involve uncertainty if the distribution of at least one random variable is unknown. The difficulties of dealing with risk and uncertainty in programming problems have been discussed in the literature since the 1950s.

Suppose that a firm producing three products at nonnegative activity levels of x_1, x_2, and x_3 earns per unit net profits of c_1, c_2, and c_3 respectively on the outputs; and all the firm's available labor b_L and capital b_K are employed to maximize net profits P; $P = c_1x_1 + c_2x_2 + c_3x_3$. Furthermore, suppose that at least d_1 units and no more than d_2 units of product 1 must be produced; $d_1 \leq x_1 \leq d_2$. To produce a single unit of the ith product ($i = 1, 2, 3$), a_{iL} units of labor and a_{iK} units of capital are required. Therefore, $a_{1L}x_1 + a_{2L}x_2 + a_{3L}x_3 \leq b_L$ and $a_{1K}x_1 + a_{2K}x_2 + a_{3K}x_3 \leq b_K$.

The mathematical programming problem the firm must solve is

$$\begin{aligned}
\text{maximize} \quad & P = c_1x_1 + c_2x_2 + c_3x_3 \\
\text{subject to} \quad & a_{1L}x_1 + a_{2L}x_2 + a_{3L}x_3 \leq b_L \\
& a_{1K}x_1 + a_{2K}x_2 + a_{3K}x_3 \leq b_K \\
& d_1 \leq x_1 \leq d_2 \\
& x_i \geq 0 \qquad i = 1, 2, 3
\end{aligned}$$

Although it has been assumed throughout previous chapters, it would be unusual for the coefficients of any programming model to be known with exact certainty. Some or all of the a's, b's, c's, and d's in the problem in the previous paragraph may be random variables rather than constants.

Each c_j is a per unit net profit determined from net revenues and costs for the jth output; however, the final selling prices and production costs may be unknown when the production decisions are made, and the net profit on each product may be a random variable with an assumed or estimated mean and variance. Furthermore, because of quality control and other unpredictable aspects, the exact required inputs per unit of output, a_{iL} or a_{iK} respectively, may not be known with certainty. The firm's experience may provide very reasonable estimates for the a's, and the likely variation around these estimates may also be available; but again the assumed constants may in fact be random variables. Furthermore, the minimum and maximum output levels for product 1, d_1 and d_2 respectively, may be subject to some variation from customer's choices that are not known or controlled by the firm. It is also possible that the entrepreneur does not know exactly how many labor and capital hours, b_L and b_K respectively, will be available during any production period because of employee absenteeism and machine breakdowns, but values for b_L and b_K may be estimated or approximated.

Random elements often appear in programming models because the analyst must assume some specific information on the environment in which the business firm or national economy operates. Any relationship that has been estimated using regression involves a random element. Often this random element is an additive error term with an expected value of zero, as assumed in least squares regression models. Even if all the available data show a particular cost or input availability to be the same for a long period, possible errors in measurement could have occurred and not been observed; an illness, an unexpected labor problem, or an unprecedented machine error could occur at any time.

In solving a stochastic programming model, the goal is usually to determine the expected value or the probability distribution of the value of the objective function; Vajda [1961, Ch. 11] called these alternatives expected value problems and distribution problems respectively. Since the value of the objective function is some function of all the constants and variables in the model, the value of the objective function is a random variable in any stochastic programming model.

Expected value problems are often more easily solved because the expected values of the random variables that appear in the problem provide sufficient information for the analyst to solve the model. However, the solutions to expected value problems are sometimes considered to be inadequate because so little information about each random element is employed. An expected value approach may be the only possible one if the means of the random variables are available and no other data exist, but this is not usually the case.

Given a reasonably sized sample on any random variable, the probability distribution of the random variable can usually be approximated by a member of the Pearson family of curves. Among the Pearson curves are nearly all the well-known statistical distributions. (The method for selecting the most appropriate member of the Pearson family is described in Appendix C.) Using the Pearson method will enable an analyst to determine the necessary probability distributions of variables and parameters to eventually determine the probability distribution of the value of the objective function. As will be seen in some of the sections in this chapter, this is not always a simple task.

In this chapter only linear and quadratic problems under risk and uncertainty are considered. Problems in which uncertainty, and therefore an unknown probability distribution, appears must be solved as expected value models; or the Pearson method is applied to determine the necessary distribution. The difficulties encountered when random variables exist in complex nonlinear programming models are beyond the scope of this presentation.

The most common method for dealing with random variables in programming models is through certainty equivalents. Under some specific assumptions, optimizing the expected value of the objective function provides the same activity levels as optimizing the objective function with the activity levels being replaced by the expected activity levels. For instance, if only the c's and b's were random variables in the production problem described earlier in this section, the same results are found whether $E(P)$ or $c_1E(x_1) + c_2E(x_2) + c_3E(x_3)$ is optimized, yet $E(P) \neq c_1E(x_1) + c_2E(x_2) + c_3E(x_3)$ since the c's may be random variables. These and further aspects of certainty equivalents are the subjects considered in Section 10.2.

The active and passive approaches to stochastic programming are presented in Sections 10.3 and 10.4. Each of these is a distribution model in which the probability distributions of the random variables are assumed or estimated using the Pearson method, and the distribution of the value of the objective function is to be derived. The active and passive approaches are similar except that an exogenous resource allocation matrix is an additional assumption in the active approach. By considering all possible allocation matrices, the active approach provides a series of possible variations on any set of observations on the random variables in the model.

Simple and complex chance-constrained programming models are presented in Section 10.5, although most of the applications of this technique involve the simple constraints. If, in the sample problem presented above, the capital constraint, $a_{1K}x_1 + a_{2K}x_2 + a_{3K}x_3 \leq b_K$, is now permitted to be violated α percent of the time, the capital constraint is chance-constrained as $Pr(a_{1K}x_1 + a_{2K}x_2 + a_{3K}x_3 \leq b_K) \geq 1 - \alpha$, so that the original constraint is not violated at least $1 - \alpha$ percent. Assuming that the random variables are all normally distributed or follow some other well-defined distribution allows replacing the chance constraint by a deterministic inequality that is similar to the chance-constrained inequality.

Charnes and Cooper, who were the first to present chance-constrained programming models, have suggested three models that have different objective functions and constraints of the simple type illustrated in the previous paragraph. The three alternatives are maximizing the expected value of the objective function (the *E* model), minimizing the generalized mean square of the objective function (the *V* model), or maximizing the probability that the aspiration level or goal of the objective function is attained (the *P* model). In each instance, the problem can be recast into a deterministic, nonstochastic model of a type that has been analyzed in earlier chapters.

This chapter is concluded with a brief examination of the dynamic aspects of optimizing under uncertainty. It will be obvious that such problems are hardly manageable unless only simple cases are considered.

10.2 UNCERTAINTY AND CERTAINTY EQUIVALENCE

The most popular method for handling uncertainty in linear and quadratic programming problems is by determining certainty equivalents to the stochastic problem. Simon [1956] and Theil [1957, 1964] have been among the most active participants in this effort.

The Quadratic Model

Theil considers the quadratic programming problem[1]

$$\text{optimize} \quad w(\mathbf{x},\mathbf{y}) = \mathbf{a}'\mathbf{x} + \mathbf{b}'\mathbf{y} + (1/2)(\mathbf{x}'\mathbf{A}\mathbf{x} + \mathbf{y}'\mathbf{B}\mathbf{y} + \mathbf{x}'\mathbf{C}\mathbf{y} + \mathbf{y}'\mathbf{C}'\mathbf{x}) \qquad (10.1)$$

$$\text{subject to} \quad \mathbf{y} = \mathbf{R}\mathbf{x} + \mathbf{s} \qquad (10.2)$$

1. Throughout this chapter the constraints $\mathbf{x} \geq 0$ and $\mathbf{y} \geq 0$ are not explicitly included for the quadratic programming problems. The reader is presumed to know how to include these constraints via the Lagrangian multiplier techniques explained in Chapter 5.

where $\mathbf{a}, \mathbf{x} = m \times 1$ vectors
$\mathbf{b}, \mathbf{y}, \mathbf{s} = n \times 1$ vectors
$\mathbf{R}$ = an $n \times m$ matrix
$\mathbf{A}$ = an $m \times m$ symmetric matrix
$\mathbf{B}$ = an $n \times n$ symmetric matrix
$\mathbf{C}$ = an $m \times n$ matrix

Substituting (10.2) into (10.1) reduces $w(\mathbf{x}, \mathbf{y})$ to be a function of only $\mathbf{x}$, $w(\mathbf{x}, \mathbf{Rx} + \mathbf{s})$.

$$w(\mathbf{x}, \mathbf{Rx} + \mathbf{s}) = \mathbf{b}'\mathbf{s} + (1/2)\,\mathbf{s}'\mathbf{Bs} + \mathbf{k}'\mathbf{x} + (1/2)\,\mathbf{x}'\mathbf{Kx} \qquad (10.3)$$
$$\mathbf{K} = \mathbf{A} + \mathbf{R}'\mathbf{BR} + \mathbf{CR} + \mathbf{R}'\mathbf{C}'$$
$$\mathbf{k}' = \mathbf{a}' + \mathbf{b}'\mathbf{R} + \mathbf{s}'(\mathbf{BR} + \mathbf{C}')$$
$$\mathbf{k} = \mathbf{a} + \mathbf{R}'\mathbf{b} + (\mathbf{R}'\mathbf{B}' + \mathbf{C})\mathbf{s}$$

Note that $\mathbf{k}$ is an $m \times 1$ vector since $\mathbf{a}$ is $m \times 1$; $\mathbf{R}'\mathbf{b}$ is an $m \times n$ times an $n \times 1$; $(\mathbf{R}'\mathbf{B}' + \mathbf{C})$ is an $m \times n$ times an $n \times n$ plus an $m \times n$, which gives an $m \times n$; since $\mathbf{s}$ is an $n \times 1$, $(\mathbf{R}'\mathbf{B}' + \mathbf{C})\mathbf{s}$ is an $m \times n$ times an $n \times 1$. Therefore, $\mathbf{k}$ is the sum of three $m \times 1$ vectors, which gives an $m \times 1$ vector.

Also, $\mathbf{K}$ is a symmetric $m \times m$ matrix since $\mathbf{A}$ is a symmetric $m \times m$ matrix; $\mathbf{B}$, which is a symmetric $n \times n$ matrix, is premultiplied by $\mathbf{R}'$, an $m \times n$ matrix, and postmultiplied by $\mathbf{R}$; $\mathbf{CR} + \mathbf{R}'\mathbf{C}'$ is the sum of an $m \times m$ matrix and its transpose, which gives a symmetric $m \times m$ matrix; and the sum of $m \times m$ symmetric matrices is a symmetric $m \times m$ matrix.

One necessary condition to determine an optimal vector of $\mathbf{x}$'s (where there are no random variables in the model) is that $w(\mathbf{x}, \mathbf{Rx} + \mathbf{s})$ is differentiated with respect to $\mathbf{x}$ and the result set equal to zero and solved for $\mathbf{x}$. The other necessary condition is that $\mathbf{K}$ is negative semidefinite to maximize w or positive semidefinite to minimize w. Together these necessary conditions are sufficient for $\mathbf{x} = \mathbf{x}^0$ to optimize $w(\mathbf{x}, \mathbf{y})$ subject to $\mathbf{y} = \mathbf{Rx} + \mathbf{s}$. (Throughout the remainder of this chapter second-order conditions are not specified in detail. It is presumed that convex functions are minimized and concave functions are maximized.)

Differentiating (10.3) with respect to $\mathbf{x}$ provides

$$\frac{\partial w(\mathbf{x}, \mathbf{Rx} + \mathbf{s})}{\partial \mathbf{x}} = \mathbf{k}' + \mathbf{Kx} = 0$$

Then

$$\mathbf{x}^0 = -\mathbf{K}^{-1}\mathbf{k} \qquad (10.4)$$

$$\mathbf{y}^0 = -\mathbf{R}\mathbf{K}^{-1}\mathbf{k} + \mathbf{s} \tag{10.5}$$

$$w^0 = w(\mathbf{x}^0, \mathbf{R}\mathbf{x}^0 + \mathbf{s}) = \mathbf{b}'\mathbf{s} + (1/2)\,\mathbf{s}'\mathbf{B}\mathbf{s} - (1/2)\,\mathbf{x}^{0\prime}\mathbf{K}\mathbf{x}^0 \tag{10.6}$$

or

$$w^0 = w(\mathbf{x}^0, \mathbf{R}\mathbf{x}^0 + \mathbf{s}) = \mathbf{b}'\mathbf{s} + (1/2)\,\mathbf{s}'\mathbf{B}\mathbf{s} - (1/2)\,\mathbf{k}'\mathbf{K}^{-1}\mathbf{k} \tag{10.7}$$

where $\mathbf{x}^0$ = the optimal level of $\mathbf{x}$
$\mathbf{y}^0$ = the optimal level of $\mathbf{y}$
w^0 = the optimal value of $w(\mathbf{x}, \mathbf{R}\mathbf{x} + \mathbf{s}) = w(\mathbf{x}, \mathbf{y})$

Since $\mathbf{K}$ is an $m \times m$ symmetric matrix, $\mathbf{K}^{-1}$ has the same characteristics. For future reference $\mathbf{K}^{-1}$ is defined as

$$\mathbf{K}^{-1} = \begin{bmatrix} K_{11}^{-1} & \cdots & K_{1m}^{-1} \\ \vdots & & \\ K_{m1}^{-1} & \cdots & K_{mm}^{-1} \end{bmatrix}$$

For most of these considerations Theil assumes that the elements of the matrices $\mathbf{A}, \mathbf{B}, \mathbf{C}$, and $\mathbf{R}$ are known constants. When the elements of $\mathbf{a}, \mathbf{b}$, and $\mathbf{s}$ are assumed to be constants, the optimal solutions are as given by (10.4), (10.5), (10.6), and (10.7). Note that the optimal levels of $\mathbf{x}$ and $\mathbf{y}$ ($\mathbf{x}^0$ and $\mathbf{y}^0$ respectively) are *linear* combinations of $\mathbf{a}, \mathbf{b}$, and $\mathbf{s}$. When these three vectors are assumed to be vectors of random variables, the expected value of $\mathbf{x}^0$, denoted by $\mathrm{E}(\mathbf{x}^0)$, has the same form as $\mathbf{x}^0$ in (10.4), except that the random variables in the solution are replaced by their expected values. The same result holds for the expected value of $\mathbf{y}^0$, $\mathrm{E}(\mathbf{y}^0)$.

$$\mathrm{E}(\mathbf{x}^0) = \mathrm{E}(-\mathbf{K}^{-1}\mathbf{k}) = -\mathbf{K}^{-1}\,\mathrm{E}(\mathbf{k}) \tag{10.8}$$

$$\begin{aligned} \mathrm{E}(\mathbf{y}^0) &= \mathrm{E}(-\mathbf{R}\mathbf{K}^{-1}\mathbf{k} + \mathbf{s}) = -\mathbf{R}\mathbf{K}^{-1}\,\mathrm{E}(\mathbf{k}) + \mathrm{E}(\mathbf{s}) \\ \mathrm{E}(\mathbf{k}) &= \mathrm{E}(\mathbf{a}) + \mathbf{R}'\mathrm{E}(\mathbf{b}) + (\mathbf{R}'\mathbf{B}' + \mathbf{C})\mathrm{E}(\mathbf{s}) \end{aligned} \tag{10.9}$$

Besides these results, Theil [1964] shows that

$$\begin{aligned} E(w^0) &= \mathrm{E}(\mathbf{b}'\mathbf{s}) + (1/2)\,\mathrm{E}(\mathbf{s}'\mathbf{B}\mathbf{s}) - (1/2)\,\mathrm{E}(\mathbf{k}'\mathbf{K}^{-1}\mathbf{k}) \\ E(w^0) &= \mathrm{E}(\mathbf{b}')\mathrm{E}(\mathbf{s}) + (1/2)\,\mathrm{E}(\mathbf{s}')\mathbf{B}\mathrm{E}(\mathbf{s}) - (1/2)\,\mathrm{E}(\mathbf{k}')\mathbf{K}^{-1}\,\mathrm{E}(\mathbf{k}) \\ &\quad + [\mathrm{E}(\mathbf{b}'\mathbf{s}) - \mathrm{E}(\mathbf{b}')\mathrm{E}(\mathbf{s})] + (1/2)\,[\mathrm{E}(\mathbf{s}'\mathbf{B}\mathbf{s}) - \mathrm{E}(\mathbf{s}')\mathbf{B}\mathrm{E}(\mathbf{s})] \\ &\quad - (1/2)\,[\mathrm{E}(\mathbf{k}'\mathbf{K}^{-1}\mathbf{k} - \mathrm{E}(\mathbf{k}')\,\mathbf{K}^{-1}\,\mathrm{E}(\mathbf{k})] \end{aligned} \tag{10.10}$$

or

$$E(w^0) = \mathbf{E(b')E(s)} + (1/2)\,\mathbf{E(s')BE(s)} - (1/2)\,\mathbf{E(k')K^{-1}E(k)} + \text{Cov}\,(\mathbf{b's}) + (1/2)\,\text{Cov}\,(\mathbf{s'Bs}) - (1/2)\,\text{Cov}\,(\mathbf{k'K^{-1}k})$$

Theil assumes constant variances and covariances. Therefore, the difference between $E[w(\mathbf{x}^0, \mathbf{y}^0)]$ and $w[\mathbf{E}(\mathbf{x}^0), \mathbf{E}(\mathbf{y}^0)]$ is a nonnegative constant.

The importance of this result should not be overlooked. The decision to optimize the expected value of a quadratic objective function having random variables, namely $E[w(\mathbf{x}, \mathbf{y})]$, provides the same set of optimal allocations, $\mathbf{x} = \mathbf{x}^0$ and $\mathbf{y} = \mathbf{y}^0$, as the decision to optimize $w[\mathbf{E}(\mathbf{x}), \mathbf{E}(\mathbf{y})]$. Optimizing objective functions with random variables by finding the optimal value of the expected value of that objective function is common.

These are the main static results that Theil [1964, pp. 52–58] derives for the certainty equivalent of the deterministic problem. Theil extends the results (in an appendix) to show the effect of permitting all the elements of $\mathbf{a}, \mathbf{b}, \mathbf{s}, \mathbf{A}, \mathbf{B}, \mathbf{C}$, and $\mathbf{R}$ to be random variables. The difficulties become much greater. Besides considering the cases where all the parameters are random variables in the static problem, Theil describes the dynamic aspects. To study these problems, the matrices and vectors of (10.1) and (10.2) are time-partitioned. A new set of variables is introduced for each time period that is within the scope of the problem. This formulation is described in Section 10.6.

Example 10.1

Theil [1964, Ch. 3] has illustrated his analysis with an application to an econometric model estimated by Klein [1950, Ch. 3]. The problem is to select optimal values for the variables to minimize the squared deviations between actual and target levels of six variables. The objective function is restricted by Klein's model. Van den Bogaard and Theil [1959] have studied macrodynamic policymaking in the United States from 1933 to 1936. The variables in the model are

Y = net national income
C = consumption
I = net investment
G = government expenditure
P = wage bill paid by private sources

W = government wage bill
K = capital stock
π = profits
T = indirect taxes
D = distribution variable
t = time in years

The behavioral equations of the model are

$$C = \alpha_0 + \alpha_1 \pi + \alpha_2 \pi_{t-1} + \alpha_3 (W + P) + u_C \qquad (10.11)$$

$$I = \beta_0 + \beta_1 \pi + \beta_2 \pi_{t-1} + \beta_3 K_{t-1} + u_I \qquad (10.12)$$

$$P = \gamma_0 + \gamma_1 (Y + T - W) + \gamma_2 (Y + T - W)_{t-1} + \gamma_3 (t - 1933) + u_P \qquad (10.13)$$

The variables u_C, u_I, and u_p are stochastic error terms for the consumption function (10.11), the investment function (10.12), and the private wage function (10.13) respectively. All the unsubscripted variables are assumed to represent 1933 levels; π_{t-1} and K_{t-1} are the respective values for profits and capital stock for 1932. The parameter estimates supplied by Theil [1964, p. 78] are given in Table 10.1.

Table 10.1. Parameter estimates

Estimate	α_1	β_1	γ_1
0	16.786	17.784	1.599
1	0.020	0.231	0.420
2	0.225	0.546	0.164
3	0.800	−0.146	0.135

The definitional equations for the model are

$$Y + T = C + I + G \qquad (10.14)$$

$$Y = \pi + P + W \qquad (10.15)$$

$$K = I + K_{t-1} \qquad (10.16)$$

$$D = P - \rho \pi \qquad (10.17)$$

The value of ρ has been set at 2.0 and D at 0.0 so that $\pi/P = 1/2$, the approximate ratio before the depression.

The purpose is presumed to be to minimize a quadratic objective function such as (10.1), where the coefficients of the linear terms are zeros. Thus the goal is to minimize

$$w = -(1/2)[(W - W^*)^2 + (T - T^*)^2 + (G - G^*)^2 + (C - C^*)^2 + (I - I^*)^2 + (D - D^*)^2]$$

subject to the model described by (10.11)–(10.17). The predetermined target values that Theil [1964, p. 78] suggests are $C^* = 49.690$, $I^* = -3.100$, $D^* = 11.250$, $W^* = 5.038$, $T^* = 7.396$, and $G^* = 10.438$ in billions of 1934 dollars.

The random variables occur as the error terms in the estimated econometric model. According to the usual assumption in statistics, $E(u_C) = E(u_I) = E(u_P) = 0$.

Applying the notation used for (10.1)–(10.10), the vectors and matrices are as follows:

$$\mathbf{a} = \mathbf{0} \qquad \mathbf{b} = \mathbf{0} \qquad \mathbf{A} = \mathbf{B} = -\mathbf{I} \qquad \mathbf{C} = \mathbf{0}$$

$$w = -(1/2)\mathbf{x}'\mathbf{x} - (1/2)\mathbf{y}'\mathbf{y}$$

$$\mathbf{x} = \begin{bmatrix} W - W^* \\ T - T^* \\ G - G^* \end{bmatrix} \qquad \mathbf{y} = \begin{bmatrix} C - C^* \\ I - I^* \\ D - D^* \end{bmatrix}$$

$$\mathbf{R} = \begin{bmatrix} 0.666 & -0.188 & 0.671 \\ -0.052 & -0.296 & 0.259 \\ 0.285 & 2.358 & -1.427 \end{bmatrix}$$

$$\mathbf{s} = \begin{bmatrix} -5.393 \\ -3.704 \\ -0.729 \end{bmatrix} + \begin{bmatrix} 1.671 & 0.671 & 1.148 \\ 0.259 & 1.259 & -0.089 \\ -1.427 & -1.427 & 2.217 \end{bmatrix} \begin{bmatrix} u_C \\ u_I \\ u_P \end{bmatrix}$$

$$\mathbf{K} = -(\mathbf{I} + \mathbf{R}'\mathbf{R}) = \begin{bmatrix} -1.527 & -0.563 & -0.026 \\ -0.563 & -6.684 & 3.568 \\ -0.026 & 3.568 & -3.553 \end{bmatrix}$$

$$\mathbf{k} = -\mathbf{R}'\mathbf{s} = -\mathbf{R}' \begin{bmatrix} -5.393 \\ -3.704 \\ -0.729 \end{bmatrix}$$

$$-\mathbf{R}' \begin{bmatrix} 1.671 & 0.671 & 1.148 \\ 0.259 & 1.259 & -0.089 \\ -1.427 & -1.427 & 2.217 \end{bmatrix} \begin{bmatrix} u_C \\ u_I \\ u_P \end{bmatrix}$$

$$\mathbf{E(k)} = -\mathbf{R}'\mathbf{E(s)} = -\mathbf{R}' \begin{bmatrix} -5.393 \\ -3.704 \\ -0.729 \end{bmatrix} = \begin{bmatrix} 3.608 \\ -0.393 \\ 3.536 \end{bmatrix}$$

From (10.8) and (10.9)

$$\mathbf{E(x^0)} = \mathbf{K}^{-1}\mathbf{E(k)} = \begin{bmatrix} 2.107 \\ 0.618 \\ 1.600 \end{bmatrix} = \begin{bmatrix} W - W^* \\ T - T^* \\ G - G^* \end{bmatrix}$$

$$\mathbf{E(y^0)} = \mathbf{RE(x^0)} + \mathbf{E(s)} = \begin{bmatrix} -0.824 \\ -2.146 \\ -2.450 \end{bmatrix} = \begin{bmatrix} C - C^* \\ I - I^* \\ D - D^* \end{bmatrix}$$

Thus, the optimal values for the variables in the macrodynamic model for 1933 are

$$\begin{aligned} E(C^0) &= C^* - 0.824 = 48.866 \\ E(I^0) &= I^* - 2.146 = -5.246 \\ E(D^0) &= D^* - 2.450 = 8.800 \\ E(W^0) &= W^* + 2.107 = 7.145 \\ E(G^0) &= G^* + 1.600 = 12.038 \\ E(T^0) &= T^* + 0.618 = 8.014 \end{aligned}$$

Note that there is nothing unusual about the negative value for $E(I^0)$ since this is the expected optimal net investment. Theil also illustrates his approach to certainty equivalents with an application to a microeconomic model. Theil considers the production model suggested by Holt et al. [1960].

Another variation of certainty equivalents that should be considered has been mentioned by van Moeseke [1964, p. 36]. In the notation of (10.1) and (10.2) van Moeseke suggests studying the problem

$$\begin{aligned} &\text{maximize} && E[w(\mathbf{x}, \mathbf{y})] - m\sigma[w(\mathbf{x}, \mathbf{y})] \\ &\text{subject to} && -\mathbf{Rx} + \mathbf{y} = \mathbf{s} \end{aligned}$$

where $E[w(\mathbf{x}, \mathbf{y})]$ = the expected value of the objective function

$\sigma[w(\mathbf{x}, \mathbf{y})]$ = the standard deviation of the objective function

m = a positive risk parameter determined by using the normal probability density function

These applications should provide the reader with more insight into the technique that has been reviewed and should suggest other possible applications.

The Linear Model

If **A**, **B**, and **C** are null matrices, (10.18) and (10.19) provide a linear programming problem with n slack variables that comprise the vector **y** and m activities in the vector **x**. The problem is

$$\text{optimize} \quad w(\mathbf{x}, \mathbf{y}) = \mathbf{a}'\mathbf{x} + \mathbf{b}'\mathbf{y} \tag{10.18}$$
$$\text{subject to} \quad -\mathbf{Rx} + \mathbf{y} = \mathbf{s} \tag{10.19}$$

The problem can be written in a more general linear programming framework. If

$$\mathbf{d}' = [a_1, \ldots, a_m; b_1, \ldots, b_n]$$
$$\mathbf{z}' = [x_1, \ldots, x_m; y_1, \ldots, y_n]$$
$$\mathbf{s}' = [s_1, \ldots, s_n]$$
$$\mathbf{D} = \mathbf{R} - \mathbf{I}_n = (d_{ij}) \qquad i = 1, \ldots, n; j = 1, \ldots, m + n$$
$$d_{ij} = r_{ij} \qquad i = 1, \ldots, n; j = 1, \ldots, m$$
$$d_{ij} = -1 \qquad i = 1, \ldots, n; j = m + i$$
$$d_{ij} = 0 \qquad i = 1, \ldots, n; j = m + k \text{ for all } k \neq i$$

the problem is

$$\text{optimize} \quad w(\mathbf{z}) = \mathbf{d}'\mathbf{z}$$
$$\text{subject to} \quad \mathbf{Dz} = \mathbf{s}$$

Note that $\mathbf{Dz} = \mathbf{Iy} - \mathbf{Rx} = \mathbf{s}$ can be studied as the set of inequalities $-\mathbf{Rx} \leq \mathbf{s}$, where the vector **y** is a set of slack variables and **I** is the identity matrix.

Two results from the complete description method of linear programming can be applied to the stochastic problem. First, an optimal solution will be one of the possible combinations of n nonzero and m zero-valued variables. Second, there are at most $(m + n)!/m!n!$ possibilities to be studied. Each of these possible solutions can be written as $\mathbf{D}^*\mathbf{z}^* = \mathbf{s}$ and $\mathbf{z}^* = (\mathbf{D}^*)^{-1}\mathbf{s}$,

where $\mathbf{z}^*$ has n of the $m + n$ elements of **z**

$\mathbf{D}^*$ = an $n \times n$ matrix formed from within **D** according to which elements of **z** are in $\mathbf{z}^*$

$(\mathbf{D}^*)^{-1}$ = the inverse matrix of $\mathbf{D}^*$
$\mathbf{s}$ = $\mathbf{s}$ as described previously

There are $(m + n)!/m!n!$ cases for which the objective function must be evaluated. For each instance there is a $w(\mathbf{z}^*)$, where

$$w(\mathbf{z}^*) = (\mathbf{d}^*)'(\mathbf{D}^*)^{-1}\mathbf{s}$$

The vector $(\mathbf{d}^*)'$ has n elements. The elements of $\mathbf{d}^*$ are those from $\mathbf{d}$ that are coefficients of the nonzero variables in $\mathbf{z}$ (the elements of $\mathbf{z}^*$).

If the only random variables in the problem are members of the vectors $\mathbf{d}$ and $\mathbf{s}$ (or $\mathbf{a}$, $\mathbf{b}$, and $\mathbf{s}$), $E[w(\mathbf{z}^*)]$ can be considered for each of the cases. Then the particular allocation $\mathbf{z}^*$ that optimizes $E[w(\mathbf{z}^*)]$ can be identified.

Several limitations of this procedure should be noted. First, the procedure is not simple for a problem in which any elements of the matrix $\mathbf{R}$ are random variables. Second, if the ith element of $\mathbf{d}^*$ and the ith element of $\mathbf{s}$ are both random variables, the expected value of a product of random variables is required. This can be found simply if the two random variables are independent or if they have a known covariance. [If u and v are random variables, $E(uv) = \text{Cov}(uv) + E(u)E(v)$. If $\text{Cov}(uv) = 0$, $E(uv) = E(u)E(v)$.] Otherwise more complicated methods are required.

Example 10.2

As an application of certainty equivalence to linear programming, consider the firm that produces and sells three products. When the decision is required on how much should be produced, the selling prices are not known specifically; at most, their probability density functions might be assumed. Two inputs are applied in the production process. The firm's manager has learned the approximate amounts of the resources that will be available. However, the exact levels that will be available cannot be known until just before production begins. Thus both the output prices and levels of available resources are not constant as is often assumed.

Table 10.2. Marginal Products

Products:	1	2	3
Resource 1	4.0	1.0	2.0
Resource 2	1.0	0.0	1.0

Table 10.3. Alternative Solutions

Case	x_1^0	x_2^0	x_3^0	y_1^0	y_2^0	w^0
1	0	0	0	$50 + \epsilon_1$	$25 + \epsilon_2$	0
2	$25 + \epsilon_2$	$-50 + \epsilon_1 - 4\epsilon_2$	0	0	0	$25p_1 - 50p_2 + p_1\epsilon_1 + p_2\epsilon_1 - 4p_2\epsilon_2$
3	$0.5\epsilon_1 - \epsilon_2$	0	$25 - 0.5\epsilon_1 + 2\epsilon_2$	0	0	$0.5p_1\epsilon_1 - p_1\epsilon_2 + 25p_3 - 0.5p_3\epsilon_1 + 2p_3\epsilon_2$
4	$25 + \epsilon_2$	0	0	$-25 + \epsilon_1 - 4\epsilon_2$	0	$25p_1 + p_1\epsilon_2$
5	$12.5 + 0.25\epsilon_1$	0	0	0	$12.5 - 0.25\epsilon_1 + \epsilon_2$	$12.5p_1 + 0.25p_1\epsilon_1$
6	0	$\epsilon_1 - 2\epsilon_2$	$25 + \epsilon_2$	0	0	$p_2\epsilon_1 - 2p_2\epsilon_2 + 25p_3 + p_3\epsilon_2$
7	0	—	0	—	0	—
8	0	$50 + \epsilon_1$	0	0	$25 + \epsilon_2$	$50p_2 + p_2\epsilon_2$
9	0	0	$25 + \epsilon_2$	$\epsilon_1 - 2\epsilon_2$	0	$25p_3 + p_3\epsilon_2$
10	0	0	$25 + 0.5\epsilon_1$	0	$-0.5\epsilon_1 + \epsilon_2$	$25p_3 + 0.5p_3\epsilon_1$

The programming problem requires maximizing the firm's revenue, $w = p_1x_1 + p_2x_2 + p_3x_3$, where p_1, p_2, and p_3 are random variables. There are two restrictions on the objective function. The approximate number of available units of the two inputs are estimated to be 50 and 25 respectively. The marginal products for the resources and products are given in Table 10.2. The constraints are

$$\begin{aligned} 4x_1 + x_2 + 2x_3 &\leq 50 + \epsilon_1 \\ x_1 \quad\quad + x_3 &\leq 25 + \epsilon_2 \\ x_i \geq 0 \quad\quad i &= 1, 2, 3 \end{aligned}$$

If y_1 and y_2 are the slack variables and unemployed resources for the first two constraints, the model becomes

$$\begin{aligned} 4x_1 + x_2 + 2x_3 + y_1 &= 50 + \epsilon_1 \\ x_1 \quad\quad + x_3 + y_2 &= 25 + \epsilon_2 \\ x_i, y_j \geq 0 \quad\quad i &= 1, 2, 3; j = 1, 2 \end{aligned}$$

where the goal is to maximize $w = p_1x_1 + p_2x_2 + p_3x_3$. The uncertainty concerning the exact amount of each resource that will be available is reflected by including ϵ_1 and ϵ_2, which are random variables.

The stochastic linear programming problem has five random variables. These are the p's and the ϵ's. With use of the complete description method, ten alternative solutions are to be investigated. These are indicated in Table 10.3. Note that case 7 cannot be solved for a unique set of results.

Table 10.4. Results of Applying Expected Values of Table 10.3

Case	$E(x_1^0)$	$E(x_2^0)$	$E(x_3^0)$	$E(y_1^0)$	$E(y_2^0)$	$E(w^0)$
1	0	0	0	56.25	27.5	0
2	27.5	–	0	0	0	–
3	0.625	0	26.875	0	0	818.75
4	27.5	0	0	–	0	–
5	14.1625	0	0	0	13.3375	238.25
6	0	1.25	27.25	0	0	830
7	0	–	0	–	0	–
8	0	56.25	0	0	27.5	562.5
9	0	0	27.5	–	0	–
10	0	0	28.125	0	–	–

Suppose that the firm's manager has underestimated the average of the available two resources by 15% and 10% respectively, and $E(\epsilon_1) = 6.25$ and $E(\epsilon_2) = 2.5$. Also, let the expected values of the prices be $E(p_1) = 20$, $E(p_2) = 10$, and $E(p_3) = 30$. Assuming that the prices and predicted resource levels are not related, $E(p_i \epsilon_j) = E(p_i)E(\epsilon_j), i = 1, 2, 3$ and $j = 1, 2$. The results of applying these expected values to Table 10.3 are presented in Table 10.4. A dash is used where the result would provide a negative value for an element of x or y, or would provide a value that is not unique.

Such applications of certainty equivalence to linear programming are not plentiful in the literature of linear programming. Yet few if any of the coefficients found in programming problems are known with certainty. Many persons estimate the parameters of a model and then proceed to apply the model as a set of constraints, while ignoring the statistical error terms in the programming model.

As indicated by Fox, Sengupta, and Thorbecke [1972, Ch. 9], several other results and features are to be considered along with the certainty equivalents delineated in this section. One feature is the derivation of optimal levels for **x** and **y** when the random variables are replaced by other measures of their averages. The mode and median have been studied for the triangular distribution of the random variables in particular. When there is only one random variable to be considered and it has a symmetric unimodal distribution function, the mean, median, and mode are the same.

A second feature of the method of certainty equivalents is that the information to be collected or the sample to be taken is minimal. The density function for the random variable need not be known or assumed. Only an estimate of the population mean is essential, and the arithmetic mean of a set of observations from a random sample provides a statistically unbiased estimate of the population mean. In this sense certainty equivalents to a stochastic programming problem do not presume the knowledge or assumption of any random variable's density function.

Two difficulties of the method of certainty equivalents are very much related to the two features cited above. First, one might prefer that more than one estimate of a random variable be considered when decisions are conceived. Second, unless the mean and mode of a distribution are the same, even the most likely occurrence is not represented by the expected value. Therefore, particular attention must be given to the application of this method to problems with random variables whose probability density functions are skewed.

Table 10.5. Special Probability Distributions

Name	Functional Form	Restrictions	Mean	Variance
Normal	$f(x) = (2\pi b^2)^{-1/2} \exp\{-(x - a)^2/2b^2\}$	$b > 0$ $-\infty < x < \infty$	a	b^2
Standard normal	$g(z) = (2\pi)^{-1/2} \exp\{-z^2/2\}$	$-\infty < z < \infty$	0	1
Exponential	$f(x) = (1/\theta) e^{-x/\theta}$	$\theta > 0$ $x > 0$	θ	θ^2
Gamma	$f(x) = \dfrac{x^a e^{-x/b}}{\Gamma(a + 1) b^{a+1}}$	$x > 0$ $a > -1$ $b > 0$	$b(a + 1)$	$b^2(a + 1)$
Beta	$f(x) = \dfrac{\Gamma(a + b + 2) x^a (1 - x)^b}{\Gamma(a + 1)\Gamma(b + 1)}$	$0 < x < 1$ $a > -1$ $b > -1$	$\dfrac{a + 1}{a + b + 2}$	$\dfrac{(a + 1)(b + 1)}{(a + b + 2)^2(a + b + 3)}$
Chi-square	$f(x) = \dfrac{x^{(v/2)-1} e^{-x/2}}{2^{v/2}\Gamma(v/2)}$	$x > 0$ $v = 1, 2, \ldots$	v	$2v$
Binomial	$f(x) = C(n.x) p^x (1 - p)^{n-x}$	$x = 0, 1, \ldots, n$ $0 < p < 1$ $n = 1, 2, \ldots$	np	$np(1 - p)$
Poisson	$f(x) = e^{-\lambda}\lambda^x/x!$	$\lambda > 0$ $x = 0, 1, \ldots$	λ	λ
Negative binomial	$f(x) = C(x + r - 1, r - 1) p^r q^x$	$x = 0, 1, \ldots$ $0 < p < 1$ $q = 1 - p$	rq/p	rq/p^2
Geometric	$f(x) = pq^x$	$x = 0, 1, \ldots$ $0 < p < 1$ $q = 1 - p$	q/p	q/p^2

Source: B. Ostle, *Statistics in Research*, 3rd ed. Iowa State Univ. Press, Ames, 1975.

In some special cases the activity levels that optimize $E[w(\mathbf{x},\mathbf{y})]$ provide a more general result. Suppose that the variance of $w(\mathbf{x},\mathbf{y})$, denoted by $V[w(\mathbf{x},\mathbf{y})]$, can be written as a function of $E[w(\mathbf{x},\mathbf{y})]$, $V[w(\mathbf{x},\mathbf{y})] = f\{E[w(\mathbf{x},\mathbf{y})]\}$. If f is a monotonic increasing function, activity levels that minimize $E[w(\mathbf{x},\mathbf{y})]$ also minimize $V[w(\mathbf{x},\mathbf{y})]$, while levels that maximize $E[w(\mathbf{x},\mathbf{y})]$ maximize $V[w(\mathbf{x},\mathbf{y})]$. If f is a monotonic decreasing function, minimizing $E[w(\mathbf{x},\mathbf{y})]$ gives a maximum $V[w(\mathbf{x},\mathbf{y})]$; maximizing $E[w(\mathbf{x},\mathbf{y})]$ provides a minimum $V[w(\mathbf{x},\mathbf{y})]$.

The significance of the conclusions in the previous paragraph should not be overlooked. If f is monotonic increasing, we should select the objective function $w(\mathbf{x},\mathbf{y})$ to be a loss or cost function and minimize its expected value and variance. For a monotonic decreasing f, let $w(\mathbf{x},\mathbf{y})$ be a profit function and maximize the expected value. Then the variance of the profits is also minimized. It may also be useful to keep in mind that $\min w(\mathbf{x},\mathbf{y}) = \max [-w(\mathbf{x},\mathbf{y})]$.

Examples where f is a monotonic increasing function occur when $w(\mathbf{x},\mathbf{y})$ is distributed as any of the statistical distributions in Table 10.5. Unfortunately, there are no known cases where f can be shown to be a monotonic decreasing function.

The possibility of selecting an objective function that involves the mean and variance has already been suggested. Sengupta and Tintner [1963] have studied the problem,

$$\text{maximize} \quad z = w_1(x_1 + x_2) - w_2[V(I)]^{1/2}$$

where

$$V(I) = \sigma_1^2 x_1^2 + \sigma_2^2 x_2^2 + 2\rho\sigma_1\sigma_2 x_1 x_2$$

w_1 and w_2 are constants, and ρ is the correlation coefficient between x_1 and x_2. A variation of the problem has been considered by Markowitz [1959] in his portfolio model. Portfolio models have been discussed in Section 5.8.

10.3 PASSIVE APPROACH TO STOCHASTIC PROGRAMMING

The first study of stochastic programming appeared when Tintner [1955] extended a linear programming application from agricultural economics. Tintner and Sengupta have since labeled Tintner's earliest effort as the passive approach. That title distinguishes the cases described in this section from those of Section 10.4, where the active approach to stochastic programming is delineated.

For both the passive and active approaches to stochastic programming, two cases will be considered. These are the quadratic programming problem of (10.1) and (10.2) and the linear programming problem

$$\text{optimize} \quad w(\mathbf{x}, \mathbf{y}) = \mathbf{a}'\mathbf{x} + \mathbf{b}'\mathbf{y} \tag{10.20}$$
$$\text{subject to} \quad \mathbf{y} = \mathbf{R}\mathbf{x} + \mathbf{s} \tag{10.21}$$

Both of these approaches are among those Vajda [1961] calls distribution techniques. The statistical distribution of $w(\mathbf{x}, \mathbf{y})$ is to be derived. Usually, all the random variables in the problem are assumed to be statistically independent. Then the joint density function of these random variables is the product of the density functions of the individual random variables. Throughout most of this section and Section 10.4 the density functions are assumed to be known or determined using the Pearson method of moments that is presented in detail in Appendix C. If the random variables are not independent, orthogonal transformations can usually be applied to provide a set of independent random variables.

For the quadratic programming problem, $w(\mathbf{x}, \mathbf{y})$ can be transformed as in Section 10.2; $w(\mathbf{x}, \mathbf{y}) = w(\mathbf{x}, \mathbf{R}\mathbf{x} + \mathbf{s})$. Thus it is sufficient to derive the statistical distribution of $w(\mathbf{x}, \mathbf{R}\mathbf{x} + \mathbf{s})$, i.e., (10.3). For the simplest case, deriving the density function of $w(\mathbf{x}, \mathbf{R}\mathbf{x} + \mathbf{s})$ can be completed without the use of higher mathematics; this will be illustrated here. For the more complex cases, the use of characteristic functions is often necessary. Throughout the remainder of this section, only simple cases are described.

It is assumed that there are no stochastic elements in the coefficient matrices: $\mathbf{A}, \mathbf{B}, \mathbf{C}$, and $\mathbf{R}$. Therefore, $\mathbf{K} = \mathbf{A} + \mathbf{R}'\mathbf{B}\mathbf{R} + \mathbf{C}\mathbf{R} + \mathbf{R}'\mathbf{C}'$ is a nonstochastic matrix of constants. The vectors $\mathbf{a}, \mathbf{b}$, and $\mathbf{s}$ are presumed to contain independent standard normally distributed random variables. Thus all these random variables have a mean of zero and a standard deviation of one. This is the simplifying assumption that is required to make the passive approach to stochastic quadratic programming problems manageable without requiring complicated mathematics. The role that this assumption plays in the derivation of the probability density function of $w(\mathbf{x}, \mathbf{R}\mathbf{x} + \mathbf{s})$ should become clear momentarily.

Since

$$k_j = a_j + \sum_{i=1}^{n} r_{ji} b_i + \sum_{i=1}^{n} r_{ij} b_{ji} s_i + \sum_{i=1}^{n} c_{ji} s_i \qquad j = 1, \ldots, m$$

k_j is a linear combination of normally distributed random variables. The mean of each k_j is zero:

$$E(k_j) = E(a_j) + \sum_{i=1}^{n} r_{ji}E(b_i) + \sum_{i=1}^{n} r_{ij}b_{ji}E(s_i) + \sum_{i=1}^{n} c_{ji}E(s_i) = 0$$

since each of these expected values is zero. The variance of an element of $\mathbf{k}$, $V(k_j)$, is derived as

$$V(k_j) = V(a_j) + \sum_{i=1}^{n} r_{ji}^2 V(b_i) + \sum_{i=1}^{n} \left(r_{ij}b_{ji}\right)^2 V(s_i)$$

$$+ \sum_{i=1}^{n} c_{ji}^2 V(s_i) \qquad j = 1,\ldots,m$$

$$V(k_j) = 1 + \sum_{i=1}^{n} r_{ji}^2 + \sum_{i=1}^{n} \left(r_{ij}b_{ji}\right)^2 + \sum_{i=1}^{n} c_{ji}^2 \qquad j = 1,\ldots,m$$

since $V(a_j) = V(b_i) = V(s_i) = 1$ for all values of i and j. Let

$$k_j^* = k_j/\sigma_j \text{ where } \sigma_j = [V(k_j)]^{1/2} \qquad j = 1,\ldots,m$$

so that k_j^* is a standard normal random variable having a zero mean and standard deviation of one.[2] Let $\mathbf{k}^*$ be the vector of these elements. Then, $(k_j^*)^2$ has a chi-square statistical distribution with one degree of freedom.

According to Sengupta, Tintner, and Millham [1963], "The (passive and active approaches) assume that in almost all possible situations, i.e., for almost all possible variations of the parameters, the conditions of the simple nonstochastic linear program are fulfilled and the maximum achieved." Thus $w(\mathbf{x}, \mathbf{Rx} + \mathbf{s})$ is to be optimized assuming no random variables as a first step. The optimal decision vector $\mathbf{x}^0 = -\mathbf{K}^{-1}\mathbf{k}$ is then substituted into $w(\mathbf{x}, \mathbf{Rx} + \mathbf{s})$ and the statistical distribution of w is to be studied henceforth.

From (10.7) recall that

$$w^0(\mathbf{x}^0, \mathbf{Rx}^0 + \mathbf{s}) = \mathbf{b's} + (1/2)\mathbf{s'Bs} - (1/2)\mathbf{k'K}^{-1}\mathbf{k}$$

$$= \sum_{i=1}^{n} b_i s_i + (1/2)\sum_{i=1}^{n}\sum_{k=1}^{n} s_i b_{ik} s_k$$

$$- (1/2)\sum_{j=1}^{m}\sum_{h=1}^{m} k_j K_{jh}^{-1} k_h$$

2. Suppose one or some of these random variables is normally distributed with a mean other than zero and/or a standard deviation other than one. Then $E(k_j)$ is not zero. Then let $k_j^* = [k_j - E(k_j)]/\sigma_j$.

Substituting $k_j = k_j^* \sigma_j$ leaves

$$w^0(\mathbf{x}^0, \mathbf{Rx}^0 + \mathbf{s}) = \sum_{i=1}^{n} b_i s_i + (1/2) \sum_{i=1}^{n} \sum_{k=1}^{n} s_i b_{ik} s_k$$
$$- (1/2) \sum_{j=1}^{m} \sum_{h=1}^{m} k_j^* \sigma_j K_{jh}^{-1} \sigma_h k_h^*$$

The summation form of this relation is provided to show that $w^0(\mathbf{x}^0, \mathbf{Rx}^0 + \mathbf{s})$ has a chi-square distribution. The first term is the sum of products of random variables that have independent standard normal density functions. Therefore, $\sum_{i=1}^{n} b_i s_i$ is distributed as a chi-square random variable with n degrees of freedom. The second term contains constants, 1/2 and b_{ik}, $i, k = 1, \ldots, n$, and random variables that are independent and distributed as standard normals. The product $s_i s_k$, $i, k = 1, \ldots, n$, has a chi-square distribution; therefore, so does $(1/2)\sum_{i=1}^{n}\sum_{k=1}^{n} s_i b_{ik} s_k$. The third term also consists of the sum of constants multiplied by independently distributed standard normal random variables, since the elements of $\mathbf{K}^{-1}$ are not stochastic and the standard deviations σ_j, $j = 1, \ldots, m$, may be presumed constant. Thus $w^0(\mathbf{x}^0, \mathbf{Rx}^0 + \mathbf{s})$ consists of a linear combination of chi-square distributed random variables, and such a combination also has a chi-square distribution.

To find the number of degrees of freedom associated with $w^0(\mathbf{x}^0, \mathbf{Rx}^0 + \mathbf{s})$, take the expected value of $w^0(\mathbf{x}^0, \mathbf{Rx}^0 + \mathbf{s})$. The expected value of a random variable distributed as a chi-square density function is the number of degrees of freedom associated with that chi-square distribution. If the numerical value of $E[w^0(\mathbf{x}^0, \mathbf{Rx}^0 + \mathbf{s})]$ is determined, this is the number of degrees of freedom associated with $w^0(\mathbf{x}^0, \mathbf{Rx}^0 + \mathbf{s})$. Fortunately, this is the only parameter that must be specified.

Example 10.3. A Production Model

Consider the production problem studied in Section 10.2:

$$\begin{aligned} \text{maximize} \quad & w = p_1 x_1 + p_2 x_2 + p_3 x_3 \\ \text{subject to} \quad & 4x_1 + x_2 + 2x_3 \le s_1 \\ & x_1 + x_3 \le s_2 \\ & x_i \ge 0 \quad i = 1, 2, 3 \end{aligned}$$

Suppose that the price of each product is a linear function of the quantity produced; $p_i = a_i + \alpha_i x_i$, $i = 1, 2, 3$, where a_i is a random variable and α_i is a constant. Let $\alpha_1 = -0.125$, $\alpha_2 = -1.00$, and $\alpha_3 = -0.25$.

Adding slack variables y_1 and y_2 transforms the constraints to be

$$y_1 = -4x_1 - x_2 - 2x_3 + s_1 \qquad y_2 = -x_1 - x_3 + s_2$$

The slack variables in the constraints represent the unused amounts of the two resources. Suppose that a unit of an unused resource can be sold as scrap or exchanged for partial credit on a new supply of the resource. Then each unit of y_1 or y_2 provides a positive contribution to the firm's revenue. Suppose that these per unit earnings are denoted by b_1 and b_2 respectively. Then the problem is

$$\begin{aligned} \text{maximize} \quad & w = p_1x_1 + p_2x_2 + p_3x_3 + b_1y_1 + b_2y_2 \\ \text{subject to} \quad & y_1 = -4x_1 - x_2 - 2x_3 + s_1 \\ & y_2 = -x_1 - x_3 + s_2 \\ & p_1 = a_1 - 0.125x_1 \qquad p_2 = a_2 - 1.0x_2 \\ & p_3 = a_3 - 0.25x_3 \end{aligned}$$

Then

$$\begin{aligned} w = {} & a_1x_1 + a_2x_2 + a_3x_3 - 0.5x_1^2 - 1.0x_2^2 - 0.5x_3^2 \\ & -4b_1x_1 - b_1x_2 - 2b_1x_3 + s_1b_1 - b_2x_1 - b_2x_3 + s_2b_2 \end{aligned}$$

and

$$\begin{aligned} \frac{\partial w}{\partial x_1} &= a_1 - 0.25x_1 - 4b_1 - b_2 = 0 \\ x_1 &= a_1 - 4b_1 - b_2 \\ \frac{\partial w}{\partial x_2} &= a_2 - 2x_2 - b_1 = 0 \\ x_2 &= 0.5a_2 - 0.5b_1 \\ \frac{\partial w}{\partial x_3} &= a_3 - 0.5x_3 - 2b_1 - b_2 = 0 \\ x_3 &= a_3 - 2b_1 - b_2 \\ w^0 &= 0.5a_1^2 + 0.25a_2^2 + 0.5a_3^2 + 10.25b_1^2 \\ &\quad + b_2^2 - 4a_1b_1 - a_1b_2 - 0.5a_2b_1 \\ &\quad - 2a_3b_1 - a_3b_2 + 6b_1b_2 + b_1s_1 + b_2s_2 \\ E(w^0) &= 12 \qquad V(w^0) = 24 \\ w^0 \sim h(w) &= w^{24.5}e^{-w/2}/2^{25.5}\Gamma(25.5) \qquad w > 0 \end{aligned}$$

Thus, the probability density function of w^0 has been derived using the solution to the nonstochastic programming problem as Sengupta and associates have suggested. It should be repeated that the simplicity of the analysis required in this example is a result of assuming that the elements of **a**, **b**, and **s** are standard normally independently distributed random variables and that the elements of the matrices **A**, **B**, **C**, and **R** are nonstochastic constants.

Example 10.4. A Linear Macroeconomic Model

Both the active and passive approaches to stochastic linear programming depend heavily upon the features of the complete description method for solving linear programming problems. Two important results from linear programming employed in solving stochastic linear problems are: (1) an optimal solution to a linear programming problem occurs at a corner point on the feasible space and (2) there are a finite number of these corner points to be investigated.

Recall the linear programming problem (10.20) and (10.21). Consider the macroeconomic programming problem where gross national product (GNP) is to be maximized in quarter t, Y_t, subject to a simplified version of Mahalanobis's two-sector planning model [1955]. Period t is presumed to be at least two quarters in the future from when this problem is solved. The two sectors are the consumption goods sector c and the investment goods sector i. The constraints provide for investment in time t, I_t and consumption in time t, C_t to be linear functions of the respective levels of the variables for each sector during the previous period $t - 1$. Thus

$$I_t = \lambda_{1i} I_{1t-1} + \lambda_{2i} I_{2t-1} \tag{10.22}$$

$$C_t = \lambda_{1c} C_{1t-1} + \lambda_{2c} C_{2t-1} \tag{10.23}$$

The intent is to determine the lambdas that reflect the marginal values for the changes in the total levels of investment and consumption with respect to a change in the level of investment or consumption in the previous period in the particular sector:

$$\lambda_{ji} = \frac{\partial I_t}{\partial I_{jt-1}} \qquad j = 1, 2$$

$$\lambda_{jc} = \frac{\partial C_t}{\partial C_{jt-1}} \qquad j = 1, 2$$

It will be assumed that there are some boundaries on the λ's and that GNP is desired to be the same in periods t and $t - 1$. (The plan is presumed to be formulated prior to period $t - 1$, and time is in quarters.) The lambdas are constrained as

$$\lambda_{1i} \geq s_1 \tag{10.24}$$

$$\lambda_{2i} \geq s_2 \tag{10.25}$$

$$\lambda_{1c} \leq s_3 \tag{10.26}$$

$$\lambda_{2c} \leq s_4 \tag{10.27}$$

$$\lambda_{1i} + \lambda_{1c} = 1 \tag{10.28}$$

$$\lambda_{2i} + \lambda_{2c} = 1 \tag{10.29}$$

$$\lambda_{ji} \geq 0, \lambda_{jc} \geq 0 \qquad j = 1, 2 \tag{10.30}$$

The linear programming problem is

$$\text{maximize} \quad Y_t = C_t + I_t \tag{10.31}$$

subject to the model described by (10.22)–(10.30). A number of substitutions can be performed to simplify the problem. Equations (10.22) and (10.23) can be substituted into (10.31) so that the goal is

$$\text{maximize} \quad Y_t = \lambda_{1i} I_{1t-1} + \lambda_{2i} I_{2t-1} + \lambda_{1c} C_{1t-1} + \lambda_{2c} C_{2t-1} \tag{10.32}$$

Then λ_{1i} and λ_{2i} can be replaced by $1 - \lambda_{1c}$ and $1 - \lambda_{2c}$ respectively in (10.32) and (10.24)–(10.27). The resulting problem is

$$\begin{aligned} \text{maximize} \quad & Y_t = \lambda_{1c}(C_{1t-1} - I_{1t-1}) + \lambda_{2c}(C_{2t-1} - I_{2t-1}) + I_{t-1} \\ \text{subject to} \quad & 1 - \lambda_{1c} \geq s_1 \\ & 1 - \lambda_{2c} \geq s_2 \\ & \lambda_{1c} \leq s_3 \qquad \lambda_{2c} \leq s_4 \\ & \lambda_{jc} \geq 0 \qquad j = 1, 2 \end{aligned}$$

Table 10.6. Probability Density Functions

Case	λ_{1c}	λ_{2c}	y_1	y_2	y_3	y_4	w
1	0	0	$1 - s_1$	$1 - s_2$	s_3	s_4	0
2	0	–	0	–	–	–	–
3	0	$1 - s_2$	$1 - s_1$	0	s_3	$s_2 + s_4 - 1$	$a_2(1 - s_2)$
4	0	–	–	–	0	–	–
5	0	s_4	$1 - s_1$	$1 - s_2 - s_4$	s_3	0	$a_2 s_4$
6	$1 - s_1$	0	0	$1 - s_2$	$s_1 + s_3 - 1$	s_4	$a_1(1 - s_1)$
7	–	0	–	0	–	–	–
8	s_3	0	$1 - s_1 - s_3$	$1 - s_2$	0	s_4	$a_1 s_3$
9	–	0	–	–	–	0	–
10	$1 - s_1$	$1 - s_2$	0	0	$s_1 + s_3 - 1$	$s_2 + s_4 - 1$	$a_1(1 - s_1) + a_2(1 - s_2)$
11	–	–	0	–	0	–	–
12	$1 - s_1$	s_4	0	$1 - s_2 - s_4$	$s_1 + s_3 - 1$	0	$a_1(1 - s_1) + a_2 s_4$
13	s_3	$1 - s_2$	$1 - s_1 - s_3$	0	0	$s_2 + s_4 - 1$	$a_1 s_3 + a_2(1 - s_2)$
14	–	–	–	0	–	0	–
15	s_3	s_4	$1 - s_1 - s_3$	$1 - s_2 - s_4$	0	0	$a_1 s_3 + a_2 s_4$

Note: $E(a_i) = E(s_j) = 0, E(1 - s_j) = 1$
$\text{Cov}(a_i s_j) = 0, E(a_i s_j) = E(a_i) E(s_j) = 0$
$E[a_i(1 - s_j)] = E(a_i) - E(a_i s_j) = 0$

Introducing four slack variables and simplifying, the constraints become

$$\lambda_{1c} + y_1 = 1 - s_1 \tag{10.33}$$

$$\lambda_{2c} + y_2 = 1 - s_2 \tag{10.34}$$

$$\lambda_{1c} + y_3 = s_3 \tag{10.35}$$

$$\lambda_{2c} + y_4 = s_4 \tag{10.36}$$

Suppose that $I_{t-1}(I_{t-1} = I_{1t-1} + I_{2t-1})$ is exogenous to this model. Then maximizing $w = Y_t - I_{t-1}$ maximizes Y_t. Furthermore, let a_1, a_2, s_1, s_2, s_3, and s_4 be random variables, where

$$a_1 = C_{1t-1} - I_{1t-1} \qquad a_2 = C_{2t-1} - I_{2t-1}$$

The s_i's are boundary points. The function to be maximized is $w = a_1\lambda_{1c} + a_2\lambda_{2c}$, when restricted by (10.33)–(10.36).

Since six variables and four constraints remain, there are at most 15 possibilities to be considered. Six of these cases are redundant; this is indicated by the use of a dash in Table 10.6. The other nine solutions are provided in this table. There are also conditions that could require certain cases to be ignored, depending on the values of s_1, s_2, s_3, and s_4. These conditions are provided in Table 10.7.

Table 10.7. Conditions Requiring Cases to Be Ignored

Condition	Cases to be ignored
$s_1 + s_3 < 1$	6, 10, 12
$s_1 + s_3 > 1$	8, 13, 15
$s_2 + s_4 < 1$	3, 10, 13
$s_2 + s_4 > 1$	5, 12, 15

Assuming density functions for the a's and s's, we can determine the probability density function of w for each case in Table 10.6. The use of characteristic functions is avoided if the a's and s's are assumed to be independent standard normal random variables. If s is a standard normal random variable, $1 - s$ is normally distributed with a mean of one and variance of one. Thus, w has a chi-square distribution for each case. For each of cases 3, 5, 6, 8, 10, 12, 13, and 15 the distribution has one degree of freedom.

This completes the second case to illustrate the passive approach to stochastic programming. The linear problem has afforded the opportunity to illustrate that the boundary points of the variables are sometimes random variables.

Before continuing into Section 10.4, it should be pointed out that there is a method for finding the probability density function of $w(\mathbf{x}, \mathbf{y})$ when the narrow assumptions made here are not valid. Suppose that many of the elements of **a**, **b**, **s**, **A**, **B**, **C**, and **R** are random variables and that a sample of size N is collected for each random variable. For each observation, optimal values for $\mathbf{x}, \mathbf{y}$, and w (labeled $\mathbf{x}^0, \mathbf{y}^0$, and w^0 respectively) are determined; then N values of w^0 are generated. The Pearson method of moments can then be applied to these N observations, and the relevant type of density function for w^0 can be selected. This is valid for both the linear and quadratic programming problems and is also employed in the active approach to stochastic programming that will be described next.

10.4 ACTIVE APPROACH TO STOCHASTIC PROGRAMMING

The aim when using the active approach to stochastic programming is the same as that of the passive approach. The statistical distribution of the objective function is to be derived; however, the technique is not the same. The cases to be studied in this section are the same as those considered in Section 10.3. These are the quadratic programming problem, whose objective function is (10.1), and the linear programming problem, whose objective function is (10.18).

For the active approach to stochastic programming, the set of constraints is $-\mathbf{RX} \leq \mathbf{SU}$,

where $\mathbf{R}$ = the same $n \times m$ matrix of coefficients assumed before
$\mathbf{X}$ = a diagonal $m \times m$ matrix such that $X_{jj} = x_j$
$\mathbf{S}$ = a diagonal $n \times n$ matrix such that $S_{ii} = s_i$, and
$\mathbf{U}$ = an $n \times m$ matrix of elements u_{ij} such that $0 \leq u_{ij} \leq 1$, and $\sum_{j=1}^{m} u_{ij} = 1$ for $i = 1, \ldots, n$

The u_{ij}'s are controlled exogenously; u_{ij} is the percentage of the ith resource allocated to the jth activity. The amount of that resource to be allocated to the jth activity is $s_i u_{ij}$.

In the active approach, the derivation of the probability density function of the objective function depends on the distributions of the random variables as well as the selected allocation matrix (the elements of **U**). The constraints being considered here are

$$-r_{ij}x_j \leq s_i u_{ij} \qquad i = 1, \ldots, n; j = 1, \ldots, m$$

Therefore, nm slack variables are required. The vectors **b** and **y** are now $nm \times 1$, and the matrix **B** is $nm \times nm$. The matrix **C** is $m \times nm$. Subscripts and summation signs will be used often in this section. This complicates the form of a few equations but enables us to keep track of the numerous slack variables.

A different slack variable must be added to each constraint, even though the same x_j appears in various constraints. The constraints could be written as

$$-r_{ij} + y_p = s_i u_{ij} \qquad i = 1, \ldots, n; j = 1, \ldots, m; p = 1, \ldots, nm$$

However, the subscript p can be eliminated because $p = j + (i - 1)m$. Thus the constraints are written as

$$-r_{ij}x_j + y_{j+(i-1)m} = s_i u_{ij} \qquad i = 1, \ldots, n; j = 1, \ldots, m$$

The Quadratic Model

The quadratic programming problem is

$$\text{optimize} \quad w(x,y) = \sum_{j=1}^{m} a_j x_j + \sum_{p=1}^{nm} b_p y_p + \frac{1}{2} \sum_{j=1}^{m} \sum_{h=1}^{m} a_{jh} x_j x_h + \frac{1}{2} \sum_{p=1}^{nm} \sum_{q=1}^{nm} b_{pq} y_p y_q + \sum_{j=1}^{m} \sum_{p=1}^{nm} c_{jp} x_j y_p \tag{10.37}$$

$$\text{subject to} \quad y_{j+(i-1)m} = r_{ij} x_j + s_i u_{ij} \qquad i = 1, \ldots, n; j = 1, \ldots, m \tag{10.38}$$

Note that two new subscripts are implemented in this formulation. Both p and q range from 1 to nm to handle the elements of **b**, **y**, **B**, and the columns of **C**.

To determine an optimum, the usual substitution is made so that the value of the objective function w would be determined by the values in **x** and **U**. Then w is differentiated with respect to each element of **x**. The result is

$$\frac{\partial w}{\partial x_j} = a_j + \sum_{i=1}^{n} b_{j+(i-1)m} r_{ij} + \frac{1}{2} \sum_{h=1}^{m} (a_{hj} + a_{jh}) x_h$$
$$+ \frac{1}{2} \sum_{p=1}^{nm} \sum_{i=1}^{n} [b_{j+(i-1)m,p} + b_{p,j+(i-1)m}] r_{ij} y_{j+(i-1)m}$$
$$+ \sum_{p=1}^{nm} c_{jp} y_p + \sum_{h=1}^{m} \sum_{i=1}^{n} c_{h,j+(i-1)m} r_{ij} x_h \qquad j = 1, \ldots, m$$

or

$$\frac{\partial w}{\partial x_j} = a_j + \sum_{i=1}^{n} b_{j+(i-1)m} r_{ij} + \sum_{h=1}^{m} a_{jh} x_h$$
$$+ \sum_{p=1}^{nm} \sum_{j=1}^{n} b_{j+(i-1)m,p} r_{ij} y_{j+(i-1)m} + \sum_{p=1}^{nm} c_{jp} y_p$$
$$+ \sum_{h=1}^{m} \sum_{i=1}^{n} c_{h,j+(i-1)m} r_{ij} x_h \qquad j = 1, \ldots, m \qquad (10.39)$$

since **A** and **B** are symmetric matrices. Given a value for x_j, $y_{j+(i-1)m}$ is only a function of r_{ij}, s_i, and u_{ij} for $i = 1, \ldots, n$ and $j = 1, \ldots, m$. Since there are m such derivations to be completed, and thus m x's, (10.39) represents m linear equations in m unknowns for each set of u_{ij}'s to be considered.

Setting $\partial w/\partial x_j = 0$, for $j = 1, \ldots, m$ gives $\mathbf{x} = -\mathbf{G}^{-1}\mathbf{F}$

where $\mathbf{G}$ = an $m \times m$ matrix with elements g_{jh}

$$g_{jh} = a_{jh} + \sum_{i=1}^{n} [c_{j,h(i-1)m} r_{ih} + c_{h,j+(i-1)m} r_{ij}] \quad h = 1, \ldots, m; j = 1, \ldots, m$$

$\mathbf{G}^{-1}$ = the inverse matrix of $\mathbf{G}$

g_{jh}^{-1} = the element in row j and column h of the matrix $\mathbf{G}^{-1}$

$\mathbf{F}$ = an $m \times 1$ vector with elements f_j, $j = 1, \ldots, m$

such that

$$f_j = a_j + \sum_{i=1}^{n} b_{j+(i-1)m} r_{ij} + \sum_{i=1}^{n} r_{ij} s_i u_{ij} \left[\sum_{p=1}^{nm} b_{j+(i-1)m,p} \right]$$
$$+ \sum_{h=1}^{m} \sum_{i=1}^{n} c_{j,h+(i-1)m} s_i u_{ih} \qquad j = 1, \ldots, m$$

Thus

$$x_h^0 = -\sum_{j=1}^{m} g_{jh}^{-1} f_j \qquad h = 1, \ldots, m$$

and

$$y_{j+(i-1)m}^0 = r_{ij}\left(\sum_{h=1}^{m} g_{jh}^{-1} f_h\right) + s_i u_{ij} \qquad i = 1, \ldots, n; j = 1, \ldots, m$$

Note that g_{jh}^{-1} is a function of the elements in the matrices **A**, **C**, and **R**, and f_j is a function of the matrices **B**, **C**, **R**, and **U** and the vectors **a**, **b**, and **s**.

Now consider the statistical distribution of $w(\mathbf{x},\mathbf{y})$. Again it is assumed that all the elements of the matrices **A**, **B**, **C**, and **R** are non-stochastic. If the elements of the vectors **a**, **b**, and **s** are standard independent normally distributed random variables, $w(\mathbf{x},\mathbf{y})$ has a chi-square distribution.

This is the same conclusion shown for the passive approach to stochastic programming when the objective function is a quadratic function. The difference in this instance is the inclusion of the exogenously controlled allocation parameters u_{ij}, $i = 1, \ldots, n$, $j = 1, \ldots, m$.

Example 10.5

Consider the following example that parallels the production model studied in the previous section. Let $n = 2$, $m = 3$.

$$\mathbf{R} = \begin{bmatrix} 0.5 & 1.0 & 0.0 \\ 1.0 & 0.0 & 1.0 \end{bmatrix}$$

$$\mathbf{A} = \begin{bmatrix} -0.5 & 0 & 0 \\ 0 & -1.0 & 0 \\ 0 & 0 & -1.0 \end{bmatrix} \qquad \mathbf{B} = \begin{bmatrix} -1 & 0 & 0 & 0 & 0 & 0 \\ 0 & -2 & 0 & 0 & 0 & 0 \\ 0 & 0 & -3 & 0 & 0 & 0 \\ 0 & 0 & 0 & -2 & 0 & 0 \\ 0 & 0 & 0 & 0 & -1 & 0 \\ 0 & 0 & 0 & 0 & 0 & -2 \end{bmatrix}$$

C is a 3×6 null matrix. Then

$$\mathbf{G}^{-1} = \begin{bmatrix} -2 & 0 & 0 \\ 0 & -1 & 0 \\ 0 & 0 & -1 \end{bmatrix}$$

and

$$
\begin{aligned}
f_1 &= a_1 + 0.5b_1 + b_4 + 0.5s_1u_{11} \\
f_2 &= a_2 + b_2 + 2s_1u_{12} \\
f_3 &= a_3 + b_6 \\
x_1 &= 2a_1 + b_1 + 2b_4 + s_1u_{11} \\
x_2 &= a_2 + b_2 + 2s_1u_{12} \\
x_3 &= a_3 + b_6 \\
y_1 &= a_1 + 0.5b_1 + b_4 + 1.5s_1u_{11} \\
y_2 &= a_2 + b_2 + 3s_1u_{12} \\
y_3 &= s_1u_{13} \\
y_4 &= 2a_1 + b_1 + 2b_4 + s_1u_{11} + s_2u_{21} \\
y_5 &= s_2u_{22} \\
y_6 &= a_3 + b_6 + s_2u_{23} \\
\sum_{j=1}^{3} u_{ij} &= 1 \qquad i = 1, 2 \\
w^0 &= 3.5a_1^2 + 0.5a_2^2 + 0.5a_3^2 + 2b_1^2 + 0.5b_2^2 + 3.5b_4^2 \\
&\quad + 0.5b_6^2 + 3.5a_1b_1 + 7a_1b_4 + a_2b_2 + a_3b_6 \\
&\quad + 3.5b_1b_4 + a_1s_1(5.5u_{11}) + a_1s_2(4u_{21}) \\
&\quad + a_2s_1(4u_{12}) + a_2s_2(2u_{12}) + a_3s_2(2u_{23}) \\
&\quad + b_1s_1(0.5u_{11}) + b_1s_2(u_{21}) + b_2s_1(3u_{12}) \\
&\quad - b_2s_2(2u_{12}) - b_3s_1(u_{13}) + b_4s_1(5.5u_{11}) \\
&\quad + b_4s_2(4u_{21} - u_{12}) - b_5s_2(u_{22}) + b_6s_2(u_{23}) \\
&\quad + s_1s_2(2u_{11}u_{21}) + s_1^2(2.375u_{11}^2 + 10u_{12}^2 \\
&\quad + 1.5u_{13}^2) + s_2^2(u_{21}^2 + .5u_{22}^2 + u_{23}^2)
\end{aligned}
$$

For each set of elements of **U**, w^0 has a chi-square distribution, so the probability density function of w^0 is determined. However, this can be concluded only because of the specific assumption that the random variables are independent and have standard normal distributions. Furthermore, the matrices **A**, **B**, **C**, and **R** are presumed to contain no stochastic elements; otherwise **G** would include stochastic elements, and $\mathbf{G}^{-1}$ could not be specified.

The more general case would be to let random variables appear where they may and collect M sample observations on each. Then for each specific set of u_{ij}'s N values are generated for w^0. If M different sets of u_{ij}'s are to be considered, there are NM values of w^0 and NM observations to which the Pearson method of distribution determination is applied.

The active approach to the stochastic linear programming problem has been delineated in some detail by Sengupta,

Tintner, and associates in references listed in the Bibliography. The problem is

$$\begin{aligned}\text{optimize} \quad & w(\mathbf{x}, \mathbf{y}) = \mathbf{a}'\mathbf{x} + \mathbf{b}'\mathbf{y} \\ \text{subject to} \quad & -\mathbf{RX} \leq \mathbf{SU}\end{aligned}$$

where all the matrices and vectors are those that have been defined for the quadratic problem in this section, (10.37), and (10.38). Recall that there are nm constraints and $m + nm$ variables (m x's and nm y's). Thus $(m + nm)!/nm!m!$ possible cases are to be considered via the complete description method of linear programming. Many of these alternatives provide solutions that are not feasible because one or more of the x's or y's are negative. The point is simply illustrated for the case where $n = 2$ and $m = 3$. Then

$$\begin{aligned}-r_{11}x_1 + y_1 &= s_1u_{11} & \quad -r_{21}x_1 + y_4 &= s_1u_{21} \\ -r_{12}x_2 + y_2 &= s_2u_{12} & \quad -r_{22}x_2 + y_5 &= s_2u_{22} \\ -r_{13}x_3 + y_3 &= s_3u_{13} & \quad -r_{23}x_3 + y_6 &= s_3u_{23}\end{aligned}$$

since y_j and y_{j+m} are functions of s_j, y_j and y_{j+m} are linearly related. There would be 84 cases to be investigated in this instance.

Example 10.6

To illustrate the results, the simplest case is considered here. Let $n = 1$ and $m = 2$. The problem is

$$\begin{aligned}\text{optimize} \quad & w(\mathbf{x}, \mathbf{y}) = a_1x_1 + a_2x_2 + b_1y_1 + b_2y_2 \\ \text{subject to} \quad & -r_{11}x_1 + y_1 = s_1u_{11} \\ & -r_{12}x_2 + y_2 = s_2u_{12}\end{aligned}$$

The cases generated using the complete description method are provided in Table 10.8. Again assume that **R** is a matrix of constants. If the elements of **a**, **b**, and **s** are independent standard normal random variables, w^0 is distributed as a chi-square density function for each set of u_{ij}'s.

The alternative to such a narrow group of assumptions is to collect a sample of N observations on each random variable. For each of M alternative U's there are N values of w^0,

Table 10.8. Cases Using Complete Description Method

Case	x_1	x_2	y_1	y_2	w
1	0	0	s_1u_{11}	s_2u_{12}	$b_1s_1u_{11} + b_2s_2u_{12}$
2	0	–	0	–	–
3	0	$\dfrac{-s_2u_{12}}{r_{12}}$	s_1u_{11}	0	$\dfrac{-a_2s_2u_{12}}{r_{12}} + b_1s_1u_{11}$
4	$\dfrac{-s_1u_{11}}{r_{11}}$	0	0	s_2u_{12}	$\dfrac{-a_1s_1u_{11}}{r_{11}} + b_2s_2u_{12}$
5	–	0	–	0	–
6	$\dfrac{-s_1u_{11}}{r_{11}}$	$\dfrac{-s_2u_{12}}{r_{12}}$	0	0	$\dfrac{-a_1u_{11}}{r_{11}} - \dfrac{s_2u_{12}}{r_{12}}$

providing NM in all. The Pearson method has been applied to such problems by Sengupta and Tintner in numerous references listed in the Bibliography. Their computations will not be duplicated here.

In many of these same references numerous theorems on stochastic linear programming are proved. Most of the theorems relate the numerical values of the objective functions for the active and passive approaches. The key result that dominates the conclusions for the linear problems follows.

Consider a particular point in the parameter space of the random variables and denote it by k. At k let $w_p^{(k)}$ and $w_a^{(k)}$ be the optimal values of $w(\mathbf{x}, \mathbf{y})$ under the passive and active approaches respectively, and let $w^{(k)}$ be any other value of $w(\mathbf{x}, \mathbf{y})$. Then if $w(\mathbf{x}, \mathbf{y})$ is maximized, $w^{(k)} \leq w_a^{(k)} \leq w_p^{(k)}$ and if $w(x, y)$ is minimized, $w^{(k)} \geq w_a^{(k)} \geq w_p^{(k)}$. Furthermore, expected values do not change the inequalities.

These results are what one might expect without engaging in the complex mathematics required to prove them. The main differences between the active and passive approaches is the implementation of the allocation matrix U in the active approach. The best choice of a set of u_{ij}'s might let $w_a^{(k)} = w_p^{(k)}$, and any less desirable set of u_{ij}'s would cause $w_a^{(k)} < w_p^{(k)}$. Furthermore, since $w^{(k)}$ can be any value of $w(\mathbf{x}, \mathbf{y})$, it cannot be made more extreme than an optimum.

10.5 CHANCE-CONSTRAINED PROGRAMMING AND DETERMINISTIC EQUIVALENTS

Chance-constrained programming has been introduced into the stochastic programming literature mainly through the expositions of Charnes and Cooper. They suggest the E, V, and P models described in Section 10.1. In each instance the stochastic programming problem is transformed into a "deterministic" (nonstochastic) convex programming problem. Only the E model will be studied here in detail. The V and P models are interesting alternatives, but they have not led to many business and economics applications.

In the E model the expected value of the objective function $E[w(\mathbf{x}, \mathbf{y})]$ is to be maximized. For the V model the intent is to minimize $E[w(\mathbf{x}, \mathbf{y}) - w^*(\mathbf{x}^*, \mathbf{y}^*)]^2$; this is the variance of the value of the objective function if the optimal value and expected value of the objective function are the same. According to Charnes and Cooper [1963, p. 30], a generalized mean square error is minimized when the V model is solved and $w^*(\mathbf{x}^*, \mathbf{y}^*)$ is a desired or target value of $w(\mathbf{x}, \mathbf{y})$. The portfolio selection model studied by Markowitz [1959] is cited by Charnes and Cooper [1963] as a special case of the V model, when $w(\mathbf{x}, \mathbf{y})$ is linear. Simon's principle of satisficing [1957, Ch. 5] motivated Charnes and Cooper to consider the P model. For this case, the probability that $w(\mathbf{x}, \mathbf{y})$ is greater than or equal to a "satisfying level" $w^*(\mathbf{x}^*, \mathbf{y}^*)$ is to be maximized.

It should be noted that Roy [1952] has previously described a problem similar to the P model. Roy attempts to find a set of activity levels that would minimize the probability of the occurrence of a disaster. He assumes the first two moments of the joint density function of the variable activities to be known. The analysis proceeds by identifying activity combinations that avoid a worse outcome than d, the disaster. Next Roy establishes the activity levels that provide the smallest chance that d actually occurs.

Simple Chance Constraints

The cases for both the linear and quadratic objective functions are studied here. These have been specified by (10.18) and (10.1) respectively. When delineating certainty equivalents in this chapter, $E[w(\mathbf{x}, \mathbf{y})]$ has been optimized. The difference in the chance-constrained model appears in the constraints. Recall that the constraints can be considered as $-\mathbf{Rx} \leq \mathbf{s}$ or $-\sum_{j=1}^{m} r_{ij}x_j \leq s_i$, $i = 1, \ldots, n$, when the elements of the vector $\mathbf{y}$ are assumed to be slack variables.

In place of these constraints, consider

$$Pr\left(-\sum_{j=1}^{m} r_{ij}x_j \leq s_i\right) \geq \alpha_i \qquad i = 1, \ldots, n \tag{10.40}$$

or

$$Pr(-\mathbf{Rx} \leq \mathbf{s}) \geq \boldsymbol{\alpha} \tag{10.41}$$

to restrain the objective function. Each of the s_i's is presumed to be a random variable having a normal distribution with a mean of $\bar{s}_i$ and a standard deviation of σ_i, where

$$\sigma_i = [E(s_i - \bar{s}_i)^2]^{1/2} > 0 \qquad i = 1, \ldots, n$$

The values of the r_{ij}'s are presumed to be constants. The value of α_i is the probability that the inequality of the ith constraint,

$$-\sum_{j=1}^{m} r_{ij}x_j \leq s_i \tag{10.42}$$

is not violated. The chance that $-\sum_{j=1}^{n} r_{ij}x_j > s_i$ must be less than or equal to $1 - \alpha_i$. Naturally, α_i is expected to be close to one. It will be assumed that $\alpha_i > 0.50$ for each constraint, $i = 1, \ldots, n$.

Equation (10.40) is transformed by subtracting $\bar{s}_i$ from each side of (10.42) and dividing the result by σ_i.[3] Then

$$\left(-\sum_{j=1}^{m} r_{ij}x_j - \bar{s}_i\right)\Big/\sigma_i \leq (s_i - \bar{s}_i)/\sigma_i \qquad i = 1, \ldots, n$$

and

$$Pr\left[\left(-\sum_{j=1}^{m} r_{ij}x_j - \bar{s}_i\right)\Big/\sigma_i \leq (s_i - \bar{s}_i)/\sigma_i\right] \geq \alpha_i \qquad i = 1, \ldots, n$$

Let $z_i = (s_i - \bar{s}_i)/\sigma_i$, $i = 1, \ldots, n$, and note that z_i is a standard normal random variable. Then

$$Pr\left[\left(-\sum_{j=1}^{m} r_{ij}x_j - \bar{s}_i\right)\Big/\sigma_i \leq z_i\right] \geq \alpha_i \qquad i = 1, \ldots, n$$

and (10.40) can be replaced by

$$F\left[\left(-\sum_{j=1}^{m} r_{ij}x_j - \bar{s}_i\right)\Big/\sigma_i\right] \geq \alpha_i \qquad i = 1, \ldots, n \tag{10.43}$$

where F is the cumulative density function for the standard normal

3. This exposition follows that of Hillier and Lieberman [1967, Ch. 15].

variable, z_i. Given a value for α_i, the value of z_i (let it be denoted by z_i^*) can be determined using a table of the cumulative standard normal distribution.

$$F(z_i^*) = \int_{z_i^*}^{\infty} (2\pi)^{-1/2} e^{-t^2/2}\, dt \qquad i = 1,\ldots,n$$

so that (10.43) is specified. The inverse function of F, F^{-1}, is used to form

$$\left(-\sum_{j=1}^{m} r_{ij} x_j - \bar{s}_i\right) \Big/ \sigma_i \leq F^{-1}(\alpha_i) \qquad i = 1,\ldots,n$$

or

$$-\sum_{j=1}^{m} r_{ij} x_j - \bar{s}_i \leq \sigma_i F^{-1}(\alpha_i) \qquad i = 1,\ldots,n$$

since $\sigma_i > 0$.

Let $F^{-1}(\alpha_i) = -K_i < 0$, $i = 1,\ldots,n$. Since $\alpha_i > 0.50$ and all this area of a standard normal variable occurs in the negative quadrant of the curve, $F^{-1}(\alpha_i)$ is negative. Substituting $-K_i$ for $F^{-1}(\alpha_i)$ leaves

$$-\sum_{j=1}^{m} r_{ij} x_j \leq \bar{s}_i - \sigma_i K_i \qquad i = 1,\ldots,n$$

Let $s_i^* = \bar{s}_i - \sigma_i K_i$, $i = 1,\ldots,n$, and $\mathbf{s}^*$ be the $n \times 1$ vector whose ith element s_i^*. Then $-\mathbf{Rx} \leq \mathbf{s}^*$ represents a set of n inequalities in which there are no stochastic elements. Adding the vector of slack variables $\mathbf{y}$ leaves

$$\mathbf{y} = \mathbf{Rx} + \mathbf{s}^* \tag{10.44}$$

which is a minor variation of $\mathbf{y} = \mathbf{Rx} + \mathbf{s}$.

For the E model the problem is to optimize $E[w(\mathbf{x}, \mathbf{y})]$ subject to the constraints of (10.44). Neither the linear nor the quadratic problem requires further consideration here. Each of these cases has been considered in Section 10.2. The result for the E model in chance-constrained programming, with (10.40) as the constraints, would be even simpler than the models studied in Section 10.2 since the random variables in $\mathbf{s}$ have already been replaced by $\mathbf{s}^*$, a vector of constants.

Example 10.7

Van de Panne and Popp [1963] have solved the deterministic equivalent of a simple linear chance-constrained programming model:

$$\begin{aligned}
&\text{minimize} \quad \psi = 24.55X_1 + 26.75X_2 + 39.00X_3 + 40.50X_4 \\
&\text{subject to} \\
&\qquad X_1 + X_2 + X_3 + X_4 = 1 \\
&\qquad 2.3X_1 + 5.6X_2 + 11.4X_3 + 1.3X_4 \geq 5 \\
&\qquad P(a_1X_1 + a_2X_2 + a_3X_3 + a_4X_4 \geq 21) \geq 0.95 \\
&\qquad X_j \geq 0 \quad j = 1, 2, 3, 4
\end{aligned}$$

Each a_j is an independent normally distributed random variable, $j = 1, 2, 3, 4$, and

$$\begin{aligned}
E(a_1) &= 12.0 & V(a_1) &= 0.28 \\
E(a_2) &= 11.9 & V(a_2) &= 0.19 \\
E(a_3) &= 41.8 & V(a_3) &= 20.5 \\
E(a_4) &= 52.1 & V(a_4) &= 0.62 \\
\mathrm{Cov}(a_i, a_j) &= 0, & i,j &= 1, 2, 3, 4;\ i \neq j
\end{aligned}$$

To solve the problem, the chance constraint is reformulated into a deterministic inequality using the assumptions that the a_j's are independent normally distributed random variables and the given means and standard deviations of the a_j's.

Let $q = a_1X_1 + a_2X_2 + a_3X_3 + a_4X_4$, then

$$\begin{aligned}
E(q) &= 12X_1 + 11.9X_2 + 41.8X_3 + 52.1X_4 \\
V(q) &= 0.28X_1^2 + 0.19X_2^2 + 20.5X_3^2 + 0.62X_4^2
\end{aligned}$$

If q is a normally distributed random variable with a mean of $E(q)$ and a standard deviation of $[V(q)]^{1/2}$, $P(q \geq q^*) \geq 0.95$ can be written as $E(q) - 1.64[V(q)]^{1/2} \geq 21$.

The chance constraint can now be replaced by

$$E(q) - 1.64[V(q)]^{1/2} \geq 21$$

or

$$\begin{aligned}
&12X_1 + 11.9X_2 + 41.8X_3 + 52.1X_4 \\
&\qquad - 1.64(0.28X_1^2 + 0.19X_2^2 + 20.5X_3^2 + 0.62X_4^2)^{1/2} \geq 21
\end{aligned}$$

The deterministic equivalent to the chance-constrained programming model is

$$\begin{array}{ll} \text{minimize} & \psi = 24.55X_1 + 26.75X_2 + 39.00X_3 + 40.50X_4 \\ \text{subject to} & X_1 + X_2 + X_3 + X_4 = 1 \\ & 2.3X_1 + 5.6X_2 + 11.1X_3 + 1.3X_4 \geq 5 \\ & 12.0X_1 + 11.9X_2 + 41.8X_3 + 52.1X_4 \\ & -1.64(0.28X_1^2 + 0.19X_2^2 + 20.5X_3^2 + 0.62X_4^2)^{1/2} \geq 21 \\ & X_j \geq 0 \quad j = 1, 2, 3, 4 \end{array}$$

Van de Panne and Popp have solved this nonlinear programming model. They find that $X_1^0 = 0.64$, $X_2^0 = 0$, $X_3^0 = 0.31$, $X_4^0 = 0.05$, and $\psi^0 = 29.89$.

Complete Chance Constraints

The set of constraints assumed by Charnes and Cooper [1963] is not only (10.41); they also assume that

$$\mathbf{x} = \mathbf{D}\mathbf{s} \tag{10.45}$$

where $\mathbf{D}$ is an $m \times n$ matrix of variables to be determined. Substituting (10.45) into (10.41) leaves $\Pr(-\mathbf{R}\mathbf{D}\mathbf{s} \leq \mathbf{s}) \geq \boldsymbol{\alpha}$. Let the ith row of the matrix $-\mathbf{R}$ be denoted by $-\mathbf{R}_i$. Then

$$Pr(-\mathbf{R}_i\mathbf{D}\mathbf{s} \leq s_i) \geq \alpha_i \qquad i = 1, \ldots, n$$
$$Pr[-(\mathbf{R}_i\mathbf{D}\mathbf{s} - \mathbf{R}_i\mathbf{D}\mathbf{u}) - \mathbf{R}_i\mathbf{D}\mathbf{u} \leq (s_i - u_i) + u_i] \geq \alpha_i \qquad i = 1, \ldots, n$$
$$Pr[-\mathbf{R}_i\mathbf{D}(\mathbf{s} - \mathbf{u}) - (s_i - u_i) \leq \mathbf{R}_i\mathbf{D}\mathbf{u} + u_i] \geq \alpha_i \qquad i = 1, \ldots, n$$
$$Pr[-\mathbf{R}_i\mathbf{D}\hat{\mathbf{s}} - \hat{s}_i \leq u_i + \mathbf{R}_i\mathbf{D}\mathbf{u}] \geq \alpha_i \qquad i = 1, \ldots, n$$

where $u_i = E(s_i)$
$\mathbf{u} = E(\mathbf{s})$
$\hat{s}_i = s_i - E(s_i)$
$\hat{\mathbf{s}} = \mathbf{s} - E(\mathbf{s})$

The approach that follows is analogous to that implemented earlier in this section. Therefore, little descriptive information is provided. Let

$$\sigma_i = [E(-\mathbf{R}_i\mathbf{D}\hat{\mathbf{s}} - \hat{s}_i)^2]^{1/2} = [E(\mathbf{R}_i\mathbf{D}\hat{\mathbf{s}} + \hat{s}_i)^2]^{1/2} > 0 \qquad i = 1, \ldots, n$$

Then

$$Pr[(-\mathbf{R}_i\mathbf{d}\hat{\mathbf{s}} - \hat{s}_i)/\sigma_i \leq (u_i + \mathbf{R}_i\mathbf{D}\mathbf{u})/\sigma_i] \geq \alpha_i \qquad i = 1, \ldots, n$$

Assuming

$$z_i = (\mathbf{R_i D\hat{s}} + \hat{s}_i)/\sigma_i \qquad i = 1,\ldots,n$$

z_i is a standard normal random variable, and

$$Pr[(-u_i - \mathbf{R_i Du})/\sigma_i \le z_i] \ge \alpha_i \qquad i = 1,\ldots,n$$

Thus

$$F[(-u_i - \mathbf{R_i Du})/\sigma_i] \ge \alpha_i \qquad i = 1,\ldots,n$$
$$(-u_i - \mathbf{R_i Du})/\sigma_i \le F^{-1}(\alpha_i) = -K_i < 0 \qquad i = 1,\ldots,n$$

where F is the cumulative standard normal distribution, and $K_i > 0$. Simplifying gives

$$-u_i - \mathbf{R_i Du} \le \sigma_i K_i \qquad i = 1,\ldots,n$$

or

$$-u_i - \mathbf{R_i Du} \le -K_i[E(\mathbf{R_i D\hat{s}} + \hat{s}_i)^2]^{1/2} \le 0 \qquad i = 1,\ldots,n$$

Let v_i be a new variable such that

$$-u_i - \mathbf{R_i Du} \le -v_i \le -K_i[E(\mathbf{R_i D\hat{s}} + \hat{s}_i)^2]^{1/2} \le 0 \qquad i = 1,\ldots,n$$

or

$$u_i + \mathbf{R_i Du} \ge v_i \ge K_i[E(\mathbf{R_i D\hat{s}} + \hat{s}_i)^2]^{1/2} \ge 0 \qquad i = 1,\ldots,n \tag{10.46}$$

The inequalities of (10.46) can be split into

$$u_i + \mathbf{R_i Du} - v_i \ge 0 \qquad i = 1,\ldots,n \tag{10.47}$$

and

$$v_i^2 - K_i^2 E(\mathbf{R_i Ds} + \hat{s}_i)^2 \ge 0 \qquad i = 1,\ldots,n \tag{10.48}$$

if the second and third terms are squared to remove the square root. These constraints involved only nonstochastic variables. These are the elements of $\mathbf{D}$ and v_i, $i = 1,\ldots,n$. Let $\mathbf{V}$ denote the $n \times 1$ vector, whose ith element is v_i, and let $\mathbf{V}^2$ denote the $n \times n$ diagonal matrix, whose element in row i and column 1 is v_i^2. Then the inequalities (10.47) and (10.48) are

$$-\mathbf{RDu} + \mathbf{V} \le \mathbf{u} \tag{10.49}$$

$$\mathbf{D'GD} + \mathbf{HD} + \mathbf{V}^2 \geq \mathbf{J} \tag{10.50}$$

where **G** and **H** are matrices and **J** is a vector; all of their elements are constants. This is true since the square of $\hat{s}_i$ has a chi-square distribution whose expected value is one. The expected value of a linear combination of chi-square variables is also a constant. Thus the constraints given by (10.41) and (10.45) have been reduced to an equivalent set in which there are no random variables. There are $2n$ constraints of which n are linear and n are quadratic. The number of slack variables that will be required is $2n$.

Finally, the new objective functions for the linear and quadratic problems must be considered. Let the n slack variables for (10.49) be denoted by $\mathbf{y}^s$ and the n slack variables for (10.50) by $\mathbf{y}^S$. Then $w(\mathbf{x}, \mathbf{y}) = \mathbf{a}'\mathbf{x} + (\mathbf{b}^s)'\mathbf{y}^s + (\mathbf{b}^S)'\mathbf{y}^S$ for the linear case, where $\mathbf{b}^s$ and $\mathbf{b}^S$ are the vectors of coefficients for $\mathbf{y}^s$ and $\mathbf{y}^S$ respectively. Adding the slack variables to the constraints leaves

$$-\mathbf{RDu} + \mathbf{V} + \mathbf{y}^s = \mathbf{u} \tag{10.51}$$

$$\mathbf{D'GD} + \mathbf{HD} + \mathbf{V}^2 - \mathbf{y}^S = \mathbf{J} \tag{10.52}$$

Substituting (10.45), (10.51), and (10.52) into the linear objective function leaves

$$\begin{aligned} w(\mathbf{D}, \mathbf{V}) = {} & \mathbf{a'Ds} + (\mathbf{b}^s)'\mathbf{u} + (\mathbf{b}^s)'\mathbf{RDu} - (\mathbf{b}^s)'\mathbf{V} + (\mathbf{b}^S)'\mathbf{D'GD} \\ & + (\mathbf{b}^S)'\mathbf{HD} + (\mathbf{b}^S)'\mathbf{V}^2 - (\mathbf{b}^S)'\mathbf{J} \end{aligned}$$

The E model requires that $E[w(\mathbf{D}, \mathbf{V})]$ be optimized. Since the first term of $w(\mathbf{D}, \mathbf{V})$ is the only one that includes a random variable, $E[w(\mathbf{D}, \mathbf{V})]$ is formed by replacing **s** by its expected value; $E(\mathbf{s}) = \mathbf{u}$.

$$\begin{aligned} E[w(\mathbf{D}, \mathbf{V})] = {} & \mathbf{a'Du} + (\mathbf{b}^s)'\mathbf{u} + (\mathbf{b}^s)'\mathbf{RDu} - (\mathbf{b}^s)'\mathbf{V} + (\mathbf{b}^S)'\mathbf{D'GD} \\ & + (\mathbf{b}^S)'\mathbf{HD} + (\mathbf{b}^S)'\mathbf{V}^2 - (\mathbf{b}^S)'\mathbf{J} \end{aligned}$$

Thus the stochastic linear programming problem is transformed into a convex programming problem with a quadratic objective function and sets of linear and quadratic constraints. This type of model can be solved using the techniques and algorithms presented in Chapters 3 and 4.

For the stochastic quadratic programming problem, the transformed constraints are the same as, (10.51) and (10.52). Suppose for this case

$$\begin{aligned} w(\mathbf{x}, \mathbf{y}) &= \mathbf{a}'\mathbf{x} + \mathbf{b}^s\mathbf{y} + (1/2)\mathbf{x}'\mathbf{A}\mathbf{x} \\ &= \mathbf{a}'\mathbf{x} + (\mathbf{b}^s)'\mathbf{y}^s + (\mathbf{b}^S)'\mathbf{y}^S + (1/2)\mathbf{x}'\mathbf{A}\mathbf{x} \end{aligned}$$

(The matrices **B** and **C** are null matrices.) Then

$$\begin{aligned} w(\mathbf{D}, \mathbf{V}) = {} & \mathbf{a}'\mathbf{D}\mathbf{s} + (\mathbf{b}^s)'\mathbf{u} + (\mathbf{b}^s)'\mathbf{R}\mathbf{D}\mathbf{u} - (\mathbf{b}^s)'\mathbf{V} + (\mathbf{b}^S)'\mathbf{D}'\mathbf{G}\mathbf{D} \\ & + (\mathbf{b}^S)'\mathbf{H}\mathbf{D} + (\mathbf{b}^S)'\mathbf{V}^2 - (\mathbf{b}^S)'\mathbf{J} + (1/2)\mathbf{s}'\mathbf{D}'\mathbf{A}\mathbf{D}\mathbf{s} \end{aligned}$$

$$\begin{aligned} E[w(\mathbf{D}, \mathbf{V})] = {} & \mathbf{a}'\mathbf{D}\mathbf{u} + (\mathbf{b}^s)'\mathbf{u} + (\mathbf{b}^s)'\mathbf{R}\mathbf{D}\mathbf{u} - (\mathbf{b}^s)'\mathbf{V} + (\mathbf{b}^S)'\mathbf{D}'\mathbf{G}\mathbf{D} \\ & + (\mathbf{b}^S)'\mathbf{H}\mathbf{D} + (\mathbf{b}^S)'\mathbf{V}^2 - (\mathbf{b}^S)'\mathbf{J} + (1/2)\mathbf{D}'\boldsymbol{\psi}\mathbf{D} \qquad (10.53) \end{aligned}$$

where $\boldsymbol{\psi}$ is a symmetric $m \times m$ variance-covariance matrix of the elements such that $\psi_{jh} = a_{jh}E(s_i s_k)$, $i, k = 1, \ldots, n$. The quadratic problem is reduced to another deterministic convex programming problem, i.e., optimize (10.53) subject to (10.51) and (10.52), which can be solved using the results of Chapters 3 and 4.

In summary, the linear and quadratic chance-constrained programming problems are to be transformed into nonstochastic problems for the E model. Two variations of the constraint sets have been considered. The first approach, suggested by Hillier and Lieberman [1967], is virtually the same as that of certainty equivalence. The addition of a second set of constraints, $\mathbf{x} = \mathbf{D}\mathbf{s}$, requires that the constraints be transformed into a convex programming problem for both the original linear and quadratic objective functions. When $\mathbf{x} = \mathbf{D}\mathbf{s}$ is presumed, a nonlinear programming problem must be solved.

Numerous extensions and applications of chance-constrained programming appeared in the 1960s. Hillier [1967] has included zero-one variables in such a problem. Naslund [1966] has studied problems in capital budgeting under uncertainty and he [1964] has also shown that under some conditions chance-constrained programming problems can be formulated as problems in variational calculus. Charnes et al. [1964, 1965] have considered the relationship between chance-constrained programming and both critical path analysis and two-stage programming.

10.6 DYNAMICS OF OPTIMIZING UNDER UNCERTAINTY

Programming problems that require the specific consideration of time can be classified as recursive, multistage, or dynamic. Recursive and dynamic problems have been described in Chapters 7 and 8, but multistage analyses have not. The multistage approach is considered here to indicate the complexity of dealing with random variables in dynamic optimization problems.

The simplest example of a multistage programming problem is one that involves only two stages. The two-stage programming problem has been compared to both the chance-constrained problem by Madansky [1963, p. 106] and the active approach to stochastic programming by Fox, Sengupta, and Thorbecke [1966, p. 279]. The presentation is for the multistage model, with the two-stage problem being cited as a special case.

In a similar manner to Dantzig's presentation [1963], the structure of the multistage programming problem can be illustrated as follows:

$$\begin{aligned} \text{optimize} \quad & w(\mathbf{x}^1, \mathbf{x}^2, \ldots, \mathbf{x}^T \mid \epsilon^2, \epsilon^3, \ldots, \epsilon^T) \\ \text{subject to} \quad & \mathbf{s}^1 = \mathbf{R}_1^1\mathbf{x}^1 \\ & \mathbf{s}^2 = \mathbf{R}_1^2\mathbf{x}^1 + \mathbf{R}_2^2\mathbf{x}^2 \\ & \mathbf{s}^3 = \mathbf{R}_1^3\mathbf{x}^1 + \mathbf{R}_2^3\mathbf{x}^2 + \mathbf{R}_3^3\mathbf{x}^3 \\ & \quad\vdots \\ & \mathbf{s}^T = \mathbf{R}_1^T\mathbf{x}^1 + \mathbf{R}_2^T\mathbf{x}^2 + \cdots + \mathbf{R}_T^T\mathbf{x}^T \end{aligned}$$

The superscripts identify within which of the T stages the vector or matrix is to be considered. It is assumed that $\mathbf{s}^1$ is a vector of known constants; $\mathbf{s}^i$, for each i between 2 and T, is a vector of random variables. In the ith stage the elements of $\mathbf{s}^i$ are functions of the random variable ϵ^i, where ϵ^i is an observation from a multivariate probability density function. The $\mathbf{R}_j^i$'s are matrices of known constants for $i,j = 1, \ldots, T$. The problem is to choose $\mathbf{x}^i$ to satisfy the set of constraints for the ith stage, given that all the constraints are satisfied for the previous stages.

Dantzig [1963] suggests the following solution process. Choose $\mathbf{x}^1$ so that $\mathbf{s}^1 = \mathbf{R}_1^1\mathbf{x}^1$ is satisfied. Then $\mathbf{x}^2$ must be such that $\mathbf{s}^2 = \mathbf{R}_1^2\mathbf{x}^1 + \mathbf{R}_2^2\mathbf{x}^2$, given that $\mathbf{s}^1 = \mathbf{R}_1^1\mathbf{x}^1$; $\mathbf{x}^3$ must solve $\mathbf{s}^3 = \mathbf{R}_1^3\mathbf{x}^1 + \mathbf{R}_2^3\mathbf{x}^2 + \mathbf{R}_3^3\mathbf{x}^3$ while maintaining $\mathbf{s}^1 = \mathbf{R}_1^1\mathbf{x}^1$ and $\mathbf{s}^2 = \mathbf{R}_1^2\mathbf{x}^2$. Finally, $\mathbf{x}^T$ is chosen so that $\mathbf{s}^T = \sum_{i=1}^T \mathbf{R}_i^T\mathbf{x}^i$, assuming that the equalities for all the previous stages are maintained. Dantzig shows that a T-stage problem can be reduced to a $(T-1)$ st stage problem in a manner that is analogous to the methodology of dynamic programming. The basis for this reduction is that the solution in the Tth stage is completed assuming optimal levels for $\mathbf{x}^1, \mathbf{x}^2, \ldots, \mathbf{x}^{T-1}$.

The two-stage problem is

$$\text{optimize} \quad w(\mathbf{x}^1, \mathbf{x}^2 \mid \epsilon^2)$$

$$\text{subject to} \quad \mathbf{s}^1 = \mathbf{R}_1^1 \mathbf{x}^1$$
$$\mathbf{s}^2 = \mathbf{R}_1^2 \mathbf{x}^1 + \mathbf{R}_2^2 \mathbf{x}^2$$

The similarity of this problem to that of (10.1) and (10.2) or (10.18) and (10.19) is apparent if $\mathbf{x}^1 = \mathbf{x}$, $\mathbf{x}^2 = \mathbf{y}$, $\mathbf{R}_1^2 = \mathbf{R}$, $\mathbf{s}^2 = \mathbf{s}$, and $\mathbf{R}_2^2 = \mathbf{I}$ (where $\mathbf{I}$ is the appropriate sized identity matrix); $\mathbf{s}^1 = \mathbf{R}_1^1\mathbf{x}$ is simply an additional set of constraints on the decision variables. The objective function could be linear or quadratic.

According to Dantzig, the existence of a convex objective function permits the reduction of the T-stage problem to a $(T - 1)$st stage problem, then to a $(T - 2)$nd stage problem, etc., and finally to a single-stage problem. The difficulty is in finding these convex functions for each of these T reductions. The importance of their existence is that a local optimum solution will be a global optimum for the problem in which the convex functions are guaranteed to exist.

Theil [1964, Ch. 4] extends his quadratic problem described by (10.1) and (10.2) to dynamic circumstances by considering each variable and parameter over T time periods. He includes a new set of variables and parameters for each time period and redefines the matrices and vectors as follows:

- $\mathbf{a}$ and $\mathbf{x}$ become $mT \times 1$ vectors
- $\mathbf{b}$, $\mathbf{y}$, and $\mathbf{s}$ become $nT \times 1$ vectors
- $\mathbf{R}$ becomes an $nT \times mT$ matrix
- $\mathbf{A}$ becomes an $mT \times mT$ symmetric matrix
- $\mathbf{B}$ becomes an $nT \times nT$ symmetric matrix
- $\mathbf{C}$ becomes an $mT \times nT$ matrix

$$\begin{array}{lll} \mathbf{a} = [a_j(t)] & \mathbf{b} = [b_i(t)] & \mathbf{A} = [a_{jh}(t,\tau)] \\ \mathbf{x} = [x_j(t)] & \mathbf{s} = [s_i(t)] & \mathbf{B} = [b_{ik}(t,\tau)] \\ \mathbf{y} = [y_i(t)] & \mathbf{R} = [r_{ij}(t,\tau)] & \mathbf{C} = [c_{jk}(t,\tau)] \end{array}$$

t and $\tau = 1, \ldots, T$, j and $h = 1, \ldots, m$, i and $k = 1, \ldots, n$.

Therefore, the requirement is to optimize w, where

$$\begin{aligned} w = \sum_{t=1}^{T} w(t) = \sum_{t=1}^{T} \Bigg[& \sum_{j=1}^{m} a_j(t)x_j(t) + \sum_{i=1}^{m} b_i(t)y_i(t) \Bigg] \\ & + (1/2)\sum_{t=1}^{T}\sum_{\tau=1}^{T} \Bigg[\sum_{j=1}^{m}\sum_{h=1}^{m} a_{jh}(t,\tau)x_j(t)x_h(\tau) \\ & + \sum_{i=1}^{n}\sum_{k=1}^{n} b_{ik}(t,\tau)y_i(t)y_k(\tau) + \sum_{j=1}^{m}\sum_{k=1}^{n} c_{jk}(t,\tau)x_j(t)y_k(\tau) \\ & + \sum_{j=1}^{m}\sum_{k=1}^{n} c_{kj}(t,\tau)x_j(t)y_k(\tau) \Bigg] \end{aligned} \tag{10.54}$$

subject to

$$y_i(t) = \sum_{\rho=1}^{t} \sum_{j=1}^{m} r_{ij}(t, \rho), x_j(\rho) + s_i(t) \quad (10.55)$$

for $i = 1, \ldots, n$ and $t = 1, \ldots, T$. For the linear problem, $a_{jh} = b_{ik} = c_{jk} = 0$ for all of the i's, j's, k's, and h's so that both the objective function and constraints are linear forms.

Upon first glance, we might conclude that the quadratic and linear dynamic problems are analogous to the static problems studied earlier, the major difference being that there are mT x's and nT y's to be determined. Unfortunately, this is not the case because the optimal levels for $w(t)$ for each $x_j(t)$ and for each $y_i(t)$ are very much dependent on the levels selected for $w(\rho)$, $y_i(\rho)$, and $x_j(\rho)$ for all $i = 1, \ldots, n$; $j = 1, \ldots, m$; and $\rho = 1, 2, \ldots, t - 1$. The role of the parameter values of $a_{jh}(t, \tau)$, $b_{ik}(t, \tau)$, $c_{jk}(t, \tau)$, and $r_{ij}(t, \tau)$ is of particular importance for $\tau \neq t$.

Setting $a_{jh}(t, \tau) \neq 0$ implies that there is a definite effect on $w(t)$, $x_j(t)$, and $y_i(t)$ from $x_j(\rho)$ for all i, j, and ρ as defined before. The $b_{ik}(t, \tau) \neq 0$ relate $y_i(\rho)$ to the $y_i(t)$, $x_j(t)$, and $w(t)$. The $c_{ji}(t, \tau) \neq 0$ link the $y_i(\tau)$ to the $x_j(t)$ for all i, j, t, and τ. In the linear case all these coefficients are zero, but there are still the $r_{ij}(t, \tau)$ to consider, since these parameters appear in both the quadratic and linear problems. The $r_{ij}(t, \tau)$ connect $y_i(t)$, for all i, to the $x_j(\tau)$ for all τ prior to t.

For completeness two more sets of restrictions should be considered. One is needed to link the $x_j(t)$ among themselves, while the other links the $y_i(t)$ among themselves.

$$x_j(t) = \sum_{h=1}^{m} \sum_{\rho=1}^{t-1} d_{jh} x_h(\rho) \qquad j = 1, \ldots, m; t = 2, \ldots, T \quad (10.56)$$

$$y_i(t) = \sum_{k=1}^{n} \sum_{\rho=1}^{t-1} g_{ik} y_k(\rho) \qquad i = 1, \ldots, n; t = 2, \ldots, T \quad (10.57)$$

$$D = [d_{jh}] \qquad j, h = 1, \ldots, m$$

$$G = [g_{ik}] \qquad i, k = 1, \ldots, n$$

Both the difficulties and importance of the dynamic problems must be recognized, for these problems allow for consideration of past history (recent and distant) in present and future decision making and planning. On the other hand, the dynamic problems require the greatest number of assumptions when the parameters are to be estimated.

Assumptions must be made about the statistical errors in the variables, due to the independence or dependence between $x_j(t)$ and $x_h(\rho)$ when $j \neq h$ and $t \neq \rho$. Similar assumptions are required for the y's. However complex these issues are for dynamic optimization problems, note that consideration of random variables has not been emphasized. Using the Pearson method, it is possible to determine probability density functions for each $x_j^0(t), y_i^0(t)$, and $w^0(t)$ for $i = 1, \ldots, n$; $j = 1, \ldots, m$; and $t = 1, \ldots, T$, when the elements of $\mathbf{a}, \mathbf{b}, \mathbf{s}, \mathbf{A}, \mathbf{B}, \mathbf{C}$, and $\mathbf{R}$ are random variables.

One simplifying assumption is that all the random variables have stationary distributions. If $q(t)$ is a random variable with a stationary density function $f[q(t)]$, then the distribution of the random variable is not permitted to shift over time. When this assumption is valid, the problem of (10.54)–(10.57) with random variables is somewhat simpler than it would be otherwise. Unfortunately, these random variables might have stationary distributions only for a short planning horizon. The elements of $\mathbf{a}, \mathbf{b}, \mathbf{A}, \mathbf{B}$, and $\mathbf{C}$ are coefficients in the objective function and are likely to vary with the thoughts of the policymaker, whether he is the production manager of a firm or the planner for a central government. Furthermore, every change of government introduces a new set of goals or new ordering of priorities. The elements of $\mathbf{R}$ are even less likely to be stationary random variables because the contribution or affect of an x_j upon a y_i in one period is very much dependent on previous levels of x_j.

At this point it should be clear that the dynamic problems with random elements are complex, even if there are few variables to be studied during a short time horizon. Unfortunately, not many of these models have been analyzed successfully.

PROBLEMS

10.1. Consider the quadratic programming problem

$$\begin{aligned} \text{maximize} \quad & w = \mathbf{c}'\mathbf{x} + \mathbf{x}'\mathbf{A}\mathbf{x} \\ \text{subject to} \quad & \mathbf{R}\mathbf{x} \leq \mathbf{s} \\ & \mathbf{x} \geq \mathbf{k} \end{aligned}$$

Identify which vectors of the problem would include random variables for each of the common sources of randomness described in Section 10.1.

10.2. Suppose the following is required:

$$\begin{aligned} \text{maximize} \quad & w = a_1x_1 + a_2x_2 \\ \text{subject to} \quad & 5x_1 + 4x_2 \le b_1 \\ & x_1 + x_2 \le b_2 \\ & 4x_1 + 5x_2 \le b_3 \\ & x_1, x_2 \ge 0 \end{aligned}$$

where a_1, a_2, b_1, b_2, and b_3 are random variables. Assume that:

$$E(a_1) = 3 \qquad E(a_2) = 4$$
$$E(b_1) = 10 \qquad E(b_2) = 5 \qquad E(b_3) = 10$$
$$E(a_ib_j) = E(a_i)\,E(b_j) \qquad i = 1, 2; j = 1, 2, 3$$

Find x_1 and x_2 such that $E(w)$ is a maximum. Find x_1 and x_2 such that $E(w)$ has a minimum positive value. What is the minimum $E(w)$ if $x_1 > 0$ or $x_2 > 0$ or both are positive?

10.3. Consider problem

$$\begin{aligned} \text{maximize} \quad & w = a_1x_1 + a_2x_2 - 2x_1^2 \\ \text{subject to} \quad & 5x_1 + 4x_2 \le b_l \\ & x_1, x_2 \ge 0 \end{aligned}$$

Assume that a_1, a_2, and b_1 are normally distributed independent random variables, and the variance of each is 1.0. If $E(a_1) = 8$, $E(a_2) = 12$, and $E(b_1) = 10$, find x_1 and x_2 that maximize $E(w)$.

10.4. Using the model, assumptions, and data in Problem 3, what is the statistical distribution of w, and what is the value of the parameter that specifies the distribution of w?

10.5. Consider the problem

$$\begin{aligned} \text{maximize} \quad & w = 4x_1 + 3x_2 \\ \text{subject to} \quad & 2x_1 + 5x_2 \le 10u_{11} + 15u_{12} \\ & 5x_1 + 2x_2 \le 3u_{21} + 20u_{22} \\ & u_{11} + u_{12} = 1 \\ & u_{21} + u_{22} = 1 \\ & x_1, x_2 \ge 0 \end{aligned}$$

Find the alternative values of w if $u_{11} = 0.6$ and $u_{22} = 0.5$. What are the values of x_1 and x_2 that provide a maximum value of w?

10.6. Using the model given in Problem 2, find x_1 and x_2 such that $E(w)$ has a minimum positive value when

$$Pr(5x_1 + 4x_2 \leq b_1) \geq 0.95$$
$$Pr(x_1 + x_2 \leq b_2) \geq 0.90$$
$$Pr(4x_1 + 5x_2 \leq b_3) \geq 0.95$$

Assume that $V(a_1) = 2$, $V(a_2) = 3$, $V(b_i) = 1$, $i = 1, 2, 3$. All the random variables are normally distributed and independent. Formulate the deterministic programming model to minimize $V(w)$.

A PROOF OF THE CRITERION OF OPTIMIZATION IN THE SIMPLEX METHOD

The problem is to

$$\begin{aligned} \text{maximize} \quad & z = d'w \\ \text{subject to} \quad & Dw = b \\ & w \geq 0 \end{aligned}$$

which can be written as one to

$$\text{maximize} \quad z = \bar{d}'\bar{w} + d^{*\prime}w^* \tag{A.1}$$

$$\text{subject to} \quad \bar{D}\bar{w} + D^*w^* = b \tag{A.2}$$

$$\bar{w}, w^* \geq 0$$

at any simplex stage, where

$$[d]' = [\bar{d} \;\; d^*]'$$

$$[D] = [\bar{D} \;\; D^*]$$

$$w = \begin{bmatrix} \bar{w} \\ w^* \end{bmatrix}$$

where $\bar{w}$ = the variables not in the solution at this stage

w^* = the included variables

$\bar{d}$ = the coefficients in d whose variables are not included in the solution

$\bar{D}$ = the vectors of D that are coefficients of variables not included in the solution

d^* = the coefficients in d whose variables are in the solution
D^* = the vectors of D that are coefficients of included variables

From (A.2)

$$w^* = (D^*)^{-1}b - (D^*)^{-1}\bar{D}\bar{w}$$

Substituting this result into (A.1) leaves

$$\begin{aligned} z &= (\bar{d})'\bar{w} + (d^*)'[(D^*)^{-1}b - (D^*)^{-1}\bar{D}\bar{w}] \\ &= (d^*)'(D^*)^{-1}b - [(d^*)'(D^*)^{-1}\bar{D} - (\bar{d})']\bar{w} \end{aligned}$$

Let

$$\begin{aligned} \hat{z} &= (d^*)'(D^*)^{-1}b \\ \tilde{z} &= (d^*)'(D^*)^{-1}\bar{D}\bar{w} - (\bar{d})'\bar{w} \end{aligned}$$

Both $\hat{z}$ and $\tilde{z}$ are scalars, and $z = \hat{z} - \tilde{z}$. If $\tilde{z} < 0$, $-\tilde{z} > 0$, and $\hat{z} - \tilde{z} > \hat{z}$. Then $\hat{z}$ is not an optimum. If $\tilde{z} > 0$, $-\tilde{z} < 0$, and $\hat{z} - \tilde{z} < \hat{z}$. Therefore, $\hat{z}$ is an optimum. For this case set $\bar{w} = 0$, so that $z = \hat{z}$.

B CONVEXITY OF FUNCTIONS AND SETS

A function $f(\mathbf{x})$ is said to be convex if for any two points $\mathbf{x}_1$ and $\mathbf{x}_2$ and for all λ, $0 \leq \lambda \leq 1$,

$$f[\lambda \mathbf{x}_2 + (1 - \lambda)\mathbf{x}_1] \leq \lambda f(\mathbf{x}_2) + (1 - \lambda) f(\mathbf{x}_1) \tag{B.1}$$

A function $f(\mathbf{x})$ is said to be concave if $-f(\mathbf{x})$ is convex. A function $f(\mathbf{x})$ is said to be strictly convex if (B.1) is a strict inequality.

Intuitively, a function $f(\mathbf{x})$ is convex if a line segment joining any two points $[\mathbf{x}_1, f(\mathbf{x}_1)]$, $[\mathbf{x}_2, f(\mathbf{x}_2)]$ on the surface of $f(\mathbf{x})$ lies on or above the surface. If $\mathbf{x}$ is univariate, the line formed by joining *any* two points on the curve $f(\mathbf{x})$ must be everywhere on or above the curve, as indicated in Figure B.1. The definition in (B.1) for convexity of a function may be used directly to prove a function is convex. Consider the following example problems.

Example B.1

Suppose that there are m functions $f_i(\mathbf{x})$, $i = 1, 2, \ldots, m$, of an n vector $\mathbf{x}$ and the first s functions are convex and the last $m - s$ are concave functions. Suppose further that there are m constant multipliers α_i, $i = 1, 2, \ldots, m$, such that $\alpha_i > 0$ for $i = 1, 2, \ldots, s$ and $\alpha_i < 0$ for $i = s + 1, \ldots, m$. Show that $F(\mathbf{x}) = \sum_{i=1}^{m} \alpha_i f_i(\mathbf{x})$ is a convex function.

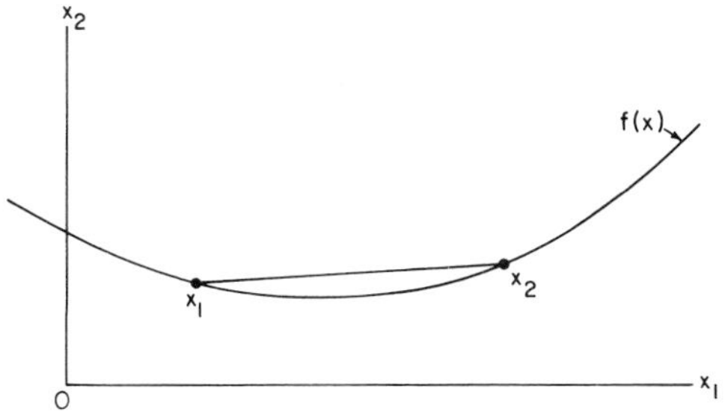

Fig. B.1.

Define the point $\hat{\mathbf{x}} = \lambda\mathbf{x}_2 + (1 - \lambda)\mathbf{x}_1$, where $\mathbf{x}_1$ and $\mathbf{x}_2$ are any two points chosen in the domain of the functions $f_i(\mathbf{x})$. Since $f_i(\mathbf{x})$, $i = 1, 2, \ldots, s$, are convex and $f_i(\mathbf{x})$, $i = s + 1, \ldots, m$, are concave, it follows from (B.1) that

$$f_i(\hat{\mathbf{x}}) \leq \lambda f_i(\mathbf{x}_2) + (1 - \lambda) f_i(\mathbf{x}_1) \qquad i = 1, 2, \ldots, s \qquad \text{(B.2)}$$

$$f_i(\hat{\mathbf{x}}) \geq \lambda f_i(\mathbf{x}_2) + (1 - \lambda) f_i(\mathbf{x}_1) \qquad i = s + 1, \ldots, m \qquad \text{(B.3)}$$

Now multiplying both sides of the inequalities in (B.2) and (B.3) by α_i results in

$$\alpha_i f_i(\hat{\mathbf{x}}) \leq \lambda \alpha_i f_i(\mathbf{x}_2) + (1 - \lambda)\alpha_i f_i(\mathbf{x}_1) \qquad i = 1, 2, \ldots, s \quad \text{(B.4)}$$

$$\alpha_i f_i(\hat{\mathbf{x}}) \leq \lambda \alpha_i f_i(\mathbf{x}_2) + (1 - \lambda)\alpha_i f_i(\mathbf{x}_1) \qquad i = s + 1, \ldots, m \quad \text{(B.5)}$$

Notice that the sense of the inequality sign in (B.5) has changed from (B.3) since α_i, $i = s + 1, \ldots, m$, are negative. Summing the left and right sides of (B.4) and (B.5) gives

$$\sum_{i=1}^{m} \alpha_i f_i(\hat{\mathbf{x}}) \leq \sum_{i=1}^{m} [\lambda \alpha_i f_i(\mathbf{x}_2) + (1 - \lambda)\alpha_i f_i(\mathbf{x}_1)]$$

$$= \lambda \sum_{i=1}^{m} \alpha_i f_i(\mathbf{x}_2) + (1 - \lambda) \sum_{i=1}^{m} \alpha_i f_i(\mathbf{x}_1)$$

so that $F(\hat{\mathbf{x}}) \leq \lambda F(\mathbf{x}_2) + (1 - \lambda) F(\mathbf{x}_1)$, which is what is to be shown. Thus $F(\mathbf{x})$ is a convex function of $\mathbf{x}$.

Fortunately, there is an easier way to demonstrate that a

function is convex. The rules are broken down into the following three cases for convenience only.

CASE 1

Let $f(x)$ be a function of a single variable such that $x \in S$ and S is a subset of the real line. Assuming that $f(x)$ is continuous and possesses a second derivative over S, then $f(x)$ is convex in this region if and only if

$$\frac{\partial^2 f(x)}{\partial x^2} \geq 0$$

Example B.2

$$\begin{aligned} f(x) &= 3x^2 + 2x + 5 \\ \frac{\partial f(x)}{\partial x} &= 6x + 2 \\ \frac{\partial^2 f(x)}{\partial x^2} &= 6 > 0 \end{aligned}$$

Thus $f(x)$ is a convex function over the entire real line.

Example B.3

$$\begin{aligned} f(x) &= x^3 - 5x^2 + 2x + 8 \\ \frac{\partial f(x)}{\partial x} &= 3x^2 - 10x + 2 \\ \frac{\partial^2 f(x)}{\partial x^2} &= 6x - 10 \end{aligned}$$

Notice that $\partial^2 f(x)/\partial x^2$ will be greater than or equal to zero if $x \geq 10/6$. Therefore, $f(x)$ will be convex for $x \geq 10/6$ and concave for $x < 10/6$.

CASE 2

Let $f(\mathbf{x})$ be a function of two variables such that $(x_1, x_2) \in S \subseteq E^2$ and that the second partial derivations of $f(\mathbf{x})$ exists over S. Compute

$$\Delta = \frac{\partial^2 f}{\partial x_1^2} \frac{\partial^2 f}{\partial x_2^2} - \left(\frac{\partial^2 f}{\partial x_1 \partial x_2}\right)^2$$

If $\Delta > 0$ and $\dfrac{\partial^2 f}{\partial x_1^2} > 0$ and $\dfrac{\partial^2 f}{\partial x_2^2} > 0$, then f is convex

$\Delta > 0$ and $\dfrac{\partial^2 f}{\partial x_1^2} < 0$ and $\dfrac{\partial^2 f}{\partial x_2^2} < 0$, then f is concave

$\Delta < 0$, f is neither convex nor concave; the function $f(x)$ is a saddle

$\Delta = 0$, this test for convexity fails and the definition of convexity must be used

Example B.4

$$f(x_1, x_2) = 2x_1^2 + 4x_2^2$$

$$\frac{\partial f}{\partial x_1} = 4x_1 \qquad \frac{\partial^2 f}{\partial x_1^2} = 4 \qquad \frac{\partial^2 f}{\partial x_1 \partial x_2} = 0$$

$$\frac{\partial f}{\partial x_2} = 8x_2 \qquad \frac{\partial^2 f}{\partial x_2^2} = 8$$

$$\Delta = (4)(8) - 0^2 = 32 > 0$$

Thus f is convex over the entire Euclidean E^2 space.

Example B.5

$$f(x_1, x_2) = x_1^2 - x_2^2$$

$$\frac{\partial f}{\partial x_1} = 2x_1 \qquad \frac{\partial^2 f}{\partial x_1^2} = 2 \qquad \frac{\partial^2 f}{\partial x_1 \partial x_2} = 0$$

$$\frac{\partial f}{\partial x_2} = -2x_2 \qquad \frac{\partial^2 f}{\partial x_2^2} = -2$$

$$\Delta = (2)(-2) - 0^2 = -4$$

Since $\Delta < 0$, f is a saddle function. The saddle point occurs at $(0, 0)$.

CASE 3

Let $f(\mathbf{x})$ be a function of n variables where all second-order derivatives exist, and let S be the subspace of the Euclidean

n-space where this condition holds. Define by Δ the determinant of the Hessian matrix of second-order partial derivatives; i.e.,

$$\Delta = |\mathbf{H}| = \begin{vmatrix} \frac{\partial^2 f}{\partial x_1^2} & \frac{\partial^2 f}{\partial x_1 \partial x_2} & \cdots & \frac{\partial^2 f}{\partial x_1 \partial x_n} \\ \frac{\partial^2 f}{\partial x_1 \partial x_2} & \frac{\partial^2 f}{\partial x_2^2} & \cdots & \frac{\partial^2 f}{\partial x_2 \partial x_n} \\ \frac{\partial^2 f}{\partial x_1 \partial x_n} & \frac{\partial^2 f}{\partial x_2 \partial x_n} & \cdots & \frac{\partial^2 f}{\partial x_n^2} \end{vmatrix}$$

Let the principal minors of the matrix $\mathbf{H}$ be given by

$$\Delta_1 = \frac{\partial^2 f}{\partial x_1^2} \qquad \Delta_2 = \begin{vmatrix} \frac{\partial^2 f}{\partial x_1^2} & \frac{\partial^2 f}{\partial x_1 \partial x_2} \\ \frac{\partial^2 f}{\partial x_1 \partial x_2} & \frac{\partial^2 f}{\partial x_2^2} \end{vmatrix}$$

$$\Delta_3 = \begin{vmatrix} \frac{\partial^2 f}{\partial x_1^2} & \frac{\partial^2 f}{\partial x_1 \partial x_2} & \frac{\partial^2 f}{\partial x_1 \partial x_3} \\ \frac{\partial^2 f}{\partial x_1 \partial x_2} & \frac{\partial^2 f}{\partial x_2^2} & \frac{\partial^2 f}{\partial x_2 \partial x_3} \\ \frac{\partial^2 f}{\partial x_1 \partial x_3} & \frac{\partial^2 f}{\partial x_2 \partial x_3} & \frac{\partial^2 f}{\partial x_3^2} \end{vmatrix}, \ldots, \Delta_n$$

The function is convex if

$$\Delta_1 > 0 \qquad \Delta_2 > 0, \ldots \qquad \Delta_n > 0 \tag{B.6}$$

The function is concave if

$$\Delta_1 < 0 \qquad \Delta_2 > 0 \qquad \Delta_3 < 0, \ldots \tag{B.7}$$

Otherwise the test fails. That is, $f(\mathbf{x})$ may be convex or concave if (B.6) or (B.7) are not satisfied, but the definition of convexity must be used to show this.

Example B.6

$$f(x_1, x_2, x_3) = x_1^2 + x_2^2 + 2x_3^2 + x_1 x_2 - 3x_1 + 2x_2 + 10$$

$$\frac{\partial f}{\partial x_1} = 2x_1 + x_2 - 3 \qquad \frac{\partial^2 f}{\partial x_1^2} = 2$$

$$\frac{\partial f}{\partial x_2} = 2x_2 + x_1 + 2 \qquad \frac{\partial^2 f}{\partial x_2^2} = 2$$

$$\frac{\partial f}{\partial x_3} = 4x_3 \qquad \frac{\partial^2 f}{\partial x_3^2} = 4$$

$$\frac{\partial^2 f}{\partial x_1 \partial x_2} = 1 \qquad \frac{\partial^2 f}{\partial x_1 \partial x_3} = 0 \qquad \frac{\partial^2 f}{\partial x_2 \partial x_3} = 0$$

$$\Delta = \begin{vmatrix} 2 & 1 & 0 \\ 1 & 2 & 0 \\ 0 & 0 & 4 \end{vmatrix} = 12 > 0 \qquad \Delta_1 = 2 > 0$$

$$\Delta_2 = \begin{vmatrix} 2 & 1 \\ 1 & 2 \end{vmatrix} = 3 > 0$$

Thus f is a convex function of **x** over the entire Euclidean 3-space.

Two important theorems on convex sets will now be given without proof.

Theorem B.1 If $f(\mathbf{x})$ is a convex function, then the set of points described by $\{\mathbf{x}/f(\mathbf{x}) \leq b\}$ is convex.

Corollary B.1 If $f(\mathbf{x})$ is a concave function, then the set of points described by $\{\mathbf{x}/f(\mathbf{x}) \geq b\}$ is convex.

Theorem B.2 The intersection of convex sets is convex.

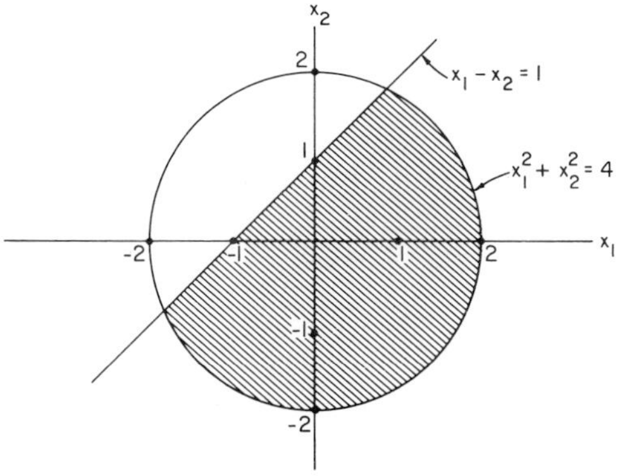

Fig. B.2.

Example B.7

Consider the convex functions $f(x_1, x_2) = x_1^2 + x_2^2$ and $h(x_1, x_2) = x_2 - x_1$. The intersection of the sets

$$\mathbf{X}_1 = \{(x_1, x_2)/f(x_1, x_2) \leq 4\}$$
$$\mathbf{X}_2 = \{(x_1, x_2)/h(x_1, x_2) \leq 1\}$$

is convex. The region is shown in Figure B.2.

C PEARSON CURVES

From among the Pearson family, one type of curve is to be selected as the appropriate density function for a sample of observations on a random variable. Elderton and Johnson [1969] show that among the Pearson curves are the exponential and normal distributions. Also included in this family are close approximations to the binomial, Poisson, chi-square, and gamma distributions. Elderton and Johnson picture the multitude of U-shaped, J-shaped, and "cocked hat shaped" cases that are alternatives in the Pearson family. Kendall [1947, pp. 139–42] proves that the incomplete beta function and rectangular distribution are members of the Pearson family. The gamma and exponential functions are special cases of the incomplete beta distribution.

Solving the differential equation,

$$\frac{dy}{dw} = \frac{(w - a)y}{b_0 + b_1 w + b_2 w^2} \tag{C.1}$$

which is the limiting case of the hypergeometric distribution, provides one particular member of the Pearson family. The parameters of (C.1) determine which type of curve appears. Members of the Pearson family are unimodal, since $dy/dw = 0$ if $w = a$, and touches the w axis at the upper and lower limits on w, since $dy/dw = 0$ if $y = 0$.

If

$$\begin{aligned} x &= w - a \\ B_0 &= b_0 + b_1 a + b_2 a^2 \\ B_1 &= b_1 + 2b_2 a \\ B_2 &= b_2 \end{aligned}$$

(C.1) is transformed into

$$\frac{d(\log y)}{dx} = \frac{x}{B_0 + B_1 x + B_2 x^2} \qquad \text{(C.2)}$$

From the first four moments of the observations on the random variable, x, a, b_0, b_1, and b_2 are determined. These parameters specify B_0, B_1, and B_2.

$$a = \frac{-\mu_3(\mu_4 + 3\mu_2^2)}{A}$$

$$b_0 = \frac{-\mu_2(4\mu_2\mu_4 - 3\mu_3^2)}{A}$$

$$b_1 = \frac{-\mu_3(\mu_4 + 3\mu_2^2)}{A}$$

$$b_2 = \frac{-(2\mu_2\mu_4 - 3\mu_3^2 - 6\mu_2^3)}{A}$$

$$A = 10\mu_4\mu_2 - 18\mu_2^3 - 12\mu_3^2$$

$$\mu_j = \sum_{i=1}^{n} (w_i - \overline{w})^i$$

$$E(w_i) = \overline{w}, N \text{ observations on } w$$

Since (C.2) is a transformed function of (C.1), the particular member of the Pearson family is determined by the two roots of $B_0 + B_1 x + B_2 x^2 = 0$. The alternatives are given in Table C.1.

Table C.1. Characteristics to Identify the Type of Pearson Curve

Characteristics of $B_0 + B_1 x + B_2 x^2$	Type of curve
Roots: real and different signs	I
Roots: complex conjugates	IV
Roots: real and same sign	VI
Roots: one approaches $\pm\infty$, $B_2 = 0$	III
Roots: equal and same sign	V
Roots: same absolute value and opposite sign	
$B_1 = 0$, B_2, and B_0 opposite sign	II
$B_1 = B_2 = a = 0$	normal
$B_1 = 0$, B_2, and B_0 same sign	VII
$B_0 = 0, B_1 > 0$	VIII
$B_0 = 0, B_1 < 0$	IX
$B_0 = B_2 = 0$	X
$B_0 = B_1 = 0$	XI

D CHARACTERISTIC ROOTS IN CONJUGATE PAIRS

Here we give proof that if characteristic roots are complex conjugate numbers, then the coefficients of integration are also complex conjugate numbers. Let

$$y = A_1 e^{\lambda_1 t} + A_2 e^{\lambda_2 t} \qquad \lambda_1 = a + bi, \lambda_2 = a - bi$$

Therefore

$$\dot{y} = \lambda_1 A_1 e^{\lambda_1 t} + \lambda_2 A_2 e^{\lambda_2 t}$$

but

$$y = A_1 e^{at} e^{bit} + A_2 e^{at} e^{-bit} = e^{at}[A_1 e^{bit} + A_2 e^{-bit}]$$

and

$$\dot{y} = (a + bi)A_1 e^{at} e^{bit} + (a - bi)A_2 e^{at} e^{-bit}$$

$$\begin{bmatrix} y \\ \dot{y} \end{bmatrix} = \begin{bmatrix} e^{at} e^{bit} & e^{at} e^{-bit} \\ (a + bi)e^{at} e^{bit} & (a - bi)e^{at} e^{-bit} \end{bmatrix} \begin{bmatrix} A_1 \\ A_2 \end{bmatrix}$$

$$\begin{bmatrix} A_1 \\ A_2 \end{bmatrix} = \begin{bmatrix} \dfrac{-(a - bi)e^{-at} e^{-bit}}{2bi} & \dfrac{e^{-at} e^{-bit}}{2bi} \\ \dfrac{(a + bi)e^{-at} e^{bit}}{2bi} & \dfrac{-e^{-at} e^{bit}}{2bi} \end{bmatrix} \begin{bmatrix} y \\ \dot{y} \end{bmatrix}$$

$$A_1 = \left[\left(\frac{a \sin bt}{2b} + \frac{1}{2}\cos bt\right) e^{-at} y - \frac{\sin bt}{2b} e^{-at} \dot{y}\right]$$

$$- i\left[\left(\frac{\sin bt}{2} - \frac{a \cos bt}{2b}\right) e^{-at} y + \frac{\cos bt}{2b} e^{-at} \dot{y}\right] = c - di$$

$$A_2 = \left[\left(\frac{a \sin bt}{2b} + \frac{1}{2}\cos bt\right) e^{-at} y - \frac{\sin bt}{2b} e^{-at} \dot{y}\right]$$

$$+ i \left[\left(\frac{\sin bt}{2} - \frac{a \cos bt}{2b}\right) e^{-at} y + \frac{\cos bt}{2b} e^{-at} \dot{y}\right] = c + di$$

BIBLIOGRAPHY

Abadie, J. (ed.). 1967. *Nonlinear Programming.* North-Holland Publ. Co., Amsterdam, Netherlands.

Agin, N. 1966. Optimum seeking with branch and bound. *Manage. Sci.* 13: 176–85.

Allen, R. G. D. 1956. *Mathematical Analysis for Economists,* 1st ed. St. Martin's, New York.

______. 1960. *Mathematical Economics,* 2nd. ed. St. Martin's, New York.

Apostol, T. M. 1957. *Mathematical Analysis.* Addison-Wesley, Reading, Mass.

Arrow, K. J., and L. Hurwicz. 1956. Reduction of constrained maxima to saddle-point problems. Proc. 3rd Berkeley Symposium on Mathematical Statistics and Probability 5:1–20, University of California Press, Berkeley.

Arrow, K. J., L. Hurwicz, and H. Uzawa. 1950. *Studies in Linear and Nonlinear Programming.* Stanford University Press, Stanford.

______. 1961. Constraint qualifications in maximization problems. *Nav. Res. Logist. Quart.* 8:175–91.

Arrow, K. J., S. Karlin, and H. Scharf. 1958. *Studies in the Mathematical Theory of Inventory and Production,* Stanford University Press, Stanford.

Asaadi, J. 1973. Computational comparison of some nonlinear programs. *Math. Program.* 4:144–54.

Balas, E. 1965. An additive algorithm for solving linear programs with zero-one variables. *J. Oper. Res. Soc. Am.* 13:517–46.

______. 1967. *Duality in Discrete Programming.* Graduate School of Industrial Administration, Carnegie-Mellon University, Pittsburgh.

Balinski, M. L. 1965. Integer programming: Methods, uses, computation. *Manage. Sci.* 12:253–313.

Balinski, M. L., and W. J. Baumol. 1968. The dual in nonlinear programming and its economic interpretation. *Rev. Econ. Stud.* 35:237–56.

Bawa, V. S. 1973. On chance-constrained programming problems with joint constraints. *Manage. Sci.* 19:1326–31.

Beale, E. M. L. 1955. Cycling in the dual-simplex algorithm. *Nav. Res. Logist. Quart.* 2:269–76.

Beale, E. M. L., and R. E. Small. 1965. Mixed integer programming by branch and bound techniques. Proc. IFIP Congress, Vol. 2, New York.

Beckmann, M. J. 1968. *Dynamic Programming and Economic Decisions.* Springer-Verlag, New York.

Bellman, R. 1957. *Dynamic Programming.* Princeton University Press, Princeton.

Bellman, R. E., and S. E. Dreyfus. 1962. *Applied Dynamic Programming.* Princeton University Press, Princeton.

Bellmore, M., H. J. Greenberg, and J. J. Jarvis. 1970. Generalized penalty function concepts in mathematical optimization. *Oper. Res.* 18:229–52.

Benichow, M., J. M. Gauther, P. Girodet, G. Hentges, G. Ribiere, and O. Vincent. 1971. Experiments in mixed-integer linear programming. *Math. Program.* 1:76–94.

Bolza O. 1961. *Lectures on the Calculus of Variations.* Dover, New York. (Originally published by University of Chicago Press, Chicago, 1904.)

Boot, J. C. G. 1964. *Quadratic Programming.* North-Holland Publ. Co., Amsterdam, Netherlands.

———. 1967. *Mathematical Reasoning in Economics and Management Science.* Prentice-Hall, Englewood Cliffs, N.J.

Boulding, K. E., and W. A. Spivey. 1960. *Linear Programming and the Theory of the Firm.* Macmillan, New York.

Bracken, J., and G. P. McCormick. 1968. *Selected Applications of Nonlinear Programming.* Wiley, New York.

Braitsch, R. J., Jr. 1972. A computer comparison of four quadratic programming algorithms. *Manage. Sci.* 18:632–43.

Carroll, C. W. 1961. The created response surface technique for optimizing nonlinear restrained systems. *Oper. Res.* 9:169–84.

Charnes, A., and W. W. Cooper. 1959. Chance-constrained programming. *Manage. Sci.* 6:73–79.

———. 1961. *Management Models and Industrial Applications of Linear Programming.* Wiley, New York.

———. 1962. Chance constraints and normal deviates. *J. Am. Stat. Assoc.* 57:134–48.

———. 1963. Deterministic equivalents for optimizing and satisficing under chance constraints. *Oper. Res.* 11:18–39.

Charnes, A., W. W. Cooper, and G. L. Thompson. 1964. Critical path analyses via chance-constrained and stochastic programming. *Oper. Res.* 12: 460–70.

———. 1965. Constrained generalized medians and hypermedians as deterministic equivalents for two-stage programs under uncertainty. *Manage. Sci.* 12:83–112.

Colville, A. 1968. A comparative study on nonlinear programming codes. IBM Tech. Rept. 329-2949. White Plains, N.Y.

Cooper, L., and D. Steinberg. 1970. *Introduction to Methods of Optimization.* Saunders, Philadelphia.

Dakin, R. J. 1965. A tree-search algorithm for mixed-integer programming problems. *Comput. J.* 8:250–55.

Danø, S. 1960. *Linear Programming in Industry.* Springer, Vienna, Austria.

Dantzig, G. B. 1960. On the significance of solving linear programming problems with some integer variables. *Econometrica* 28:30–44.

______. 1963. *Linear Programming and Extensions.* Princeton University Press, Princeton.

Davis, H. T. 1960. *Introduction to Nonlinear Differential and Integral Equations.* U.S. Atomic Energy Commission, Washington, D.C.

Day, R. H. 1963. *Recursive Programming and Production Response.* North-Holland Publ. Co., Amsterdam, Netherlands.

______. 1967. A microeconomic model of business growth, decay, and cycle. *Unternehmensforschung* 11:1–20.

______. (ed.). 1975. *Stimulating Economic Behavior: The Recursive Programming Approach.* North-Holland Publ. Co., Amsterdam, Netherlands.

Day, R. H., and E. H. Tinney. 1969a. Cycles, phases and growth in a generalized cobweb theory. *Econ. J.* 79:90–108.

______. 1969b. A dynamic von Thunen model. *Geograph. Anal.* 1:137–51.

Dorfman, R. 1951. *Applications of Linear Programming to the Theory of the Firm.* University of California Press, Berkeley.

______. 1969. An economic interpretation of optimal control theory. *Am. Econ. Rev.* 59:817–31.

Dorfman, R., P. A. Samuelson, and R. M. Solow. 1958. *Linear Programming and Economic Analysis.* McGraw-Hill, New York.

Dorn, W. S. 1960. Duality in quadratic programming. *Quart. Appl. Math.* 18:155–62.

______. 1961. Self-dual quadratic programs. *J. Soc. Ind. Appl. Math.* 9:51–54.

Dresher, M. 1961. *Games of Strategy: Theory and Applications.* Prentice-Hall, Englewood Cliffs, N.J.

Duffin, R. J., E. L. Peterson, and C. Zener. 1967. *Geometric Programming.* Wiley, New York.

Dykstra, O. 1971. The augmentation of experimental data to maximize $|X'X|$. *Technometrics* 13:682–88.

Elderton, W. P., and N. L. Johnson. 1969. *Systems of Frequency Curves.* Cambridge University Press, Cambridge, England.

Elsgolc, L. E. 1961. *Calculus of Variations.* Pergamon Press, New York.

Evans, G. C. 1964. *Functionals and Their Applications.* Dover, New York.

Fiacco, A. V., and G. P. McCormick. 1964. The sequential unconstrained minimization technique for nonlinear programming. *Manage. Sci.* 10: 360–66.

______. 1968. *Nonlinear Programming, Sequential Unconstrained Minimization Technique.* Wiley, New York.

Ford, L. R., Jr., and D. R. Fulkerson. 1962. *Flows in Networks.* Princeton University Press, Princeton.

Fox, K., J. K. Sengupta, and E. Thorbecke. 1972. *The Theory of Quantitative Economic Policy,* 2nd rev. ed. North Holland Publ. Co., Amsterdam, Netherlands.

Frank, M., and P. Wolfe. 1956. An algorithm for quadratic programming. *Nav. Res. Logist. Quart.* 3:95–109.

Gale, D. 1960. *The Theory of Linear Economic Models.* McGraw-Hill, New York.

Gandolfo, G. 1972. *Mathematical Methods and Models in Economic Dynamics.* North-Holland Publ. Co., Amsterdam, Netherlands.

Garfinkle, S., and G. Nemhauser. 1972. *Integer Programming.* Wiley, New York.
Gass, S. I. 1969. *Linear Programming Methods and Applications.* 3rd ed. McGraw-Hill, New York.
Geoffrion, A. H. 1969. An improved implicit enumeration approach for integer programming. *Oper. Res.* 17:137–51.
Gerber, F. 1974. Mathematical programming book reviews: 1965–1972. *Manage. Sci.* 20:875–95.
Glass, H., and L. Cooper. 1965. Sequential search: A method for solving constrained optimization problems. *J. Assoc. Comput. Mach.* 12:71–82.
Glover, F. 1968. A new foundation for a simplified primal integer programming algorithm. *J. Oper. Res. Soc. Am.* 16:727–40.
———. 1972. Cut search methods in integer programming. *Math. Program.* 3:86–100.
Glover, F., and E. Woolsey. 1974. Converting the 0–1 polynomial programming problem to a 0–1 linear program. *Oper. Res.* 22:180–82.
Goldberg, S. 1961. *Introduction to Difference Equations.* Wiley, New York.
Goldberger, A. S. 1964. *Econometric Theory.* Wiley, New York.
Gomory, R. E. 1960. An algorithm for the mixed integer problem. RAND Rept. P-1885, Stanford.
———. 1963. An algorithm for integer solutions to linear programs. In *Recent Advances in Mathematical Programming,* R. L. Graves and P. Wolfe (eds.). McGraw-Hill, New York.
———. 1969. Some polyhedra related to combinatorial problems. *Linear Algebra, and Its Applications* 2:451–558.
Gomory, R. E., and W. J. Baumol. 1960. Integer programming and pricing. *Econometrica* 28:521–50.
Gould, F. J., and J. W. Tolle. 1972. Geometry of optimality conditions and constraint qualifications. *Math. Program.* 2:1–18.
Greenberg, H. 1971. *Integer Programming.* Academic Press, New York.
Greenstadt, J. 1972. A ricocheting gradient method for nonlinear optimization. *J. SIAM Appl. Math.* 14:429–45.
Grinold, R. C. 1974. A generalized discrete dynamic programming model. *Manage. Sci.* 20:1092–1103.
Gunther, P. 1955. Use of linear programming in capital budgeting. *Oper. Res.* 3:219–24.
Hadley, G. 1961. *Linear Algebra.* Addison-Wesley, Reading, Mass.
———. 1962. *Linear Programming.* Addison-Wesley, Reading, Mass.
———. 1964. *Nonlinear and Dynamic Programming.* Addison-Wesley, Reading, Mass.
Hamilton, W. F., and M. A. Moses. 1973. An optimization model for corporate financial planning. *Oper. Res.* 21:677–92.
Hartley, H. O. 1961. Nonlinear programming by the simplex methods. *Econometrica* 29:223–37.
Hartley, H. O., and R. R. Hocking. 1963. Convex programming by tangential approximation. *Manage. Sci.* 9:600–12.
Hartley, H. O., and R. C. Pfaffenberger. 1971. Statistical control of optimization. In *Optimizing Methods in Statistics,* J. S. Rastagi (ed.). Academic Press, New York.
Hartley, H. O., and P. G. Ruud. 1969. Computer optimization of second order

response surface designs. In *Statistical Computation,* R. C. Milton and J. A. Nelder (eds.). Academic Press, New York.

Hartley, H. O., R. R. Hocking, and W. P. Cooke. 1967. Least squares fit of definite quadratic forms by convex programming. *Manage. Sci.* 13:913–25.

Heady, E. O., and W. Candler. 1958. *Linear Programming Methods.* Iowa State University Press, Ames.

Henderson, J. M. 1959. The utilization of agricultural land, a theoretical and empirical inquiry. *Rev. Econ. Stat.* 41:242–60.

Henderson, J. M., and R. E. Quandt. 1958. *Microeconomic Theory: A Mathematical Approach.* McGraw-Hill, New York.

Hestenes, M. R. 1966. *Calculus of Variations and Optimal Control Theory.* Wiley, New York.

Higgins, G. F. 1966. Game theory and sequential analysis applied to firm behavior in an oligopolistic market structure. Department of Economics, Iowa State University, Ames.

Hillier, F. S. 1967. Change-constrained programming with 0–1 or bounded continuous decision variables. *Manage. Sci.* 14:34–57.

Hillier, F. S., and C. J. Lieberman. 1967. *Introduction to Operations Research,* Holden-Day, San Francisco.

Hitchcock, F. L. 1941. Distribution of a product from several sources to numerous liabilities. *J. Math. Phys.* 20:224–30.

Hohn, F. E. 1958. *Elementary Matrix Algebra.* Macmillan, New York.

Holt, C. C., F. Modigliani, J. F. Muth, and H. A. Simon. 1960. *Planning Production, Inventories, and Work Force.* Prentice-Hall, Englewood Cliffs, N.J.

Hooke, R., and T. A. Jeeves. 1961. Direct search solution of numerical and statistical problems. *J. Assoc. Comput. Mach.* 8:212–20.

Howard, R. A. 1960. *Dynamic Programming and Markov Processes.* M.I.T. Press, Cambridge, Mass.

______. 1966. Dynamic programming. *Manage. Sci.* 12:317–48.

Hu, T. C. 1969. *Integer Programming and Network Flows.* Addison-Wesley, Reading, Mass.

Hu, T. C., and S. M. Robinson. 1973. *Mathematical Programming.* Academic Press, New York.

Ignizio, J. P. 1971. On the establishment of standards for comparing algorithmic performance. *TIMS Interfaces* 2:8–11.

International Business Machine Corporation. 1967. Mathematical Programming System/360. IBM Corp. Tech. Publ. Dept., White Plains, New York.

Intriligator, M. D. 1971. *Mathematical Optimization and Economic Theory.* Prentice-Hall, Englewood Cliffs, N.J.

Jagonnathan, R. 1974. Chance-constrained programming with joint constraints. *Oper. Res.* 22:358–72.

Kelley, J. E., Jr. 1960. The cutting plane method for solving convex programs. *J. Soc. Ind. Appl. Math.* 8:703–12.

Kendall, M. G. 1947. *The Advanced Theory of Statistics.* Vol. 1. Charles Griffin, London, England.

Klein, L. W. 1950. *Economic Fluctuations in the United States: 1921–1941.* Cowles Commission Monogr. 11. Wiley, New York.

Koopmans, T. C. (ed.). 1951. *Activity Analysis of Production and Allocation.* Cowles Commission Monogr. 13. Wiley, New York.

Koopmans, T. C., and A. F. Bausch. 1959. Selected topics on economics involving mathematical reasonings. *Soc. Ind. Appl. Math. Rev.* 1:79–149.

Koot, R. S., and D. A. Walker. 1970. Economic growth and stability since the employment act of 1946. *Quart. Rev. Econ. Bus.* 10:7–18.

———. 1971. Pearson curve fitting to evaluate the performance of a firm. *Rivista Internationale di Scienze Economiche e Commerciale* 18:487–97.

Kowalik, I., M. R. Osborne, and D. M. Ryan. 1969. A new method for constrained optimization problems. *Oper. Res.* 17:973–83.

Kuhn, H. W. 1957. Lectures on the theory of games. *Ann. Math. Stud.,* No. 37. Princeton University Press, Princeton.

Kuhn, H. W., and A. W. Tucker. 1959. Nonlinear programming. In Proc. 2nd Berkeley Symposium on Mathematical Statistics and Probability, pp. 481–92. University of California Press, Berkeley.

Kwak, N. K. 1973. *Mathematical Programming with Business Applications.* McGraw-Hill, New York.

Lancaster, K. 1968. *Mathematical Economics.* Macmillan, New York.

Lefscheta, S. 1956. Linear and nonlinear oscillations. In *Modern Mathematics for the Engineer,* Edwin F. Beckenbach, (ed.). McGraw-Hill, New York.

Lemke, C. E., and K. Spielberg. 1967. Direct search algorithm for zero-one and mixed-integer programming. *J. Oper. Res. Soc. Am.* 15:892–915.

Leontief, W. W. 1936. Quantitative input and output relations in the economic systems of the United States. *Rev. Econ. Stat.* 18:105–25.

———. 1953. *Studies in the Structure of the American Economy.* Oxford University Press, New York.

———. 1958. Theoretical note of time preference productivity of capital, stagnation and economic growth. *Am. Econ. Rev.* 48:105–11.

———. 1959. Time-preference and economic growth, reply. *Am. Econ. Rev.* 49:1041–43.

Luce, R. D., and H. Raiffa. 1957. *Games and Decisions, Introduction and Critical Survey.* Wiley, New York.

Luenberger, D. G. 1973. *Introduction to Linear and Nonlinear Programming.* Addison-Wesley, Reading, Mass.

Madansky, A. 1962. Methods of solution of linear programs under uncertainty. *Oper. Res.* 10:463–70.

Mahalanobis, P. C. 1955. Approach of operational research to planning in India. *Sankhya* 16:3–62.

Mangasarian, O. L. 1969. *Nonlinear Programming.* McGraw-Hill, New York.

Markowitz, H. 1956. The optimization of a quadratic function subject to linear constraints. *Nav. Res. Logist. Quart.* 3:111–33.

———. 1959. *Portfolio Selection.* Cowles Foundation for Research in Economics at Yale University, Monogr. 16. Wiley, New York.

McCormick, G. P. 1971. Penalty function versus nonpenalty function methods for constrained nonlinear programming problems. *Math. Program.* 1:217–38.

McCormick, G. P., W. C. Mylander III, and A. V. Fiacco. 1965. Computer program implementing the sequential unconstrained minimization technique for nonlinear programming. Research Analysis Corporation, McLean, Va.

McKinsey, J. C. C. 1962. *Introduction to the Theory of Games.* McGraw-Hill, New York.

McMillan, C. 1970. *Mathematical Programming.* Wiley, New York.

Miernyk, W. H. 1965. *The Elements of Input-Output Analysis.* Random House, New York.

Morse, M. 1934. *The Calculus of Variations in the Large.* American Mathematical Society, New York.

Naslund, B. 1964. Decisions under risk: Economic applications of chance constrained programming. ONR Res. Memo. 134. Graduate School of Industrial Administration, Carnegie Institute of Technology, Pittsburgh.

———. 1966. A model of capital budgeting under risk. *J. Bus.* 39:257–71.

Naslund, B., and A. Whinston. 1962. A model of multi-period investment under uncertainty. *Manage. Sci.* 8:184–200.

Nemhauser, G. L. 1966. *Introduction to Dynamic Programming.* Wiley, New York.

Nemhauser, G. L., and Z. Ullmann. 1969. Discrete dynamic programming and capital budgeting. *Manage. Sci.* 15:494–505.

Orden, A. 1956. The transshipment problem. *Manage. Sci.* 2:276–85.

Oshima, H. T. 1957. Share of government in gross national product for various countries. *Am. Econ. Rev.* 48:381–90.

Ostle, B. *Statistics in Research*, 3rd ed., 1975. Iowa State University Press, Ames.

Pekelman, D., and S. Sen. 1974. Mathematical programming models for the determination of attribute weights. *Manage. Sci.* 20:1217–29.

Phillips, A. 1955. The tableau économique as a simple Leontief model. *Quart. J. Econ.* 69:137–44.

Pindyck, R. S. 1973. *Optimal Planning for Economic Stabilization.* American Elsevier, New York.

Plane, D., and McMillan, C. 1971. *Discrete Optimization.* Prentice-Hall, Englewood Cliffs, N.J.

Ponstein, J. 1967. Seven kinds of convexity. *Soc. Ind. Appl. Math. Rev.* 9:115–19.

Powell, M. J. D. 1972. Recent advances in unconstrained optimization. *Math. Program.* 1:26–57.

———. 1973. On search directions for minimization algorithms. *Math. Program.* 4:193–201.

Programming. *Oper. Res.* 21:1–394.

Ramsey, F. P. 1928. A mathematical theory of saving. *Econ. J.* 38:543–59.

Reiter, S., and Rice, D. R. 1966. Discrete optimizing solution procedures for linear and nonlinear integer programming problems. *Manage. Sci.* 12:829–50.

Robinson, S. M. 1973. Computable error bounds for nonlinear programming. *Math. Program.* 5:235–42.

Rosen, J. B. 1960. The gradient projection method for nonlinear programming: Part I, linear constraints. *J. Soc. Ind. Appl. Math.* 8:181–217.

———. 1961. The gradient projection method for nonlinear programming: Part II, nonlinear constraints. *J. Soc. Ind. Appl. Math.* 9:514–32.

Roy, A. D. 1952. Safety first and the holding of assets. *Econometrica* 20:431–49.

Saaty, T. L., and J. Bram. 1964. *Nonlinear Mathematics.* McGraw-Hill, New York.

Samuelson, P. A. 1963. *Foundations of Economic Analysis.* Harvard University Press, Cambridge, Mass.

———. 1969. Lifetime portfolio selection by dynamic stochastic programming. *Rev. Econ. Stat.* 51:239–46.

Sengupta, J. K. 1964. On the relative stability and optimality of consumption in aggregative growth models. *Economica* 31:34–50.

———. 1966. The stability of truncated solutions of stochastic linear programming. *Econometrica* 34:77–104.

———. 1969. Safety first rules under chance-constrained programming. *Oper. Res.* 17:112–32.

———. 1970. Stochastic linear programming with chance-constraints. *Int. Econ. Rev.* 11:287–304.

———. 1973. *Stochastic Programming: Methods and Applications.* American Elsevier, New York.

Sengupta, J. K., and K. A. Fox. 1969. *Optimization Techniques in Quantitative Economic Models.* North-Holland Publ. Co., Amsterdam, Netherlands.

Sengupta, J. K., and T. K. Kumar. 1965. An application of sensitivity analysis to a linear programming problem. *Unternehmensforschung* 9:18–36.

Sengupta, J. K., and G. Tintner. 1963. On some economic models of development planning. *Econ. Int.* 16:34–50.

Sengupta, J. K., and D. A. Walker. 1964. On the empirical specification of optimal economic policy. *J. Manchester School Soc. Sci.* 32:215–37.

Sengupta, J. K., C. Millham, and G. Tintner. 1965. On the stability of solutions under error on stochastic linear programming. *Metrika* 9:47–60.

Sengupta, J. K., G. Tintner, and C. Millham. 1963. On some theorems of stochastic linear programming with applications. *Manage. Sci.* 10:143–59.

Sengupta, J. K., G. Tintner, and B. Morrison. 1963. Stochastic linear programming with applications to economic models. *Economica* 30:262–76.

Shapiro, D. F., and R. L. Well. 1972. Management science: A view from nonlinear programming. *Comm. ACM* 15:542–49.

Sharpe, W. F. 1970. *Portfolio Theory and Capital Markets.* McGraw-Hill, New York.

Shubik, M. 1961. Objective functions and models of corporate optimization. *Quart. J. Econ.* 75:345–75.

Simon, H. A. 1956. Dynamic programming under uncertainty with a quadratic criterion function. *Econometrica* 24:74–81.

———. 1957. *Models of Man.* Wiley, New York.

Smith, D. V. 1973. Decision rules in chance-constrained programming: Some experimental comparisons. *Manage. Sci.* 19:688–702.

Spang, H. A. 1962. A review of minimization techniques for nonlinear functions. *J. Soc. Ind. Appl. Math.* 4:343–65.

Stigler, G. J. 1945. The cost of subsistence. *J. Farm Econ.* 27:303–14.

Stocker, D. C. 1969. A comparative study of nonlinear programming codes. Unpubl. M.S. thesis, University of Texas, Austin.

Stong, R. E. 1965. A note on the sequential unconstrained minimization technique for nonlinear programming. *Manage. Sci.* 12:142–44.

Struble, R. 1962. *Nonlinear Differential Equations.* McGraw-Hill, New York.

Theil, H. 1957. A note on certainty equivalence on dynamic planning. *Econometrica* 25:346–49.

———. 1964. *Optimal Decision Rules for Government and Industry.* North Holland Publ. Co., Amsterdam, Netherlands.

Tintner, 1955. Stochastic linear programming with applications to agricultural economics. Second Symposium in Linear Programming Vol. 1. National Bureau of Standards, Washington, D.C.

______. 1960. A note on stochastic linear programming. *Econometrica* 28: 490–95.

Tomlin, J. A. 1971. An improved branch and bound method for integer programming. *Oper. Res.* 19:1070–75.

Vajda, S. 1958. *Readings in Linear Programming.* Wiley, New York.

______. 1961. *Mathematical Programming.* Addison-Wesley, Reading, Mass.

______. 1967. Nonlinear programming and duality. In *Nonlinear Programming,* J. Abadie, (ed.). North-Holland Publ. Co., Amsterdam, Netherlands.

______. 1973. *Parametric Programming.* Academic Press, New York.

Van den Bogaard, P. J. M., and H. Theil. 1959. Macrodynamic Policy-Making: Economy of the United States, 1933–1938. *Metroeconomica* 11:149–67.

Van de Panne, C. 1968. Programming with a quadratic constraint. *Manage. Sci.* 12:798–815.

Van de Panne, C., and W. Popp. 1963. Minimum-cost cattle feed under probabilistic protein constraints. *Manage. Sci.* 9:405–30.

Van Moeseke, P. 1964. Duality theorems in convex homogeneous programming. *Metroeconomica* 16:32–40.

Von Neumann, J., and O. Morgenstern. 1964. *Theory of Games and Economic Behavior,* 3rd ed. Science Editions, Wiley, New York.

Wagner, H. M. 1969. *Principles of Operations Research.* Prentice-Hall, Englewood Cliffs, N.J.

Walker, D. A. 1972. A recursive programming approach to bank asset management. *J. Financ. Quant. Anal.*, pp. 2055–75.

Watters, L. J. 1967. Reduction of integer polynomial programming problems to zero-one linear programming problems. *Oper. Res.* 15:1171–74.

Weingartner, H. 1963. *Mathematical Programming and the Analysis of Capital Budgeting Problems.* Prentice-Hall, Englewood Cliffs, N.J.

White, D. J. 1969. *Dynamic Programming.* Holden-Day, San Francisco.

White, W. W. 1966. On a group theoretic approach to linear integer programming. ORC Report 66–27. University of California, Berkeley.

Wilde, D. J. 1962. Differential calculus in nonlinear programming. *Oper. Res.* 10:764–73.

Wilde, D. J., and C. S. Beightler. 1967. *Foundations of Optimization.* Prentice-Hall, Englewood Cliffs, N.J.

Williams, J. B. 1954. *The Complete Strategist.* McGraw-Hill, New York.

______. 1967. The path to equilibrium. *Quart. J. Econ.* 81:241–55.

Wolfe, P. 1959. The simplex method for quadratic programming. *Econometrica* 27:382–98.

______. 1961. A duality theorem for nonlinear programming. *Quart. Appl. Math.* 19:239–44.

______. 1967. Methods of nonlinear programming. In *Nonlinear Programming,* J. Abadie (ed.). North-Holland Publ. Co., Amsterdam, Netherlands.

Wood, M. K. 1951. Representation in a linear model of non-linear growth curves in the aircraft industry. In *Activity Analysis of Production and Allocation,* Tjalling C. Koopmans (ed.). Wiley, New York.

Young, R. D. 1968. A simplified primal (all-integer) integer programming algorithm. *J. Oper. Res. Soc. Am.* 16:750–82.

Zangwill, W. I. 1967. The convex simplex method. *Manage. Sci.* 14:221–39.

______. 1967. Nonlinear programming via penalty functions. *Manage. Sci.* 13:344–58.

Zangwill, W. I. 1968. *Nonlinear Programming*. Prentice-Hall, Englewood Cliffs, N.J.

Zellner, A. 1963. Decision rules for economic forecasting. *Econometrica* 31: 111–31.

Zoutendijk, G. 1959. Maximizing a function in a convex region. *J. Roy. Stat. Soc.* Ser. B, 21:338–55.

INDEX